Nonlinear Option Pricing

T0341062

CHAPMAN & HALL/CRC
Financial Mathematics Series

Aims and scope:
The field of financial mathematics forms an ever-expanding slice of the financial sector. This series aims to capture new developments and summarize what is known over the whole spectrum of this field. It will include a broad range of textbooks, reference works and handbooks that are meant to appeal to both academics and practitioners. The inclusion of numerical code and concrete real-world examples is highly encouraged.

Series Editors

M.A.H. Dempster
Centre for Financial Research
Department of Pure
Mathematics and Statistics
University of Cambridge

Dilip B. Madan
Robert H. Smith School
of Business
University of Maryland

Rama Cont
Department of Mathematics
Imperial College

Published Titles

American-Style Derivatives; Valuation and Computation, *Jerome Detemple*

Analysis, Geometry, and Modeling in Finance: Advanced Methods in Option
  Pricing, *Pierre Henry-Labordère*

Computational Methods in Finance, *Ali Hirsa*

Credit Risk: Models, Derivatives, and Management, *Niklas Wagner*

Engineering BGM, *Alan Brace*

Financial Modelling with Jump Processes, *Rama Cont and Peter Tankov*

Interest Rate Modeling: Theory and Practice, *Lixin Wu*

Introduction to Credit Risk Modeling, Second Edition, *Christian Bluhm,*
  Ludger Overbeck, and Christoph Wagner

An Introduction to Exotic Option Pricing, *Peter Buchen*

Introduction to Stochastic Calculus Applied to Finance, Second Edition,
  Damien Lamberton and Bernard Lapeyre

Monte Carlo Methods and Models in Finance and Insurance, *Ralf Korn, Elke Korn,*
  and Gerald Kroisandt

Monte Carlo Simulation with Applications to Finance, *Hui Wang*

Nonlinear Option Pricing, *Julien Guyon and Pierre Henry-Labordère*

Numerical Methods for Finance, *John A. D. Appleby, David C. Edelman,*
  and John J. H. Miller

Option Valuation: A First Course in Financial Mathematics, *Hugo D. Junghenn*

Portfolio Optimization and Performance Analysis, *Jean-Luc Prigent*

Quantitative Fund Management, *M. A. H. Dempster, Georg Pflug,*
  and Gautam Mitra

Risk Analysis in Finance and Insurance, Second Edition, *Alexander Melnikov*

Proposals for the series should be submitted to one of the series editors above or directly to:
CRC Press, Taylor & Francis Group

Chapman & Hall/CRC FINANCIAL MATHEMATICS SERIES

Nonlinear Option Pricing

Julien Guyon

Pierre Henry-Labordère

CRC Press
Taylor & Francis Group
Boca Raton London New York

CRC Press is an imprint of the
Taylor & Francis Group, an **informa** business

A CHAPMAN & HALL BOOK

First published in paperback 2024

First published 2014
by CRC Press
2385 NW Executive Center Drive, Suite 320, Boca Raton FL 33431

and by CRC Press
4 Park Square, Milton Park, Abingdon, Oxon, OX14 4RN

CRC Press is an imprint of Taylor & Francis Group, LLC

© 2014, 2024 Taylor & Francis Group, LLC

Publisher's Note
The publisher has gone to great lengths to ensure the quality of this reprint but points out that some imperfections in the original copies may be apparent.

ISBN: 978-1-4665-7033-7 (hbk)
ISBN: 978-1-03-291939-3 (pbk)
ISBN: 978-0-429-10149-6 (ebk)

DOI: 10.1201/b16332

**Visit the Taylor & Francis Web site at
http://www.taylorandfrancis.com**

**and the CRC Press Web site at
http://www.crcpress.com**

To our beloved ones

Contents

Preface

The field of mathematics is very wide and it is not easy to predict what happens next, but I can tell you it is alive and well. Two general trends are obvious and will surely persist. In its pure aspect, the subject has changed, much for the better I think, by moving to more concrete problems. In both its pure and applied aspects, an equally beneficial shift to nonlinear problems can be seen. Most mathematical questions suggested by Nature are genuinely nonlinear, meaning very roughly that the result is not proportional to the cause, but varies with it as the square or the cube, or in some more complicated way. The study of such questions is still, after two or three hundred years, in its infancy. Only a few of the simplest examples are understood in any really satisfactory way. I believe this direction will be a principal theme in the future.

— Henry P. McKean, in *Some Mathematical Coincidences*, May 2003

The valuation of European options in the Black-Scholes paradigm evolves organically *linear* second order parabolic partial differential equations (PDEs). The optionality in American-style options, or market imperfections not taken into account in the Black-Scholes framework, such as illiquidity, transaction costs, uncertain volatilities, uncertain default intensities, different rates for borrowing and lending, and counterparty risk lead in turn to *nonlinear* parabolic PDEs. So do the calibrations of local stochastic volatility models and local correlation models to market prices of vanilla options. Some examples of nonlinear PDEs arising in quantitative finance are listed in Table 1. All of them are investigated in this book.

In practice, analytical solutions are not available and we must therefore rely on numerical methods. For a small number of underlyings, these nonlinear PDEs can be solved efficiently with finite difference schemes. However, PDEs suffer from the curse of dimensionality. When pricing complex financial derivatives, we often deal with a large number of underlyings, be they asset values, path-dependent variables, or hidden factors such as a stochastic volatility. Hence we must turn to probabilistic methods that allow for Monte Carlo techniques. However, the classical Feynman-Kac's probabilistic representation is only relevant for linear PDEs: it is not applicable for these nonlinear PDEs. Hence one needs *new advanced probabilistic methods*. Note

Table 1: Examples of nonlinear PDEs arising in option pricing. $\Sigma(\Gamma)$ denotes the volatility in the nonlinear PDE $\partial_t u + \frac{1}{2}\Sigma(\partial_x^2 u)^2 x^2 \partial_x^2 u = 0$, $u(T,x) = g(x)$.

Transaction costs	$\Sigma(\Gamma)^2 = \sigma^2 + \sqrt{\frac{2}{\pi}\frac{k\sigma}{\sqrt{\delta t}}}\mathrm{sign}(\Gamma)$
Illiquid market	$\Sigma(\Gamma)^2 = \frac{\sigma^2}{(1-\epsilon x\Gamma)^2}$
American option	$\max\left(\partial_t u + \frac{1}{2}\sigma^2 x^2 \partial_x^2 u, g - u\right) = 0$
Uncertain default rate model	$\partial_t u + \frac{1}{2}\sigma^2 x^2 \partial_x^2 u + \lambda(t,u)(u^D - u) = 0$
Uncertain volatility model	$\Sigma(\Gamma)^2 = \overline{\sigma}^2 \mathbf{1}_{\Gamma \geq 0} + \underline{\sigma}^2 \mathbf{1}_{\Gamma < 0}$
Local stochastic volatility model	$-\partial_t p + \frac{1}{2}\partial_x^2(\sigma^2(t,x,p)p(t,x)) = 0$
Local correlation model	$-\partial_t p + \frac{1}{2}\partial_{x_1}^2(\sigma_1^2 p) + \frac{1}{2}\partial_{x_2}^2(\sigma_2^2 p)$ $+\partial_{x_1 x_2}(\rho(t,x_1,x_2,p)\sigma_1\sigma_2 p) = 0$
Counterparty risk	$\partial_t u + \frac{1}{2}\sigma^2 x^2 \partial_x^2 u + \lambda(t)(u^+ - u) = 0$

that in most cases these numerical methods can easily be parallelized using, for example, graphics processing units (GPUs).

The purpose of this book is to present and compare various numerical methods for solving high-dimensional nonlinear problems arising in option pricing. Its originality lies in the following points. Firstly, although some of these advanced methods are already scattered across various recent articles and books, to the best of our knowledge, this is the first monograph dedicated to nonlinear option pricing. We present many different methods in the same book, when usually they are described separately. This allows us to compare their efficiency. Secondly, certain methods suggested in this book are original contributions to the field. Our original contributions include regression methods and dual methods for pricing chooser options (Chapter 6), the Monte Carlo approaches for pricing in the uncertain lapse and mortality model (Chapter 8) and in the uncertain volatility model (Chapter 9), the Markovian projection method and the particle method for calibrating local stochastic volatility models to market prices of vanilla options, with or without stochastic interest rates (Chapter 11), the $a + b\lambda$ technique for building local correlation models that calibrate to market prices of vanilla options on a basket (Chapter 12), and a new stochastic representation of solutions of some nonlinear PDEs based on marked branching diffusions (Chapter 13). While we have already presented some of these new techniques in various articles, it is the first time that we publish our results on chooser options (Chapter 6) and on the uncertain lapse and mortality model (Chapter 8). Thirdly, we focus on general mathematical tools rather than on specific financial questions. The main advantage in doing so is that the tools can be straightforwardly used by readers to solve their own (nonlinear) problems. Finally, we strove to make this book reasonably comprehensive, and to find a right balance between ideas, mathematical theory, and numerical implementations. We devote ample space to the theory: the relevant mathematical notions are introduced, the important results are given, and some proofs are either detailed or sketched when needed, or when it is instructive. But the main focus is deliberately on ideas and on numerical examples, which we believe help a lot in understanding the tools and building intuition. In this respect, this book is meant to be a practitioner's guide: all the mathematical methods that are introduced are illustrated on practical nonlinear option pricing problems. In layman's terms, these problems have been considered in our careers as quantitative analysts.

In order to guide the reader, here is a general description of the book's chapters, which can be read independently:

• Chapter 1 is a quick recap of classical linear PDEs arising in mathematical finance. In particular, we focus on probabilistic representations of linear PDEs using Feynman-Kac's formula, including the cases of Dirichlet options and Neumann options. This enables us to recall key notions in option pricing such as super-replication, convex duality, and completeness.

• Stochastic representations of linear PDEs allow us to switch from finite difference schemes to Monte Carlo. In Chapter 2, we recall the principle of the Monte Carlo method and focus on various issues related to it such as reduction of variance, bias of discretization schemes for stochastic differential equations (SDEs), and Romberg's extrapolation. Throughout this book, we will review existing and suggest new Monte Carlo methods for solving nonlinear PDEs.

• In Chapter 3, we explore a variety of advanced models and pricing concepts that are not often covered in monographs: complete market models, utility indifference pricing, quantile hedging, and P&L variance minimization. These hedging frameworks lead to nonlinear pricing operators which, for Markovian models, are connected to nonlinear parabolic PDEs.

• Chapter 4 is an overview of basic properties of nonlinear second order parabolic PDEs. We explain that due to the comparison principle, these equations naturally show up in finance. Many nonlinear PDEs arising in finance are of the Hamilton-Jacobi-Bellman type. For readers not familiar with stochastic control, we present a crash course on this subject highlighting the main notions and results such as verification theorems and viscosity solutions.

• In Chapter 5, as an appetizer for the rest of the book, we present a collection of examples of nonlinear PDEs arising in quantitative finance. They arise for instance when one considers the pricing of American-style options, the uncertain volatility model, the uncertain lapse and mortality model, transaction costs, illiquid markets, different rates for borrowing and lending, counterparty risk valuation, or the pricing under delta and gamma constraints. Some of these examples, together with probabilistic methods for solving the PDE in high dimension, are studied in detail in the next chapters.

• Chapter 6 focuses purely on American-style options. Their pricing involves a special variety of nonlinear PDEs called *variational inequalities*. We begin by recalling some finite difference methods for solving these PDEs. We thus present regression methods such as the Longstaff-Schwartz and Tsitsiklis-Van Roy algorithms as well as the parametric approach and the dual approach introduced by Rogers. Then, these algorithms are generalized when we consider chooser options. Chooser options are a generalization of American options. They give the holder the right but not the obligation to change his/her payoff at some discrete dates. Some pricing examples—a multi-asset convertible bond, the restrikable put, the "passport option"—are looked at in detail and illustrate primal and dual methods.

• Chapter 7 investigates backward stochastic differential equations (in short BSDEs). These SDEs differ from classical forward SDEs in that we impose a terminal condition. One of the key features of BSDEs is that they provide a probabilistic representation of solutions of nonlinear parabolic PDEs, generalizing the Feynman-Kac formula. For example, the PDE arising in the uncertain lapse and mortality model (introduced in Chapter 8) is connected to first order BSDEs; and the PDE arising in the uncertain volatility model (introduced in Chapter 9) is connected to second order BSDEs.

• Chapter 8 deals with the uncertain lapse and mortality model. This model is used for reinsurance deals. Since there is no market to hedge the lapse risk (the risk that an insurance policy subscriber may cancel his/her policy before maturity) or to hedge the mortality risk, it is natural to price conservatively, using adequate ranges for lapse and mortality rates. The nonlinear PDE solving the problem is simple, in that the nonlinearity affects none of the differential terms. We first present the PDE method, and then a Monte Carlo method that naturally extends the Longstaff-Schwartz algorithm. We also make a link with first order BSDEs.

• In Chapter 9, we consider the pricing of options in the uncertain volatility model as introduced by Avellaneda *et al.* [36] and Lyons [159]. We suggest two probabilistic methods to solve the problem when the number of variables prevents us from using a finite difference scheme. The first method requires a parameterization of the optimal covariance matrix and consists of a series of backward low-dimensional optimizations. The second method relies heavily on a recently established connection between second order BSDEs and fully nonlinear second order parabolic PDEs.

• In Chapter 10, we introduce nonlinear McKean SDEs. McKean SDEs will be used in Chapters 11 and 12 to calibrate models to market smiles. These SDEs differ from classical Itô diffusions in that the drift and volatility coefficients are allowed to depend on the (unknown) marginal distribution of the process. We review classical existence results for a simple nonlinear SDE, the so-called McKean-Vlasov process. We then present the particle algorithm, an elegant stochastic simulation of such processes. We emphasize the link with mean-field approximations in statistical physics.

• In Chapter 11, we focus on the calibration of local stochastic volatility models to market smiles. The dynamics of an asset in such a model is described by a nonlinear McKean SDE. Existence of this nonlinear McKean SDE is not in anyway obvious. We argue numerically that this nonlinear diffusion does not exist when the smile produced by the naked stochastic volatility model is very far from the market smile. We also consider the case of stochastic interest rates. The calibration of such models is challenging and requires advanced tools such as the particle method, Malliavin calculus, or the Markovian projection technique.

• In Chapter 12 we investigate the calibration of local correlation models to market smiles of vanilla options written on a basket of assets. We review the two calibration methods that have been suggested so far in the literature, and introduce a new general technique which uses the particle method to produce a whole family of calibrated local correlation models. We also show how the new method generalizes (i) to models that combine stochastic interest rates, stochastic dividend yield, local stochastic volatility, and local correlation; and (ii) to single-asset path-dependent volatility models.

• One of the main issues of BSDEs, introduced in Chapter 7, is that they require the computation of conditional expectations, usually by regression methods. This could be a hard task, in particular for multi-asset options. In Chapter 13 we introduce a new, promising, method based on branching diffusions for solving high-dimensional nonlinear PDEs which does not require estimations of conditional expectations. We first apply this stochastic representation to the pricing of counterparty risk. Then we briefly show how this method can be extended to fully nonlinear PDEs.

Exercises have been included at the end of almost each chapter. This allows readers to check their understanding. Most of the time, these problems originate from discussions we have had with traders.

The idea for this book originated in the presentation given by one of the authors at the conference Numerical Methods in Finance held at Ecole des Ponts in Paris in April 2009 [127]. Originally an oral and partial version was given as a mini-course by one of the authors at the conference Research in Options that took place in Rio de Janeiro in November 2009 [119]. From the authors' perspective, this book is a good illustration of quantitative analyst's skills: a mix between analytical and numerical expertise.

Book audience

This book is aimed mainly at practitioners working in quantitative finance. As a consequence, we assume familiarity with classical stochastic analysis and Black-Scholes option pricing. The reader needing a reminder on these matters could consult [20], [7], and Chapter 1 in [11]. Regarding the new tools introduced, the book is self-contained. This book may also be of interest for academic researchers and PhD students working in the field of financial mathematics.

Acknowledgments

Most of this research was done while we were colleagues at Société Générale. The content of Chapter 12 was developed when Julien was working at Bloomberg LP. We would like to thank the members of the Global Markets Quantitative Research team at Société Générale for many helpful discussions: Lorenzo Bergomi, Philippe Carpentier, Alexandre Charitopoulos, Stéphane Crapanzano, Adel Dellali, Constantin Denuelle, Olivier Drevillon, Benoît Humez, Rémi Monsarrat, and Julien Tijou.

We say a special thank-you to Lorenzo Bergomi for sharing with us during many years his insights and ideas on quantitative finance in general and on stochastic volatility modeling in particular. The reader is warmly encouraged to consult his forthcoming monograph on this subject. Julien is also grateful to Bruno Dupire for sharing so many ideas and creating a stimulating research environment at Bloomberg.

Pierre would like to thank his collaborators at Ecole Polytechnique, in particular Alfred Galichon, Xiaolu Tan, and Nizar Touzi. Julien owes a particular debt to his PhD advisor, Jean-François Delmas, who guided his first steps in the world of research. He warmly thanks his colleagues at Bloomberg: Bruno Dupire, Chakri Cherukuri, Sylvain Corlay, Xin Cui, Ivailo Dimov, Bryan Liang, Alexey Polishchuk, and Arun Verma for carefully reading preliminary versions of this book and suggesting improvements. He is grateful to Oleg Kovrizhkin and Sylvain Corlay for providing him with the FX data of Chapter 12, and to Ivailo Dimov for producing Figure 12.18.

We also both wish to thank Frédéric Abergel, Aurélien Alfonsi, Marco Avellaneda, Vlad Bally, Pascal Delanoe, Jean-François Delmas, Nicole El Karoui, Jim Gatheral, Emmanuel Gobet, Benjamin Jourdain, Bernard Lapeyre, Fabio Mercurio, Gilles Pagès, Chris Rogers, and Denis Talay for fruitful discussions. We also pay tribute to the memory of Paul Malliavin who pointed out to us the efficiency of the Clark-Ocone formula while we were working on Section 11.7.6.

We would also like to thank Ms. Sarah Ramsay for her very careful proofreading. We fully acknowledge that all remaining errors are ours. We are grateful in advance to all the readers who will report them via email.

Eventually, we would like to express our gratitude to our families, Julien, Véronique, Aurélien, Emma, and Florian, for their support and patience, especially during the cyclical periods when this book project consumed many evenings and weekends.

Pierre and Julien dedicate this book to their parents. Julien has special thoughts for his father Xavier, his late friend Pierrette Sentenac, and Annick Mahieux, his math teacher in mathématiques supérieures at Lycée Henri IV in Paris, who introduced him many years ago to the incredibly beautiful world of mathematics.

About the Authors

Julien Guyon is a member of the Quantitative Research Group at Bloomberg LP, New York. Before joining Bloomberg, Julien worked in the Global Markets Quantitative Research team at Société Générale in Paris (2006–2012). He was also a visiting professor at Université Paris 7 and at Ecole des Ponts, where he taught mathematics of finance in master programs. Julien holds a Ph.D. in probability theory and statistics from Ecole des Ponts (Paris). He graduated from Ecole Polytechnique (Paris), Université Paris 6, and Ecole des Ponts. His main current research interests include numerical probabilistic methods and volatility and correlation modeling.

Pierre Henry-Labordère works in the Global Markets Quantitative Research team at Société Générale. After receiving his Ph.D. at Ecole Normale Supérieure (Paris) in the theory of superstrings, he joined the Theoretical Physics department at Imperial College London before moving to finance in 2004. Since 2011, Pierre has also been an associate researcher at Centre de Mathématiques Appliquées, Ecole Polytechnique. He was the recipient of the 2013 Quant of the Year award from *Risk Magazine*.

Symbol Description

$\mathbb{Q}$	Probability measure, usually a risk-neutral measure.
$\mathbb{P}^{\text{hist}}$	Historical or real probability measure.
$\mathbb{Q}^T$	Forward measure associated with maturity T.
$\mathbb{Q}^f$	Risk-neutral measure associated to a foreign currency.
$\mathbb{E}^{\mathbb{P}}$	Expectation with respect to the measure $\mathbb{P}$.
$\text{Var}^{\mathbb{P}}$	Variance with respect to the measure $\mathbb{P}$.
$\mathbb{P} \sim \mathbb{Q}$	Equivalent probability measures.
$\frac{d\mathbb{Q}}{d\mathbb{P}}$	Radon-Nikodym derivative of the measure $\mathbb{Q}$ with respect to the measure $\mathbb{P}$.
$\mathbb{P}_t$	Law of X_t under $\mathbb{P}$.
$\mathbb{P}_t^N$	Empirical measure of the $(X_t^{i,N})_{1 \le i \le N}$.
$\mathcal{F}_t$	Filtration.
$Y \in \mathcal{F}_t$	Y $\mathcal{F}_t$-measurable.
X_t	Vector of all market variables, or asset price, depending on the context.
S_t	Asset price.
W_t, Z_t	Brownian motion under the measure $\mathbb{Q}$.
W_t^T, Z_t^T	Brownian motion under the measure $\mathbb{Q}^T$.
W_t^f, Z_t^f	Brownian motion under the measure $\mathbb{Q}^f$.
$\langle X, Y \rangle_t$	Covariation of the processes X and Y.
$\langle X \rangle_t$	$\langle X, X \rangle_t$.
$f(t, X_t, Y_t, Z_t)$	Driver of a 1-BSDE.
$f(t, X_t, Y_t, Z_t, \Gamma_t)$	Driver of a 2-BSDE.
$D_s^W X_t$	Malliavin derivative of X_t w.r.t. W at time s.
N_t	For a branching diffusion, number of particles alive at t.
$z_t^1, \ldots, z_t^{N_t}$	The particles alive at t.
Ω_k	Number of branchings of type k on $[0, T]$.
$\xrightarrow{(d)}$	Convergence in distribution.
$\mathcal{N}(m, \sigma^2)$	Gaussian distribution with mean m and variance σ^2.
$\#p$	Number of paths in a Monte Carlo run.
a.s.	Almost surely.
r.v.	Random variable.
i.i.d.	Independent and identically distributed.
w.r.t.	With respect to.
r.h.s.	Right-hand side.
l.h.s.	Left-hand side.
T	Maturity.
K	Strike.
$\mathcal{C}(T, K)$	Market value of a call option of strike K and maturity T.
B_t	Money-market account.
P_{st}	Time-s value of a bond of maturity t.
D_{tT}	Discount factor between the dates t and T.
$\tilde{Y}_t$	Discounted value $D_{0t} Y_t$ of a price process Y_t.
f_t	Forward value of the asset X_t (for a fixed maturity T).
r_t	Instantaneous interest rate.
$\sigma(t, S_t)$	Local volatility of the asset S.
$\sigma_{\text{Dup}}(t, S_t)$	Dupire local volatility of the asset S implied by the market prices of vanilla options.

$\sigma(t, S_t, X_t)$ Path-dependent volatility of the asset S.

a_t Stochastic volatility.

$\sigma_\mathrm{r}(t, r_t)$ Instantaneous volatility of the rate r_t.

$\sigma_\mathrm{P}^T(t)$ Instantaneous volatility of the bond P_{tT}.

$\rho(t, S_t^1, \dots, S_t^N)$ Local correlation of the assets $S^1, \dots, S^N$.

ρ^hist Historical correlation.

F_T $\mathcal{F}_T$-measurable payoff.

$g(X_T)$ Payoff of a vanilla option.

$L_i(t)$ Value at t of the Libor fixing at T_{i-1} and operating between T_{i-1} and T_i.

$T_{\beta(t)-1}$ The nearest future Libor fixing date.

$\mathcal{B}_t(F_T)$ Buyer's super-replication price.

$\mathcal{S}_t(F_T)$ Seller's super-replication price.

$\mathcal{T}_D$ The set of all $(\mathcal{F}_t)$-stopping times τ such that, $\mathbb{P}^\mathrm{hist}$-a.s., $\tau \in D$.

$\mathcal{T}_{tT}$ $\mathcal{T}_D$ with $D = [t, T]$.

$\mathcal{S}_t$ Snell envelope.

pdv_t Path-dependent variables.

ddv_t Decision-dependent variables.

cddv_t Continuous decision-dependent variables.

dddv_t Discrete decision-dependent variables.

λ_t^D Instantaneous mortality rate.

λ_t^L Instantaneous lapse rate.

$\mathbb{R}_+$ $\{x \in \mathbb{R} \mid x \geq 0\}$.

$\mathbb{R}_+^*$ $\{x \in \mathbb{R} \mid x > 0\}$.

$a \wedge b$ $\min(a, b)$.

$a \vee b$ $\max(a, b)$.

$\mathbf{1}_{x>0}$ Heaviside function: 1 if $x > 0$, and 0 otherwise.

$\delta(x)$ Delta Dirac function.

$\delta_K(x)$ $\delta(x - K)$.

δ_{ij} Kronecker delta symbol: $\delta_{ij} = 1$ if $i = j$, and 0 otherwise.

$\mathcal{S}^n$ Space of real n-dimensional symmetric matrices.

bp Basis point: $\times 10^{-4}$.

BS$(\cdot)$ Price of a call option in the Black-Scholes model.

GBM Geometric Brownian motion.

PDE Partial differential equation.

SDE Stochastic differential equation.

BSDE Backward stochastic differential equation.

ELMM Equivalent local martingale measure.

HJB Hamilton-Jacobi-Bellman.

LV Local volatility.

LVM Local volatility model.

SV Stochastic volatility.

SVM Stochastic volatility model.

LSV Local stochastic volatility.

LSVM Local stochastic volatility model.

SIR-LSVM Stochastic interest rate–Local stochastic volatility model.

ULMM Uncertain lapse and mortality model.

UVM Uncertain volatility model.

CVA Credit Valuation adjustment.

$[\underline{\lambda}, \overline{\lambda}]$ Range of the uncertain parameter λ.

V_t Quadratic variation of $\ln X_t$.

M_t $\max_{0 \leq s \leq t} X_s$.

m_t $\min_{0 \leq s \leq t} X_s$.

$Du, D_x u$ Gradient of $u(t, x)$ w.r.t. the space variables x.

$D^2 u, D_x^2 u$ Hessian matrix of $u(t, x)$ w.r.t. the space variables x.

$H(t, x, u, Du, D^2 u)$ Hamiltonian.

H^* Parabolic envelope of H.

$L^\infty([0,T] \times \mathbb{R}^d)$ Bounded functions on $[0,T] \times \mathbb{R}^d$.

$C_b(\mathbb{R}^d)$ Bounded continuous functions on $\mathbb{R}^d$.

$C_b^k(\mathbb{R}^d)$ Bounded k times continuously differentiable functions on $\mathbb{R}^d$ with bounded derivatives.

$C_b^\infty(\mathbb{R}^d)$ Infinitely differentiable functions with bounded derivatives of any order.

$C_{\text{pol}}^\infty(\mathbb{R}^d)$ Infinitely differentiable functions with polynomially growing derivatives of any order.

$C^{1,2}$ Functions of (t,x) that are one time continuously differentiable in t and two times continuously differentiable in x.

$\|f\|_{\text{Lip}}$ Lipschitz constant of the Lipschitz-continuous function f.

$\bar{\mathcal{D}}$ Closure of the domain $\mathcal{D}$.

$\partial \mathcal{D}$ Boundary of the domain $\mathcal{D}$.

g^{conc} Concave envelope of g.

g^{conv} Convex envelope of g.

List of Figures

List of Tables

Chapter 1

Option Pricing in a Nutshell

> Tell me what the true price of this exotic option is.
> — A structurer at Société Générale[1]

In this first chapter, we recall basic results on stochastic representations of solutions of linear parabolic PDEs. This allows us to revisit key notions of option pricing and set our notations. Most of the material presented here is standard and can be found in various classical references. Yet we also highlight several important notions which are not often covered, and were not covered in [11], such as super-replication, pricing in incomplete models, and convex duality, as well as Dirichlet options and Neumann options.

1.1 The super-replication paradigm

1.1.1 Models of financial markets

Let us consider a filtered probability space $\left(\Omega, (\mathcal{F}_t)_{0 \leq t \leq T}, \mathbb{P}^{\text{hist}}\right)$. Here $\mathbb{P}^{\text{hist}}$ is the historical or real probability measure under which we model our market. A market model is defined by an n-dimensional stochastic differential equation (SDE)

$$dX_t^i = b_i(t, X_t)\, dt + \sum_{j=1}^{d} \sigma_{i,j}(t, X_t)\, dW_t^j, \qquad i \in \{1, \ldots, n\} \qquad (1.1)$$

and by another positive stochastic process B_t, called the money-market account, representing the value of cash, which satisfies

$$dB_t = r_t B_t\, dt, \qquad B_0 = 1$$

i.e.,

$$B_t = \exp\left(\int_0^t r_s\, ds\right)$$

---

[1]Despite our best efforts, we have never been able to answer this question (see Section 1.1.6).

r_t is the short term interest rate. It is adapted to $\mathcal{F}_t$, which is the (natural) filtration generated by the d-dimensional uncorrelated standard Brownian motion $\{W_t^j\}_{1 \leq j \leq d}$. In order to ensure that SDE (1.1) admits a unique strong solution (see e.g., [13]), we assume that b and σ satisfy:

Assum(SDE): The functions b and σ are Lipschitz-continuous in x uniformly in t, and satisfy a linear growth condition: there exists a positive constant C such that for all $t \geq 0$, $x, y \in \mathbb{R}^n$,

$$|b(t, x) - b(t, y)| + |\sigma(t, x) - \sigma(t, y)| \leq C|x - y|$$
$$|b(t, x)| + |\sigma(t, x)| \leq C(1 + |x|)$$

We set

$$D_{tu} \equiv B_t B_u^{-1} = \exp\left(-\int_t^u r_s \, ds\right)$$

which is the discount factor from date u to date t. Throughout the book, we will denote by $\tilde{Y}_t \equiv D_{0t} Y_t$ the discounted value of any price process Y_t. Certain market components X^i may not be sold or bought in the market, such as the short term interest rate, or a stochastic volatility. Throughout this book, a market component X^i that can be sold and bought in the market is called an "asset."

1.1.2 Self-financing portfolios

Let us assume that we have a portfolio consisting of m assets, say $X_t^1, \ldots, X_t^m$, and the money-market account B_t. It is convenient to use the notation X^0 for B. The portfolio at a time t is composed of Δ_t^i assets X_t^i and Δ_t^0 units of X_t^0 (cash). The Δ_t^i's must be $\mathcal{F}_t$-measurable, i.e., we cannot look into the future. The portfolio value π_t is

$$\pi_t \equiv \sum_{i=0}^m \Delta_t^i X_t^i \tag{1.2}$$

As time passes, we can readjust the allocations Δ_t^i, but no cash is ever injected or removed from the portfolio: between t and $t + dt$, the variation in the portfolio value is only due to the variation of the values of the assets, i.e.,

$$d\pi_t = \sum_{i=0}^m \Delta_t^i \, dX_t^i \tag{1.3}$$

We then speak of a *self-financing portfolio*. In terms of discounted values, this reads

$$d\tilde{\pi}_t = \sum_{i=0}^m \Delta_t^i \, d\tilde{X}_t^i = \sum_{i=1}^m \Delta_t^i \, d\tilde{X}_t^i \tag{1.4}$$

because for any price process Y_t, $d\tilde{Y}_t = D_{0t}(dY_t - r_t Y_t \, dt)$, concluding that[2]

$$\tilde{\pi}_t = \pi_0 + \sum_{i=1}^{m} \int_0^t \Delta_s^i \, d\tilde{X}_s^i \qquad (1.5)$$

We may also write this as

$$\tilde{\pi}_t = \pi_0 + \int_0^t \Delta_s \cdot d\tilde{X}_s$$

where $\cdot$ denotes the usual scalar product in $\mathbb{R}^m$. As a technical condition, we need to introduce the notion of admissible portfolio:

DEFINITION 1.1 Admissible portfolio $(\Delta_t, 0 \le t \le T)$ *defines an admissible portfolio if $\tilde{\pi}_t$ is bounded from below for all t $\mathbb{P}^{\text{hist}}$-a.s., i.e., there exists $M \in \mathbb{R}$ such that*

$$\mathbb{P}^{\text{hist}} (\forall t \in [0, T], \ \tilde{\pi}_t \ge M) = 1$$

1.1.3 Arbitrage and arbitrage-free models

Let us now introduce the notion of arbitrage. An arbitrage is a self-financing strategy that is worth zero initially and yields a positive gain without any risk.

DEFINITION 1.2 Arbitrage *A self-financing admissible portfolio is called an arbitrage if the corresponding value process π_t satisfies $\pi_0 = 0$ and*

$$\pi_T \ge 0 \quad \mathbb{P}^{\text{hist}} - \text{a.s} \qquad \text{and} \qquad \mathbb{P}^{\text{hist}}(\pi_T > 0) > 0$$

Arbitrageurs are a special kind of trader. Their role is precisely to detect and take full advantage of arbitrage opportunities as soon as they appear in the market. This impacts market prices: arbitrage opportunities tend to disappear as soon as they arise. Absence of arbitrage opportunities is therefore a natural modeling assumption. The next lemma gives a sufficient condition under which we exclude arbitrage opportunities in our market model.

LEMMA 1.1 Sufficient condition excluding arbitrage
Suppose there exists a measure $\mathbb{Q}$ on $(\Omega, \mathcal{F}_T)$ such that[3] $\mathbb{Q} \sim \mathbb{P}^{\text{hist}}$ and such that, for all asset X^i, the discounted price process $\{\tilde{X}_t^i\}_{t \in [0,T]}$ is a local mar-

---

[2]The stochastic integral is well-defined with the condition $\int_0^t (\Delta_s^i)^2 \, d\langle X \rangle_s < \infty$ $\mathbb{P}^{\text{hist}}$-a.s. $\tilde{\pi}_t$ is then a local martingale.

[3]$\mathbb{P}$ is equivalent to $\mathbb{Q}$ (denoted by $\mathbb{P} \sim \mathbb{Q}$) if and only if $\forall A \in \mathcal{F}_T$, $\mathbb{P}(A) = 0 \iff \mathbb{Q}(A) = 0$.

tingale with respect to $\mathbb{Q}$.[4] *Then the market $\{X_t\}_{t\in[0,T]}$ has no arbitrage.*

Note that the assumption of Lemma 1.1 bears only on *assets X^i only*, not on non-tradable components of X, such as instantaneous interest rates, instantaneous stochastic volatility, etc.

PROOF   Let us suppose that there exists an arbitrage. From (1.5) and the fact that Δ defines an admissible portfolio, $\tilde{\pi}_t$ is a lower bounded local martingale w.r.t. $\mathbb{Q}$. By a standard result (see Exercise 1.3.1), $\tilde{\pi}_t$ is thus a $\mathbb{Q}$-supermartingale and hence $\mathbb{E}^{\mathbb{Q}}[\tilde{\pi}_T] \leq \tilde{\pi}_0 = 0$. But since $\tilde{\pi}_T \geq 0$ $\mathbb{P}^{\mathrm{hist}}$-a.s. and $\mathbb{Q} \sim \mathbb{P}^{\mathrm{hist}}$, we have $\tilde{\pi}_T \geq 0$ $\mathbb{Q}$-a.s. Hence $\tilde{\pi}_T = 0$ $\mathbb{Q}$-a.s., so $\tilde{\pi}_T = 0$ $\mathbb{P}^{\mathrm{hist}}$-a.s., which yields $\pi_T = 0$ $\mathbb{P}^{\mathrm{hist}}$-a.s. This contradicts the fact that $\mathbb{P}^{\mathrm{hist}}(\pi_T > 0) > 0$.
□

This simple lemma invites one to introduce the notion of an equivalent local martingale measure (in short ELMM):

DEFINITION 1.3 Equivalent local martingale measure   *Any measure $\mathbb{Q} \sim \mathbb{P}^{\mathrm{hist}}$ on $(\Omega, \mathcal{F}_T)$ such that, for all asset X^i, the discounted price process $\{\tilde{X}_t^i\}_{t\in[0,T]}$ is a local martingale w.r.t. $\mathbb{Q}$ is called an equivalent local martingale measure (ELMM). An ELMM is also called a risk-neutral measure.*

Lemma 1.1 now reads: if there exists an equivalent local martingale measure, then the market has no arbitrage opportunities. Throughout this book we will always assume that we are in a situation where there exists at least one equivalent local martingale measure.

REMARK 1.1 Drift of an asset under an equivalent local martingale measure   Let us consider an asset X^i. Under an ELMM $\mathbb{Q}$, $\{\tilde{X}_t^i\}_{t\in[0,T]}$ is an $(\mathcal{F}_t)$-local martingale, hence has zero drift. As a consequence, X_t^i has drift $r_t X_t^i$:

$$dX_t^i = \exp\left(\int_0^t r_s\, ds\right)(d\tilde{X}_t^i + r_t \tilde{X}_t^i\, dt)$$

Hence the drift of an *asset* X^i under an ELMM is $r_t X_t^i$. This is not the case for non-tradable components of the market such as interest rates, stochastic volatility, etc.
□

---

[4]Throughout this book, unless stated otherwise, martingales and local martingales are considered with respect to the filtration $(\mathcal{F}_t)_{0 \leq t \leq T}$.

1.1.4 Super-replication

Let us assume that, at time t, we buy and delta-hedge a European option[5] written on m assets, say $X_t^1, \ldots, X_t^m$, with maturity T and payoff F_T, at the price z. In general, the payoff F_T is a function of the paths $(X_t^i, 0 \le t \le T)$ followed by the prices of the m assets between times 0 and T. The final value of the buyer's portfolio, discounted at time 0, is

$$
\begin{aligned}
\tilde{\pi}_T^B &= -D_{0t}z + \sum_{i=1}^m \int_t^T \Delta_s^i \, d\tilde{X}_s^i + D_{0T}F_T \\
&= -D_{0t}z + \int_t^T \Delta_s \cdot d\tilde{X}_s + D_{0T}F_T
\end{aligned}
$$

We can then define the buyer's super-replication price at time t as the greatest price z such that the value of the buyer's portfolio $\tilde{\pi}_T^B$ is $\mathbb{P}^{\text{hist}}$-a.s. nonnegative. To be precise, we introduce the following:

DEFINITION 1.4 Buyer's price

$$
\mathcal{B}_t(F_T) = \sup \left\{ z \in \mathcal{F}_t \; \middle| \; \text{there exists an admissible portfolio } \Delta \text{ such that} \right.
$$

$$
\left. \tilde{\pi}_T^B \equiv -D_{0t}z + \int_t^T \Delta_s \cdot d\tilde{X}_s + D_{0T}F_T \ge 0 \; \mathbb{P}^{\text{hist}} - a.s. \right\} \quad (1.6)
$$

The price z must be $\mathcal{F}_t$-measurable, denoted by $z \in \mathcal{F}_t$, i.e., we cannot look into the future. Similarly, we can define the seller's super-replication price as:

DEFINITION 1.5 Seller's price

$$
\mathcal{S}_t(F_T) = \inf \left\{ z \in \mathcal{F}_t \; \middle| \; \text{there exists an admissible portfolio } \Delta \text{ such that} \right.
$$

$$
\left. \tilde{\pi}_T^S \equiv D_{0t}z + \int_t^T \Delta_s \cdot d\tilde{X}_s - D_{0T}F_T \ge 0 \; \mathbb{P}^{\text{hist}} - a.s. \right\} \quad (1.7)
$$

THEOREM 1.1 Arbitrage-free bounds
Assume that there exists an equivalent local martingale measure $\mathbb{Q} \sim \mathbb{P}^{\text{hist}}$. We then have

$$
\mathcal{B}_t(F_T) \le \mathbb{E}^{\mathbb{Q}}[D_{tT}F_T|\mathcal{F}_t] \le \mathcal{S}_t(F_T)
$$

---

[5]American options will be treated in Chapter 6.

PROOF   From the definition of $\mathcal{B}_t(F_T)$, we assume that there exists an admissible portfolio Δ such that

$$-D_{0t}z + \int_t^T \Delta_s \cdot d\tilde{X}_s + D_{0T}F_T \geq 0 \qquad \mathbb{P}^{\text{hist}} - a.s.$$

As seen previously (see proof of Lemma 1.1 and Exercise 1), $\int_0^t \Delta_s \cdot d\tilde{X}_s$ is a $\mathbb{Q}$-supermartingale, so $\mathbb{E}^{\mathbb{Q}}[\int_t^T \Delta_s \cdot d\tilde{X}_s | \mathcal{F}_t] \leq 0$ and we deduce that

$$D_{0t}z \leq \mathbb{E}^{\mathbb{Q}}[D_{0T}F_T | \mathcal{F}_t]$$

i.e.,

$$z \leq \mathbb{E}^{\mathbb{Q}}[D_{tT}F_T | \mathcal{F}_t]$$

Taking the supremum over z, we get $\mathcal{B}_t(F_T) \leq \mathbb{E}^{\mathbb{Q}}[D_{tT}F_T | \mathcal{F}_t]$. The inequality $\mathbb{E}^{\mathbb{Q}}[D_{tT}F_T | \mathcal{F}_t] \leq \mathcal{S}_t(F_T)$ can be derived similarly. $\Box$

These bounds can be sharpened by assuming that the market is complete. In order to define what it means for a market to be complete, we need to introduce the notion of attainable payoff:

DEFINITION 1.6 Attainable payoff   *A payoff F_T is said to be attainable (at time 0) if there exists an admissible portfolio Δ and a real number z such that $z + \int_0^T \Delta_s \cdot d\tilde{X}_s - D_{0T}F_T = 0$ $\mathbb{P}^{\text{hist}}$-a.s. and $\int_0^t \Delta_s \cdot d\tilde{X}_s$ is a (true) $\mathbb{Q}$-martingale with $\mathbb{Q} \sim \mathbb{P}^{\text{hist}}$.*

Stated otherwise, a payoff F_T is said to be attainable if we can generate the wealth F_T at time T from an initial wealth z by trading only in the underlying assets and the cash in a self-financing way. The portfolio Δ and the real number z are unique because if $z' + \int_0^T \Delta_s' \cdot d\tilde{X}_s - D_{0T}F_T = 0$, then

$$\int_0^T (\Delta_s' - \Delta_s) \cdot d\tilde{X}_s = z - z'$$

and since $\tilde{X}$ is a (nontrivial) $\mathbb{Q}$-local martingale, this yields $\Delta_s' = \Delta_s$, hence $z' = z$. As a consequence, there is a unique fair price at $t = 0$ for an attainable payoff, and this price is z. The assumption that $\int_0^t \Delta_s \cdot d\tilde{X}_s$ is a (true) $\mathbb{Q}$-martingale guarantees that $z = \mathbb{E}^{\mathbb{Q}}[D_{0T}F_T]$, so we have a formula to compute the price z from the payoff F_T. See Theorem 1.2 below.

DEFINITION 1.7 Complete market   *A market is said to be complete when every payoff is attainable at time 0.*

THEOREM 1.2
For a complete market, for any ELMM $\mathbb{Q}$ and any payoff F_T, we have

$$\mathcal{B}_t(F_T) = \mathcal{S}_t(F_T) = \mathbb{E}^{\mathbb{Q}}[D_{tT}F_T|\mathcal{F}_t]$$

PROOF By completeness, we can find a real number z and an admissible portfolio Δ such that $z + \int_0^T \Delta_s \cdot d\tilde{X}_s - D_{0T}F_T = 0$ $\mathbb{P}^{\text{hist}}$-a.s., hence $\mathbb{Q}$-a.s. As a consequence,

$$D_{0t}z_t + \int_t^T \Delta_s \cdot d\tilde{X}_s - D_{0T}F_T = 0 \qquad \mathbb{Q}\text{--a.s.}$$

with $z_t = D_{0t}^{-1}\left(z + \int_0^t \Delta_s \cdot d\tilde{X}_s\right)$ being $\mathcal{F}_t$-measurable. Since $\int_0^t \Delta_s \cdot d\tilde{X}_s$ is a $\mathbb{Q}$-martingale, we have $D_{0t}z_t = \mathbb{E}^{\mathbb{Q}}[D_{0T}F_T|\mathcal{F}_t]$, i.e., $z_t = \mathbb{E}^{\mathbb{Q}}[D_{tT}F_T|\mathcal{F}_t]$, whence $\mathcal{B}_t(F_T) \geq \mathbb{E}^{\mathbb{Q}}[D_{tT}F_T|\mathcal{F}_t]$. Combined with the opposite inequality $\mathcal{B}_t(F_T) \leq \mathbb{E}^{\mathbb{Q}}[D_{tT}F_T|\mathcal{F}_t]$ (see Theorem 1.1), we get $\mathcal{B}_t(F_T) = \mathbb{E}^{\mathbb{Q}}[D_{tT}F_T|\mathcal{F}_t]$. We proceed similarly for the other equality. ⬜

With incomplete markets or markets with friction, perfect replication is no longer possible. In the theorem below, we state that the buyer's super-replication price is given by the infimum of expectations of the discounted payoff over the set ELMM of equivalent local martingale measures.

THEOREM 1.3 Super-replication price [100, 151]
The buyer's super-replication price and the seller's super-replication price are given by

$$\mathcal{B}_t(F_T) = \inf_{\mathbb{Q}\in\text{ELMM}} \mathbb{E}^{\mathbb{Q}}[D_{tT}F_T|\mathcal{F}_t]$$

$$\mathcal{S}_t(F_T) = \sup_{\mathbb{Q}\in\text{ELMM}} \mathbb{E}^{\mathbb{Q}}[D_{tT}F_T|\mathcal{F}_t]$$

Here, we give a formal proof using convex duality (minimax principle) as explained in [177]. For simplicity, we assume that $t = 0$. We only look at the seller's super-replication price—the case of the buyer's super-replication price is treated similarly by replacing F_T by $-F_T$. For readers not familiar with the minimax principle, we think it is more important in a first step to understand the formal proof than to go along the lines of the rigorous proof (see [100, 151]), which involves the optional decomposition theorem.

PROOF (formal) As seen in (1.4), the portfolio Δ is self-financing if and only if

$$d\tilde{\pi}_t = \Delta_t \cdot d\tilde{X}_t \equiv \sum_{i=1}^{m} \Delta_t^i \, d\tilde{X}_t^i$$

Note that in general, the vector X_t that describes the market at date t is not only made of the assets $X_t^1, \dots, X_t^m$, but has extra non-asset components $X_t^{m+1}, \dots, X_t^n$ that may describe stochastic interest rates, stochastic volatility, etc. Using the dynamics (1.1) for X_t, we get that the portfolio Δ is self-financing if and only if

$$d\tilde{\pi}_t = \sum_{i=1}^{m} \Delta_t^i D_{0t} \left((b_i(t, X_t) - r_t X_t^i) \, dt + \sum_{j=1}^{d} \sigma_{i,j}(t, X_t) \, dW_t^j \right)$$

$$= D_{0t} \Delta_t \cdot ((b(t, X_t) - r_t X_t) \, dt + \sigma(t, X_t) \, dW_t) \qquad (1.8)$$

Recall that the scalar product $\cdot$ concerns only the first m coordinates of the vectors $b(t, X_t)$, X_t and $\sigma(t, X_t) \, dW_t$.

By introducing the penalty function U given by $U(x) = +\infty$ for $x < 0$, 0 otherwise, we rewrite the seller's price at time 0 as

$$S_0(F_T) = \inf_{\Delta \text{ adm., } \tilde{\pi}_t \text{ given by (1.8) and } \tilde{\pi}_0 = 0, \, z} \mathbb{E}^{\mathbb{P}^{\text{hist}}} \left[z + U \left(z + \tilde{\pi}_T - D_{0T} F_T \right) \right]$$

The presence of the U-term ensures that $z + \int_0^T \Delta_t \cdot d\tilde{X}_t - D_{0T} F_T \geq 0$ $\mathbb{P}^{\text{hist}}$-a.s. The basic idea is then to rewrite this constrained problem as an "inf sup" unconstrained problem, and then exchange the two operators inf and sup by formally applying a minimax principle, i.e., replacing an "inf sup" problem by a "sup inf" problem. In this respect, we introduce the Lagrange multiplier Y_t, also known as the *shadow price*, associated to the constraint (1.8) on $d\tilde{\pi}_t$, defined by

$$dY_t = A_t \, dt + Z_t \, dW_t \qquad (1.9)$$

for some processes A and Z, where Z has the dimension of W, i.e., is d-dimensional, and is such that $\int_0^t \tilde{\pi}_s Z_s \, dW_s$ and $\int_0^t D_{0s} Y_s \Delta_s \cdot (\sigma(s, X_s) \, dW_s)$ are $\mathbb{P}^{\text{hist}}$-martingales. The initial condition $Y_0 \in \mathbb{R}$ is left unspecified for the moment. We consider the stochastic integral $\int_0^T Y_t \, d\tilde{\pi}_t$. On the one hand, integration by parts gives

$$\int_0^T Y_t \, d\tilde{\pi}_t = \tilde{\pi}_T Y_T - \tilde{\pi}_0 Y_0 - \int_0^T \tilde{\pi}_t \, dY_t - \langle \tilde{\pi}, Y \rangle_T$$

On the other hand, we have, provided dynamics (1.8) holds,

$$\int_0^T Y_t \, d\tilde{\pi}_t = \int_0^T Y_t D_{0t} \Delta_t \cdot ((b(t, X_t) - r_t X_t) \, dt + \sigma(t, X_t) \, dW_t)$$

Taking the expectation in both equations above, we get that $\tilde{\pi}_t$ satisfies the two constraints (1.8) and $\tilde{\pi}_0 = 0$ if and only if for all Y

$$
\mathbb{E}^{\mathbb{P}^{\text{phist}}} \left[\tilde{\pi}_T Y_T \right.
$$
$$
\left. - \int_0^T (\tilde{\pi}_t A_t + D_{0t} \Delta_t \cdot (\sigma(t, X_t) Z_t) + D_{0t} Y_t \Delta_t \cdot (b(t, X_t) - r_t X_t)) \, dt \right] = 0
$$
(1.10)

We set

$$
H(Y) \equiv \mathbb{E}^{\mathbb{P}^{\text{phist}}} \left[z + U \left(z + \tilde{\pi}_T - D_{0T} F_T \right) + \tilde{\pi}_T Y_T \right.
$$
$$
\left. - \int_0^T (\tilde{\pi}_t A_t + D_{0t} \Delta_t \cdot (\sigma(t, X_t) Z_t) + D_{0t} Y_t \Delta_t \cdot (b(t, X_t) - r_t X_t)) \, dt \right]
$$

As this functional is linear in Y, it is easy to check that

$$
\sup_Y H(Y) = \begin{cases} \mathbb{E}^{\mathbb{P}^{\text{phist}}} [z + U (z + \tilde{\pi}_T - D_{0T} F_T)] & \text{if (1.8) holds and } \tilde{\pi}_0 = 0 \\ +\infty & \text{otherwise} \end{cases}
$$

The seller's price can then be written as a minimax problem:

$$
S_0(F_T) = \inf_{\Delta \text{ adm., } \tilde{\pi}_t \text{ unconst., } z} \sup_Y H(Y)
$$

Here the minimax principle can be applied. It consists of exchanging the inf and sup operators. Taking for granted that this principle can be invoked (the functional H should be concave (resp. convex) with respect to Y (resp. Δ, $\tilde{\pi}_t$, z), see Section 3 in [177]), we obtain the dual formulation:

$$
S_0(F_T) = \sup_Y \inf_{\Delta \text{ adm., } \tilde{\pi}_t \text{ unconst., } z} H(Y)
$$
(1.11)

Taking the infimum of $H(Y)$ over $\tilde{\pi}_T$ (the final value of $(\tilde{\pi}_t)$), we obtain a finite value if and only if $Y_T \geq 0$:

$$
\inf_{\tilde{\pi}_T} (U (z + \tilde{\pi}_T - D_{0T} F_T) + \tilde{\pi}_T Y_T) = (D_{0T} F_T - z) Y_T
$$

Indeed for $Y_T < 0$, this infimum is equal to $-\infty$. We can disregard this case since we consider the supremum over Y_T in a second step.

Similarly, by taking the infimum over $\tilde{\pi}_t$ ($t \in [0, T)$), we obtain a finite value if and only if

$$
A_t = 0
$$

and the minimization over Δ_t results in a finite value if and only if the complementary slackness condition

$$
\forall i \in \{1, \ldots, m\}, \qquad (\sigma(t, X_t) Z_t)_i + Y_t(b_i(t, X_t) - r_t X_t^i) = 0
$$

holds. By setting $Z_t \equiv -Y_t\lambda_t \equiv -Y_t(\lambda_t^1, \ldots, \lambda_t^d)$, we get from (1.9) that Y_t follows the SDE

$$dY_t = -Y_t\lambda_t \, dW_t \qquad (1.12)$$

where λ_t is a d-dimensional adapted process such that

$$\forall i \in \{1, \ldots, m\}, \qquad (\sigma(t, X_t)\lambda_t)_i = b_i(t, X_t) - r_t X_t^i \qquad (1.13)$$

Finally,

$$S_0(F_T) = \sup_{\lambda \in \mathcal{S}, y_0 \in \mathbb{R}} \inf_z \mathbb{E}^{\mathbb{P}^{\text{hist}}} [z + (D_{0T}F_T - z) \, Y_T(-\lambda; y_0)]$$

where $Y_t(-\lambda; y_0)$ is a solution to (1.12) with $Y_0 = y_0$ and

$$\mathcal{S} \equiv \{\mathcal{F}_t\text{--adapted process } (\lambda_t) \text{ such that (1.13) holds}\}$$

The minimization over z results in a finite value if and only if $\mathbb{E}^{\mathbb{P}^{\text{hist}}}[Y_T(-\lambda; y_0)] = 1$, i.e., assuming that $Y_t(-\lambda; y_0)$ is a true $\mathbb{P}^{\text{hist}}$-martingale, $y_0 = 1$:

$$S_0(F_T) = \sup_{\lambda \in \mathcal{S}} \mathbb{E}^{\mathbb{P}^{\text{hist}}} [D_{0T}F_T Y_T(-\lambda; 1)]$$

Let us introduce the probability measure $\mathbb{Q}^\lambda$ defined by

$$\frac{d\mathbb{Q}^\lambda}{d\mathbb{P}^{\text{hist}}}\Big|_{\mathcal{F}_T} = Y_T(-\lambda; 1)$$

Using Girsanov's theorem, one can verify that, if $\lambda \in \mathcal{S}$, the discounted prices of all *assets* $(\tilde{X}_t^i)$, $i \in \{1, \ldots, m\}$, are local martingales under $\mathbb{Q}^\lambda$:

$$\forall i \in \text{asset}, \qquad d\tilde{X}_t^i = D_{0t}(\sigma(t, X_t) \, dW_t')_i$$

with W' a $\mathbb{Q}^\lambda$-Brownian motion. Actually, any ELMM is of the $\mathbb{Q}^\lambda$ type, with $\lambda \in \mathcal{S}$. As a consequence,

$$S_0(F_T) = \sup_{\mathbb{Q} \in \text{ELMM}} \mathbb{E}^{\mathbb{Q}}[D_{0T}F_T]$$

which completes the proof.   □

COROLLARY 1.1 *Complete market*
A market is complete if and only if there exists a unique ELMM.

PROOF   Assume the market is complete. Let $\mathbb{Q}^0$ be an ELMM. From Theorems 1.2 and 1.3, for any payoff F_T, $\mathcal{B}_0(F_T) = S_0(F_T) = \mathbb{E}^{\mathbb{Q}^0}[D_{0T}F_T] = \inf_{\mathbb{Q} \in \text{ELMM}} \mathbb{E}^{\mathbb{Q}}[D_{0T}F_T] = \sup_{\mathbb{Q} \in \text{ELMM}} \mathbb{E}^{\mathbb{Q}}[D_{0T}F_T]$. Hence, for all F_T, for all

$\mathbb{Q} \in$ ELMM, $\mathbb{E}^{\mathbb{Q}}[D_{0T}F_T] = \mathbb{E}^{\mathbb{Q}^0}[D_{0T}F_T]$, so that ELMM $= \{\mathbb{Q}^0\}$. Conversely, if ELMM $= \{\mathbb{Q}^0\}$, then $\mathcal{B}_0(F_T) = \mathcal{S}_0(F_T) = \mathbb{E}^{\mathbb{Q}^0}[D_{0T}F_T]$. As a consequence, there exist two admissible portfolios Δ^B and Δ^S (one for the buyer and one for the seller) such that $\mathbb{P}^{\text{hist}}$-a.s., i.e., $\mathbb{Q}^0$-a.s.

$$-\mathbb{E}^{\mathbb{Q}^0}[D_{0T}F_T] + \int_0^T \Delta_t^B \cdot d\tilde{X}_t + D_{0T}F_T \geq 0$$

$$\mathbb{E}^{\mathbb{Q}^0}[D_{0T}F_T] + \int_0^T \Delta_t^S \cdot d\tilde{X}_t - D_{0T}F_T \geq 0$$

Summing those inequalities, this means that $\mathbb{Q}^0$-a.s.

$$\int_0^T (\Delta_t^B + \Delta_t^S) \cdot d\tilde{X}_t \geq 0$$

and, because the discounted prices of all *assets* $(\tilde{X}_t^i)$ are $\mathbb{Q}^0$-local martingales, this yields $\Delta_t^B = -\Delta_t^S$ $\mathbb{Q}^0$-a.s. Eventually, we have that $\mathbb{E}^{\mathbb{Q}^0}[D_{0T}F_T] + \int_0^T \Delta_t^S \cdot d\tilde{X}_t - D_{0T}F_T = 0$ $\mathbb{Q}^0$-a.s., hence $\mathbb{P}^{\text{hist}}$-a.s., and the market is complete. □

1.1.5   Complete models versus incomplete models

From (1.13), the market model (1.1) is complete if there exists a unique $\mathcal{F}_t$-adapted vector $\lambda_t \in \mathbb{R}^d$ such that

$$\forall i \in \text{asset}, \qquad b_i(t, X_t) - r_t X_t^i = \sum_{j=1}^d \sigma_{i,j}(t, X_t)\lambda_t^j \qquad (1.14)$$

The unique ELMM $\mathbb{Q}$ is then given by

$$\frac{d\mathbb{Q}}{d\mathbb{P}^{\text{hist}}}\Big|_{\mathcal{F}_T} \equiv Y_T(-\lambda; 1) = \prod_{j=1}^d e^{-\int_0^T \lambda_t^j \, dW_t^j - \frac{1}{2}\int_0^T (\lambda_t^j)^2 \, dt}$$

The unique arbitrage-free price is

$$\mathcal{B}_t(F_T) = \mathcal{S}_t(F_T) = \mathbb{E}^{\mathbb{Q}}[D_{tT}F_T | \mathcal{F}_t]$$

An inspection of (1.14) reveals that if the market is complete, then the rank of $\sigma(t, X_t)$ is equal to d a.s. (which implies that #assets $\geq d$). In the case where #assets $< d$, the market cannot be complete. Moreover, if the number of assets coincides with the number of Brownian motions, i.e., #assets $= d$, and $(\sigma_{i,j}(t, X_t))_{i \in \text{asset}, 1 \leq j \leq d}$ is invertible, then the market is complete. So, provided the volatility matrix $\sigma_{i,j}(t, X_t)$ is correctly estimated, there is a unique arbitrage-free price.

Examples of complete models that are commonly used by practitioners include Dupire's local volatility model [95]; Libor market models with local volatilities, e.g., BGM with deterministic volatilities [67]; and Markov functional models [138] (see also [11], Chapter 2).

Common examples of incomplete models are stochastic volatility models (in short SVMs). Here #assets $< d$. An example of stochastic volatility model is the double lognormal SVM, which has attracted the attention of practitioners in equity markets [113, 137]. The dynamics of the underlying, denoted by X_t, reads under a risk-neutral measure $\mathbb{Q}^0 \sim \mathbb{P}^{\text{hist}}$ as

$$\frac{dX_t}{X_t} = r_t \, dt + \sqrt{V_t} \left(\rho_{\text{XV}^0} \, dW_t^3 + \frac{\rho_{\text{XV}} - \rho \rho_{\text{XV}^0}}{\sqrt{1-\rho^2}} \, dW_t^2 + \chi \, dW_t^1 \right) \quad (1.15)$$

$$dV_t = -k(V_t - V_t^0) \, dt + \sigma V_t \left(\rho \, dW_t^3 + \sqrt{1-\rho^2} \, dW_t^2 \right) \quad (1.16)$$

$$dV_t^0 = -k(V_t^0 - V^{00}) \, dt + \sigma^0 V_t^0 \, dW_t^3 \quad (1.17)$$

with $\chi^2 = 1 - \rho_{\text{XV}^0}^2 - \frac{(\rho_{\text{XV}} - \rho \rho_{\text{XV}^0})^2}{1-\rho^2}$ and W_t^1, W_t^2, W_t^3, three uncorrelated standard $\mathbb{Q}^0$-Brownian motions. V_t is the instantaneous variance, and V_t^0 plays the role of a moving long-term average value for V_t. Neither V_t nor V_t^0 are tradable instruments. The only asset is X_t. Its drift under any ELMM is $r_t X_t$. The incompleteness of such a model can be detected as the drift terms in V_t and V_t^0 can be arbitrarily modified with a change of measure from $\mathbb{Q}^0$ to $\mathbb{Q}^1 \sim \mathbb{P}^{\text{hist}}$:

$$\frac{d\mathbb{Q}^1}{d\mathbb{Q}^0}\Big|_{\mathcal{F}_T} = \prod_{j=1}^{3} e^{-\int_0^T \lambda_t^j \, dW_t^j - \frac{1}{2}\int_0^T (\lambda_t^j)^2 \, dt}$$

such that

$$\rho_{\text{XV}^0} \lambda_t^3 + \frac{\rho_{\text{XV}} - \rho_{\text{XV}^0} \rho}{\sqrt{1-\rho^2}} \lambda_t^2 + \chi \lambda_t^1 = 0 \quad (1.18)$$

Equation (1.18) guarantees that the drift of X_t under $\mathbb{Q}^1$ is still $r_t X_t$, i.e., that $\mathbb{Q}^1$ is still an ELMM. In Exercise 4.7.2, we will obtain that the seller's super-replication (undiscounted) price at $t = 0$ of a European vanilla payoff $F_T \equiv g(X_T)$ in such an SVM is $g^{\text{conc}}(X_0)$ with g^{conc} the concave envelope of g, i.e., the smallest concave function that is greater than or equal to g. Similarly, the buyer's super-replication (undiscounted) price at $t = 0$ of a European vanilla payoff $g(X_T)$ in such an SVM is $g^{\text{conv}}(X_0)$ with g^{conv} the convex envelope of g, i.e., the largest convex function that is smaller than or equal to g. For example, for a call option $g(x) = (x-K)^+$, $g^{\text{conc}}(x) = x$ and $g^{\text{conv}}(x) = 0$. This simple example illustrates the main drawback of the super-replication framework: the buyer's and seller's super-replication prices are not close and therefore cannot be used in practice. This drawback can be circumvented by adding extra market instruments, such as vanilla options, in the superhedging strategy. This leads to tight bounds and a nice generalization of the optimal transport theory [52, 112] where now the transport is performed along a martingale measure.

An alternative approach consists of completing the market by modeling directly the dynamics of market instruments, instead of the dynamics of unobservable quantities such as a stochastic volatility. This is in line with a familiar trader's quote: "An option should be hedged with options." Like for any asset, the discounted prices of these market instruments are local martingales under any risk-neutral measure and there are no more arbitrary drifts as in (1.16)–(1.17). In Section 3.1 we will present various examples of complete market models.

1.1.6 Pricing in practice

In practice, the seller's price at time t is computed by picking out a particular ELMM $\mathbb{Q}$:

$$u_t \equiv \mathbb{E}^{\mathbb{Q}}[D_{tT}F_T|\mathcal{F}_t] \tag{1.19}$$

Under this measure $\mathbb{Q}$, the drift for an asset X^i is fixed to $b_i(t, X_t) \equiv r_t X_t^i$ (see Remark 1.1). In an incomplete market, $\mathbb{Q}$ does not necessary achieve the supremum in Theorem 1.3, and we lose the superhedging strategy paradigm. Selling options becomes a risky business. However, it seems that the idea of a "true price" (based on a "true model") is still vivid in the community of structurers and sales people (see the quote at the beginning of this chapter). In practice, picking a particular ELMM simplifies a lot the pricing problem: it becomes a *linear* problem, i.e., the price of the (European) payoff $F_T^1 + F_T^2$ equals the sum of the prices of the (European) payoffs F_T^1 and F_T^2.

We will always assume that there exists a (deterministic) function r such that $r_t = r(t, X_t)$. Then

$$D_{t_1 t_2} = \exp\left(-\int_{t_1}^{t_2} r(s, X_s)\, ds\right)$$

If there exists g such that $F_T = g(X_T)$, we speak of a vanilla option. In such a case, by the Markov property of X,

$$u_t = \mathbb{E}^{\mathbb{Q}}\left[\exp\left(-\int_t^T r(s, X_s)\, ds\right) g(X_T)\Big|\mathcal{F}_t\right] \equiv u(t, X_t)$$

is a function u of (t, X_t). Below, we recall that $u(t, x)$ is a solution to a linear second order parabolic PDE, the so-called Black-Scholes pricing PDE.

1.2 Stochastic representation of solutions of linear PDEs

1.2.1 The Cauchy problem

We denote by $\mathcal{L}$ the Itô generator of X defined under a risk-neutral measure $\mathbb{Q}$:

$$\mathcal{L} = \sum_{i=1}^{n} b_i(t,x)\partial_i + \frac{1}{2}\sum_{i,j=1}^{n}\sum_{k=1}^{d}\sigma_{i,k}(t,x)\sigma_{j,k}(t,x)\partial_{ij} \qquad (1.20)$$

We recall that if X^i is an asset, $b_i(t,x) = r(t,x)x_i$. The discounted price process

$$D_{0t}u_t = D_{0t}\mathbb{E}^{\mathbb{Q}}[D_{tT}F_T|\mathcal{F}_t] = \mathbb{E}^{\mathbb{Q}}[D_{0T}F_T|\mathcal{F}_t]$$

is a $\mathbb{Q}$-martingale. Then a straightforward application of Itô's lemma to $D_{0t}u(t,X_t) = \exp\left(-\int_0^t r(s,X_s)\,ds\right)u(t,X_t)$ shows that the pricing function u of a vanilla option, if smooth enough, satisfies the second order parabolic linear PDE:

$$\partial_t u(t,x) + \mathcal{L}u(t,x) - r(t,x)u(t,x) = 0 \qquad (1.21)$$

with the terminal condition $u(T,x) = g(x)$. Conversely, a solution u to (1.21) admits the stochastic representation (1.19) as stated by the Feynman-Kac theorem:

THEOREM 1.4 Feynman-Kac, see e.g., [13]
*Let b, σ satisfy **Assum(SDE)**, r be uniformly bounded from below and f have quadratic growth in x uniformly in t. Let $u \in C^{1,2}([0,T]\times\mathbb{R}^d)\cap C^0([0,T]\times\mathbb{R}^d)$ with quadratic growth in x uniformly in t and solution to the parabolic PDE:*

$$\partial_t u(t,x) + \mathcal{L}u(t,x) - r(t,x)u(t,x) + f(t,x) = 0$$

with terminal condition $u(T,x) = g(x)$. Then u admits the stochastic representation

$$u(t,x) = \mathbb{E}^{\mathbb{Q}}\left[\int_t^T e^{-\int_t^s r(u,X_u)du}f(s,X_s)ds + e^{-\int_t^T r(s,X_s)ds}g(X_T)\,\bigg|\,X_t = x\right]$$

REMARK 1.2 Path-dependent options Path-dependent payoffs are those payoffs for which one cannot write $F_T = g(X_T)$, i.e., whose value depends on the asset values not only at maturity T, but also at previous dates $t < T$. For instance, with one asset $X_t = X_t^1$:

- Asian options: $F_T = g\left(\frac{1}{T}\int_0^T X_t\,dt\right)$

- Barrier and lookback options: $F_T = g\left(\min_{t\in[0,T]} X_t, \max_{t\in[0,T]} X_t, X_T\right)$

- Cliquet options: $F_T = g\left(X_{T_i}/X_{T_{i-1}}, 1 \le i \le N\right)$

In some cases, one can write SDEs for the path-dependent variables and add them to the SDEs describing the market X so that one can write $F_T = g(X_T')$ for an enlarged market X', and the option price satisfies a (degenerate) PDE in the variables x'. For instance, the price of an Asian option satisfies a PDE in the variables (x, a) where $A_t = \int_0^t X_s\,ds$ and $dA_t = X_t\,dt$. For a general path-dependent payoff $F_T = g(X_{t_1}, \ldots, X_{t_n})$, depending on the spot values at the observation dates $t_1 < \cdots < t_n \equiv T$, the PDE, depending on the past spot values $Z^1, \ldots, Z^n$, can be written as

$$\partial_t u_i(t, x, Z^1, \ldots, Z^{i-1}) + \mathcal{L}u_i(t, x, Z^1, \ldots, Z^{i-1}) = 0, \qquad \forall t \in [t_{i-1}, t_i)$$

with the matching conditions at the constat dates

$$u_n(t_n, x, Z^1, \ldots, Z^{n-1}) = g(Z^1, \ldots, Z^{n-1}, x)$$
$$u_i(t_i^-, x, Z^1, \ldots, Z^{i-1}) = u_{i+1}(t_i^+, x, Z^1, \ldots, Z^{i-1}, x)$$

The final price is $u_1(0, X_0)$. ⬜

Example 1.1
Let us now derive the pricing PDE for an option that delivers $g(\tau, X_\tau)$ at time τ if $\tau < T$, or $g(T, X_T)$ at time T if $T \ge \tau$, where τ is the first time of jump of a Poisson process with deterministic intensity $\beta(t) \in \mathbb{R}_+$, independent of the filtration $(\mathcal{F}_t)$. Think of τ as a default time. Assume that $\tau > t$. Then the price of the option at time t is

$$u(t, x) = \mathbb{E}\left[\mathbf{1}_{\tau \ge T} e^{-\int_t^T r(s, X_s)ds} g(T, X_T) + \mathbf{1}_{\tau < T} e^{-\int_t^\tau r(s, X_s)ds} g(\tau, X_\tau)\,\Big|\, X_t = x\right]$$

$$= \mathbb{E}\left[g(T, X_T)e^{-\int_t^T (r(s, X_s)+\beta(s))ds}\right.$$
$$\left. + \int_t^T \beta(s)g(s, X_s)e^{-\int_t^s (r(u, X_u)+\beta(u))du}\,ds\,\Big|\, X_t = x\right]$$

Feynman-Kac's theorem states that if v is a solution to the PDE

$$\partial_t v + \mathcal{L}v + \beta(g - v) - rv = 0 \qquad (1.22)$$

with the terminal condition $v(T, x) = g(T, x)$, then $v = u$. Under the usual assumptions guaranteeing the existence and uniqueness of a solution to the above PDE, we deduce that (1.22) is the pricing PDE for such an option. Note that PDE (1.22) can also be directly deduced from Itô's formula for processes with jumps. ⬜

1.2.2 The Dirichlet problem

Let $\mathcal{D}$ be a domain in $\mathbb{R}^n$ and let us consider the Dirichlet problem: given a continuous function Φ on $\partial\mathcal{D}$, find $u \in C(\bar{\mathcal{D}})$ such that u is C^2 in $\mathcal{D}$ and

$$\mathcal{A}u = 0 \text{ in } \mathcal{D}, \qquad u = \Phi \text{ on } \partial\mathcal{D} \qquad (1.23)$$

Here, $\mathcal{A}$ is a parabolic second order differential operator of the type

$$\mathcal{A} = \sum_{i=1}^{n} \mu_i(x)\partial_i + \frac{1}{2}\sum_{i,j=1}^{n}\sum_{k=1}^{d} \sigma_{i,k}(x)\sigma_{j,k}(x)\partial_{ij} \qquad (1.24)$$

In particular, $\mathcal{A}$ does not depend on time t, and differs from the infinitesimal generator $\mathcal{L}$ of our market model X_t defined in (1.20).

THEOREM 1.5 Dirichlet problem, see e.g., [13]
Let X be a diffusion with generator $\mathcal{A}$ and $\tau_D = \inf\{t \geq 0 \mid X_t \notin \mathcal{D}\}$ denote the exit time of the domain $\mathcal{D}$. If $\mathbb{E}^{\mathbb{Q}}[\tau_D|X_0 = x] < \infty$ and X admits a unique strong solution, then the solution to (1.23) satisfies

$$u(x) = \mathbb{E}^{\mathbb{Q}}[\Phi(X_{\tau_D})|X_0 = x] \qquad (1.25)$$

Conversely, suppose that u, as given by (1.25), belongs to $C(\bar{\mathcal{D}}) \cap C^2(\mathcal{D})$. Then u satisfies (1.23).

Dirichlet options

As an example of pricing PDEs with Dirichlet boundary conditions, let us cite (generalized) *timer options*. These options were first studied in [62], then first sold by Société Générale in 2007, and have since attracted the attention of investors [179]. They pay the holder a payoff $\Phi(X, V)$ when the spot X and the variance V, defined as the quadratic variation of the log-spot, $V_t \equiv \langle \ln X \rangle_t$, reach the boundary of a domain in (X, V), say $\mathcal{D} \subset \mathbb{R}_+ \times \mathbb{R}_+$.

For simplicity, let us take zero rates, repos, and dividends. We consider a function $u(X_t, V_t)$ with u smooth enough so that we can apply Itô's formula. Since V is an increasing process, $\langle X, V \rangle_t = \langle V \rangle_t = 0$, so Itô's formula reads for $s \geq t$

$$u(X_s, V_s) = u(X_t, V_t) + \int_t^s \partial_x u(X_r, V_r)\, dX_r + \int_t^s \partial_v u(X_r, V_r)\, dV_r$$
$$+ \frac{1}{2}\int_t^s \partial_x^2 u(X_r, V_r)\, d\langle X \rangle_r$$

Now, by assuming that the spot process X_t is a continuous local martingale, we have $d\langle X \rangle_t = X_t^2 dV_t$, so

$$u(X_s, V_s) = u(X_t, V_t) + \int_t^s \partial_x u(X_r, V_r) \, dX_r$$

$$+ \int_t^s \left(\partial_v u(X_r, V_r) + \frac{1}{2} X_r^2 \partial_x^2 u(X_r, V_r) \right) dV_r$$

$$= u(X_t, V_t) + \int_t^s \partial_x u(X_r, V_r) \, dX_r + \int_t^s \mathcal{A} u(X_r, V_r) \, dV_r$$

where

$$\mathcal{A} = \partial_v + \frac{1}{2} x^2 \partial_x^2$$

If u is a solution to the Dirichlet problem (1.23) then by taking $s = \tau_{\mathcal{D}}$ we get

$$\Phi(X_{\tau_{\mathcal{D}}}, V_{\tau_{\mathcal{D}}}) = u(X_t, V_t) + \int_t^{\tau_{\mathcal{D}}} \partial_x u(X_r, V_r) \, dX_r$$

Since interest rates, repos, and dividends are zero, the above equation means that one can replicate the option payout from an initial wealth $u(X_t, V_t)$ at time t by trading only in the underlying asset and the cash. As a consequence, the fair price of the option at time t is $u(X_t, V_t)$. By assuming that $\int_0^t \partial_x u(X_r, V_r) \, dX_r$ is a true $\mathbb{Q}$-martingale, we get

$$u(X_t, V_t) = \mathbb{E}^{\mathbb{Q}} \left[\Phi(X_{\tau_{\mathcal{D}}}, V_{\tau_{\mathcal{D}}}) | \mathcal{F}_t \right] \qquad (1.26)$$

In particular, for $t = 0$, we recover (1.25):

$$u(x, v) = \mathbb{E}^{\mathbb{Q}} \left[\Phi(X_{\tau_{\mathcal{D}}}, V_{\tau_{\mathcal{D}}}) | X_0 = x, V_0 = v \right] \qquad (1.27)$$

Note that the solution u does not depend on the volatility of the asset X_t:

$$\partial_v u(x, v) + \frac{x^2}{2} \partial_x^2 u(x, v) = 0 \qquad (x, v) \in \mathcal{D} \qquad (1.28)$$

$$u(x, v) = \Phi(x, v) \qquad (x, v) \in \partial \mathcal{D}$$

As a striking consequence, both the price and hedge of timer options are model-free within the class of arbitrage-free models with continuous paths, zero rates, zero repo, and zero dividends, which makes these products very convenient for hedging. For example, we end up with the same pricing PDE whether we assume that X_t follows an arbitrary stochastic volatility model, or a local volatility model, or the Black-Scholes model, whatever the value of the Black-Scholes volatility. From Equation (1.27), we have

$$u(x, v) = \mathbb{E}^{\mathbb{Q}} [\Phi(B_{\tau_{\mathcal{D}}}^{\text{geo}}, V_{\tau_{\mathcal{D}}}) | B_0^{\text{geo}} = x] \qquad (1.29)$$

where B^{geo} is a geometric Brownian motion with volatility one under $\mathbb{Q}$ and $V_t = v + t = v + \langle \ln B^{\text{geo}} \rangle_t$.

PDE (1.28) can be solved numerically with a Monte Carlo simulation using the above stochastic representation (1.29). We observe below that for particular payoffs and domains, u admits an analytical expression.

Example 1.2 Timer calls
The holder of a timer call receives a call payoff $(X - K)^+$ when the variance reaches the target $V \geq \bar{V}$. This corresponds to the domain $\mathcal{D} = \mathbb{R}_+ \times [0, \bar{V}]$. PDE (1.28) can be seen as a heat kernel equation with the variance playing the role of the time. The solution is then given by the well known Black-Scholes formula, see (3.4), for a call option with volatility $\sigma_{\text{BS}} = 1$ and maturity $\bar{V}$:

$$u(X_0, V_0 = 0) = \text{BS}\left(\sigma_{\text{BS}}^2 T = \bar{V}, K | X_0\right)$$

$\Box$

Example 1.3 Hint on the Dambis-Dubins-Schwartz theorem
As a new example of timer options, we pay the payoff $\exp(\lambda X)$ when the quadratic variation of the spot reaches the target $V \geq \bar{V}$. Here we take $V_t \equiv \langle X \rangle_t$ and PDE (1.28) is replaced by

$$\partial_v u(x, v) + \frac{1}{2} \partial_x^2 u(x, v) = 0 \qquad \forall\, v < \bar{V}$$

$$u(x, \bar{V}) = e^{\lambda x}$$

The solution is $u(x, v) = e^{\lambda x - \frac{\lambda^2}{2}(v - \bar{V})}$, from which we deduce the identity

$$\mathbb{E}_{x,v}^{\mathbb{Q}}\left[e^{\lambda(X_{\tau_{\bar{V}}} - x)}\right] = e^{\frac{\lambda^2}{2}(\bar{V} - v)}$$

where $\tau_{\bar{V}} = \inf\{t \geq 0 \mid V_t \geq \bar{V}\}$. More generally, applying (1.26) with $t = \tau_{\bar{V}_1}$, we have that for $0 \leq \bar{V}_1 \leq \bar{V}_2 \equiv V$,

$$\mathbb{E}^{\mathbb{Q}}\left[e^{\lambda\left(X_{\tau_{\bar{V}_2}} - X_{\tau_{\bar{V}_1}}\right)} \Big| \mathcal{F}_{\tau_{\bar{V}_1}}\right] = e^{\frac{\lambda^2}{2}(\bar{V}_2 - \bar{V}_1)}$$

which means that the process $(X_{\tau_{\bar{V}}}, \bar{V} \geq 0)$ has independent and stationary increments. Since it has continuous paths, it is a $(\mathcal{F}_{\tau_{\bar{V}}})$-Brownian motion $B_{\bar{V}}$.

$\Box$

This last result, known as the Dambis-Dubins-Schwartz theorem, states that a continuous local martingale is a time-changed Brownian motion, namely, a Brownian motion evaluated at the quadratic variation of the local martingale:

THEOREM 1.6 Dambis-Dubins-Schwartz, see e.g., [13]
Let (X_t) be a continuous $(\mathcal{F}_t)$-local martingale satisfying $\lim_{t\to\infty}\langle X\rangle_t = \infty$ $\mathbb{Q}$-a.s. Define for each $0 \leq \bar{V} < \infty$ the stopping time $\tau_{\bar{V}} = \inf\{t \geq 0 \mid \langle X\rangle_t \geq \bar{V}\}$. Then the time-changed process $B_{\bar{V}} \equiv X_{\tau_{\bar{V}}}$ is a $\mathcal{G}_{\bar{V}}$-standard Brownian motion, where $\mathcal{G}_{\bar{V}} \equiv \mathcal{F}_{\tau_{\bar{V}}}$. In particular, we have $\mathbb{Q}$-a.s, $X_t = B_{\langle X\rangle_t}$.

Example 1.4 Double barrier variance calls
The holder of this option receives a call payoff on the variance, $(V - K)^+$, when the spot reaches the low barrier $\underline{x}$ or the high barrier $\bar{x}$. The domain is $\mathcal{D} = [\underline{x}, \bar{x}] \times \mathbb{R}_+$. We have for $x \in (\underline{x}, \bar{x})$ [74]:

$$u(x,v) = \int_{-\infty-\alpha i}^{\infty-\alpha i} \frac{\sqrt{\frac{x}{\bar{x}}}\sinh\left(\ln\left(\frac{x}{\underline{x}}\right)\sqrt{\frac{1}{4}-2iz}\right) - \sqrt{\frac{x}{\underline{x}}}\sinh\left(\ln\left(\frac{\bar{x}}{x}\right)\sqrt{\frac{1}{4}-2iz}\right)}{2\pi z^2 e^{i(K-v)z}\sinh\left(\ln\left(\frac{\bar{x}}{\underline{x}}\right)\sqrt{\frac{1}{4}-2iz}\right)}dz$$

where $\alpha > 0$. Any such α gives the same value for the integral. □

1.2.3 The Neumann problem

The Neumann problem is the following: given a smooth function Φ on $\partial\mathcal{D}$, find $u \in C(\bar{\mathcal{D}}) \cap C^2(\mathcal{D})$ such that

$$\mathcal{A}u = 0 \text{ in } \mathcal{D}, \qquad \frac{\partial u}{\partial\nu} = \Phi \text{ in } \partial\mathcal{D}$$

where $\mathcal{D}$ is a domain of $\mathbb{R}^n$, $\mathcal{A}$ is a parabolic second order differential operator of the type (1.24), and $\nu(x)$ denotes an inward-pointing unit vector at $x \in \partial\mathcal{D}$; $\frac{\partial u}{\partial\nu}$ should be read as $\sum_{i=1}^n \nu^i(x)\frac{\partial^i u}{\partial x^i}$. Here again we stress that time is not involved in this problem, $\mathcal{A}$ does not depend on time t, and differs from the infinitesimal generator $\mathcal{L}$ of our market model X_t given by (1.20).

Neumann options

Let us introduce the running minimum and running maximum processes: $M_t \equiv \max_{s\in[0,t]} X_s$, $m_t \equiv \min_{s\in[0,t]} X_s$. We consider an option that pays a payoff on the maximum M_t and the spot X_t when (X_t, M_t) reaches the boundary $\partial\mathcal{D}'$ of a domain $\mathcal{D}'$. Let us assume zero rates, repos, and dividends. We consider a function $u(X_t, M_t)$, with u smooth enough so that we can apply Itô's formula. Since M is an increasing process, $\langle X, M\rangle_t = \langle M\rangle_t = 0$, so Itô's formula reads for $s \geq t$

$$u(X_s, M_s) = u(X_t, M_t) + \int_t^s \partial_x u(X_r, M_r)\,dX_r + \int_t^s \partial_M u(X_r, M_r)\,dM_r$$

$$+ \frac{1}{2}\int_t^s \mathcal{A}u(X_r, M_r)\,d\langle X\rangle_r$$

where $\mathcal{A} = \partial_x^2$. If u is a solution to the problem

$$\mathcal{A}u(x, M) = 0 \qquad \{x < M\} \cap \mathcal{D}' \equiv \mathcal{D} \tag{1.30}$$

$$\partial_M u(x, M)|_{x=M} = 0 \tag{1.31}$$

$$u(x, M) = \Phi(x, M) \qquad (x, M) \in \partial\mathcal{D}' \tag{1.32}$$

then by taking $s = \tau_{\mathcal{D}'}$, we get

$$\Phi(X_{\tau_{\mathcal{D}'}}, M_{\tau_{\mathcal{D}'}}) = u(X_t, M_t) + \int_t^{\tau_{\mathcal{D}'}} \partial_x u(X_r, M_r)\, dX_r$$

Indeed, the condition $\partial_M u(x, M)|_{x=M} = 0$ ensures that $\partial_M u(X_r, M_r)\, dM_r \equiv 0$, because dM_r is always zero except when the spot equals the maximum, and when $dM_r \neq 0$, it is the integrand $\partial_M u(X_r, M_r)$ that vanishes. Since interest rates, repos, and dividends are zero, the above equation means that one can replicate the option payout from an initial wealth $u(X_t, M_t)$ at time t by trading only in the underlying asset and the cash. As a consequence, the fair price of the option at time t is $u(X_t, M_t)$. Note that Problem (1.30)-(1.32) is not a pure Neumann problem. It is a mixed Dirichlet and Neumann problem. Indeed, the conditions at the boundaries of the domain $\mathcal{D} = \{x < M\} \cap \mathcal{D}'$ are twofold:

- At the boundary of $\{x < M\}$, the condition is of Neumann type.

- At the boundary of $\mathcal{D}'$, the condition is of Dirichlet type.

Obviously, the same derivation holds with the running minimum instead of the running maximum, with $\{x < M\}$ replaced by $\{x > m\}$.

Example 1.5 Lookback Neumann options
Let us take $\mathcal{D}' = (\underline{x}, \overline{x}) \times \mathbb{R}_+$. Here, Equation (1.32) reads

$$u(\underline{x}, M) = \Phi(\underline{x}, M) \tag{1.33}$$

$$u(\overline{x}, \overline{x}) = \Phi(\overline{x}, \overline{x}) \tag{1.34}$$

as $M = \overline{x}$ when the spot reaches the upper barrier. Then, a straightforward computation gives

$$u(x, M) = \left(-\int_{\overline{x}}^M \frac{\partial_y \Phi(\underline{x}, y)}{y - \underline{x}}\, dy + \frac{\Phi(\overline{x}, \overline{x}) - \Phi(\underline{x}, \overline{x})}{\overline{x} - \underline{x}}\right)(x - \underline{x}) + \Phi(\underline{x}, M)$$

and therefore

$$u(X_0, X_0) = \left(-\int_{\overline{x}}^{X_0} \frac{\partial_y \Phi(\underline{x}, y)}{y - \underline{x}}\, dy + \frac{\Phi(\overline{x}, \overline{x}) - \Phi(\underline{x}, \overline{x})}{\overline{x} - \underline{x}}\right)(X_0 - \underline{x}) + \Phi(\underline{x}, X_0)$$

$$\square$$

Example 1.6 Digital Neumann options and Azéma-Yor martingales

Let us consider $\Phi(x, M) = \mathbf{1}_{x \geq \zeta}$ (with $\zeta > 0$) and $\mathcal{D}' = \{(x, M) \mid x \geq g(M)\}$ with g an increasing function such that $g(M) < M$. Equations (1.30)–(1.32) read

$$\partial_x^2 u(x, M) = 0 \qquad g(M) < x < M \tag{1.35}$$

$$\partial_M u(x, M)|_{x=M} = 0 \tag{1.36}$$

$$u(g(M), M) = \mathbf{1}_{g(M) \geq \zeta} \tag{1.37}$$

Using Equations (1.35) and (1.37) we get

$$u(x, M) = a(M)(x - g(M)) + \mathbf{1}_{g(M) \geq \zeta}$$

$u(X_t, M_t)$ is an example of an Azéma-Yor's martingale (see [42] and Exercise 3). Using the remaining equation (1.36), we get

$$a(M) = \alpha e^{\int_0^M \frac{g'(y)}{y - g(y)} dy} - \frac{\mathbf{1}_{g(M) \geq \zeta}}{g^{-1}(\zeta) - \zeta} e^{\int_{g^{-1}(\zeta)}^M \frac{g'(y)}{y - g(y)} dy}$$

The integration constant α is fixed by requiring that $u(0, 0) = 0$: when the asset (a nonnegative martingale) starts from 0, it stays at 0 and the option price is zero. Since $g(0) < 0$, this yields $\alpha = 0$, whence the price of the Neumann digital call:

$$\mathbb{E}^{\mathbb{Q}}[\mathbf{1}_{X_{\tau_{\mathcal{D}'}} \geq \zeta}] = \left(1 - \frac{X_0 - g(X_0)}{g^{-1}(\zeta) - \zeta} e^{\int_{g^{-1}(\zeta)}^{X_0} \frac{g'(y)}{y - g(y)} dy}\right) \mathbf{1}_{g(X_0) \geq \zeta} \tag{1.38}$$

with $\tau_{\mathcal{D}'} = \inf\{t \geq 0 \mid X_t \leq g(M_t)\}$.

Azéma and Yor [42] have considered such stopping times to build a simple solution to the Skorokhod embedding problem: Given a probability distribution μ such that $\int x \, d\mu(x) = X_0$, find a stopping time τ such that X_τ has distribution μ. One can build g so that $\tau_{\mathcal{D}'}$ is a solution:

$$g_\mu^{-1}(\zeta) = \frac{\int_\zeta^\infty x \, d\mu(x)}{\int_\zeta^\infty d\mu(x)} \tag{1.39}$$

□

Example 1.7 Range timer options

Define the range process as $R_t \equiv M_t - m_t$ and the stopping time $\tau_{\mathcal{D}'} = \inf\{t \geq 0 \mid R_t \geq L\}$ where $L > 0$ is a given threshold. A range timer option pays the holder a payoff $\Phi(x)$ on the spot when the range reaches the level L. Assume zero interest rates, repos, and dividends. We consider a function $u(X_t, m_t, M_t)$, with u smooth enough so that we can apply Itô's formula.

Since m and M are monotonous processes, $\langle X, M \rangle_t = \langle X, m \rangle_t = \langle M, m \rangle_t = \langle M \rangle_t = \langle m \rangle_t = 0$, so Itô's formula reads for $s \geq t$

$$u(X_s, m_s, M_s) = u(X_t, m_t, M_t) + \int_t^s \partial_x u(X_r, m_r, M_r)\, dX_r$$

$$+ \int_t^s \partial_m u(X_r, m_r, M_r)\, dm_r + \int_t^s \partial_M u(X_r, m_r, M_r)\, dM_r$$

$$+ \frac{1}{2} \int_t^s \mathcal{A}u(X_r, m_r, M_r)\, d\langle X \rangle_r$$

where $\mathcal{A} = \partial_x^2$. If u is a solution to the problem

$$\mathcal{A}u(x, m, M) = 0 \qquad m < x < M \tag{1.40}$$
$$\partial_M u(x, m, M)|_{x=M} = 0 \tag{1.41}$$
$$\partial_m u(x, m, M)|_{x=m} = 0 \tag{1.42}$$
$$u(m, m, L + m) = \Phi(m) \tag{1.43}$$
$$u(M, M - L, M) = \Phi(M) \tag{1.44}$$

then by taking $s = \tau_{\mathcal{D}'}$, we get

$$\Phi(X_{\tau_{\mathcal{D}'}}) = u(X_t, m_t, M_t) + \int_t^{\tau_{\mathcal{D}'}} \partial_x u(X_r, m_r, M_r)\, dX_r$$

Indeed, the conditions $\partial_m u(x, m, M)|_{x=m} = 0$ and $\partial_M u(x, M)|_{x=M} = 0$ ensure that $\partial_m u(X_r, m_r, M_r)\, dm_r$ and $\partial_M u(X_r, m_r, M_r)\, dM_r$ vanish. Since interest rates, repos, and dividends are zero, the above equation means that one can replicate the option payout from an initial wealth $u(X_t, V_t)$ at time t by trading only in the underlying asset and the cash, so the fair price of the option at time t is $u(X_t, m_t, M_t)$. Here again, the problem is not of a pure Neumann problem. It is a mixed Dirichlet and Neumann problem.

The first equation implies that the gamma of the option is zero and u is a linear function in x:

$$u(x, m, M) = a(m, M)x + b(m, M)$$

The other equations, written in a convenient form, are

$$\partial_M \left(a(m, M)M + b(m, M) \right) = a(m, M) \tag{1.45}$$
$$\partial_m \left(a(m, M)m + b(m, M) \right) = a(m, M) \tag{1.46}$$
$$a(m, L + m)m + b(m, L + m) = \Phi(m) \tag{1.47}$$
$$a(m, L + m)(L + m) + b(m, L + m) = \Phi(L + m) \tag{1.48}$$

By subtracting Equation (1.47) from Equation (1.48), we obtain

$$a(m, L + m) = L^{-1} \left(\Phi(L + m) - \Phi(m) \right) \tag{1.49}$$
$$b(m, L + m) = L^{-1} \left((L + m)\Phi(m) - m\Phi(m + L) \right) \tag{1.50}$$

Equations (1.45)–(1.46) can be integrated with respect to M and m, respectively, and we get

$$a(m, M) = a(m, L + m) - \int_{L+m}^{M} \frac{1}{y} \partial_M b(m, y) \, dy \qquad (1.51)$$

$$- \partial_m b(m, M) = \partial_m a(m, L + m) - \partial_m \int_{L+m}^{M} \frac{1}{y} \partial_M b(m, y) \, dy \qquad (1.52)$$

Differentiating (1.52) w.r.t. M, we get

$$\partial_{mM} b(m, M) = 0$$

which gives, by using (1.50),

$$\partial_m b(m, M) = L^{-1} \partial_m \left((L + m) \Phi(m) - m \Phi(m + L) \right)$$

Finally, by integrating this equation, the solution is

$$u(x, m, M) = L^{-1} \left((M - x) \Phi(M - L) + (x - m) \Phi(m + L) \right)$$
$$+ L^{-2} \left(\int_{M-L}^{m} (x - y) \Phi(y) \, dy + \int_{M}^{m+L} (x - y) \Phi(y) \, dy \right)$$

and

$$u(X_0, X_0, X_0) = L^{-2} \left(\int_{X_0-L}^{X_0} (X_0 - y) \Phi(y) \, dy + \int_{X_0}^{X_0+L} (y - X_0) \Phi(y) \, dx \right)$$

This solution was already obtained in [82] using probabilistic arguments. Our approach looks more straightforward. □

Conclusion

Stochastic representations of linear PDEs allow us to switch from a finite difference scheme method to a Monte Carlo implementation. Monte Carlo is unavoidable when the number of underlyings is large, because the computational time of finite difference methods grows exponentially with the number of variables. In the next chapter, we present the principle of the Monte Carlo method. Since time discretization is almost always needed when one implements a Monte Carlo algorithm, we also review recent results on discretization schemes.

1.3 Exercises

1.3.1 Local martingales and supermartingales

1. Recall the definition of a local martingale and of a supermartingale.

2. Prove that a local martingale that is bounded from below is a supermartingale. Hint: use Fatou's lemma.

1.3.2 Timer options on two assets

Let us consider two assets X^1 and X^2 which are modeled as continuous local martingales. A timer option on two underlyings pays $X^1 \Phi \left(\frac{X^2}{X^1} \right)$ when the quadratic variation $V_t = \langle \ln \frac{X^2}{X^1} \rangle_t$ exceeds the variance budget $\bar{V}$.

1. Prove that the fair value $u(X_t^1, X_t^2, V_t)$ can be written as

$$u(X_t^1, X_t^2, V_t) = X_t^1 U \left(V_t, \frac{X_t^2}{X_t^1} \right)$$

where $U(\cdot, \cdot)$ satisfies the one-dimensional PDE

$$\partial_v U(v, x) + \frac{1}{2} x^2 \partial_x^2 U(v, x) = 0$$
$$U(v, x) = \Phi(x) \qquad \forall v \geq \bar{V}$$

Hint: use X^1 as the numéraire.

2. Compute the solution for $\Phi(x) = (K - x)^+$. Hint: Write v as t!

1.3.3 Azéma-Yor martingales

Let us consider (X_t) a positive continuous $\mathbb{Q}$-local martingale; $M_t = \sup_{0 \leq s \leq t} X_s$, its running maximum; and $\Phi(x, M)$, a smooth function in $C^{2,1}$.

1. Prove that $\Phi(X_t, M_t)$ is a $\mathbb{Q}$-local martingale if and only if

$$\partial_x^2 \Phi(x, M) = 0 \qquad \forall x < M$$
$$\partial_M \Phi(x, M)|_{M=x} = 0$$

Hint: see Section 1.2.3 on Neumann options.

2. Derive the solution

$$\Phi(x, M) = a(M)(x - M) + \int_{x^*}^{M} a(y) \, dy \qquad (1.53)$$

where $a \in C^1$ and $x^* \in \mathbb{R}_+$ is an integration constant. $\Phi(X_t, M_t)$ is called an Azéma-Yor martingale and we have shown that this is the only (local) martingale under $\mathbb{Q}$ that can be written as a smooth function of X_t and its running maximum M_t. A stronger result that does not require the smoothness of a was proved in [168]: $\Phi(X_t, M_t)$ is a $\mathbb{Q}$-local martingale if and only if Φ is given by (1.53) with a a locally integrable function.

3. Deduce that

$$\mathbb{E}^{\mathbb{Q}}\left[a(M_T)(M_T - X_T) - \int_{X_T}^{M_T} a(y)\, dy\right] = \mathbb{E}^{\mathbb{Q}}\left[\int_{X_0}^{X_T} a(y)\, dy\right]$$

4. Apply the above equation with $a(y) = 2y$. Why is this relation useful for pricing the (exotic) option with payoff $(M_T - X_T)^2$ using T-vanilla options? How do you hedge this option?

5. Same question for the payoffs $1 - \frac{X_T}{M_T} + \ln \frac{X_T}{M_T}$ and $(M_T - B)\, \mathbf{1}_{X_T \le B \le M_T}$.

6. By assuming now that X_t is a discontinuous $\mathbb{Q}$-local martingale (in particular $dM_t(M_t - X_t)$ can be positive), show that

$$\mathbb{E}^{\mathbb{Q}}\left[a(M_T)(M_T - X_T) - \int_{X_T}^{M_T} a(y)\, dy\right] > \mathbb{E}^{\mathbb{Q}}\left[\int_{X_0}^{X_T} a(y)\, dy\right]$$

if a is an increasing function.

1.3.4 Azéma-Yor martingales and the Skorokhod embedding problem

The goal of this exercise is to prove in two different ways that a solution of the Skorokhod embedding problem is given by (1.39).

1. Derive (1.39) from Equation (1.38). Hint: Use that X_τ has distribution μ if and only if for all ζ, $\mathbb{E}^{\mathbb{Q}}[\mathbf{1}_{X_\tau \ge \zeta}] = 1 - F_\mu(\zeta)$, where F_μ is the cumulative distribution function of μ.

2. Using Exercise 3 and the optimal stopping theorem, show that for all $a \in C^0$ having compact support in $[X_0, +\infty)$,

$$\mathbb{E}^{\mathbb{Q}}\left[\int_{X_0}^{M_{\tau_{D'}}} a(u)\, du - (M_{\tau_{D'}} - g(M_{\tau_{D'}}))a(M_{\tau_{D'}})\right] = 0$$

Deduce that $\gamma(x) \equiv \mathbb{Q}(M_{\tau_{D'}} \ge x)$ satisfies

$$d\gamma(x) = -\frac{\gamma(x)\, dx}{x - g(x)}$$

Prove that a solution of the Skorokhod embedding problem is given by (1.39).

Chapter 2

Monte Carlo

> Mathematicians are like the French: You tell them something, they put it into their own language, and then it means something completely different.
>
> — Goethe

In the perspective of numerics, the Feynman-Kac theorem that we recalled in Chapter 1 allows us to switch from finite difference scheme methods to Monte Carlo implementations. Monte Carlo is more efficient with a high number of underlyings, because it hardly depends on the dimension of the problem while the computational time of finite difference methods grows exponentially with the number of variables. In this chapter, we briefly recall the principle of the Monte Carlo method and focus on various issues related to it, such as reduction of variance, bias of discretization schemes, and Romberg's extrapolation. In particular, we present new advanced results regarding the convergence of the Euler scheme, probably the most widely used discretization scheme, showing that the well known result for the rate of convergence, which assumes smooth payoffs, actually holds even for extremely irregular payoffs. We also devote a large section to the so-called Romberg extrapolation, a very simple technique for building high order discretization schemes from low order ones at almost no extra cost.

2.1 The Monte Carlo method

2.1.1 Principle of the method

As Equation (1.19) shows, the pricing of a derivative can be achieved by computing an expectation, namely, the expectation of the discounted payoff under an equivalent local martingale measure $\mathbb{Q}$. The Monte Carlo method consists of approximating the expectation $\mathbb{E}^{\mathbb{Q}}[Z]$ of a $\mathbb{Q}$-integrable random variable Z by

$$\hat{Z}_{\#p} = \frac{1}{\#p} \sum_{p=1}^{\#p} Z^{(p)}$$

where $\#p$ is large enough, and where the random variables $Z^{(p)}$ are independent and have the same distribution as Z. Indeed, the law of large numbers states that $\hat{Z}_{\#p}$ almost surely converges to $\mathbb{E}^{\mathbb{Q}}[Z]$ under $\mathbb{Q}$.

2.1.2   Sampling error, reduction of variance

The sampling error is defined as

$$\hat{Z}_{\#p} - \mathbb{E}^{\mathbb{Q}}[Z]$$

According to the central limit theorem, if Z^2 is integrable under $\mathbb{Q}$, then the sampling error is of magnitude $1/\sqrt{\#p}$:

$$\frac{\sqrt{\#p}}{\sigma}\left(\hat{Z}_{\#p} - \mathbb{E}^{\mathbb{Q}}[Z]\right) \xrightarrow{(d)} \mathcal{N}(0,1)$$

when $\#p$ tends to infinity, where $\xrightarrow{(d)}$ denotes convergence in distribution, and σ stands for the standard deviation of Z, i.e.,

$$\sigma^2 = \mathrm{Var}^{\mathbb{Q}}(Z) = \mathbb{E}^{\mathbb{Q}}\left[\left(Z - \mathbb{E}^{\mathbb{Q}}[Z]\right)^2\right]$$

This means that, for large $\#p$,

$$\hat{Z}_{\#p} \xrightarrow{(d)} \mathcal{N}\left(\mathbb{E}^{\mathbb{Q}}[Z], \frac{\mathrm{Var}^{\mathbb{Q}}(Z)}{\#p}\right)$$

Of course, in general, $\mathrm{Var}^{\mathbb{Q}}(Z)$, like $\mathbb{E}^{\mathbb{Q}}[Z]$, is unknown, and has to be estimated as well. Slutsky's theorem (see e.g., [12]) states that

$$\frac{\sqrt{\#p}}{\hat{\sigma}_{\#p}}\left(\hat{Z}_{\#p} - \mathbb{E}^{\mathbb{Q}}[Z]\right) \xrightarrow{(d)} \mathcal{N}(0,1)$$

when $\#p$ tends to infinity, where

$$\hat{\sigma}_{\#p}^2 = \frac{1}{\#p}\sum_{p=1}^{\#p}\left(Z^{(p)} - \hat{Z}_{\#p}\right)^2$$

In order to improve the accuracy of the Monte Carlo method, we may find another random variable Y that we can simulate under $\mathbb{Q}$, such that

$$\mathbb{E}^{\mathbb{Q}}[Y] = \mathbb{E}^{\mathbb{Q}}[Z] \qquad \text{and} \qquad \mathrm{Var}^{\mathbb{Q}}(Y) < \mathrm{Var}^{\mathbb{Q}}(Z)$$

In fact, we may even change measures: we may look for another probability measure $\mathbb{Q}'$ and another random variable Y that we can simulate under $\mathbb{Q}'$, such that

$$\mathbb{E}^{\mathbb{Q}'}[Y] = \mathbb{E}^{\mathbb{Q}}[Z] \qquad \text{and} \qquad \mathrm{Var}^{\mathbb{Q}'}(Y) < \mathrm{Var}^{\mathbb{Q}}(Z)$$

This is known in the literature as *reduction of variance*. There is a huge number of variance reduction techniques that gave rise to a vast literature:

- **Control variate methods** consist of building a random variable Z' close to Z in the sense that $\mathrm{Var}^{\mathbb{Q}}(Z - Z') < \mathrm{Var}^{\mathbb{Q}}(Z)$ and such that $\mathbb{E}^{\mathbb{Q}}[Z']$ is known. Then one picks $Y = Z - Z'$ and adds $\mathbb{E}^{\mathbb{Q}}[Z']$ to the Monte Carlo result. The closer Z' is to Z, the more accurate the Monte Carlo method. Ideally, $Z' = Z$, but $\mathbb{E}^{\mathbb{Q}}[Z]$ is not known—if it was known, one would not turn to a Monte Carlo procedure to compute it! Obviously, one can introduce a real number β, take $Y = Z - \beta Z'$, add $\beta \mathbb{E}^{\mathbb{Q}}[Z']$ to the Monte Carlo result, and search for the value of β that minimizes the variance of Y. This value is given by the slope of the straight line $z = \alpha + \beta z'$ that best approximates (in the mean square sense) Z as a function of Z', i.e., by the slope of the linear regression of Z over Z', or explicitly

$$\beta = \frac{\mathrm{Cov}^{\mathbb{Q}}(Z, Z')}{\mathrm{Var}^{\mathbb{Q}}(Z')}$$

Note that $\mathrm{Cov}^{\mathbb{Q}}(Z, Z')$ is usually unknown. It can be estimated in a pilot run, i.e., a first independent Monte Carlo run that usually makes use of a small number of samples.

- **Importance sampling methods** consist of building a probability measure $\mathbb{Q}'$ equivalent to $\mathbb{Q}$. Then

$$\mathbb{E}^{\mathbb{Q}'}\left[Z \frac{d\mathbb{Q}}{d\mathbb{Q}'} \right] = \mathbb{E}^{\mathbb{Q}}[Z]$$

so one simulates $Y = Z \frac{d\mathbb{Q}}{d\mathbb{Q}'}$ under $\mathbb{Q}'$. The latter must be chosen such that $\mathrm{Var}^{\mathbb{Q}'}(Z \frac{d\mathbb{Q}}{d\mathbb{Q}'}) < \mathrm{Var}^{\mathbb{Q}}(Z)$ and we know how to simulate Z and $\frac{d\mathbb{Q}}{d\mathbb{Q}'}$ under $\mathbb{Q}'$.

- **Antithetic variate methods** basically apply when Z is a function of some symmetric random variable G, i.e., when $Z = f(G)$, where G and $-G$ have the same distribution. Then one picks $Y = (f(G) + f(-G))/2$, hoping that $\mathrm{Var}^{\mathbb{Q}}((f(G) + f(-G))/2) < \mathrm{Var}^{\mathbb{Q}}(f(G))$. This happens, for instance, when f is monotonous.

- **Stratified sampling methods** consist of introducing a random variable A that can take a finite number of values $a_1, \ldots, a_m$ with known probabilities $q_i = \mathbb{Q}(A = a_i)$, and such that we know how to simulate Z given $A = a_i$ under $\mathbb{Q}$. Since

$$\mathbb{E}^{\mathbb{Q}}[Z] = \sum_{i=1}^{m} q_i \mathbb{E}^{\mathbb{Q}}[Z|A = a_i]$$

a stratified sampling estimator of Z is given by

$$\hat{Z}_{\#p_1, \ldots, \#p_m} = \sum_{i=1}^{m} q_i \frac{1}{\#p_i} \sum_{p=1}^{\#p_i} Z_i^{(p)}$$

where, for each i, the $Z_i^{(p)}$'s are i.i.d. and have the same distribution as Z given $A = a_i$, and where $\sum_{i=1}^m \#p_i = \#p$. The variance of this (unbiased) estimator is

$$\operatorname{Var}^{\mathbb{Q}}\left(\hat{Z}_{\#p_1,\ldots,\#p_m}\right) = \sum_{i=1}^m \frac{q_i^2}{\#p_i}\sigma_i^2$$

where $\sigma_i^2 = \operatorname{Var}^{\mathbb{Q}}(Z|A = a_i)$. Taking $\#p_i = q_i\#p$ yields

$$\operatorname{Var}^{\mathbb{Q}}\left(\hat{Z}_{\#p_1,\ldots,\#p_m}\right) = \frac{1}{\#p}\sum_{i=1}^m q_i\sigma_i^2 \leq \frac{\sigma^2}{\#p} = \operatorname{Var}^{\mathbb{Q}}(\hat{Z}_{\#p})$$

hence the reduction of variance. The sample sizes that minimize the variance of the stratified estimator are given by

$$\#p_i = \frac{q_i\sigma_i}{\sum_{j=1}^m q_j\sigma_j}\#p$$

The σ_j's are usually unknown, but they can be estimated in a pilot run.

- **Adaptive methods**, more involved and more recent, aim at offering generic variance reduction. They may consist, for example, of building good control variates dynamically within the Monte Carlo procedure. More generally, these methods use the information available in the already drawn random samples to optimize control variates, importance sampling, or stratified sampling. For instance, instead of running a pilot run, one may estimate the optimal β of the control variate method, or the optimal $\#p_i$'s of the stratified sampling method within the Monte Carlo procedure. This usually produces biased estimators. [146] is a nice overview of adaptive variance reduction techniques in finance.

A lot of books and monographs deal with variance reduction, for instance [7, 16], so we will not go into further detail.

2.1.3   Discretization error, reduction of bias

It is often not possible to simulate exactly the random variable Z, at least in a reasonably short time. This is the case in general when $Z = f(X_T)$ is a function of X_T, the solution at time T of an SDE (however, see Remark 2.1). We are particularly interested in this situation, because of the Feynman-Kac representation theorem, see Theorem 1.4. What we are always able to do in such a case is to generate a random variable $\bar{Z}$ close to Z. For instance, we may pick $\bar{Z} = f(X_T^n)$ where X^n is the Euler scheme with time step T/n. Or alternatively, just name your favorite discretization scheme. We then estimate $\mathbb{E}^{\mathbb{Q}}[\bar{Z}]$ *via* Monte Carlo. The global error is the sum of the sampling error, also called the Monte Carlo error, and the discretization error, also called bias:

$$\frac{1}{\#p}\sum_{p=1}^{\#p}\bar{Z}^{(p)} - \mathbb{E}^{\mathbb{Q}}[Z] = \left\{\frac{1}{\#p}\sum_{p=1}^{\#p}\bar{Z}^{(p)} - \mathbb{E}^{\mathbb{Q}}[\bar{Z}]\right\} + \left\{\mathbb{E}^{\mathbb{Q}}[\bar{Z}] - \mathbb{E}^{\mathbb{Q}}[Z]\right\}$$

The sampling error is given by the central limit theorem. It has the magnitude of the square root of $\mathrm{Var}^{\mathbb{Q}}(Z)/\#p$, because $\mathrm{Var}^{\mathbb{Q}}(\bar{Z})$ should be close to $\mathrm{Var}^{\mathbb{Q}}(Z)$. The discretization error $\mathbb{E}^{\mathbb{Q}}[\bar{Z}] - \mathbb{E}^{\mathbb{Q}}[Z]$ depends on the particular scheme used. We will see in Section 2.2 that for the Euler scheme, this bias due to time discretization is of magnitude T/n, the size of the time step:

$$\mathbb{E}^{\mathbb{Q}}[f(X_T^n)] - \mathbb{E}^{\mathbb{Q}}[f(X_T)] = c/n + O\left(1/n^2\right) \tag{2.1}$$

Practical interest in such an expansion must be emphasized. It tells us how n and $\#p$ should be chosen to reach a given accuracy in the estimation of $\mathbb{E}^{\mathbb{Q}}[f(X_T)]$. In a time of order $n\#p$, one gets an error of order $1/\sqrt{\#p} + 1/n$. Given a tolerance $\varepsilon \ll 1$, in order to minimize the time of calculus, one should then choose $\#p = O\left(n^2\right)$ and then one obtains a result in a time of order $1/\varepsilon^3$.

We have seen how reduction variance techniques allow us to lessen the sampling error. There also exist some techniques to reduce the bias. The most well known technique is the Romberg extrapolation, also known as the Richardson extrapolation. Its principle is straightforward. If one runs $\#p$ independent copies $(X_T^{2n,(p)}, X_T^{n,(p)})$ of the couple (X_T^{2n}, X_T^n), which still requires a time of order $n\#p$, then computing

$$\frac{1}{\#p}\sum_{p=1}^{\#p}\left(2f(X_T^{2n,(p)}) - f(X_T^{n,(p)})\right)$$

one gets an estimate of $\mathbb{E}^{\mathbb{Q}}[f(X_T)]$ whose accuracy is of order $1/\sqrt{\#p} + 1/n^2$, since (2.1) implies that

$$\mathbb{E}^{\mathbb{Q}}\left[2f(X_T^{2n}) - f(X_T^n)\right] = \mathbb{E}^{\mathbb{Q}}[f(X_T)] + O(1/n^2)$$

Given a tolerance $\varepsilon \ll 1$, one should now choose $\#p = O\left(n^4\right)$ and then one gets a result in a time of order $1/\varepsilon^{5/2}$. For more on the Romberg extrapolation, see Section 2.3.

It is often believed that the bias due to the Euler discretization is of order $1/n$ only when the payoff f is smooth, or non-smooth but bounded. Now, usual payoffs in finance are neither smooth nor bounded—think of call options, straddles, strangles, etc. Actually, under the usual uniform ellipticity assumption, the Euler discretization error can be expanded in powers of $\Delta t \equiv T/n$, for a much wider range of functions f—in fact, even for f's not being functions, like tempered distributions (see [9, 118]). As these results are recent and not often mentioned in the literature, we will review these properties of the Euler bias in greater detail in Section 2.2. Then Section 2.3 will be devoted to the reduction of bias.

REMARK 2.1   When the SDE has dimension one, meaning that its solution X and the driving Brownian motion are scalars, we can exactly simulate X_T; see [60] and Chapter 13, Section 13.6.4, for an alternative but related method. However, finite difference schemes are more efficient than the Monte Carlo method in this one-dimensional case.   ⬜

2.2   Euler discretization error

To make the statements precise, we must introduce some notation. Let $d, r \geq 1$ be two integers. Let $(\Omega, \mathcal{F}, \mathbb{Q})$ be a probability space on which lives an r-dimensional Brownian motion W. We denote by $\mathcal{F}_t = \sigma(W_s, 0 \leq s \leq t)$, the filtration generated by W. Let us give two functions $b : \mathbb{R}^d \to \mathbb{R}^d$ and $\sigma : \mathbb{R}^d \to \mathbb{R}^{d \times r}$. We use (column) vector and matrix notations, so that $b(x)$ should be thought of as a vector of size d and $\sigma(x)$ as a matrix of size $d \times r$. We denote transposition by a star and define a $d \times d$ matrix-valued function by putting $a = \sigma\sigma^*$. For a multi-index $\alpha \in \mathbb{N}^d$, $|\alpha| = \alpha_1 + \cdots + \alpha_d$ is its length and ∂^α is the differential operator $\partial^{|\alpha|}/\partial x_1^{\alpha_1} \cdots \partial x_d^{\alpha_d}$. Equipping $\mathbb{R}^d$ with the Euclidian norm $\|\cdot\|$, we denote by

- $C^\infty_{\mathrm{pol}}(\mathbb{R}^d)$ the set of infinitely differentiable functions $f : \mathbb{R}^d \to \mathbb{R}$ with polynomially growing derivatives of any order, i.e., such that for all $\alpha \in \mathbb{N}^d$, there exists $c \geq 0$ and $q \in \mathbb{N}$ such that for all $x \in \mathbb{R}^d$,

$$|\partial^\alpha f(x)| \leq c\,(1 + \|x\|^q) \tag{2.2}$$

- $C^\infty_b(\mathbb{R}^d)$ the set of infinitely differentiable functions $f : \mathbb{R}^d \to \mathbb{R}$ with bounded derivatives of any order, i.e., such that $\partial^\alpha f \in \mathrm{L}^\infty(\mathbb{R}^d)$ for all $\alpha \in \mathbb{N}^d$.

We shall make use of the following assumptions:

(A) For all $i \in \{1, \ldots, d\}$ and $j \in \{1, \ldots, r\}$, b_i and $\sigma_{i,j}$ belong to $C^\infty_{\mathrm{pol}}(\mathbb{R}^d)$ and have bounded first derivatives.

(B) For all $i \in \{1, \ldots, d\}$ and $j \in \{1, \ldots, r\}$, b_i and $\sigma_{i,j}$ belong to $C^\infty_b(\mathbb{R}^d)$.

(C) There exists $\eta > 0$ such that for all $x, \xi \in \mathbb{R}^d$,

$$\xi^* a(x)\xi \geq \eta\|\xi\|^2 \tag{2.3}$$

(C) is known as the *uniform ellipticity condition.*

It is well known that, given $x \in \mathbb{R}$, the hypothesis (A) guarantees the existence and the $\mathbb{Q}$-almost sure uniqueness of a solution $X^x = (X_t^x, t \geq 0)$ of the SDE

$$X_t^x = x + \int_0^t b(X_s^x) \, ds + \int_0^t \sigma(X_s^x) \, dW_s \tag{2.4}$$

The fact that here b and σ do not depend on t is only for the sake of simplicity and is not crucial at all.

2.2.1 Weak error

Let us fix a time horizon $T > 0$. Without loss of generality, we can and do assume that $T = 1$. In view of the Monte Carlo method, we need to estimate the law of X_1^x. To do so, the most natural idea is to approach X^x by its Euler scheme of order $n \geq 1$, say $X^{n,x} = (X_t^{n,x}, t \geq 0)$, defined as follows. We consider the regular subdivision $\mathfrak{S}_n = \{0 = t_0^n < t_1^n < \cdots < t_{n-1}^n < t_n^n = 1\}$ of the interval $[0,1]$, i.e., $t_k^n = k/n$, and we put $X_0^{n,x} = x$ and, for all $k \in \{0, \ldots, n-1\}$ and $t \in [t_k^n, t_{k+1}^n]$,

$$X_t^{n,x} = X_{t_k^n}^{n,x} + b\left(X_{t_k^n}^{n,x}\right)(t - t_k^n) + \sigma\left(X_{t_k^n}^{n,x}\right)(W_t - W_{t_k^n}) \tag{2.5}$$

Then the random variable $X_1^{n,x}$ can be exactly simulated and should be close in law to X_1^x. Precisely, we measure the weak error between $X_1^{n,x}$ and X_1^x by the quantities

$$\Delta_1^n f(x) = \mathbb{E}^{\mathbb{Q}}[f(X_1^{n,x})] - \mathbb{E}^{\mathbb{Q}}[f(X_1^x)]$$

and we try to find *the largest space of test functions* f for which, for each x, there exists a constant $C_1 f(x)$ such that

$$\Delta_1^n f(x) = C_1 f(x)/n + O\left(1/n^2\right) \tag{2.6}$$

2.2.2 Results concerning smooth or bounded payoffs

Using Itô expansions, D. Talay and L. Tubaro [190] have shown that (2.6) holds when $f \in C_{\text{pol}}^\infty(\mathbb{R}^d)$ under condition

(B') The b_i's and the $\sigma_{i,j}$'s are infinitely differentiable functions with bounded derivatives of any order ≥ 1.

Hypothesis (B') is almost (B), but in (B') the functions b_i and $\sigma_{i,j}$ are not supposedly bounded themselves. Using Malliavin calculus, V. Bally and D. Talay [44] extended this result to the case of measurable and bounded f's, with the extra hypothesis that X is uniformly hypoelliptic.

2.2.3 Results for general payoffs

In [9, 118], it is shown that, under the assumptions of regularity of coefficients and uniform ellipticity, one can give a precise meaning to $\Delta_1^n f(x) = \mathbb{E}^{\mathbb{Q}}[f(X_1^{n,x})] - \mathbb{E}^{\mathbb{Q}}[f(X_1^x)]$ when f is any tempered distribution, and it is proved that expansions in powers of $1/n$ remain valid in this extremely general setting. This comes as a consequence of the following theorem:

THEOREM 2.1
Under (B) and (C),

(i) *for all $t \in (0, 1]$ and $x \in \mathbb{R}^d$, X_t^x has a density $p(t, x, \cdot)$ and $p \in \mathcal{G}(\mathbb{R}^d)$;*

(ii) *for all $t \in (0, 1]$, $x \in \mathbb{R}^d$ and $n \geq 1$, $X_t^{n,x}$ has a density $p_n(t, x, \cdot)$ and $(p_n, n \geq 1)$ is a bounded sequence in $\mathcal{G}(\mathbb{R}^d)$;*

(iii) *there exists $\pi \in \mathcal{G}_1(\mathbb{R}^d)$ and a bounded sequence $(\pi_n, n \geq 1)$ in $\mathcal{G}_4(\mathbb{R}^d)$ such that for all $n \geq 1$,*

$$p_n - p = \pi/n + \pi_n/n^2 \tag{2.7}$$

In this theorem, the spaces $\mathcal{G}(\mathbb{R}^d)$ and $\mathcal{G}_l(\mathbb{R}^d)$ are spaces of Gaussian-like functions. To be precise, for $l \in \mathbb{Z}$, we first define $\mathcal{G}_l(\mathbb{R}^d)$ as the set of all measurable functions $\pi : (0, 1] \times \mathbb{R}^d \times \mathbb{R}^d \to \mathbb{R}$ such that

- for all $t \in (0, 1]$, $\pi(t, \cdot, \cdot)$ is infinitely differentiable,

- for all $\alpha, \beta \in \mathbb{N}^d$, there exist two constants $c_1 \geq 0$ and $c_2 > 0$ such that for all $t \in (0, 1]$ and $x, y \in \mathbb{R}^d$,

$$\left| \partial_x^\alpha \partial_y^\beta \pi(t, x, y) \right| \leq c_1 t^{-(|\alpha| + |\beta| + d + l)/2} \exp\left(-c_2 \|x - y\|^2 / t\right) \tag{2.8}$$

We say that a subset $\mathcal{B} \subset \mathcal{G}_l(\mathbb{R}^d)$ is bounded if, in (2.8), c_1 and c_2 can be chosen independently on $\pi \in \mathcal{B}$. We also introduce the space $\mathcal{G}(\mathbb{R}^d)$ defined in the same way as $\mathcal{G}_l(\mathbb{R}^d)$ with (2.8) replaced by the following two conditions:

$$\left| \partial_x^\alpha \partial_y^\beta \pi(t, x, y) \right| \leq c_1 t^{-(|\alpha| + |\beta| + d)/2} \exp\left(-c_2 \|x - y\|^2 / t\right), \tag{2.9}$$

$$\left| \partial_x^\alpha \left(\pi\left(t, x, x + y\sqrt{t}\right) \right) \right| \leq c_1 t^{-d/2} \exp\left(-c_2 \|y\|^2\right) \tag{2.10}$$

We say that a subset $\mathcal{B} \subset \mathcal{G}(\mathbb{R}^d)$ is bounded if, in (2.9) and (2.10), c_1 and c_2 can be chosen independently on $\pi \in \mathcal{B}$.

The function π in Equation (2.7) is the principal *functional* error term. It can be expressed in terms of p by

$$\pi(t, x, y) = \frac{1}{2} \int_0^t \int_{\mathbb{R}^d} p(s, x, z) L_2^*(p(t - s, \cdot, y))(z) \, dz \, ds \tag{2.11}$$

where the differential operator L_2^* is explicitly given in terms of the functions a and b by

$$- L_2^* = \sum_{i=1}^{d} \left(b \cdot Db_i + \frac{1}{2} \operatorname{tr} \left(aD^2 b_i \right) \right) \partial_i$$

$$+ \sum_{i,j=1}^{d} \left(\frac{1}{2} b \cdot Da_{i,j} + a_j \cdot Db_i + \frac{1}{4} \operatorname{tr} \left(aD^2 a_{i,j} \right) \right) \partial_{ij} + \frac{1}{2} \sum_{i,j,k=1}^{d} a_k \cdot Da_{i,j} \partial_{ijk}$$

$$(2.12)$$

Here, $\cdot$, a_k, tr, D, and D^2, respectively, stand for the inner product in $\mathbb{R}^d$, the k-th column of a, the trace of a matrix, the gradient vector, and the Hessian matrix.

Using Theorem 2.1, one can now prove [9, 118] that if X is uniformly elliptic, the expansion (2.6) is valid in the very general case when f is a tempered distribution. Let us denote by $\mathcal{S}(\mathbb{R}^d)$ Schwartz's space, i.e., the space of infinitely differentiable functions $\varphi : \mathbb{R}^d \to \mathbb{R}$ such that $x \mapsto x^\alpha \partial^\beta \varphi(x) \in L^\infty(\mathbb{R}^d)$ for all $\alpha, \beta \in \mathbb{N}^d$ (x^α stands for $x_1^{\alpha_1} \cdots x_d^{\alpha_d}$), and let us denote by $\mathcal{S}'(\mathbb{R}^d)$ the space of tempered distributions. The seminorms $(N_q, q \in \mathbb{N})$ are defined on $\mathcal{S}(\mathbb{R}^d)$ by

$$N_q(\varphi) = \sum_{|\alpha| \leq q, |\beta| \leq q} \sup_{x \in \mathbb{R}^d} \left| x^\alpha \partial^\beta \varphi(x) \right|$$

and the order $\#S$ of $S \in \mathcal{S}'(\mathbb{R}^d)$ is the smallest integer q such that there is a $c \geq 0$ such that $|\langle S, \varphi \rangle| \leq c N_q(\varphi)$ for all $\varphi \in \mathcal{S}(\mathbb{R}^d)$. Note that whenever $\pi \in \mathcal{G}_l(\mathbb{R}^d)$, $\pi(t, x, \cdot)$ and $\pi(t, \cdot, y)$ belong to $\mathcal{S}(\mathbb{R}^d)$. More precisely, for $\mathcal{B} \subset \mathcal{G}_l(\mathbb{R}^d)$ bounded, there exists $c \geq 0$ such that for all $\pi \in \mathcal{B}$, $t \in (0, 1]$ and $x, y \in \mathbb{R}^d$,

$$N_q(\pi(t, x, \cdot)) \leq c t^{-(d+l+q)/2} \left(1 + \|x\|^q \right)$$
$$N_q(\pi(t, \cdot, y)) \leq c t^{-(d+l+q)/2} \left(1 + \|y\|^q \right)$$

Applying a tempered distribution S to (2.7), t and x or t and y being fixed, we immediately deduce from Theorem 2.1:

THEOREM 2.2
Under (B) and (C), for all $S \in \mathcal{S}'(\mathbb{R}^d)$, there exists $c \geq 0$ such that for all $n \geq 1$, $t \in (0, 1]$ and $x, y \in \mathbb{R}^d$,

$$\langle S, p_n(t, x, \cdot) \rangle - \langle S, p(t, x, \cdot) \rangle = \frac{1}{n} \langle S, \pi(t, x, \cdot) \rangle + r_n'(t, x)$$

$$\langle S, p_n(t, \cdot, y) \rangle - \langle S, p(t, \cdot, y) \rangle = \frac{1}{n} \langle S, \pi(t, \cdot, y) \rangle + r_n''(t, y)$$

and

$$|r_n'(t,x)| + |r_n''(t,x)| \leq cn^{-2}t^{-(d+4+\#S)/2}\left(1 + \|x\|^{\#S}\right)$$

Let us define $\mathbb{E}^{\mathbb{Q}}[S(Y)]$ by $\langle S, p_Y \rangle$ when $S \in \mathcal{S}'(\mathbb{R}^d)$ and Y is a random variable with density $p_Y \in \mathcal{S}(\mathbb{R}^d)$. Note that, when S is a measurable and polynomially growing function, this definition coincides with the usual expectation. We have then proved that, under (B) and (C), (2.6) is valid for f's being only tempered distributions, and not only for $t = 1$, but also for any time $t \in (0,1]$, and we have even determined exactly the way the $O(1/n^2)$ remainder depends on t, f, and x. Precisely, this remainder grows slower than $\|x\|^{\#f}$ as x tends to infinity, and explodes slower than $t^{-(\#f+d+4)/2}$ as t tends to 0.

We can now state an easy consequence of Theorem 2.1, including an application to financial markets. We denote by $\mathcal{E}_\mu$, for $\mu \in (0,2)$, the set of all measurable functions $f : \mathbb{R}^d \to \mathbb{R}$ such that there exists $c_1, c_2 \geq 0$ such that for all $y \in \mathbb{R}^d$,

$$|f(y)| \leq c_1 \exp\left(c_2 \|y\|^\mu\right)$$

Theorem 2.1 leads to:

PROPOSITION 2.1
Under (B) and (C), for all $\alpha \in \mathbb{N}^d$, $\mu \in (0,2)$, and $f \in \mathcal{E}_\mu$, there exists $c_1, c_2 \geq 0$ such that for all $n \geq 1$, $t \in (0,1]$ and $x \in \mathbb{R}^d$,

$$\partial_x^\alpha \mathbb{E}^{\mathbb{Q}}[f(X_t^{n,x})] - \partial_x^\alpha \mathbb{E}^{\mathbb{Q}}[f(X_t^x)] = \frac{1}{n}\int_{\mathbb{R}^d} f(y)\partial_x^\alpha \pi(t,x,y)\,dy + r_n(t,x) \quad (2.13)$$

with

$$|r_n(t,x)| \leq c_1 n^{-2} t^{-(|\alpha|+4)/2} \exp\left(c_2 \|x\|^\mu\right)$$

This result can now be used in the context of financial markets.

2.2.4 Application to option pricing and hedging

Let $S^v = (S^{v,1}, \ldots, S^{v,d})$ be a basket of assets satisfying

$$\frac{dS_t^{v,i}}{S_t^{v,i}} = \mu_i(S_t^v)\,dt + \sum_{j=1}^r \sigma_{i,j}(S_t^v)\,dW_t^j, \qquad S_0^{v,i} = v^i > 0$$

with $\mu, \sigma \in C_b^\infty(\mathbb{R}^d)$ and σ satisfying (C). Given a measurable and polynomially growing function ϕ, we try to estimate the price, Price $= \mathbb{E}^{\mathbb{Q}}[\phi(S_t^v)]$; the deltas, $\text{Delta}_i = \partial_v^{e_i} \mathbb{E}^{\mathbb{Q}}[\phi(S_t^v)]$; and the gammas, $\text{Gamma}_{i,j} = \partial_v^{e_i+e_j} \mathbb{E}^{\mathbb{Q}}[\phi(S_t^v)]$ of the European option of maturity t and payoff ϕ (($e_1, \ldots, e_d$) is the canonical base of $\mathbb{R}^d$). For convenience, let us set $x = \ln v$ (i.e., $x^i = \ln v^i$) and $X_t^{x,i} = \ln(S_t^{v,i})$. X is then the solution to (2.4) with $b = \mu - \|\sigma\|^2/2 \in C_b^\infty(\mathbb{R}^d)$, where $\|\sigma\|_i^2(x) = \sum_{j=1}^r \sigma_{i,j}^2(x)$. If we set $\exp(x) = (\exp(x^1), \ldots, \exp(x^d))$ and $f(x) = \phi(\exp(x))$, we define a function $f \in \mathcal{E}_1$ and, since $\text{Price} = \mathbb{E}^{\mathbb{Q}}[f(X_t^x)]$, (2.13) leads to

$$\text{Price}^n - \text{Price} = C_t^{\text{Price}}\phi(v)/n + O\left(n^{-2}t^{-2}\exp\left(c_2\|\ln v\|\right)\right)$$

where Price^n stands for the approximated price $\mathbb{E}^{\mathbb{Q}}[f(X_t^{n,x})]$ and

$$C_t^{\text{Price}}\phi(v) = \int_{(\mathbb{R}_+^*)^d} \phi(u)\frac{\pi(t, \ln v, \ln u)}{u_1 \cdots u_d}\,du$$

Besides, if we set

$$\text{Delta}_i^n = \partial_v^{e_i}\mathbb{E}^{\mathbb{Q}}[f(X_t^{n,\ln v})], \qquad \text{Gamma}_{i,j}^n = \partial_v^{e_i+e_j}\mathbb{E}^{\mathbb{Q}}[f(X_t^{n,\ln v})]$$

then (2.13) shows that

$$\text{Delta}^n - \text{Delta} = C_t^{\text{Delta}}\phi(v)/n + O\left(n^{-2}t^{-5/2}\exp\left(c_2\|\ln v\|\right)\right)$$
$$\text{Gamma}^n - \text{Gamma} = C_t^{\text{Gamma}}\phi(v)/n + O\left(n^{-2}t^{-3}\exp\left(c_2\|\ln v\|\right)\right)$$

where

$$C_t^{\text{Delta}}\phi(v)_i = \frac{1}{v_i}\int_{(\mathbb{R}_+^*)^d}\phi(u)\frac{\partial_2^{e_i}\pi(t, \ln v, \ln u)}{u_1 \cdots u_d}\,du$$

and

$$C_t^{\text{Gamma}}\phi(v)_{i,j}$$
$$= \frac{1}{v_i v_j}\int_{(\mathbb{R}_+^*)^d}\phi(u)\frac{\partial_2^{e_i+e_j}\pi(t, \ln v, \ln u) - \mathbf{1}_{\{i=j\}}\partial_2^{e_i}\pi(t, \ln v, \ln u)}{u_1 \cdots u_d}\,du$$

Eventually, we have proved that applying the Euler scheme of order n to the logarithm of the underlying leads to approximations of the price, the deltas and the gammas that converge to the true price, deltas, and gammas with speed $1/n$, at least when the drift and volatility of the underlying satisfy (B) and (C), which in the context of financial markets is not a restricting hypothesis. Note that the principal part of the error explodes as t tends to 0 as $t^{-1/2}$ for the prices, t^{-1} for the deltas, and $t^{-3/2}$ for the gammas.

2.3   Romberg extrapolation

The main message of Section 2.2 is that, when one wants to estimate $\mathbb{E}^{\mathbb{Q}}[f(X_T)]$, the discretization error induced by the Euler scheme is of order the time step $\Delta t \equiv T/n$. So is the Milstein discretization error. As a consequence, if one uses either scheme plus a Monte Carlo method to estimate $\mathbb{E}^{\mathbb{Q}}[f(X_T)]$, in a time of order $n\#p$, one gets an error of order $1/\sqrt{\#p} + 1/n$, where $\#p$ stands for the number of independent copies of X_T^n generated by the Monte Carlo procedure. Given a tolerance $\varepsilon \ll 1$, in order to minimize the time of calculus, one should then choose $\#p = O\left(n^2\right)$, which gets a result in a time of order $1/\varepsilon^3$.

To diminish this time, one can implement a higher order discretization scheme, such as the Ninomiya-Victoir scheme [166]. Imagine we use a scheme of order $\alpha > 1$, meaning that the discretization error is of magnitude Δt^α. Then given a tolerance $\varepsilon \ll 1$, in order to minimize the time of calculus, one should now choose $\#p = O\left(n^{2\alpha}\right)$, which gets a result in a time of order $\varepsilon^{-2-1/\alpha}$. The bigger α, the bigger the time step we can use, and hence the shorter the computation time. Shorter time steps would be useless, because the induced increase in accuracy would be hidden by the sampling error. Note that the computation time cannot be less than ε^{-2}, because of the central limit theorem.

However, high order discretization schemes, for which $\alpha = 2$, 3, or more, are quite involved, meaning that each time step costs much more time than the Euler scheme, and the risk of code error—or operational risk—is considerably larger. There is a simple way to reduce, at almost no cost, the discretization error, and hence the total computation time of the Monte Carlo procedure: the Romberg extrapolation, which we will now explain in detail. We begin with the standard extrapolation method, before investigating more involved versions of the Romberg extrapolation. The efficiency of the extrapolation will be illustrated by various numerical results.

2.3.1   Standard Romberg extrapolation

Following [169], let us assume that we were able to build a discretization scheme $\bar{X}$ whose induced error can be expanded in powers of the time step

$$\forall f \in V, \ \forall R \geq 1, \ \mathbb{E}^{\mathbb{Q}}[f(X_T)] = \mathbb{E}^{\mathbb{Q}}[f(\bar{X}_T)] + \sum_{k=1}^{R-1} \frac{c_k}{n^k} + O(n^{-R}) \quad (2.14)$$

for a large space V of functions. In Section 2.2, we proved that the above expansion holds under regularity assumptions on the coefficients of the SDE for the Euler scheme, for $R = 2$, when V is a space of measurable functions with exponential growth. The expansion is actually valid for any order $R \geq 2$, but one has to go through tedious computations to prove it. Let $X^{(1)} = X$ and $W^{(1)} = W$. The standard L^2 Monte Carlo error squared is worth

$$\left\| \mathbb{E}^{\mathbb{Q}}[f(X_T^{(1)})] - \frac{1}{\#p} \sum_{p=1}^{\#p} f\left(\left(\bar{X}_T^{(1)} \right)^{(p)} \right) \right\|_2^2$$

$$= \left| \mathbb{E}^{\mathbb{Q}}[f(X_T^{(1)})] - \mathbb{E}^{\mathbb{Q}}[f(\bar{X}_T^{(1)})] \right|^2 + \left\| \mathbb{E}^{\mathbb{Q}}[f(\bar{X}_T^{(1)})] - \frac{1}{\#p} \sum_{p=1}^{\#p} f\left(\left(\bar{X}_T^{(1)} \right)^{(p)} \right) \right\|_2^2$$

$$= \frac{c_1^2}{n^2} + \frac{\text{Var}^{\mathbb{Q}}\left(f(\bar{X}_T^{(1)}) \right)}{\#p} + O\left(n^{-3} \right)$$

We can easily reduce this error. To this end, let $\left(W^{(2)}, X^{(2)} \right)$ be a copy in distribution of $\left(W^{(1)}, X^{(1)} \right)$, and let $\bar{X}^{(2)}$ be the scheme with time step $T/(2n)$. Then

$$\mathbb{E}^{\mathbb{Q}}[f(X_T)] = \mathbb{E}^{\mathbb{Q}}[2f(\bar{X}_T^{(2)}) - f(\bar{X}_T^{(1)})] - \frac{1}{2}\frac{c_2}{n^2} + O\left(n^{-3} \right)$$

and the L^2 error squared becomes

$$\left\| \mathbb{E}^{\mathbb{Q}}[f(X_T)] - \frac{1}{\#p} \sum_{p=1}^{\#p} \left(2f\left(\left(\bar{X}_T^{(2)} \right)^{(p)} \right) - f\left(\left(\bar{X}_T^{(1)} \right)^{(p)} \right) \right) \right\|_2^2$$

$$= \frac{c_2^2}{4n^4} + \frac{\text{Var}^{\mathbb{Q}}\left(2f(\bar{X}_T^{(2)}) - f(\bar{X}_T^{(1)}) \right)}{\#p} + O\left(n^{-5} \right)$$

We say we performed a (Richardson-)Romberg extrapolation of order 2. In order to minimize the asymptotic variance[1] $\text{Var}^{\mathbb{Q}}\left(2f(X_T^{(2)}) - f(X_T^{(1)}) \right)$, it is enough to choose $W^{(2)} = W^{(1)}$, whence $X^{(2)} = X^{(1)}$, see [169]. Then,

$$\text{Var}^{\mathbb{Q}}\left(2f(X_T^{(2)}) - f(X_T^{(1)}) \right) = \text{Var}^{\mathbb{Q}}\left(f(X_T^{(1)}) \right)$$

which is the variance of the non-extrapolated scheme. We have hence reduced the bias of the Monte Carlo estimator without increasing its (asymptotic) variance. Note that constants c_k are hard to estimate *a priori*, so for a given time step T/n, $\frac{c_2^2}{4n^4}$ may be bigger than $\frac{c_1^2}{n^2}$!

---

[1]This is indeed the asymptotic variance if f is continuous with polynomial growth, or if f is measurable and bounded and the diffusion is uniformly elliptic; see hypothesis (C) in Section 2.2.

2.3.2   Iterated Romberg extrapolation

A natural iteration of the Romberg extrapolation consists of using the schemes with time steps $T/(4n)$, say $\bar{X}^{(3)}$, then $T/(8n)$, say $\bar{X}^{(4)}$, etc. For instance,

$$\mathbb{E}^{\mathbb{Q}}[f(X_T)] = \mathbb{E}^{\mathbb{Q}}[2f(\bar{X}_T^{(3)}) - f(\bar{X}_T^{(2)})] - \frac{1}{8}\frac{c_2}{n^2} + O\left(n^{-3}\right)$$

$$\mathbb{E}^{\mathbb{Q}}[f(X_T)] = \mathbb{E}^{\mathbb{Q}}[2f(\bar{X}_T^{(2)}) - f(\bar{X}_T^{(1)})] - \frac{1}{2}\frac{c_2}{n^2} + O\left(n^{-3}\right)$$

so

$$\mathbb{E}^{\mathbb{Q}}[f(X_T)] = \frac{4}{3}\mathbb{E}^{\mathbb{Q}}[2f(\bar{X}_T^{(3)}) - f(\bar{X}_T^{(2)})] - \frac{1}{3}\mathbb{E}^{\mathbb{Q}}[2f(\bar{X}_T^{(2)}) - f(\bar{X}_T^{(1)})] + O\left(n^{-3}\right)$$

i.e.,

$$\mathbb{E}^{\mathbb{Q}}[f(X_T)] = \mathbb{E}^{\mathbb{Q}}\left[\frac{8}{3}f(\bar{X}_T^{(3)}) - 2f(\bar{X}_T^{(2)}) + \frac{1}{3}f(\bar{X}_T^{(1)})\right] + O\left(n^{-3}\right)$$

If we choose $W^{(3)} = W^{(2)} = W^{(1)}$, the asymptotic variance of the new estimator is equal to that of the basic estimator, that is, $\mathrm{Var}^{\mathbb{Q}}(f(X_T))$. Again, we reduced the asymptotic bias of the estimator without increasing its asymptotic variance. We generally observe that coefficient c_k grows exponentially with k, so that, for a given time step, the above third order remainder may be greater than $\frac{c_2}{2n^2}$, or even than $\frac{c_1}{n}$. This estimator ($R = 3$) needs $4n\#p$ Gaussian increments computations. More generally, the estimator induced by $R - 2$ natural iterations of the Romberg extrapolation needs $\#p \times 2^{R-1}n$ Gaussian increment computations.

We can also consider another type of iteration of the Romberg extrapolation, as suggested by G. Pagès in [169]. We now build linear combinations of schemes $\bar{X}^{(r)}$ with time steps $T/n, T/(2n), T/(3n), \ldots, T/(Rn)$ instead of $T/n, T/(2n), T/(4n), \ldots, T/(2^{R-1}n)$. By solving a linear system of equations, we get

$$\mathbb{E}^{\mathbb{Q}}[f(X_T)] = \mathbb{E}^{\mathbb{Q}}\left[\sum_{r=1}^{R}\alpha_r f\left(\bar{X}_T^{(r)}\right)\right] + \frac{\tilde{c}_R}{n^R} + O\left(n^{-R-1}\right)$$

with

$$\sum_{r=1}^{R}\alpha_r = 1, \quad \alpha_r = (-1)^{R-r}\frac{r^R}{r!(R-r)!}$$

If we pick $W^{(r)} = \cdots = W^{(2)} = W^{(1)}$, the asymptotic variance of the new estimator is

$$\mathrm{Var}^{\mathbb{Q}}\left(\sum_{r=1}^{R}\alpha_r f\left(X_T^{(r)}\right)\right) = \left(\sum_{r=1}^{R}\alpha_r\right)^2 \mathrm{Var}^{\mathbb{Q}}\left(f\left(X_T\right)\right) = \mathrm{Var}^{\mathbb{Q}}\left(f\left(X_T\right)\right)$$

Table 2.1: One-year implied volatility.

K/X_0	0.3	0.5	0.7	0.8	0.9	1
1Y implied vol (in %)	86.96	74.56	64.79	60.78	57.42	54.68
K/X_0	1.1	1.2	1.3	1.5	2.0	
1Y implied vol (in %)	52.52	50.85	49.51	47.31	42.51	

It is equal to that of the basic estimator. Pagès [169] shows that this choice is never optimal when $R \geq 3$. The advantage of such an iteration only appears for large R's: we only need to compute $O(R^2 n \# p)$ Gaussian increments, instead of $2^{R-1} n \# p$. For $R = 3$, one must compute $4n \# p$ Gaussian increments, like in the natural iteration; for $R = 4$, $6n \# p$, instead of $8n \# p$. Now, it is advised not to use large values of R, because of the growth of the absolute value of the coefficients $|\alpha_r|$. After examining the numerical experiments below, we recommend using $R = 2$ (the standard Romberg extrapolation), or $R = 3$. Since in the latter case there is no decrease in the number of Gaussian increments to be computed, we recommend using the natural iteration, because it uses a uniform splitting in 4 of a time step $(k\Delta t, (k+1)\Delta t)$, whereas the Pagès iteration uses the nonuniform splitting

$$(k\Delta t, (k+1)\Delta t) = (k\Delta t, (k+1/3)\Delta t) \cup ((k+1/3)\Delta t, (k+1/2)\Delta t)$$
$$\cup ((k+1/2)\Delta t, (k+2/3)\Delta t) \cup ((k+2/3)\Delta t, (k+1)\Delta t)$$

resulting in a bigger value of $\max_{1 \leq r \leq 3} |\alpha_r|$, which is not desirable.

2.3.3 Numerical experiments with vanilla options

In this section, we price vanilla options with maturity $T = 1$ under a local volatility model. The time-homogeneous local volatility function is inferred from the one-year implied volatility given in Table 2.1.

Interest rates and repos are zero, and the underlying delivers no dividend. We plot the estimator of the price and its standard deviation as functions of the time step (1/2, 1/4, 1/8, etc.), respectively, for $R = 1$ (classical Euler scheme), $R = 2$ (classical Romberg extrapolation), $R = 3$ (iterated Romberg extrapolation with time steps T/n, $T/(2n)$ and $T/(4n)$), and $R = 4$ (iterated Romberg extrapolation with time steps T/n, $T/(2n)$, $T/(4n)$, and $T/(8n)$). We use $\#p = 2^{18}$ paths in the Monte Carlo sampling.

Put

We first price a put option with strike 80%; see Figures 2.1 and 2.2. The estimators of the price for $R = 3$ and $R = 4$ have very small bias, even for large time steps. However, the standard deviation of the price estimator for

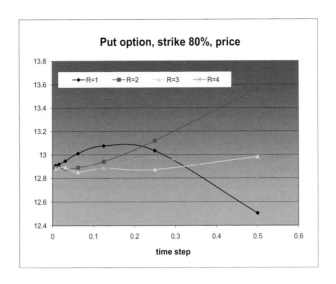

Figure 2.1: Price of the put option in percent as a function of the time step. $K = 80\%$.

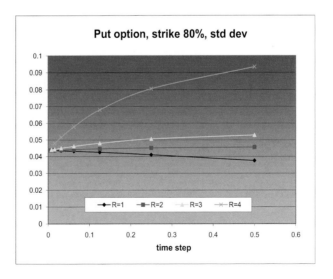

Figure 2.2: Standard deviation of the estimator of the price of the put option as a function of the time step. $K = 80\%$.

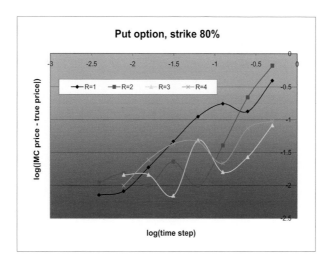

Figure 2.3: Log-error on the price of the put option as a function of the time step. $K = 80\%$.

$R = 4$ is much larger than that for $R = 1$, 2, or 3: the variance of the estimator is not yet controlled by the asymptotic variance. The waves on the price curve for $R = 4$ are evidence of a large variance. Even for desirable time steps, such as $1/8$ or $1/16$, the standard deviation for $R = 4$ is more than 1.5 times those for $R = 1$, 2, or 3. For those time steps, the basic estimator $(R = 1)$ shows a clear bias (17 and 11 bps, resp. 4 and 2.5 standard deviations), whereas the estimators for $R = 2$ (4 and 1 bps) and $R = 3$ (2 and 5 bps) lie within the Monte Carlo error zone.

In Figure 2.3, in order to check how the bias of the price estimator depends on R, we plot

$$\log_{10}(1/n) \mapsto \log_{10} |\text{Estimator}_R(1/n) - \text{Exact Price}|$$

for various values of R. For $R = 1$, we distinguish three different regimes:

- for large time steps, the asymptotic regime $(n \to \infty)$ is not reached;

- for time steps smaller than $1/8$, we enter into the asymptotic regime: as expected, the graph is affine with a slope close to 1;

- for time steps smaller than $1/128$, the asymptotic regime is hidden by the Monte Carlo sampling error.

For $R = 2$ we only see the last two regimes, with slope 2 in the affine regime, as expected. As for $R = 3$ and $R = 4$, we only see the last regime: for all

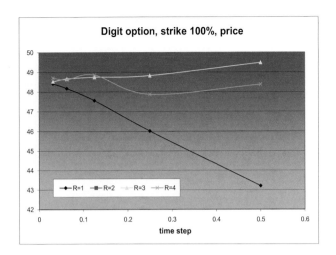

Figure 2.4: Price of the digital call option in percent as a function of the time step. $K = 100\%$.

time steps smaller than $1/2$, the price estimator seems to stand within the Monte Carlo error domain—this domain is broader for $R = 3$ or $R = 4$ than for $R = 1$ or $R = 2$, as shown in Figure 2.2.

Digital option

Figures 2.4 and 2.5 show the estimated price and standard deviation for the European digital call option with strike 100%. They lead to similar conclusions.

These experiments on vanilla options show that the estimator with $R = 2$ gives excellent results: when the time step decreases, the bias is quickly absorbed in the Monte Carlo error, with a minimal increase in variance.

2.3.4 Numerical experiments with path-dependent options

Using iterated conditional expectations, one can easily see that expansion (2.14) remains valid for the Euler scheme for discretely monitored path-dependent options. One just needs to slightly adapt the scheme by adding the discrete observation dates involved in the payout of the option. However, if there are a lot of observation dates, the asymptotic regime may be visible only for very small time steps, that is, time steps smaller than the typical length of time between two observation dates. Here we consider a path-dependent option with a one-year maturity, and with 4 uniformly spread constatation

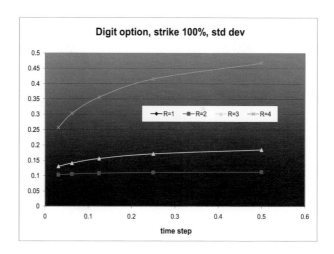

Figure 2.5: Standard deviation of the price of the digital call option as a function of the time step. $K = 100\%$. The graph for $R = 1$, hardly visible, is almost constant at 0.1.

dates t_i, i.e., one every 3 months. As in the previous section, we price under a local volatility model, with a time-homogeneous local volatility function inferred from the one-year implied volatility given in Table 2.1.

Lookback put (put on minimum over time)

This option delivers $\left(K - \min_{i \in \{1,2,3,4\}} X_{t_i}/X_{t_0}\right)_+$. The estimated price and standard deviation are shown in Figures 2.6 and 2.7. Like for vanilla options, for $R = 1$, we observe a linear behavior of the bias of the estimator as a function of the time step. We also distinguish the quadratic behavior of the estimator for $R = 2$. This estimator has the most desirable properties: no visible bias for $\Delta t = 1/8$ and $\#p = 2^{18}$, and almost no increase in variance, compared to $R = 1$.

Of course, computational time matters a lot. For instance, in Figure 2.8, we plot the computational time (in arbitrary units) as a function of the time step for $R = 1, 2, 3, 4$, in the case of the lookback put option. Similar graphs were obtained for the vanilla options tested in the previous section. It takes less time to get an unbiased result (i.e., the bias is hidden by the sampling error) with $R = 2$ and $\Delta t = 1/8$ than it takes to get a biased result with $R = 1$ and $\Delta t = 1/64$. Hence, we definitely advise using the Romberg extrapolation with $R = 2$ and a reasonably large time step.

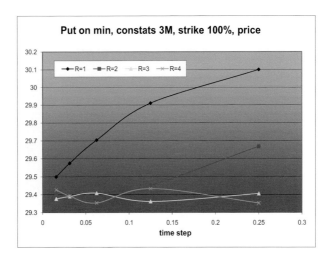

Figure 2.6: Price of the lookback put option in percent as a function of the time step. $K = 80\%$.

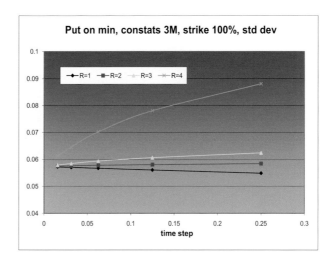

Figure 2.7: Standard deviation of the estimator of the price of the lookback put option as a function of the time step. $K = 80\%$.

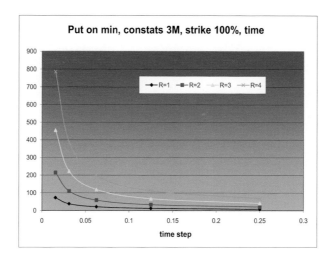

Figure 2.8: Computation time of the estimator of the price of the lookback put option as a function of the time step. $K = 80\%$.

Conclusion

Monte Carlo will be our preferred method for solving PDEs in high dimension, but it has to be adapted to handle nonlinearity. Throughout this book, we will review existing and suggest new "nonlinear" Monte Carlo methods, i.e., Monte Carlo methods for solving nonlinear PDEs. But before that, in the next chapter, we present new advanced models for financial markets, as well as recent hedging concepts that lead to nonlinear pricing operators.

Chapter 3

Some Excursions in Option Pricing

> Be bold and mighty forces will come to your aid.
>
> — Goethe

In this chapter, we explore a variety of advanced models and pricing concepts that were not covered in [11] and are not often covered in monographs. We first present various examples of complete market models, which try to directly specify the dynamics of vanilla option prices in an arbitrage-free consistent way. Although not all examples are fully conclusive, we believe it is worth explaining them to point out where the difficulties are. Then, in order to illustrate that replication and super-replication, presented in Chapter 1, are not the only pricing paradigms, we introduce three alternatives: utility indifference pricing, quantile hedging, and P&L variance minimization, which are connected to nonlinear pricing operators (see Exercises 4.7.2 and 4.7.3).

3.1 Complete market models

In this section, for the ease of presentation, we will assume zero interest rates, no repo or dividends. Let us consider for each maturity date $T \in (0, \infty)$ a European vanilla payoff $\Phi(X_T)$. Its market value at time t is denoted Φ_t^T. The no-arbitrage condition dictates that the processes Φ_t^T, X_t are local martingales under a risk-neutral measure

$$dX_t = \sigma_t X_t \, dW_t^0, \qquad X_{t=0} = X_0 \tag{3.1}$$

$$d\Phi_t^T = Z_t^T . dW_t, \qquad \Phi_{t=0}^T = \Phi_0^T \tag{3.2}$$

Here, W_t denotes a multi-dimensional Brownian motion whose first component is W_t^0, or possibly a random field parameterized by T. We complement these equations by specifying the terminal condition at each maturity date T

$$\Phi_T^T = \Phi(X_T) \tag{3.3}$$

As always, we have assumed that there exists at least one risk-neutral measure. We assume that

$$\begin{cases} \forall T \geq 0, \ Z_t^T . \lambda_t = 0 \\ \sigma_t \lambda_t^0 = 0 \end{cases} \Longleftrightarrow \lambda_t = \lambda_t^0 = 0$$

so we have exactly one risk-neutral measure: the market is complete (see Section 1.1.5). Note that in the situation where W has finite dimension d, the volatility matrix has d columns and *an infinity* of rows, and the above assumption means that at least d rows are linearly independent, which is not requiring much.

For convenience, we may change coordinates and write Φ_t^T in terms of a new process Y_t^T such that $\Phi_t^T = \Psi(X_t, Y_t^T)$ for a deterministic function Ψ; Y_t^T is not necessarily a driftless process. In Sections 3.1.1 to 3.1.5, we look at some examples of complete market models.

3.1.1   Modeling the dynamics of implied volatility: Schönbucher's market model

Here we follow [180]. We take $\Phi(x) = (x - K)^+$, a call with fixed strike K, and we parameterize its market value by its implied volatility $\sigma_t^{T,K}$ defined by

$$\Phi_t^T = BS((\sigma_t^{T,K})^2(T - t), K | X_t)$$

$BS(\sigma^2 T, K | X)$ denotes the Black-Scholes formula with variance $\sigma^2 T$, strike K, and spot X:

$$BS(\sigma^2 T, K | X) = X N(d_+) - K N(d_-) \tag{3.4}$$

with

$$d_\pm = \frac{\ln \frac{X}{K}}{\sigma \sqrt{T}} \pm \frac{\sigma \sqrt{T}}{2}$$

$N(x) = \frac{1}{\sqrt{2\pi}} \int_{-\infty}^x e^{-\frac{y^2}{2}} dy$ is the standard normal cumulative distribution function.

A straightforward application of Itô's formula shows that Φ_t^T is a local martingale if and only if the implied volatility dynamics is given by

$$d\sigma_t^{T,K} = u_t^{T,K} \, dt + v_t^{T,K} . dW_t \tag{3.5}$$

with the drift $u_t^{T,K}$ linked to the volatilities $\sigma_t^{T,K}$ and $v_t^{T,K}$ by

$$\sigma_t^{T,K} u_t^{T,K}(T-t) = \frac{1}{2}\left((\sigma_t^{T,K})^2 - \sigma_t^2\right) - \frac{1}{2}(v_t^{T,K})^2 \frac{(\ln \frac{X_t}{K})^2 - \frac{1}{4}(\sigma_t^{T,K})^4(T-t)}{(\sigma_t^{T,K})^2}$$

$$+ \sigma_t v_t^{T,K,0} \frac{\ln \frac{X_t}{K} - \frac{1}{2}(\sigma_t^{T,K})^2(T-t)}{\sigma_t^{T,K}} \tag{3.6}$$

For $K = X_T$, $t = T$, we get

$$\sigma_T^{T,X_T} = \sigma_T \tag{3.7}$$

For each volatility $v_t^{T,K}$ of the implied volatility, (3.6) defines the only acceptable arbitrage-free drift $u_t^{T,K}$ as a function of σ_t, $\sigma_t^{T,K}$, and $v_t^{T,K}$. It is not at all obvious to choose $v_t^{T,K}$ so that $\lim_{t \to T} \sigma_t^{T,K} \sqrt{T - t} = 0$—to ensure condition (3.3).

The complicated form of Equation (3.6) shows that the implied volatility does not seem to be the right coordinate to work with, although it is a very natural one. Below, we introduce a market model which entails much simpler algebra, and has the great advantage of admitting a low-dimensional Markovian representation.

3.1.2 Modeling the dynamics of log-contract prices: Bergomi's variance swap model

Let us take $\Phi(x) = \ln x$ a log-contract payoff. In the Black-Scholes model with time-dependent volatility $\sigma(t)$, we have (left as an exercise to the reader)

$$\Phi_t^T = \ln X_t - \frac{1}{2} \int_t^T \sigma(s)^2 \, ds$$

Let us set

$$\Phi_t^T = \ln X_t - \frac{1}{2} \int_t^T \xi_t^s \, ds$$

for a family of processes $(\xi_t^u, u \geq t)$. Φ_t^T is a local martingale under the risk-neutral measure if and only if

$$\sigma_t^2 = \xi_t^t$$

and the ξ_t^s are local martingales:

$$dX_t = X_t \sqrt{\xi_t^t} \, dW_t^0$$
$$d\xi_t^s = \beta_t^s . dW_t$$

$\xi_t^s = \mathbb{E}^{\mathbb{Q}}[\xi_s^s | \mathcal{F}_t] = \mathbb{E}^{\mathbb{Q}}[\sigma_s^2 | \mathcal{F}_t] \geq 0$ represents the forward instantaneous variance at time s, as seen at time $t \leq s$. If ξ_t^s behaves nicely when $s \to T$, condition (3.3) is automatically satisfied. We can trace back the creation of the notion of forward variance, forward instantaneous variance, and variance swaps to Dupire [94]. Bergomi [57] uses this framework to build a variance swap model that admits a low-dimensional Markovian representation, and which is now widely used in the industry:

$$d\xi_t^T = \sigma(t, T, \xi_t^T) \, dW_t \tag{3.8}$$

PROPOSITION 3.1

SDE (3.8) admits a two-dimensional Markovian representation, i.e., $\xi_t^T = f(T, X_t^0, X_t^1)$, depending on two one-dimensional Itô processes X_t^0, X_t^1 if and only if there exists Φ such that for all t, T, x

$$\partial_t \sigma(t, T, x) + \frac{1}{2}\sigma(t, T, x)^2 \partial_x^2 \sigma(t, T, x) = \Phi(t)\sigma(t, T, x) \qquad (3.9)$$

PROOF   Using Stratonovich calculus, SDE (3.8) reads as the time-homogeneous SDE

$$d\xi_t^T = -\frac{1}{2}\sigma(u, T, \xi_t^T)'\sigma(u, T, \xi_t^T)\, dt + \sigma(u, T, \xi_t^T) \diamond dW_t$$
$$du = dt$$

Here the prime denotes the (Gâteaux-Fréchet) functional derivative with respect to ξ^T. The corresponding vector fields are

$$V_0 \equiv -\frac{1}{2}\int dT\, \sigma'(u, T, \xi_T)\sigma(u, T, \xi_T)\frac{\delta}{\delta \xi_T} + \partial_u$$
$$V \equiv \int dT\, \sigma(u, T, \xi_T)\frac{\delta}{\delta \xi_T}$$

We have

$$[V_0, V] = \int dT \left(\partial_u \sigma(u, T, \xi_T) + \frac{1}{2}\sigma(u, T, \xi_T)^2 \sigma''(u, T, \xi_T)\right)\frac{\delta}{\delta \xi_T}$$

From Theorem 8.1 in [11], we have a Markov representation of dimension 2 if and only if there exists a function Φ such that

$$[V_0, V] = \Phi(u)V$$

which is equivalent to (3.9).   □

Example 3.1   One-factor Bergomi model [57]
This model corresponds to the case where

$$\sigma(t, T, x) = \omega e^{-k(T-t)}x$$

The dynamics of the instantaneous forward variance reads

$$\xi_t^T = \xi_0^T \exp\left(\omega e^{-k(T-t)}X_t - \frac{\omega^2}{2}e^{-2k(T-t)}\mathbb{E}^{\mathbb{Q}}[X_t^2]\right)$$

with

$$X_t^0 \equiv X_t = \int_0^t e^{-k(t-s)}dW_s, \quad \mathbb{E}^{\mathbb{Q}}[X_t^2] = \frac{1 - e^{-2kt}}{2k}, \quad X_t^1 = t$$

We speak of a one-factor model because the Itô process X_t^1 is deterministic: it is just the time t; X_t^0 is an Ornstein-Uhlenbeck process: $dX_t = -kX_t\,dt + dW_t$.
▯

Example 3.2 Two-factor Bergomi model [57]
The dynamics for the two-factor Bergomi model reads

$$dX_t = X_t\sqrt{\xi_t^t}\,dW_t$$
$$\xi_t^T = \xi_0^T f^T(t, x_t^T)$$
$$f^T(t, x) = \exp(2\sigma x - 2\sigma^2 h(t, T))$$
$$x_t^T = \alpha_\theta\left((1-\theta)e^{-k_Y(T-t)}Y_t + \theta e^{-k_Z(T-t)}Z_t\right)$$
$$\alpha_\theta = \left((1-\theta)^2 + \theta^2 + 2\rho_{YZ}\theta(1-\theta)\right)^{-1/2}$$
$$dY_t = -k_Y Y_t\,dt + dW_t^Y$$
$$dZ_t = -k_Z Z_t\,dt + dW_t^Z$$

where

$$h(t, T) = (1-\theta)^2 e^{-2k_Y(T-t)}\mathbb{E}^{\mathbb{Q}}\left[Y_t^2\right] + \theta^2 e^{-2k_Z(T-t)}\mathbb{E}^{\mathbb{Q}}\left[Z_t^2\right]$$
$$+ 2\theta(1-\theta)e^{-(k_Y+k_Z)(T-t)}\mathbb{E}^{\mathbb{Q}}\left[Y_t Z_t\right]$$
$$\mathbb{E}^{\mathbb{Q}}\left[Y_t^2\right] = \frac{1 - e^{-2k_Y t}}{2k_Y}$$
$$\mathbb{E}^{\mathbb{Q}}\left[Z_t^2\right] = \frac{1 - e^{-2k_Z t}}{2k_Z}$$
$$\mathbb{E}^{\mathbb{Q}}\left[Y_t Z_t\right] = \rho_{YZ}\frac{1 - e^{-(k_Y+k_Z)t}}{k_Y + k_Z}$$

This model, commonly used by practitioners, is a variance swap curve model that admits a two-dimensional (three-dimensional if we consider time as one of the Markov processes) Markovian representation.
▯

3.1.3 Modeling the dynamics of power-payoff prices: an HJM-like model

As a generalization of Bergomi's model, we introduce the p-payoff market model and show its link with the Heath-Jarrow-Morton forward rate curve framework in fixed income (see e.g., [1], Chapter 5).
We take $\Phi(x) = x^p$, the so-called p-payoff, or power-payoff, with $p \in \mathbb{R}\backslash\{0, 1\}$ a fixed number. In the Black-Scholes model with time-dependent volatility $\sigma(t)$, we have

$$\Phi_t^T = X_t^p \exp\left(\frac{1}{2}p(p-1)\int_t^T \sigma(s)^2\,ds\right)$$

Let us set

$$\Phi_t^T = X_t^p \exp\left(\frac{1}{2}p(p-1)Y_t^T\right)$$

with $Y_t^T = \int_t^T \xi_t^s \, ds$. We write

$$d\xi_t^T = \mu(t,T)\, dt + \nu(t,T).\, dW_t, \qquad \xi_{t=0}^T = \xi_0^T$$

where $(\mu(t,s), 0 \le t \le s)_{s \ge 0}$ and $(\nu(t,s), 0 \le t \le s)_{s \ge 0}$ are two families of respectively scalar-valued and vector-valued processes. Then

$$dY_t^T = \left(-\xi_t^t + \int_t^T \mu(t,s)\, ds\right) dt + \left(\int_t^T \nu(t,s)\, ds\right).\, dW_t$$

and

$$
\begin{aligned}
\frac{d\Phi_t^T}{\Phi_t^T} &= \frac{1}{2}p(p-1)\sigma_t^2\, dt + p\sigma_t\, dW_t^0 \\
&\quad + \frac{1}{2}p(p-1)\left(\left(-\xi_t^t + \int_t^T \mu(t,s)\, ds\right) dt + \left(\int_t^T \nu(t,s)\, ds\right).\, dW_t\right) \\
&\quad + \frac{1}{8}p^2(p-1)^2\left(\int_t^T \nu(t,s)\, ds\right)^2 dt + \frac{1}{2}p^2(p-1)\sigma_t \int_t^T \nu^0(t,s)\, ds
\end{aligned}
$$

By imposing that Φ_t^T is a driftless process, we obtain the condition for all $T \in [t, \infty)$:

$$
\begin{aligned}
m_t^T &\equiv \frac{1}{2}p(p-1)\sigma_t^2 + \frac{1}{2}p(p-1)\left(-\xi_t^t + \int_t^T \mu(t,s)\, ds\right) \\
&\quad + \frac{1}{8}p^2(p-1)^2\left(\int_t^T \nu(t,s)\, ds\right)^2 + \frac{1}{2}p^2(p-1)\sigma_t \int_t^T \nu^0(t,s)\, ds = 0
\end{aligned}
$$

This statement is equivalent to the conditions $m_t^t = 0$ and $\partial_T m_t^T = 0$, which imply

$$\sigma_t^2 = \xi_t^t \tag{3.10}$$

$$\mu(t,T) = -\frac{1}{2}p(p-1)\nu(t,T).\int_t^T \nu(t,s)\, ds - p\sigma_t \nu^0(t,T)$$

Hence the drift $\mu(t,T)$ can be inferred from the volatilities $\nu(t,s)$, $s \in [t,T]$, and ξ_t^T follows Heath-Jarrow-Morton-like dynamics (HJM):

PROPOSITION 3.2 Arbitrage-free dynamics of ξ_t^T
The arbitrage-free dynamics of ξ_t^T read as

$$d\xi_t^T = -\frac{1}{2}p(p-1)\nu(t,T).\left(\int_t^T \nu(t,s)\,ds\right)dt - p\sqrt{\xi_t^t}\nu^0(t,T)\,dt + \nu(t,T).\,dW_t$$

$$\xi_0^T = \frac{2}{p(p-1)}\partial_T \ln \frac{\Phi_0^T}{X_0^p} \tag{3.11}$$

Any admissible specification of the family $(\nu(t,s), 0 \le t \le s)_{s \ge 0}$ completely describes the model. A similar connection between p-payoffs and HJM was obtained independently in [72].

For the moment, we have focused only on the modeling of the term-structure of European vanilla payoffs with a fixed strike or fixed number p. What if we want to include all the strikes or all the p's? This is investigated in the next section.

3.1.4 The case of a family of power-payoffs

As a next step, we assume that our market model is calibrated to a strip of p-payoffs at each maturity date with p belonging to a subset I of $\mathbb{R}$. As shown above, the model is equivalent to an infinite collection of HJM-type models parameterized by the real number p:

$$d\xi_t^{T,(p)} = -\nu^{(p)}(t,T).\left(\frac{1}{2}p(p-1)\int_t^T \nu^{(p)}(t,s)\,ds + p\sigma_t\right)dt + \nu^{(p)}(t,T).\,dW_t$$

$$\xi_0^{T,(p)} = \frac{2}{p(p-1)}\partial_T \ln \frac{\Phi_0^T}{X_0^p} \tag{3.12}$$

with the compatibility condition (3.10)

$$\forall p \in I, \qquad \xi_t^{t,(p)} = \sigma_t^2 \tag{3.13}$$

Here, $\nu^{(p)}(t,T).\sigma_t$ means $\nu^{(p),0}(t,T)\sigma_t$. Equation (3.13) ensures that all the forward HJM curves $(\xi_t^{u,(p)}, u \ge t)$ have the same "short rate," defined as $\sigma_t^2 \equiv \xi_t^t$.

At this point, it is not in any way obvious under which conditions on $\nu^{(p)}(t,s)$ the above compatibility condition is satisfied. Furthermore, the implementation of this HJM-like model requires the simulation of the full forward curve. In order to circumvent this last problem, we choose a particular family $(\nu^{(p)}(t,T))$ of volatilities of instantaneous forward variance curves that admits a low-dimensional Markovian representation (see in [11] the link with the Frobenius theorem). This is similar to the Cheyette model as introduced in [81]. For the sake of simplicity, here W_t is a one-dimensional Brownian motion equal to W_t^0. The straightforward generalization is left to the reader.

PROPOSITION 3.3 Markovian representation

Assume that the volatility of the instantaneous forward variance curve can be written as

$$\nu^{(p)}(t,T) = \sum_{i=1}^{N} \alpha_i^{(p)}(T)\beta_i^{(p)}(t)$$

with $(\beta_i^{(p)})_{1\le i\le N}$ N positive stochastic processes and $(\alpha_i^{(p)})_{1\le i\le N}$ N positive deterministic functions of the maturity T. Then the instantaneous forward variance curve, described by (3.12), admits a Markov representation as a function of $N(N+3)/2$ state variables x_i^p, $V_{ij}^p = V_{ji}^p$, $i,j \in \{1,\ldots,N\}$, following the SDE

$$dx_i^{(p)}(t) = \left(x_i^{(p)}(t)\partial_t \ln \alpha_i^{(p)}(t) - \frac{p(p-1)}{2}\sum_{k=1}^{N} V_{ik}^{(p)}(t) - p\sqrt{\xi_t^t}\alpha_i^{(p)}(t)\beta_i^{(p)}(t)\right)dt$$

$$+\beta_i^{(p)}(t)\alpha_i^{(p)}(t)\,dW_t$$

$$dV_{ij}^{(p)}(t) = \left(\beta_i^{(p)}(t)\beta_j^{(p)}(t)\alpha_i^{(p)}(t)\alpha_j^{(p)}(t) + V_{ij}^{(p)}(t)\partial_t \ln\left(\alpha_i^{(p)}(t)\alpha_j^{(p)}(t)\right)\right)dt$$

with the initial conditions $x_i^{(p)}(0) = 0$, $V_{ij}^{(p)}(0) = 0$. The forward curve $\xi_t^{T,(p)}$ can be expressed as

$$\xi_t^{T,(p)} = \xi_0^{T,(p)} + \sum_{j=1}^{N} \frac{\alpha_j^{(p)}(T)}{\alpha_j^{(p)}(t)}\left(x_j^{(p)}(t) - \frac{p(p-1)}{2}\sum_{i=1}^{N} \frac{A_i^{(p)}(T) - A_i^{(p)}(t)}{\alpha_i^{(p)}(t)}V_{ij}^{(p)}(t)\right)$$

where $A_k^{(p)}(t) = \int_0^t \alpha_k^{(p)}(s)\,ds$.

The proposition follows from a straightforward computation. Note that the "short rate" $\xi_t^{t,(p)}$, i.e., the instantaneous variance, is given by

$$\xi_t^{t,(p)} = \xi_0^{t,(p)} + \sum_{i=1}^{N} x_i^{(p)}(t)$$

If we assume that the initial forward curve $\xi_0^{t,(p)}$ is a quadratic function in p, i.e.,

$$\xi_0^{t,(p)} = a(t) + b(t)p - \frac{p(p-1)}{2}c(t) \tag{3.14}$$

the compatibility condition $\partial_p \xi_t^{t,(p)} = 0$ holds if we take $\alpha_i^{(p)}(t)$ and $\beta_i^{(p)}(t)$ independent of p and satisfying

$$b(t) = \sum_{i=1}^N \int_0^t \alpha_i(t)\beta_i(s) \cdot \sigma_s \, ds$$

$$c(t) = \sum_{i=1}^N \int_0^t \alpha_i(t)\beta_i(s) \cdot \sum_{k=1}^N \left(\int_s^t \alpha_k(u) du \right) \beta_k(s) \, ds$$

3.1.5 Modeling the dynamics of local volatility: Carmona and Nadtochiy's market model

Following [93, 72], as another parametrization, let us now try to describe the market-implied volatility dynamics in terms of the dynamics of the corresponding local volatility function whose square is defined as the conditional expectation of the square of the stochastic volatility σ_T^2 given the spot value:

$$\sigma_t^2(T, K) \equiv \mathbb{E}^{\mathbb{Q}}[\sigma_T^2 | X_T = K, \mathcal{F}_t] = \frac{\mathbb{E}^{\mathbb{Q}}[\sigma_T^2 \delta(X_T - K) | \mathcal{F}_t]}{\mathbb{E}^{\mathbb{Q}}[\delta(X_T - K) | \mathcal{F}_t]} \qquad (3.15)$$

The stochastic volatility is defined as $\sigma_t^2 \equiv \frac{d\langle \ln X \rangle_t}{dt}$. Let $\mathcal{C}_t(T, K)$ denote the market value of a call option with strike K and maturity T at time t. From Dupire's formula (see [95]),

$$\forall t < T, \qquad \sigma_t^2(T, K) = \frac{\partial_T \mathcal{C}_t(T, K)}{\frac{1}{2} K^2 \partial_K^2 \mathcal{C}_t(T, K)}$$

Note that $\sigma_t^2(T, K)$ is known at $t = 0$ from the initial implied volatility surface. From the definition (3.15), we have

$$\sigma_t^2(t, K) = \mathbb{E}^{\mathbb{Q}}[\sigma_t^2 | X_t = K, \mathcal{F}_t] = \sigma_t^2 \mathbb{E}^{\mathbb{Q}}[1 | X_t = K, \mathcal{F}_t]$$

from which we deduce that

$$\sigma_t(t, X_t) = \sigma_t$$

Below, we formally derive the arbitrage-free dynamics of $\sigma_t(T, K)$. To this purpose, we observe that $\mathbb{E}^{\mathbb{Q}}[\sigma_T^2 \delta(X_T - K) | \mathcal{F}_t]$ and $M_t^{T,K} \equiv \mathbb{E}^{\mathbb{Q}}[\delta(X_T - K) | \mathcal{F}_t]$ are both (closed) $\mathbb{Q}$-martingales. As a consequence, $\sigma_t^2(T, K)$ should be a local martingale in the measure $\mathbb{Q}^{T,K}$ associated to the numéraire $M_t^{T,K} \equiv p_t(T, K | t, X_t)$:

$$d\sigma_t^2(T, K) = \beta_t^{T,K} . dW_t^{T,K}$$

with $W_t^{T,K}$ a $\mathbb{Q}^{T,K}$-Brownian motion. By changing the measure from $\mathbb{Q}^{T,K}$ to $\mathbb{Q}$ ($\frac{d\mathbb{Q}}{d\mathbb{Q}^{T,K}}|_{\mathcal{F}_t} = \frac{1}{M_t^{T,K}}$), we obtain (see the change of measure formula in

[11], Proposition 2.1)

$$d\sigma_t^2(T,K) = -\langle d\sigma_t^2(T,K), \frac{dM_t^{T,K}}{M_t^{T,K}} \rangle + \beta_t^{T,K}.dW_t$$

$$= -\beta_t^{T,K} \cdot \left(\sigma_t X_t \partial_{X_t} \ln p_t(T,K|t,X_t) \right.$$

$$\left. + \int_t^T dT' \int_0^\infty dK' \beta_t^{T',K'} \frac{\delta \ln p_t(T,K|t,X_t)}{\delta \sigma_t^2(T',K')} \right) dt + \beta_t^{T,K}.dW_t$$

with W a $\mathbb{Q}$-Brownian motion. Here $\beta_t^{T,K}.\sigma_t$ means $\beta_t^{T,K,0}\sigma_t$ and we have assumed that X_t admits a smooth density $p_t(T,K|t,X_t)$, which depends explicitly on X_t and implicitly on the variables $\sigma_t(T,K)$ for all T and for all K. $\frac{\delta \ln p_t(T,K|t,X_t)}{\delta \sigma_t^2(T',K')}$ indicates the functional Gâteaux-Fréchet derivative of $\ln p_t$ with respect to $\sigma_t(T',K')$. For the reader not familiar with this tool, consider that this is a classical derivative with respect to the variable $\sigma_t(T_i,K_\alpha), i \in \{1,\ldots,n\}$, $\alpha \in \{1,\ldots,N\}$, where we have discretized the maturity and the strike spaces. The integral $\int_t^T dT' \int_0^\infty dK'$ should then be replaced by a discrete sum $\sum_{T_i} \sum_{K_\alpha}$. In particular, we have $\frac{\delta \sigma_t^2(T_j,K_\beta)}{\delta \sigma_t^2(T_i,K_\alpha)} = \delta_{ij}\delta_{\alpha\beta}$. In the limit $n,N \to \infty$, we have

$$\frac{\delta \sigma_t^2(T,K)}{\delta \sigma_t^2(T',K')} = \delta(T'-T)\delta(K'-K) \tag{3.16}$$

The transition function $p_t(T,K|u,x)$ satisfies the backward Kolmogorov equation

$$\partial_u p_t(T,K|u,x) + \frac{1}{2}\sigma_t^2(u,x)\partial_x^2 p_t(T,K|u,x) = 0$$

$$\lim_{u \to T} p_t(T,K|u,x) = \delta(K-x)$$

By applying the functional derivation $\frac{\delta}{\delta\sigma_t^2(T',K')}$ on both sides of this equation (using (3.16) and Leibnitz's differentiation rule), we get

$$\partial_u \frac{\delta p_t(T,K|u,x)}{\delta\sigma_t^2(T',K')} + \frac{1}{2}\sigma_t^2(u,x)\partial_x^2 \frac{\delta p_t(T,K|u,x)}{\delta\sigma_t^2(T',K')}$$

$$+ \frac{1}{2}x^2\delta(T'-u)\delta(S-K')\partial_x^2 p_t(T,K|u,x) = 0$$

$$\lim_{u \to T} \frac{\delta p_t(T,K|u,x)}{\delta\sigma_t^2(T',K')} = 0$$

From Feynman-Kac's theorem (see Theorem 1.4 for a reminder), we conclude that

$$\frac{\delta p_t(T, K | u, x)}{\delta \sigma_t^2(T', K')} = \frac{1}{2} \mathbb{E}\Big[\int_u^T X_s^2 \partial_x^2 p_t(T, K | s, X_s) \delta(T' - s) \delta(X_s - K') \, ds \Big| X_u = x \Big]$$

$$= \frac{1}{2} K'^2 \partial_{K'}^2 p_t(T, K | T', K') p_t(T', K' | u, x) \mathbf{1}_{T' \in [u, T]}$$

$$= -\frac{\partial_{T'} p_t(T, K | T', K') p_t(T', K' | u, x)}{\sigma_t^2(T', K')} \mathbf{1}_{T' \in [u, T]}$$

from which we get the dynamics of the local volatility (a longer proof not relying on the change of measure from $\mathbb{Q}^{T,K}$ to $\mathbb{Q}$ can be found in [72]):

PROPOSITION 3.4 Arbitrage-free dynamics of $\sigma_t^2(T, K)$
The arbitrage-free dynamics of $\sigma_t^2(T, K)$ read

$$dX_t = X_t \sigma_t(t, X_t) \, dW_t^0 \qquad (3.17)$$

$$d\sigma_t^2(T, K) = \mu_t^{T,K} \, dt + \beta_t^{T,K}.dW_t \qquad (3.18)$$

with the non-trivial drift

$$\mu_t^{T,K} = -\beta_t^{T,K}.\Big(\sigma_t(t, X_t) X_t \partial_{X_t} \ln p_t(T, K | t, X_t)$$

$$- \int_t^T dT' \int_0^\infty dK' \, \partial_{T'} p_t(T, K | T', K') \frac{\beta_t^{T',K'}}{\sigma_t^2(T', K')} \frac{p_t(T', K' | t, X_t)}{p_t(T, K | t, X_t)} \Big) \qquad (3.19)$$

Example 3.3 Dupire local volatility model
By taking $\beta_t^{T,K} = 0$, we obtain Dupire's local volatility model [96] $dX_t = X_t \sigma_0(t, X_t) \, dW_t$ with

$$\sigma_0^2(T, K) = \frac{\partial_T \mathcal{C}_0(T, K)}{\frac{1}{2} K^2 \partial_K^2 \mathcal{C}_0(T, K)}$$

▯

Note that the first term in (3.19) is incorrectly missing in [93]. It is not obvious that SDE (3.18) is well-defined, in particular that the drift term behaves smoothly when $t \to T$. Furthermore, SDE (3.18) does not appear to admit a low-dimensional Markov representation.

As a conclusion, building a well-defined market model admitting a low-dimensional Markov representation (and so allowing an easy numerical simulation) is not at all obvious. Under the assumption (3.14), we have, however, produced one example of such a model. As far as we know, no explicit solutions valid for arbitrary (initial) market data are known. Note, however, that in Chapter 11 we will be able to build (incomplete) local stochastic volatility models that are perfectly calibrated to the prices of all vanilla options.

3.2 Beyond replication and super-replication

This section is a brief overview of alternative approaches for pricing options, one based on a utility function, and the other based on quantile hedging. In Exercise 4.7.3, we show that both approaches are identical under the assumption of uncertain volatilities.

3.2.1 The utility indifference price

We introduce a utility function U, which is strictly increasing and strictly concave. We consider the value $u(x,0)$ of the supremum over all admissible hedging portfolios starting from the initial capital x of the expectation of the utility of the discounted final wealth under the historical measure $\mathbb{P}^{\text{hist}}$:

$$u(x,0) \equiv \sup_{\Delta \text{ adm.}} \mathbb{E}^{\mathbb{P}^{\text{hist}}} \left[U\left(x + \int_0^T \Delta_t \, d\tilde{X}_t \right) \right]$$

Similarly, the value $u(x - \mathcal{C}, F_T)$ is defined for a claim F_T as

$$u(x - \mathcal{C}, F_T) \equiv \sup_{\Delta \text{ adm.}} \mathbb{E}^{\mathbb{P}^{\text{hist}}} \left[U\left(x - \mathcal{C} + D_{0T} F_T + \int_0^T \Delta_t \, d\tilde{X}_t \right) \right]$$

The utility indifference buyer's price, as introduced by Hodges-Neuberger [135], is the quantity $\mathcal{B}(x, F_T)$ such that[1]

$$u(x,0) = u(x - \mathcal{B}(x, F_T), F_T)$$

This means that a buyer should accept quoting a price for the claim F_T when buying and delta hedging this derivative becomes as profitable as setting up a pure delta strategy. For instance, if $U(x) = x$, then it is easy to check that $\mathcal{B}(x, F_T)$ does not depend on the initial capital x and is equal to $\mathbb{E}^{\mathbb{P}^{\text{hist}}}[D_{0T} F_T]$. For a general utility function (satisfying a technical assumption known as Inada's condition: $\lim_{x \to \infty} \frac{U(x)}{x} = 0$), a dual expression can be obtained. Proceeding as in the proof of Theorem 1.3, one can prove that

$$u(x, F_T) = \inf_{\mathbb{Q} \in \text{ELMM}} \inf_{p \geq 0} \mathbb{E}^{\mathbb{P}^{\text{hist}}} \left[p\left(x + D_{0T} F_T \right) \frac{d\mathbb{Q}}{d\mathbb{P}^{\text{hist}}} + \tilde{U}\left(p \frac{d\mathbb{Q}}{d\mathbb{P}^{\text{hist}}} \right) \right] \quad (3.20)$$

with $\tilde{U}(p) \equiv \sup_{x \in \mathbb{R}} \{U(x) - px\}$ the Legendre-Fenchel transform of U. The functions U and $\tilde{U}$ also satisfy $U(x) = \inf_{p \in \mathbb{R}} \{px + \tilde{U}(p)\}$.

---

[1] Obviously one can define the utility indifference seller's price in a similar way.

Example 3.4

For $U_\alpha(x) = 1 - e^{-\alpha x}$ with $\alpha > 0$, we have $\tilde{U}_\alpha(p) = 1 - \frac{p}{\alpha} + \frac{p}{\alpha} \ln \frac{p}{\alpha}$, and the utility indifference buyer's price is

$$\alpha \mathcal{B}_\alpha(F_T) = \inf_{\mathbb{Q} \in \text{ELMM}} \left\{ \mathbb{E}^{\mathbb{P}^{\text{hist}}} \left[\frac{d\mathbb{Q}}{d\mathbb{P}^{\text{hist}}} \ln \frac{d\mathbb{Q}}{d\mathbb{P}^{\text{hist}}} \right] + \alpha \mathbb{E}^{\mathbb{P}^{\text{hist}}} \left[\frac{d\mathbb{Q}}{d\mathbb{P}^{\text{hist}}} D_{0T} F_T \right] \right\}$$

$$- \inf_{\mathbb{Q} \in \text{ELMM}} \left\{ \mathbb{E}^{\mathbb{P}^{\text{hist}}} \left[\frac{d\mathbb{Q}}{d\mathbb{P}^{\text{hist}}} \ln \frac{d\mathbb{Q}}{d\mathbb{P}^{\text{hist}}} \right] \right\} \quad (3.21)$$

Note that due to the particular form of the exponential utility function, $\mathcal{B}_\alpha(F_T)$ does not depend on the initial capital x.

In particular, we have

$$\lim_{\alpha \to \infty} \mathcal{B}_\alpha(F_T) = \inf_{\mathbb{Q} \in \text{ELMM}} \mathbb{E}^{\mathbb{Q}}[D_{0T} F_T]$$

which corresponds to the buyer's super-replication price. This was expected as $\lim_{\alpha \to \infty} U_\alpha(x) = -\infty$ if $x < 0$, 1 otherwise, which puts us exactly in the situation of the proof of Theorem 1.3.[2]

Furthermore, we have

$$\lim_{\alpha \to 0} \mathcal{B}_\alpha(F_T) = \mathbb{E}^{\mathbb{Q}^*}[D_{0T} F_T]$$

where $\mathbb{Q}^* \in$ ELMM is, among all the risk-neutral measures, the one that minimizes the relative entropy $H\left(\mathbb{Q}|\mathbb{P}^{\text{hist}}\right)$ with respect to $\mathbb{P}^{\text{hist}}$:

$$H\left(\mathbb{Q}|\mathbb{P}^{\text{hist}}\right) = \mathbb{E}^{\mathbb{P}^{\text{hist}}} \left[\frac{d\mathbb{Q}}{d\mathbb{P}^{\text{hist}}} \ln \frac{d\mathbb{Q}}{d\mathbb{P}^{\text{hist}}} \right]$$

$\mathbb{Q}^*$ is called the minimal entropy martingale measure. ⬚

Note that in general, the utility indifference price involves a nonlinear pricing operator, associated to a nonlinear PDE: if F_T^1 and F_T^2 are two payoffs,

$$\mathcal{B}(x, F_T^1 + F_T^2) \leq \mathcal{B}(x, F_T^1) + \mathcal{B}(x, F_T^2)$$

3.2.2 Quantile hedging

Quantile hedging consists of replacing Definition 1.5 of the seller's price by the following:

DEFINITION 3.1 *p-quantile seller's price*

$$\mathcal{S}_t^{(p)}(F_T) = \inf \{ z \in \mathcal{F}_t \mid \text{there exists an admissible portfolio } \Delta \text{ such that}$$

$$\mathbb{P}^{\text{hist}}[\tilde{\pi}_T^S \geq 0] \geq p \} \quad (3.22)$$

---

[2]The constant 1 in the definition of U_α plays no role here: We could have taken $U_\alpha(x) = -e^{-\alpha x}$ instead.

$p \in [0, 1]$ is interpreted as the probability of super-replicating the claim F_T under the historical measure. By definition, $\mathcal{S}_t^{(1)}(F_T) = \mathcal{S}_t(F_T)$, the super-replication price, and $\mathcal{S}_t^{(0)}(F_T) = -\infty$.

The super-replication price of an option can be very high—recall for instance that the super-replication price of a call option equals the forward (see Section 1.1.5). In the quantile hedging approach, we only impose that the payoff be super-replicated with probability p. In their seminal paper [107], Föllmer and Leukert show that the corresponding optimal strategy consists of superhedging a modified payoff provided by the Neyman-Pearson lemma:

$$\mathcal{S}_t^{(p)}(F) = \inf_{A \in \mathcal{F}_T \text{ s.t. } \mathbb{E}^{\mathbb{P}\text{hist}}[1_A|\mathcal{F}_t] \geq p} \sup_{\mathbb{Q} \in \text{ELMM}} \mathbb{E}^{\mathbb{Q}}[D_{tT}F_T 1_A|\mathcal{F}_t]$$

3.2.3   Minimum variance hedging

Another approach consists of minimizing the quadratic error of the (global) P&L over the set of all admissible strategies:

$$\inf_{z, \Delta \text{ adm.}} \mathbb{E}^{\mathbb{P}} \left[\left(-D_{0t}z + \int_t^T \Delta_s . d\tilde{X}_s + D_{0T}F_T \right)^2 \bigg| \mathcal{F}_t \right]$$

Note that this optimization problem depends on the choice of the underlying probability measure $\mathbb{P}$. Here, since the utility function is quadratic, the pricing operator is linear. Otherwise, it can be nonlinear.

If we assume that $\mathbb{P}$ is an equivalent local martingale measure, thus denoted by $\mathbb{Q}$, the optimal trading strategy is obtained by applying the so-called Kunita-Watanabe decomposition of $D_{tT}F_T$ on $\tilde{X}$ [106]:

$$\mathbb{E}^{\mathbb{Q}}[D_{tT}F_T|\mathcal{F}_t] = \mathbb{E}^{\mathbb{Q}}[D_{tT}F_T] + \int_0^t \Delta_s^* . d\tilde{X}_s + X_t^{\perp} \tag{3.23}$$

with $X_t^{\perp}$ a $\mathbb{Q}$-martingale orthogonal to $\tilde{X}_s$. The optimal delta is Δ^* and the price is given the usual pricing rule in a complete market, i.e., $\mathbb{E}^{\mathbb{Q}}[D_{tT}F_T|\mathcal{F}_t]$. Schweizer [181] introduces locally risk-minimizing hedging, which consists of minimizing the quadratic error of the P&L between t and $t + dt$. The two notions coincide if $\mathbb{P}$ is a martingale measure.

Conclusion

The hedging frameworks presented in this chapter, as well as the super-replication paradigm described in Section 1.1.4, lead to nonlinear pricing operators which, for Markovian models, are connected to nonlinear parabolic

PDEs; see Exercises 4.7.2 and 4.7.3. In the next chapter we give a short introduction to the theory of nonlinear PDEs.

3.3 Exercises

3.3.1 Super-replication and quantile hedging

Let us consider a one-period model: $S_0 = 1$, and S_T can take only three values 0.5, 1, and 1.5 at maturity T. Under the historical measure $\mathbb{P}^{\text{hist}}(S_T = 0.5) = 1/3$, $\mathbb{P}^{\text{hist}}(S_T = 1.0) = 1/3$, and $\mathbb{P}^{\text{hist}}(S_T = 1.5) = 1/3$. Assume zero interest rates.

1. Compute the super-replication price of the call option with maturity T and strike 1.

2. Compute the quantile hedging price of the same option as a function of $p \in [0, 1]$.

3.3.2 Duality for the utility indifference price

Prove Equation (3.20). Hint: proceed as in the proof of Theorem 1.3, and introduce $p \equiv Y_0$, where Y is the shadow price.

3.3.3 Utility indifference price for exponential utility

Prove Equation (3.21).

3.3.4 Minimum variance hedging for the double lognormal stochastic volatility model

Explain the Kunita-Watanabe decomposition (3.23) in the case of the double lognormal stochastic volatility model defined by Equations (1.15)–(1.17). For the sake of simplicity, take zero interest rates.

Chapter 4

Nonlinear PDEs: A Bit of Theory

La finance est décrite par des EDPs paraboliques non-linéaires et cela n'a rien à voir avec le mouvement brownien.[1]

— Pierre-Louis Lions, cours du Collège de France (2007)

In this chapter we review basic general properties of nonlinear second order parabolic PDEs. As a consequence of the comparison principle, these equations naturally show up in finance. Since many nonlinear PDEs arising in finance are of the Hamilton-Jacobi-Bellman type, we present a crash course on stochastic control theory highlighting the main notions and results such as verification theorems and viscosity solutions. Although proofs have been replaced by appropriate references such as [21, 24], we strive to give some mathematical intuition.

4.1 Nonlinear second order parabolic PDEs: Some generalities

Throughout this book we consider second order fully nonlinear partial differential equations (in short, PDEs) of the type

$$-\partial_t u(t,x) - H(t,x,u(t,x),Du(t,x),D^2u(t,x)) = 0, \quad x \in \Omega, \quad t \in [0,T) \quad (4.1)$$

with the terminal condition $u(T,x) = g(x)$. Here, Ω is an open domain of $\mathbb{R}^n$, and Du (resp. D^2u) denotes the gradient (resp. Hessian) of the solution u with respect to spatial variables x. Below, we explain why we have introduced this overall minus sign when we discuss the concept of supersolutions (resp. subsolutions) and its link with supermartingales (resp. submartingales).

$H = H(t,x,z,p,r)$, called the Hamiltonian when connected to a stochastic control problem, is defined on $[0,T) \times \Omega \times \mathbb{R} \times \mathbb{R}^n \times \mathcal{S}^n$ with $\mathcal{S}^n$ the space

---

[1] "Finance is described by nonlinear parabolic PDEs, and that has nothing to do with the Brownian motion."

of real n-dimensional symmetric matrices. Note that by inverting the time, $\tau = T - t$, the above PDE can be cast into an initial value problem

$$-\partial_\tau u + H(T - \tau, x, u, Du, D^2 u) = 0, \qquad x \in \Omega, \qquad \tau \in (0, T] \quad (4.2)$$

with the initial condition $u(\tau = 0, x) = g(x)$. As final value problems are more common in finance, we have decided to give an informal presentation of nonlinear PDEs with a terminal condition. In most mathematics books we refer to in this chapter, theorems are stated for these PDEs (4.1).

4.1.1 Parabolic PDEs

PDE (4.1) is called *parabolic* if the condition below is satisfied:

DEFINITION 4.1 Parabolic PDE   *PDE (4.1) is called parabolic if H is elliptic: $[\frac{\partial H}{\partial r_{ij}}]_{1 \leq i,j \leq n}$ is a symmetric positive semi-definite matrix.*

The sagacious reader can argue that H is not necessarily differentiable. A more general definition is: H is said to be parabolic if for all choices of t, x, z, p, r_1, r_2,

$$r_2 \geq r_1 \implies H(t, x, z, p, r_2) \geq H(t, x, z, p, r_1)$$

where $r_2 \geq r_1$ means $x^*(r_2 - r_1)x \geq 0$ for all $x \in \mathbb{R}^n$; x^* denotes the transpose of the vector x. Parabolicity will play a key role when we introduce the comparison principle.

Two important special cases of (4.1) can be highlighted. The first case is a *semilinear* second order parabolic PDE, which can be written as

$$-\partial_t u - \sum_{i,j=1}^n a_{ij}(t, x)\partial_{ij} u = f(t, x, u, Du)$$

In layman's terms, H is linear in $D^2 u$ and the coefficients a_{ij} do not depend on u and its derivatives. Parabolicity means that the symmetric matrix a is positive semi-definite. The second case is a *quasilinear* PDE, which can be written as

$$-\partial_t u - \sum_{i,j=1}^n a_{ij}(t, x, u, Du)\partial_{ij} u = f(t, x, u, Du)$$

In a *fully nonlinear* PDE, the coefficients $a_{ij}(t, x, u, Du, D^2 u)$ depend on u, its gradient Du, and its Hessian $D^2 u$.

4.1.2   What do we mean by a well-posed PDE?

A PDE is said to be well-posed if both existence and uniqueness of a solution can be established for arbitrary data (i.e., terminal condition) belonging to a specified large space of functions, such as a class of smooth functions. Moreover, the solutions must depend continuously on the data: If two terminal conditions are "close," the corresponding solutions should be "close" enough.

4.1.3   What do we mean by a solution?

We call u a *classical solution* of (4.1) if $u \in C^{1,2}([0,T] \times \Omega)$, the space of functions which are twice continuously differentiable in space, and once continuously differentiable in time. All the derivatives in (4.1) exist and are continuous in (t, x). This is certainly the most obvious definition of a solution. However, such a requirement is usually too strong. For example, the nonlinear PDE satisfied by the price in the uncertain volatility model with unbounded volatility does not have a classical solution, as the (generalized or weak) solution is not even continuous in time (we may have $u(T^-, x) > u(T, x)$). In order to have a well-posed PDE which admits a unique solution, one needs to weaken the notion of solution. It is usually easy to prove existence and uniqueness of weak solutions. In a second and important step, we can then ask if the weak solution has some regularity properties. In the context of second order nonlinear parabolic PDEs, this usually requires the use of the Giorgi-Nash-Moser theorem [104].

The (weak) notion of a viscosity solution, as first introduced by P.-L. Lions, only requires the solution to be locally bounded and seems to be the right weak condition for nonlinear second order parabolic PDEs connected to stochastic control problems (see Section 4.4).

4.1.4   Comparison principle and uniqueness

4.1.4.1   Definition and sufficient condition

We require that PDE (4.1) satisfies the comparison (or maximum) principle:

DEFINITION 4.2 Comparison principle   *If u and v are two solutions, then*

$$h(x) \equiv v(T, x) \leq g(x) \equiv u(T, x), \quad \forall x \in \Omega$$
$$\implies \quad v(t, x) \leq u(t, x), \quad \forall x \in \Omega, \ \forall t \in [0, T]$$

If a European payoff g dominates another European payoff h, then their values should satisfy the same inequality. In terms of derivatives pricing, this principle is equivalent to the absence of arbitrage opportunities. If this condition is not satisfied, we can build a (static) arbitrage opportunity consisting of holding $g - h$: the terminal wealth of this static portfolio is nonnegative and its t-value is negative.

A striking consequence of this comparison principle is a uniqueness result: if u and v are two classical (or weak, see later) solutions with the same terminal condition $g = h$, then $u(t,x) = v(t,x)$ for all $x \in \Omega$ and $t \in [0,T]$. Indeed, $h(x) \leq g(x)$ (resp. $g(x) \leq h(x)$) implies $v(t,x) \leq u(t,x)$ (resp. $u(t,x) \leq v(t,x)$).

What conditions should we impose on H for the comparison principle to hold? If $H = H(t, x, Du, D^2u)$ is independent of u, a sufficient condition is that H be parabolic. For a general Hamiltonian (depending on u), the comparison principle is satisfied if H is a proper operator:

DEFINITION 4.3 Proper   *$H(t, x, z, p, r)$ is said to be a proper operator if it is parabolic and (strictly) decreasing in z.*

PROPOSITION 4.1
PDE (4.1) satisfies the comparison principle if H is proper.

Below, we outline the proof by assuming that Ω is a compact domain to avoid difficulties at the boundary of Ω.

PROOF   Let $h \leq g$. Our objective is to prove that

$$M \equiv \max_{(t,x)\in[0,T]\times\Omega} (v(t,x) - u(t,x)) \leq 0$$

Assume that $M > 0$. The maximum is reached at a point (t_0, x_0). Since $h(x) - g(x) \leq 0$, $t_0 < T$. As a consequence,

$$u(t_0, x_0) < v(t_0, x_0)$$
$$Dv(t_0, x_0) = Du(t_0, x_0)$$
$$D^2v(t_0, x_0) \leq D^2u(t_0, x_0)$$

and

$$\partial_t v(t_0, x_0) \leq \partial_t u(t_0, x_0)$$

The above inequality is an equality if $t_0 \neq 0$. From the above inequalities and the parabolic and proper conditions, we have

$$\partial_t v(t_0, x_0) + H\left(t_0, x_0, v(t_0, x_0), Dv(t_0, x_0), D^2v(t_0, x_0)\right) \overset{\text{parabolic}}{\leq}$$

$$\partial_t u(t_0, x_0) + H\left(t_0, x_0, v(t_0, x_0), Du(t_0, x_0), D^2u(t_0, x_0)\right) \overset{\text{proper}}{<}$$

$$\partial_t u(t_0, x_0) + H\left(t_0, x_0, u(t_0, x_0), Du(t_0, x_0), D^2u(t_0, x_0)\right)$$

which implies a contradiction as u and v are two solutions ($0 < 0$).   □

Below, as an alternative proof, we check that the linearized PDE satisfies the comparison principle.

4.1.4.2 Linearization

Linearization around special solutions is an extremely important tool when studying nonlinear PDEs. Difficult problems in nonlinear PDEs often reduce to understanding specific simpler problems in linear PDEs. Linearization should certainly be used by quantitative analysts as a preliminary tool for understanding the behavior of a nonlinear PDE. Below, we explain how to prove a comparison principle by linearization of PDE (4.1) around a solution $u(t, x)$ with a terminal condition $u(T, x) = g(x)$.

Let us compute the solution, denoted with $u + \epsilon \delta u$, associated to the terminal condition $g(x) + \epsilon \delta g(x)$ at first order in ϵ. The linearized PDE around the solution u reads

$$\partial_t \delta u + \left(H'_u + H'_{Du} D + H'_{D^2 u} D^2 \right) \delta u = 0, \qquad \delta u(T, x) = \delta g(x)$$

where H'_Y denotes the first order derivative of H with respect to Y. This is a second order *linear* PDE which is well-posed if and only if the variance matrix $H'_{D^2 u}$ is positive semi-definite. This is guaranteed by the parabolic condition. By applying Feynman-Kac's formula, we get:

$$\delta u(t, x) = \mathbb{E}[\delta g(X_T) e^{\int_t^T H'_u(s, X_s, u(s, X_s), Du(s, X_s), D^2 u(s, X_s)) \, ds} | X_t = x]$$

with X_t the Itô diffusion defined with drift

$$H'_{Du}(t, X_t, u(t, X_t), Du(t, X_t), D^2 u(t, X_t))$$

and half variance

$$H'_{D^2 u}(t, X_t, u(t, X_t), Du(t, X_t), D^2 u(t, X_t))$$

u being fixed. In order to ensure that this stochastic representation is well defined, we assume that H'_u is bounded from above (see Theorem 1.4). As a consequence, if $\delta g(x) \geq 0$, then $\delta u(t, x) \geq 0$. This is the comparison principle at first order in ϵ. Here we only require that H'_u be bounded from above and not that H'_u be negative, as implied by the definition of a proper operator. This suggests that the stronger condition $H'_u < 0$ can be relaxed (see [86]):

DEFINITION 4.4 Proper II $H(t, x, z, p, r)$ *is said to be a proper operator if it is parabolic and there exists* $\gamma > 0$ *such that for all* t, x, p, r, *and* $z \leq z'$,

$$H(t, x, z', p, r) - H(t, x, z, p, r) \leq \gamma (z' - z)$$

Proposition 4.1 still holds with condition Proper II; see e.g., [5]. For H differentiable in z, by taking the limit where $z' \to z$, this implies that H'_z is bounded from above by γ.

4.1.4.3   Supersolutions, subsolutions, and the comparison principle

DEFINITION 4.5 Supersolutions and subsolutions $u \in C^{1,2}([0,T) \times \Omega)$ *is called a classical supersolution (resp. subsolution) of PDE (4.1) if*

$$- \partial_t u(t,x) - H\left(t, x, u(t,x), Du(t,x), D^2 u(t,x)\right) \geq 0 \quad (\text{resp. } \leq 0) \quad (4.3)$$

PDE (4.1) was presented with an overall minus sign because we wanted to define a supersolution as a function for which the l.h.s. is greater than 0. Note that if $u(t, X_t)$ is a supermartingale with $u \in C^{1,2}([0,T) \times \Omega)$ and X_t is the solution of an SDE, then by applying Itô's formula and using the fact that a supermartingale has a negative drift term, we get that $u(t,x)$ is a supersolution to

$$- \partial_t u - \mathcal{L}u = 0$$

with $\mathcal{L}$ the Itô generator of X. Similarly, submartingales are associated with subsolutions of PDEs.

We extend the notion of comparison principle for sub- and supersolutions:

DEFINITION 4.6 Comparison principle II   *We say that PDE (4.1) satisfies a comparison principle if for all u supersolution, and for all v subsolution, to PDE (4.1), then*

$$h(x) \equiv v(T,x) \leq g(x) \equiv u(T,x), \quad \forall x \in \Omega$$
$$\implies \quad v(t,x) \leq u(t,x), \quad \forall x \in \Omega, \ \forall t \in [0,T]$$

Following closely the proof of Proposition 4.1, we obtain

PROPOSITION 4.2
PDE (4.1) satisfies the comparison principle II if H is proper (I or II).

Assum: Throughout this book, we will assume that a comparison principle holds for PDE (4.1) (e.g., H is proper I or II).

4.2 Why is a pricing equation a parabolic PDE?

A common belief is that (linear) pricing equations deriving from profit and loss (P&L) arguments take the form of a parabolic second order PDE due to the particular properties of the Brownian motion. As explained below, the reason why nonlinear second order parabolic PDEs naturally show up in finance is far more fundamental and has nothing to do with the Brownian motion (see P.-L. Lions' quote at the beginning of the chapter). Let us proceed axiomatically and list some natural conditions that we want our (nonlinear) pricing operator to satisfy. These conditions are drawn from [31, 50]. We denote by $T_{t,t+h}$ the pricing functor that assigns to a payoff at time t its fair price $u(t + h, \cdot)$ at time $t + h$ with $h \leq 0$.[2] This can be interpreted as a causality principle. The family $(T_{t,s})_{s \leq t}$ forms a semigroup.

Causality axiom: There exists a set of applications $T_{t,t+h} : \mathcal{P} \to \mathcal{P}$ defined for all $t \geq 0$, $h \leq 0$ such that

$$u(t + h, x) = T_{t,t+h}[u(t, \cdot)](x), \qquad x \in \mathbb{R}^n$$

and $T_{t,t} = \mathrm{id}$ for all $t \geq 0$. $\mathcal{P}$ denotes a set of payoffs that we do not specify (see [31, 50]). We have

$$\frac{u(t + h, x) - u(t, x)}{h} = \frac{T_{t,t+h}[u(t, \cdot)](x) - T_{t,t}[u(t, \cdot)](x)}{h}$$

By taking the limit $h \to 0$, we obtain

$$\partial_t u(t, x) = H[u(t, \cdot)](x)$$

where H is the infinitesimal generator of the semigroup $(T_{t,s})_{s \geq t}$.

As a next condition, we impose that the operators $T_{t,t+h}$ satisfy a comparison principle in order to exclude arbitrage opportunities (see Section 4.1).

Monotonicity axiom:

$$T_{t,t+h} u \leq T_{t,t+h} v \qquad \text{if } u \leq v \text{ in } \mathbb{R}^n$$

for all functions $u, v \in \mathcal{P}$, for all $t \geq 0$, $h \leq 0$. Monotonicity corresponds to the comparison principle and implies uniqueness.

We then impose some technical conditions.

Locality axiom: If $f, g \in \mathcal{P} \cap C^\infty(\mathbb{R}^n)$ satisfy $D^\alpha f = D^\alpha g$ at a point $x \in \mathbb{R}^n$ for all multi-indices α, then

$$T_{t,t+h}[f](x) - T_{t,t+h}[g](x) = o(h) \qquad \text{when } h \to 0^-$$

---

[2]We recall that in our convention we proceed backward in time.

for all $t \geq 0$. This condition requires that the (infinitesimal) generator of the semi-group H be a local operator that depends only on derivatives of f at the point x:

$$\lim_{t \to 0} \frac{T_{t,t+h}[f](x) - f(x)}{h} = H(t, x, Df(x), D^2 f(x), \dots, D^n f(x), \dots)$$

No non-local operators are allowed.

Regularity axiom: For all $f, g \in \mathcal{P} \cap C^\infty(\mathbb{R}^n)$, for all $t \geq 0$, there exists a constant $C = C(f, g, t)$ such that for all $h \leq 0$, $\mu \in \mathbb{R}$

$$||T_{t,t+h}[f + \mu g] - T_{t,t+h}[f] - \mu T_{t,t+h}[g]||_\infty \leq C\mu h$$

This second condition further restricts H to depend only on u, Du (i.e., the so-called delta), and $D^2 u$ (i.e., the so-called gamma): $H = H(t, x, u, Du, D^2 u)$.

Cash invariance axioms are related to the way H depends on x and u. One may impose that the fair value of a financial contract which delivers a claim f and a cash amount c be equal to $T_{t,t+h}[f] + c$ (assuming, for the sake of simplicity, zero interest rates) as cash in a book does not incorporate risks. This corresponds to axiom I1 below. Alternatively, one may require axiom I2.

Cash invariance axioms: For all $f \in \mathcal{P}$, we have

$$\text{I1} \; : \; T_{t,t+h}[f + c] = T_{t,t+h}[f] + c$$
$$\text{I2} \; : \; T_{t,t+h}[f(\cdot + c)] = T_{t,t+h}[f](\cdot + c)$$

for all $t \geq 0$, $h \leq 0$ and $c \in \mathbb{R}$. The first (resp. second) invariance principle imposes that H be independent of u (resp. x). We should remark that I1 is not justified from a financial point of view due to counterparty risk (see Section 5.8). Indeed, if we buy a bond delivering 1 at a maturity T, we will receive 1 only if the seller of the bond has not defaulted.

From these axioms, we suspect that $T_{t,t+h}$ should be the semi-group associated to a nonlinear parabolic PDE (4.1) with a Hamiltonian $H(t, Du, D^2 u)$. This result is proved in the following theorem:

THEOREM 4.1 [31]
Under the axioms above, there exists a continuous function $H : [0, T) \times \mathbb{R}^n \times \mathcal{S}^n \to \mathbb{R}$ such that the function u is a (viscosity) solution to

$$- \partial_t u(t, x) - H(t, Du, D^2 u) = 0 \qquad \text{in } \mathbb{R}^n \times [0, T)$$

and H is parabolic. If we relax assumption I1, I2, we get

$$- \partial_t u(t, x) - H(t, x, u, Du, D^2 u) = 0$$

where H is a proper Hamiltonian.

All PDEs we focus on in this book are of this form.

No-arbitrage: As a final requirement, the no-arbitrage condition is equivalent to the existence of an ELMM under which the price of any asset is a local martingale (for the sake of simplicity, we take zero rates). From a PDE point of view, the solution to the pricing PDE with the terminal condition $u(T, x) = x^i$ should be $u(t, x) = x^i$, where the variable x^i corresponds to an asset. Therefore, the Hamiltonian $H(t, Du, D^2u)$ must satisfy

$$H(t, (\delta_{ij})_{1 \leq j \leq n}, 0) = 0, \qquad \forall i \text{ such that } x^i \text{ is an asset}$$

REMARK 4.1 Call–Put parity   Let us denote $c(t, x)$ (resp. $p(t, x)$), the solution to PDE (4.1) with the terminal condition $c(T, x) = (x^i - K)^+$ (resp. $p(T, x) = (K - x^i)^+$). The call-put parity $c(t, x) - p(t, x) = x^i - K$ holds if H does not depend on u and Du. Indeed, let us denote $u(t, x) = p(t, x) + x^i - K$; it is a solution of

$$\partial_t u + H(t, x, u - (x^i - K), Du - \delta_{ij}, D^2u) = 0, \qquad u(T, x) = c(T, x)$$

Since H does not depend on u and Du, this reads

$$\partial_t u + H(t, x, D^2u) = 0, \qquad u(T, x) = c(T, x)$$

Uniqueness implies that $u = c$, hence the call-put parity. ⬜

Before jumping to Hamilton-Jacobi-Bellman equations, we recall some general facts on numerical finite difference schemes.

4.3   Finite difference schemes

4.3.1   Introduction

For a small number of variables (say no more than three), PDE (4.1) can be solved using a finite difference scheme. The problem is usually posed on an unbounded domain. The numerical implementation requires localizing the problem on a bounded domain. Judicious boundary conditions have to be imposed. The numerical solution should only slightly depend on these boundary conditions. This could be checked by changing the size of the domain and/or changing the boundary condition. However, inappropriate boundary conditions on a finite grid could lead to numerical instability. After choosing appropriate boundary conditions, the finite difference scheme method consists of discretizing the solution u on a space–time grid $\mathcal{G}_h = \Delta t\{0, 1, \ldots, n_T\} \times \Delta x \mathbb{Z}^n$ where $h = (\Delta t, \Delta x)$ and $n_T \Delta t = T$:

$$u_i^n = u(t_n, x_i), \qquad \forall (t_n, x_i) \in \mathcal{G}_h$$

and of replacing the differentiation operators ∂_t, D, and D^2 by their finite difference approximations. In one dimension, one can use

$$\partial_t^{\text{fwd}} u(t_n, x_i) = \frac{u_i^{n+1} - u_i^n}{\Delta t}$$

$$D^{\text{fwd}} u(t_n, x_i) = \frac{u_{i+1}^n - u_i^n}{\Delta x}$$

$$D^{\text{bwd}} u(t_n, x_i) = \frac{u_i^n - u_{i-1}^n}{\Delta x}$$

$$D^{\text{cen}} u(t_n, x_i) = \frac{u_{i+1}^n - u_{i-1}^n}{2\Delta x}$$

$$(D^2)^{\text{cen}} u(t_n, x_i) = \frac{u_{i+1}^n + u_{i-1}^n - 2u_i^n}{\Delta x^2}$$

PDE (4.1) can then be approximated by a (nonlinear) algebraic equation with unknowns u_i^n. By using centered discrete differentiation operators, an *explicit* scheme that uses centered finite differences reads

$$\frac{u_i^{n+1} - u_i^n}{\Delta t} + H(t_{n+1}, x_i, u_i^{n+1}, D^{\text{cen}} u(t_{n+1}, x_i), (D^2)^{\text{cen}} u(t_{n+1}, x_i)) = 0,$$

$$\forall (t_{n+1}, x_i) \in \mathcal{G}_h$$

The *implicit* scheme reads

$$\frac{u_i^{n+1} - u_i^n}{\Delta t} + H(t_n, x_i, u_i^n, D^{\text{cen}} u(t_n, x_i), (D^2)^{\text{cen}} u(t_n, x_i)) = 0,$$

$$\forall (t_n, x_i) \in \mathcal{G}_h \backslash \{t = T\}$$

A θ-scheme reads

$$\frac{u_i^{n+1} - u_i^n}{\Delta t} + \theta H(t_{n+1}, x_i, u_i^{n+1}, D^{\text{cen}} u(t_{n+1}, x_i), (D^2)^{\text{cen}} u(t_{n+1}, x_i))$$

$$+ (1 - \theta) H(t_n, x_i, u_i^n, D^{\text{cen}} u(t_n, x_i), (D^2)^{\text{cen}} u(t_n, x_i)) = 0$$

These schemes can be summarized by an equation of the following form

$$\mathcal{S}_h(u_h(t, x), [u_h]_{t,x}) = 0, \qquad \forall (t, x) \in \mathcal{G}_h \backslash \{t = T\} \tag{4.4}$$

where $u_h(t, x)$ stands for an approximation of u at the point (t, x) and $[u_h]_{t,x}$ represents the values of u_h at points other than (t, x). Does the numerical solution to Equation (4.4) converge numerically to the true solution to PDE (4.1) when h goes to 0?

4.3.2 Consistency

The first step consists of replacing (4.1) by its discrete approximation (4.4). Consistency requires that the l.h.s. of (4.4) converges to the l.h.s. of PDE

(4.1) for $h = (\Delta t, \Delta x) \to 0$: For every smooth function u, we have

$$\mathcal{S}_h(u_h(t,x), [u_h]_{t,x}) - \left(\partial_t u(t,x) + H(t,x,u(t,x), Du(t,x), D^2 u(t,x))\right)$$
$$= O(\Delta t), \qquad \forall (t,x) \in \mathcal{G}_h \backslash \{t = T\}$$

When PDE (4.1) does not admit classical solutions, this smoothness condition should be relaxed. We will dwell on this when we come to discuss viscosity solutions.

4.3.3 Stability

The second step involves a stability criterion: there is no propagation of errors due to rounding error in our numerical implementation. The discrete numerical solution does not explode and converges to the true discrete solution to (4.4).

Stability (in the l_∞ norm): For every $\Delta t > 0$, the scheme has a unique solution u_h, and this solution is uniformly bounded independently of Δt:

$$\exists C \geq 0, \; \forall \Delta t > 0, \quad ||u_h||_\infty \leq C$$

4.3.4 Convergence

Does the numerical solution to Equation (4.4) converge numerically to the true solution to (4.1)? For linear PDEs, the Lax equivalence theorem states that for consistent schemes, convergence is equivalent to stability:

THEOREM 4.2 Lax's equivalence theorem
A consistent finite difference scheme for a linear PDE, for which the terminal-value problem is well posed, is convergent if and only if it is stable.

Checking the consistency condition is not difficult and involves using a Taylor expansion of (4.1) assuming enough regularity. This is done explicitly on some classical schemes for linear PDEs in the example below.

Example 4.1 Linear 1d-PDE
Let us consider the one-dimensional PDE associated to a local volatility model:

$$\partial_t u(t,x) + \frac{1}{2}\sigma^2(t,x)\partial_x^2 u(t,x) = 0, \qquad u(T,x) = g(x) \qquad (4.5)$$

The θ-scheme is specified by

$$
\mathcal{S}_h(u_i^n, [u_{i+1}^n, u_{i-1}^n, u_{i-1}^{n+1}, u_i^{n+1}, u_{i+1}^{n+1}]) = \frac{u_i^{n+1} - u_i^n}{\Delta t} + \frac{1}{2} \times
$$

$$
\left((1-\theta)\sigma^2(t_n, x_i) \frac{u_{i+1}^n + u_{i-1}^n - 2u_i^n}{\Delta x^2} + \theta\sigma^2(t_{n+1}, x_i) \frac{u_{i+1}^{n+1} + u_{i-1}^{n+1} - 2u_i^{n+1}}{\Delta x^2} \right)
$$

Here we have decided to single out u_i^n as the approximation for u on the grid and consider $[u_{i+1}^n, u_{i-1}^n, u_{i-1}^{n+1}, u_i^{n+1}, u_{i+1}^{n+1}]$ as u represented at neighborhood points.

A simple use of the Taylor expansion allows us to conclude that this scheme is consistent of order two in space and order one in time:

$$
\mathcal{S}_h(u_h(t,x), [u_h]_{t,x}) - \left(\partial_t + \frac{1}{2}\sigma^2(t,x)\partial_x^2 \right) u(t_n, x_i) = O(\Delta t + \Delta x^2)
$$

Indeed, assuming that the solution is regular enough, the Taylor expansion gives

$$
u_{i+1}^n = u_i^n + \partial_x u(t_n, x_i)\Delta x + \frac{1}{2}\partial_x^2 u(t_n, x_i)\Delta x^2 + \frac{1}{6}\partial_x^3 u(t_n, x_i)\Delta x^3
$$
$$
+ \frac{1}{24}\partial_x^4 u(t_n, x_i)\Delta x^4 + O(\Delta x^5)
$$

and

$$
u_{i-1}^n = u_i^n - \partial_x u(t_n, x_i)\Delta x + \frac{1}{2}\partial_x^2 u(t_n, x_i)\Delta x^2 - \frac{1}{6}\partial_x^3 u(t_n, x_i)\Delta x^3
$$
$$
+ \frac{1}{24}\partial_x^4 u(t_n, x_i)\Delta x^4 + O(\Delta x^5)
$$

This means that the truncation error of the second order derivative (resp. first order derivative) is of order 2 in space (resp. order 1 in time):

$$
\frac{u_{i+1}^n + u_{i-1}^n - 2u_i^n}{\Delta x^2} = \partial_x^2 u(t_n, x_i) + \frac{\Delta x^2}{12}\partial_x^4 u(t_n, x_i) + O(\Delta x^3)
$$
$$
\frac{u_{i+1}^{n+1} + u_{i-1}^{n+1} - 2u_i^{n+1}}{\Delta x^2} = \partial_x^2 u(t_n, x_i) + \frac{\Delta x^2}{12}\partial_x^4 u(t_n, x_i) + \partial_{tx^2} u(t_n, x_i)\Delta t
$$
$$
+ O(\Delta t^2 + \Delta x^3 + \Delta t \Delta x^2)
$$
$$
\frac{u_i^{n+1} - u_i^n}{\Delta t} = \partial_t u(t_n, x_i) + \frac{1}{2}\partial_t^2 u(t_n, x_i)\Delta t + O(\Delta t^2)
$$

We obtain

$$
\mathcal{S}_h = \partial_t u(t_n, x_i) + \frac{1}{2}\partial_t^2 u(t_n, x_i)\Delta t + \frac{1}{2}\partial_t \sigma^2(t_n, x_i)\partial_x^2 u(t_n, x_i)\Delta t \theta
$$
$$
+ \frac{1}{2}\sigma^2(t_n, x_i)\left(\partial_x^2 u(t_n, x_i) + \frac{\Delta x^2}{12}\partial_x^4 u(t_n, x_i) + \theta\partial_{tx^2} u(t_n, x_i)\Delta t \right)
$$
$$
+ O(\Delta t^2 + \Delta x^3 + \Delta t \Delta x^2)
$$

Note that the term proportional to θ can be rearranged using PDE (4.5) into

$$\frac{1}{2}\partial_t \left(\sigma^2(t,x)\partial_x^2 u(t,x)\right) = -\partial_t^2 u(t,x)$$

and the term in Δt is

$$\left(\frac{1}{2}-\theta\right)\partial_t^2 u(t_n,x_i)$$

We conclude that the θ-scheme is consistent:

$$\mathcal{S}_h - \left(\partial_t + \frac{1}{2}\sigma^2(t_n,x_i)\partial_x^2\right)u(t_n,x_i) = \left(\frac{1}{2}-\theta\right)\partial_t^2 u(t_n,x_i)\Delta t$$

$$+ \frac{1}{24}\sigma^2(n\Delta,x_i)\partial_x^4 u(t_n,x_i)\Delta x^2 + O(\Delta t^2 + \Delta x^3 + \Delta t\Delta x^2)$$

For $\theta = \frac{1}{2}$, the so-called Crank-Nicholson scheme, the truncation error is of order two in both space and time. It is also stable for all $\theta \leq \frac{1}{2}$. Using Lax's equivalence theorem, we get that this numerical scheme is convergent. $\square$

4.4 Stochastic control and the Hamilton-Jacobi-Bellman PDE

A typical example of nonlinear PDE (4.1) is the Hamilton-Jacobi-Bellman (in short, HJB) PDE:

$$-\partial_t u(t,x) - \sup_{\alpha \in \mathcal{A}} \mathcal{L}^\alpha u(t,x) = 0 \qquad \text{in } \Omega \times [0,T)$$

where $\mathcal{L}^\alpha u = \sum_{i,j} a_{ij}(t,x,\alpha)\partial_{ij}^2 u$ with $a \geq 0$. This equation arises naturally in stochastic control theory.

4.4.1 Introduction

In this chapter $(\Omega, (\mathcal{F}_t)_{0\leq t\leq T}, \mathbb{Q})$ denotes a probability space equipped with a d-dimensional Brownian motion W; $(\mathcal{F}_t)$ denotes the natural filtration of W. We introduce the following SDE

$$dX_t^{\alpha,i} = b_i(t,X_t^\alpha,\alpha_t)\,dt + \sum_{j=1}^d \sigma_{i,j}(t,X_t^\alpha,\alpha_t)\,dW_t^j, \qquad i \in \{1,\dots,n\} \tag{4.6}$$

where the drift b and the volatility σ depend on t, X_t^α and a control parameter α_t, which is an $\mathcal{F}_t$-adapted process valued in a domain $\mathcal{A}$, not necessarily compact. We denote

$$\mathcal{A}_{t,T} = \{(\alpha_s)_{t\leq s\leq T} \text{ adapted} \mid \forall s \in [t,T], \ \alpha_s \in \mathcal{A}\}$$

In order to ensure that SDE (4.6) admits a strong solution, we assume that b, σ satisfy:

Assum(SDE$_\alpha$): b and σ are Lipschitz-continuous functions in x uniformly in t and α, and satisfy a linear growth condition:

$$|b(t, x, \alpha) - b(t, y, \alpha)| + |\sigma(t, x, \alpha) - \sigma(t, y, \alpha)| \leq C|x - y|$$
$$|b(t, x, \alpha)| + |\sigma(t, x, \alpha)| \leq C\left(1 + |x|\right)$$

for every $t \geq 0$, $x, y \in \mathbb{R}^n$ and $\alpha \in \mathcal{A}$ where C is a positive constant. This condition ensures that

$$\mathbb{E}^{\mathbb{Q}}\left[\sup_{0 \leq s \leq T} |X_s^\alpha|^2\right] < \infty \tag{4.7}$$

A standard stochastic control problem consists of maximizing (or minimizing) a cost function J given by

$$J(t, x) \equiv \mathbb{E}^{\mathbb{Q}}\left[\int_t^T f(s, X_s^\alpha, \alpha_s)\, ds + g(X_T^\alpha)\middle| X_t^\alpha = x\right]$$

with respect to the control $\alpha \in \mathcal{A}_{t,T}$:

$$u(t, x) \equiv \sup_{\alpha \in \mathcal{A}_{t,T}} \mathbb{E}^{\mathbb{Q}}\left[\int_t^T f(s, X_s^\alpha, \alpha_s)\, ds + g(X_T^\alpha)\middle| X_t^\alpha = x\right] \tag{4.8}$$

where the initial condition is $X_t^\alpha = x$. In order to ensure that the cost function is finite, we assume that the supremum is taken over the subset $\mathcal{A}_{t,T}^*$ of $\mathcal{A}_{t,T}$ satisfying

$$\mathbb{E}^{\mathbb{Q}}\left[\int_t^T |f(s, X_s^\alpha, \alpha_s)|\, ds + |g(X_T^\alpha)|\right] < \infty \tag{4.9}$$

4.4.2   Standard form

By augmenting the state variables X_t^α with the path-dependent continuous variable

$$dZ_t = f(t, X_t^\alpha, \alpha_t)\, dt$$

our stochastic control problem can be written without loss of generality as

$$u(t, \tilde{x}) = \sup_{\alpha \in \mathcal{A}_{t,T}^*} \mathbb{E}^{\mathbb{Q}}[\tilde{g}(\tilde{X}_T^\alpha)|\tilde{X}_t^\alpha = (x, z)] - z$$

with $\tilde{X}^\alpha = (X^\alpha, Z)$ and $\tilde{g}(\tilde{x}) = g(x) + z$. We shall appeal to this standard form (i.e., $f = 0$)

$$u(t, x) \equiv \sup_{\alpha \in \mathcal{A}_{t,T}^*} \mathbb{E}^{\mathbb{Q}}[g(X_T^\alpha)|X_t^\alpha = x] \tag{4.10}$$

to review results on stochastic control. Below, as a technical condition, we assume:

Assum(g): g has quadratic growth, i.e., there exists $C \geq 0$ such that for all $x \in \mathbb{R}^n$, $|g(x)| \leq C\left(1 + |x|^2\right)$.

From the integrability result (4.7), this assumption on g implies that $\mathcal{A}^*_{t,T} = \mathcal{A}_{t,T}$.

4.4.3 Bellman's principle

The function u is solution to a nonlinear PDE called the Hamilton-Jacobi-Bellman (in short, HJB) PDE. This PDE is derived from Bellman's principle, which can be explained simply as follows: Using iterated conditional expectation, we get

$$u(t, x) = \sup_{\alpha \in \mathcal{A}_{t,T}} \mathbb{E}^{\mathbb{Q}}[\mathbb{E}^{\mathbb{Q}}[g(X_T^\alpha)|\mathcal{F}_{t+h}]|X_t^\alpha = x]$$

By definition, we have for all $\alpha \in \mathcal{A}_{t+h,T}$

$$u(t + h, X_{t+h}^\alpha) \geq \mathbb{E}^{\mathbb{Q}}[g(X_T^\alpha)|X_{t+h}] = \mathbb{E}^{\mathbb{Q}}[g(X_T^\alpha)|\mathcal{F}_{t+h}]$$

from which we deduce the inequality

$$u(t, x) \leq \sup_{\alpha \in \mathcal{A}_{t,t+h}} \mathbb{E}^{\mathbb{Q}}[u(t + h, X_{t+h}^\alpha)|X_t^\alpha = x] \tag{4.11}$$

To prove the reverse inequality, we consider an ϵ-control α_s^ϵ between $t + h$ and T satisfying (with $\epsilon > 0$):

$$u(t + h, X_{t+h}^\alpha) \leq \mathbb{E}^{\mathbb{Q}}[g(X_T^{\alpha^\epsilon})|X_{t+h}^\alpha] + \epsilon$$

where $X_T^{\alpha^\epsilon}$ is the diffusion (4.6) controlled with α^ϵ. We glue together a control $\alpha_s \in \mathcal{A}$ defined in $[t, t + h)$ with the (adapted) ϵ-control α_s^ϵ:

$$\tilde{\alpha}_s^\epsilon = \begin{cases} \alpha_s & \text{if } s \in [t, t + h) \\ \alpha_s^\epsilon & \text{if } s \in [t + h, T] \end{cases}$$

Note that it is not at all obvious that $\tilde{\alpha} \in \mathcal{A}_{t,T}$.[3] Taking this fact for granted, we have for all $\alpha \in \mathcal{A}_{t,t+h}$,

$$\begin{aligned} \mathbb{E}^{\mathbb{Q}}[u(t + h, X_{t+h}^\alpha)|X_t^\alpha = x] &\leq \mathbb{E}^{\mathbb{Q}}[\mathbb{E}^{\mathbb{Q}}[g(X_T^{\alpha^\epsilon})|X_{t+h}^\alpha]|X_t^\alpha = x] + \epsilon \\ &= \mathbb{E}^{\mathbb{Q}}[\mathbb{E}^{\mathbb{Q}}[g(X_T^{\alpha^\epsilon})|\mathcal{F}_{t+h}]|X_t^\alpha = x] + \epsilon \\ &= \mathbb{E}^{\mathbb{Q}}[\mathbb{E}^{\mathbb{Q}}[g(X_T^{\tilde{\alpha}^\epsilon})|\mathcal{F}_{t+h}]|X_t^\alpha = x] + \epsilon \\ &= \mathbb{E}^{\mathbb{Q}}[g(X_T^{\tilde{\alpha}^\epsilon})|X_t^\alpha = x] + \epsilon \leq u(t, x) + \epsilon \end{aligned}$$

---

[3]The technical proof requires difficult measurability selection arguments.

Letting $\epsilon \to 0$, we obtain that the reverse inequality of (4.11) holds as well, from which we get Bellman's principle:

$$u(t,x) = \sup_{\alpha \in \mathcal{A}_{t,t+h}} \mathbb{E}^{\mathbb{Q}}[u(t+h, X_{t+h}^\alpha)|X_t^\alpha = x] \qquad (4.12)$$

Note that this result can be generalized to any stopping time τ in $\mathcal{T}_{tT} \equiv \{\tau \in [t,T] \ \mathbb{Q}-\text{a.s.}\}$:

$$u(t,x) = \sup_{\alpha \in \mathcal{A}_{t,T}} \mathbb{E}^{\mathbb{Q}}[u(\tau, X_\tau^\alpha)|X_t^\alpha = x] \qquad (4.13)$$

and it reads in a discrete time setting as

$$u(t,x) = \sup_{\alpha_t \in \mathcal{A}} \mathbb{E}^{\mathbb{Q}}[u(t+1, X_{t+1}^\alpha)|X_t^\alpha = x] \qquad (4.14)$$

This relation is known as the dynamic programming equation. From Equation (4.12), we can now sketch the derivation of the HJB equation.

4.4.4   Formal derivation of the HJB PDE

Let us take an arbitrary constant control $\alpha_s = a$ with $a \in \mathcal{A}$ during the interval $[t, t+h]$. From (4.12), we get

$$u(t,x) \geq \mathbb{E}^{\mathbb{Q}}[u(t+h, X_{t+h}^a)|X_t^a = x]$$

By applying Itô's lemma to $u(t+h, X_{t+h}^\alpha)$ (u should be smooth enough, $u \in C^{1,2}([0,T) \times \mathbb{R}^n))$, we get

$$u(t,x) \geq \mathbb{E}^{\mathbb{Q}}\left[u(t,x) + \int_t^{t+h} (\partial_s + \mathcal{L}^a) u(s, X_s^a) \, ds + M \Big| X_t^a = x\right]$$

where $M \equiv \int_t^{t+h} Du(s, X_s^a)\sigma(s, X_s^a, a) \, dW_s$ and $\mathcal{L}^a$ is the Itô generator associated to X_t (4.6) controlled by the constant $a \in \mathcal{A}$. By assuming that M is not only a local martingale but a true martingale (or by a localization argument), we obtain

$$\mathbb{E}^{\mathbb{Q}}\left[\int_t^{t+h} (\partial_s + \mathcal{L}^a) u(s, X_s) \, ds \Big| X_t = x\right] \leq 0$$

Dividing by $h > 0$, letting $h \to 0$, and permuting the limit and the expectation using the dominated convergence theorem, we finally get

$$\partial_t u(t,x) + \mathcal{L}^a u(t,x) \leq 0, \qquad \forall a \in \mathcal{A} \qquad (4.15)$$

Then we take the optimal control $\alpha^* \in \mathcal{A}_{t,T}$ that realizes the supremum in Equation (4.12). By following a similar route, we get

$$\partial_t u(t,x) + \mathcal{L}^{\alpha_t^*} u(t,x) = 0 \qquad (4.16)$$

By combining Equations (4.15) and (4.16), we get that u, as given by Equation (4.10), is a solution to the following parabolic second order fully nonlinear PDE, the so-called HJB equation:

$$-\partial_t u(t,x) - \sup_{a\in\mathcal{A}} \mathcal{L}^a u(t,x) = 0, \qquad u(T,x) = g(x) \qquad (4.17)$$

The interpretation is clear: For a deterministic constant "control" a, the linear PDE reads

$$-\partial_t u(t,x) = \mathcal{L}^a u(t,x), \qquad u(T,x) = g(x)$$

At each time, we actually have the possibility to choose a in $\mathcal{A}$ so as to maximize $u(0,\cdot)$. Since u is known at the final date, $a \in \mathcal{A}$ must be chosen so that $-\partial_t u$ is maximized. This yields

$$-\partial_t u(t,x) = \sup_{a\in\mathcal{A}} \mathcal{L}^a u(t,x), \qquad u(T,x) = g(x)$$

which is Equation (4.17).

For our initial stochastic control problem with a source term (4.8), the HJB PDE reads

$$-\partial_t u(t,x) - \sup_{a\in\mathcal{A}}\{\mathcal{L}^a u(t,x) + f(t,x,a)\} = 0, \qquad u(T,x) = g(x) \qquad (4.18)$$

As the linear operator $\mathcal{L}^\alpha$ involves only first and second order derivatives, the Hamiltonian $H(t,x,Du,D^2u) \equiv \sup_{a\in\mathcal{A}}\{\mathcal{L}^a u(t,x) + f(t,x,a)\}$ does not depend on u. In order to get a general nonlinear PDE (4.1), we shall appeal to the following form of the stochastic control problem, which includes a discount factor:

$$u(t,x) = \sup_{\alpha\in\mathcal{A}_{t,T}} \mathbb{E}^{\mathbb{Q}}\left[g(X_T^\alpha)e^{-\int_t^T r(s,X_s,\alpha_s)\,ds} + \right.$$

$$\left. \int_t^T e^{-\int_t^s r(u,X_u,\alpha_u)du} f(s,X_s,\alpha_s)\,ds \,\middle|\, X_t = x\right] \qquad (4.19)$$

The HJB PDE reads

$$-\partial_t u(t,x) - \sup_{a\in\mathcal{A}}\{\mathcal{L}^a u(t,x) + f(t,x,a) - r(t,x,a)u(t,x)\} = 0 \qquad (4.20)$$

$$u(T,x) = g(x)$$

Note that if the control can only be chosen at discrete dates, the HJB equation is not applicable because it assumes that the control can be chosen continuously in time. In these cases, we must rely on the (discrete) Bellman's dynamic programming equation (4.14).

In practice, once the optimal control has been computed by formally taking the supremum in the above equation, we try to find a smooth solution U.

Then one can show that $U = u$ as defined by (4.10). This is the purpose of the so-called verification theorem:

THEOREM 4.3 Verification theorem [21, 24]
Let $U \in C^{1,2}([0,T) \times \mathbb{R}^n) \cap C^0([0,T] \times \mathbb{R}^n)$ be a function with quadratic growth in x, uniformly in t, i.e., there exists a constant C such that

$$\forall (t,x) \in [0,T) \times \mathbb{R}^n, \qquad |U(t,x)| \leq C(1 + |x|^2)$$

1. *Let us assume that*

$$-\partial_t U(t,x) - \sup_{a \in \mathcal{A}} \{\mathcal{L}^a U(t,x)\} \geq 0, \qquad \forall (t,x) \in [0,T) \times \mathbb{R}^n$$

$$U(T,x) \geq g(x), \qquad \forall x \in \mathbb{R}^n$$

   Then $U \geq u$ in $[0,T] \times \mathbb{R}^n$.

2. *Let us also assume that $U(T,\cdot) = g$, that for all $(t,x) \in [0,T) \times \mathbb{R}^n$ there exists $\alpha^*(t,x) \in \mathcal{A}$ such that*

$$-\partial_t U(t,x) - \sup_{a \in \mathcal{A}} \{\mathcal{L}^a U(t,x)\} = -\partial_t U(t,x) - \mathcal{L}^{\alpha^*} U(t,x) = 0$$

   and that the SDE (4.6) with the control $\alpha^* \in \mathcal{A}_{0,T}$ admits a strong solution. Then

$$U = u \qquad \text{in } [0,T] \times \mathbb{R}^n$$

It is usually very difficult to determine the regularity of U. Conditions that guarantee the smoothness of U are stated in Theorem 4.4.

THEOREM 4.4 Regularity [14]
Let us assume that

- $\sigma(t,x,a)$ *is uniformly elliptic: There is a constant $c > 0$ such that*

$$\forall (t,x,\alpha) \in [0,T] \times \mathbb{R}^n \times \mathcal{A}, \quad \zeta^{\dagger}\sigma(t,x,\alpha)\sigma^{\dagger}(t,x,\alpha)\zeta \geq c|\zeta|^2$$

- $\mathcal{A}$ *is compact*

- $b, \sigma, f \in C_b^{1,2}([0,T] \times \mathbb{R}^n)$

- $g \in C_b^3(\mathbb{R}^n)$

Then the HJB equation (4.18) has a unique solution in $C_b^{1,2}([0,T) \times \mathbb{R}^n)$.

Optimal stopping is another type of stochastic control that we introduce below. In this case, what you control is the maturity T in the cost function.

4.5 Optimal stopping problems

The stopping time control problem (in standard form) consists of solving

$$u(t, X_t) = \sup_{\tau \in \mathcal{T}_{tT}} \mathbb{E}^{\mathbb{Q}}[g(X_\tau)|\mathcal{F}_t]$$

where $\mathcal{T}_{tT}$ is the set of all stopping time τ valued in $[t, T]$ $\mathbb{Q}$-a.s. Note that $u(t, x) \geq g(x)$ for all t, x as $\{\tau = t\}$ is a possible stopping time. Formally, if it is optimal to stop at a time τ^*, the value of the cost function is $u(\tau^*, X_{\tau^*}) = g(X_{\tau^*})$. Therefore, as long as $u(t, X_t) > g(X_t)$, it is not optimal to stop, and we expect that

$$\tau^* = \inf\{s \geq t \mid u(s, X_s) = g(X_s)\} \tag{4.21}$$

be an optimal stopping time.

Formal derivation of the variational inequality

This variational inequality reads

$$\max\left(\partial_t U(t, x) + \mathcal{L}U(t, x), g(x) - U(t, x)\right) = 0, \quad U(T, x) = g(x) \tag{4.22}$$

We now show that if U is a smooth solution of this variational inequality, then $U(t, x) = u(t, x) \equiv \sup_{\tau \in \mathcal{T}_{tT}} \mathbb{E}^{\mathbb{Q}}[g(X_\tau)|X_t = x]$.

(i) By applying the optional stopping theorem to the martingale $M_s = U(s, X_s) - \int_t^s \left(\partial_t U(r, X_r) + \mathcal{L}U(r, X_r)\right) dr$ and a stopping time $\tau \in \mathcal{T}_{tT}$ (bounded by T), we get

$$\mathbb{E}^{\mathbb{Q}}[M_\tau|\mathcal{F}_t] = M_t = U(t, X_t)$$

As $\partial_t U(s, X_s) + \mathcal{L}U(s, X_s) \leq 0$ and $U(\tau, X_\tau) \geq g(X_\tau)$ from Equation (4.22), this relation implies that

$$U(t, X_t) \geq \mathbb{E}^{\mathbb{Q}}[g(X_\tau)|X_t], \qquad \forall \tau \in \mathcal{T}_{tT}$$

and

$$U(t, X_t) \geq \sup_{\tau \in \mathcal{T}_{tT}} \mathbb{E}^{\mathbb{Q}}[g(X_\tau)|\mathcal{F}_t] \tag{4.23}$$

(ii) By applying the optional stopping theorem to the martingale M_s and the stopping time $\tau^* \in \mathcal{T}_{tT}$ defined by (4.21), we get

$$\mathbb{E}^{\mathbb{Q}}[M_{\tau^*}|\mathcal{F}_t] = M_t = U(t, X_t)$$

As $\partial_t U(s, X_s) + \mathcal{L}U(s, X_s) = 0$ for all $s \in [t, \tau^*]$ and $U(\tau^*, X_{\tau^*}) = g(X_{\tau^*})$, we obtain

$$U(t, X_t) = \mathbb{E}^{\mathbb{Q}}[g(X^*_{\tau})|\mathcal{F}_t]$$

Using Equation (4.23), we eventually get that $U(t, X_t) = \sup_{\tau \in \mathcal{T}_{tT}} \mathbb{E}^{\mathbb{Q}}[g(X_\tau)|\mathcal{F}_t] \equiv u(t, X_t)$. We have also shown that the supremum over $\mathcal{T}_{tT}$ is attained by the (optimal) stopping time τ^*.

In presence of interest rates $r(t, X_t)$, u is a solution to the variational inequality

$$\max\left(\partial_t u(t, x) + \mathcal{L}u(t, x) - r(t, x)u(t, x), g(x) - u(t, x)\right) = 0 \qquad (4.24)$$
$$u(T, x) = g(x)$$

4.6   Viscosity solutions

4.6.1   Motivation

In order to justify the introduction of a notion of weak solutions of HJB equations, we consider the HJB equation

$$\partial_t u(t, x) + \frac{1}{2} \sup_{\sigma \in [0, \infty)} \sigma^2 x^2 \partial^2_x u(t, x) = 0 \qquad (4.25)$$

with terminal condition $u(T, x) = g(x)$. It corresponds to the uncertain volatility model with unbounded volatility $\underline{\sigma} = 0$, $\overline{\sigma} = \infty$. From Section 5.2 and Chapter 9, we have

$$u(t, X_t) = \sup_{[t,T]} \mathbb{E}^{\mathbb{Q}}[g(X_T)|\mathcal{F}_t]$$

where $dX_t = \sigma_t X_t \, dW_t$ and $\sup_{[t,T]}$ means that the supremum is taken over all $(\mathcal{F}_t)$-adapted volatility processes $(\sigma_s)_{t \le s \le T}$ such that for all $s \in [t, T]$, σ_s (the control) belongs to the domain $[\underline{\sigma}, \overline{\sigma}]$, here, $[0, \infty)$. We assume that g is *not* a concave function. Here, as the control is unbounded, we speak of a *singular* stochastic control problem.

From Equation (4.25), if $\partial^2_x u(t, x) > 0$, then the supremum over the control σ is reached for $\sigma = \infty$ and is infinite, so all solutions u must satisfy $\partial^2_x u(t, x) \le 0$. Then the supremum is always reached for $\sigma = 0$ and the HJB equation reduces to $\partial_t u = 0$. These conditions can be summarized into

$$\partial^2_x u \le 0$$
$$\partial_t u = 0$$
$$u(T, x) = g(x)$$

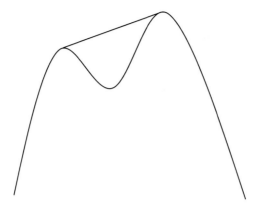

Figure 4.1: Example of a concave envelope.

This system has no solution. Indeed, the first two equations imply that u is a time-independent concave function and the last equation gives $u(t, x) = g(x)$, which was assumed to be not concave. Unless g is concave and smooth, the HJB PDE does not admit any classical (smooth) solution.

Note that by taking $\sigma_t = \underline{\sigma} = 0$, we have

$$u(t, x) \geq g(x)$$

This leads to

$$u(t, x) \geq g^{\mathrm{conc}}(x)$$

where g^{conc} is the smallest concave function greater than or equal to g (see Figure 4.1). As $g^{\mathrm{conc}}(x) \geq g(x)$, we have

$$u(t, x) \leq \sup_{[t,T]} \mathbb{E}^{\mathbb{Q}}[g^{\mathrm{conc}}(X_T)|X_t = x]$$

By applying Jensen's inequality, we obtain that

$$u(t, x) \leq g^{\mathrm{conc}}(\mathbb{E}^{\mathbb{Q}}[X_T|X_t = x])] = g^{\mathrm{conc}}(x)$$

Eventually, u coincides with the concave envelope of g:

$$u(t, x) = g^{\mathrm{conc}}(x)$$

Note that u does *not* satisfy the terminal condition: $g(x)$ differs from $g^{\mathrm{conc}}(x)$! In order to justify that $g^{\mathrm{conc}}(x)$ is indeed a (weak) solution to the HJB equation (4.25), we must rely on the theory of viscosity solutions developed by Crandall, Ishii, and Lions [86]. Viscosity solutions are considered the right concept of solutions for HJB PDEs because usually for such PDEs one can prove the existence and uniqueness of a viscosity solution. For weaker concepts of solutions, one may lose uniqueness. For stronger concepts of solutions, such as the classical concept of smooth solutions, one usually loses existence: typically, as explained above, an HJB equation admits no smooth solution. Keeping to the objective of this book, we will not give a detailed exposition of viscosity solutions, but will simply piece together some definitions and results. As suggested by the example above, the viscosity solution is not necessarily continuous (in time, in our example) and we will only impose that the function u be locally bounded.

4.6.2   Definition

The definition of a viscosity solution is motivated by the following theorem:

THEOREM 4.5
$u \in C^{1,2}([0, T) \times \Omega)$ *is a classical solution to (4.1) if and only if it satisfies* $u(T, x) = g(x)$ *and for all* $\varphi \in C^{1,2}([0, T) \times \Omega)$:

1. *If* $(t_0, x_0) \in [0, T) \times \Omega$ *is such that* $\min_{(t,x)}(u - \varphi) = u(t_0, x_0) - \varphi(t_0, x_0) \equiv 0$, *we have*

$$- \partial_t \varphi(t_0, x_0) - H(t_0, x_0, \varphi(t_0, x_0), D\varphi(t_0, x_0), D^2\varphi(t_0, x_0)) \geq 0$$

2. *If* $(t_0, x_0) \in [0, T) \times \Omega$ *is such that* $\max_{(t,x)}(u - \varphi) = u(t_0, x_0) - \varphi(t_0, x_0) \equiv 0$, *we have*

$$- \partial_t \varphi(t_0, x_0) - H(t_0, x_0, \varphi(t_0, x_0), D\varphi(t_0, x_0), D^2\varphi(t_0, x_0)) \leq 0$$

PROOF   ($\Longrightarrow$) Assume that u is a classical solution. Let us prove Assertion 1. Let $\varphi \in C^{1,2}([0, T) \times \Omega)$ such that $\min_{(t,x)}(u - \varphi) = u(t_0, x_0) - \varphi(t_0, x_0) \equiv 0$. The first and second order optimality conditions imply that

$$Du(t_0, x_0) = D\varphi(t_0, x_0)$$
$$\partial_t u(t_0, x_0) \leq \partial_t \varphi(t_0, x_0) \; (= \partial_t \varphi(t_0, x_0) \text{ if } t_0 < T)$$
$$D^2 u(t_0, x_0) \geq D^2 \varphi(t_0, x_0)$$

From the ellipticity condition on H, we deduce that

$$- \partial_t \varphi(t_0, x_0) - H(t_0, x_0, \varphi(t_0, x_0), D\varphi(t_0, x_0), D^2\varphi(t_0, x_0))$$
$$\geq - \partial_t u(t_0, x_0) - H(t_0, x_0, u(t_0, x_0), Du(t_0, x_0), D^2u(t_0, x_0)) \equiv 0$$

Assertion 2 is proved in a similar way.
($\Longleftarrow$) Take $\varphi = u$.   □

This result justifies the definition of viscosity solutions (see Definition 4.8). In the unbounded UVM example introduced above, the solution u was discontinuous in time. In particular, $u(T^-, x) = g^{\text{conc}}(x)$ and $u(T, x) = g(x)$. This leads us to define upper and lower semi-continuous functions as follows:

DEFINITION 4.7 Upper and lower semi-continuous   *A locally bounded function $u : [0, T) \times \Omega \to \mathbb{R}$ is said to be upper semi-continuous (USC in short) if*

$$(t_k, x_k) \to (t, x) \implies u(t, x) \geq \limsup_{k \to \infty} u(t_k, x_k) \equiv \lim_{k_0 \to \infty} \sup_{k \geq k_0} u(t_k, x_k)$$

and lower semi-continuous (LSC in short) if

$$(t_k, x_k) \to (t, x) \implies u(t, x) \leq \liminf_{k \to \infty} u(t_k, x_k) \equiv \lim_{k_0 \to \infty} \inf_{k \geq k_0} u(t_k, x_k)$$

A function that is both USC and LSC is continuous. Below we denote by $\overline{u}(t, x)$ (resp. $\underline{u}(t, x)$) the upper (resp. lower) semi-continuous envelope of u:

$$\overline{u}(t, x) \equiv \limsup_{(t_k, x_k) \to (t, x)} u(t_k, x_k)$$

$$\underline{u}(t, x) \equiv \liminf_{(t_k, x_k) \to (t, x)} u(t_k, x_k)$$

Eventually, viscosity solutions are defined as follows:

DEFINITION 4.8 (Discontinuous) viscosity solution   *Let $u :$ $[0, T) \times \Omega \to \mathbb{R}$ be locally bounded.*

(i) *u is said to be a viscosity supersolution (resp. subsolution) of PDE (4.1) if for all $\varphi \in C^{1,2}([0, T) \times \Omega)$ and $(t_0, x_0) \in [0, T) \times \Omega$ such that $\min_{(t,x)}(\underline{u} - \varphi) = \underline{u}(t_0, x_0) - \varphi(t_0, x_0) \equiv 0$ (resp. $\max_{(t,x)}(\overline{u} - \varphi) = \overline{u}(t_0, x_0) - \varphi(t_0, x_0) \equiv 0$), we have*

$$- \partial_t \varphi(t_0, x_0) - H(t_0, x_0, \varphi(t_0, x_0), D\varphi(t_0, x_0), D^2\varphi(t_0, x_0)) \geq 0$$
$$(\text{resp.} \ \leq 0) \quad (4.26)$$

This means that φ is a classical supersolution (resp. subsolution) of PDE (4.1) at (t_0, x_0) (see Definition 4.5).

(ii) *u is said to be viscosity solution of PDE (4.1) if it is both a viscosity subsolution and a viscosity supersolution of PDE (4.1).*

4.6.3 Viscosity solutions to HJB equations

For use below, we denote

$$\text{dom}(H) = \{(t, x, z, p, r) \in [0, T] \times \mathbb{R}^n \times \mathbb{R} \times \mathbb{R}^n \times \mathcal{S}_n \mid H(t, x, z, p, r) < \infty\}$$

and set

$$u(t, x) = \sup_{\alpha \in \mathcal{A}_{t,T}} \mathbb{E}^{\mathbb{Q}}\left[g(X_T^\alpha) e^{-\int_t^T r(s, X_s, \alpha_s)\, ds} \right.$$
$$\left. + \int_t^T e^{-\int_t^s r(u, X_u, \alpha_u)\, du} f(s, X_s, \alpha_s)\, ds \,\middle|\, X_t = x \right] \quad (4.27)$$

with X the (strong) solution to SDE (4.6) (see Equation (4.19)).

Assum(H): H is continuous on $\text{int}(\text{dom}(H))$ and there exists $G : [0, T] \times \mathbb{R}^n \times \mathbb{R} \times \mathbb{R}^n$ nondecreasing in its last argument and continuous such that

$$(t, x, z, p, r) \in \text{dom}(H) \iff G(t, x, z, p, r) \geq 0$$

THEOREM 4.6 [21]
Assume that Assum(H) holds. Then u defined by (4.27) is a viscosity solution to the HJB variational inequality: for all $(t, x) \in [0, T) \times \Omega$,

$$\min\{-\partial_t u - H(t, x, u, Du, D^2 u), G(t, x, u, Du, D^2 u)\} = 0 \quad (4.28)$$

REMARK 4.2 So far, viscosity solutions have only been defined for PDE (4.1) (see Definition 4.8). In the case of variational inequalities, Equation (4.26) is replaced by

$$\min\{-\partial_t \varphi(t_0, x_0) - H(t_0, x_0, \varphi(t_0, x_0), D\varphi(t_0, x_0), D^2\varphi(t_0, x_0)),$$
$$G(t_0, x_0, \varphi(t_0, x_0), D\varphi(t_0, x_0), D^2\varphi(t_0, x_0))\} \geq 0$$

‎□

In the above theorem, there is no uniqueness result; in particular, the terminal condition is not specified. The terminal condition is specified in the next theorem. In particular, $u(T, x) = g(x)$ is not always realized. A uniqueness result requires a comparison principle (for viscosity solutions); see e.g., [24].

THEOREM 4.7 [21]
Assume that g is continuous and lower-bounded or satisfies a linear growth condition, and that Assum(H) holds. Then u, defined by (4.27), is a viscosity

solution to

$$\min\{u(t,x) - g(x), G(t, x, u(t,x), Du(t,x), D^2u(t,x))\} = 0 \quad \text{on } \{T\} \times \Omega$$

This means that $\overline{u}(T, \cdot)$ is a viscosity supersolution to

$$\min\{\overline{u}(T,x) - g(x), G(T, x, \overline{u}(T,x), D\overline{u}(T,x), D^2\overline{u}(T,x))\} \geq 0 \quad \text{on } \Omega$$

and $\underline{u}(T, \cdot)$ is a viscosity subsolution to

$$\min\{\underline{u}(T,x) - g(x), G(T, x, \underline{u}(T,x), D\underline{u}(T,x), D^2\underline{u}(T,x))\} \leq 0 \quad \text{on } \Omega$$

If the domain $\mathcal{A}$ is bounded, i.e., if G is finite, we have $u(T, x) = g(x)$.

Example 4.2 Unbounded UVM
For the unbounded UVM where $\underline{\sigma} = 0$ and $\overline{\sigma} = \infty$ (see Section 4.6.1), $G(\partial_x^2 u) = -\partial_x^2 u$. Theorems 4.6 and 4.7 imply that u is a viscosity solution to

$$\min\{-\partial_t u, -\partial_x^2 u\} = 0 \qquad \forall (t,x) \in [0, T) \times \mathbb{R}_+$$
$$\min\{u(T,x) - g(x), -\partial_x^2 u(T,x)\} = 0 \qquad \forall x \in \mathbb{R}_+$$

The solution is unique and coincides with the concave envelope of g. ▯

When the domain is reduced to a singleton (i.e., no control), Theorem 4.6 gives a generalization of the Feynman-Kac formula. The smoothness requirement $u \in C^{1,2}([0, T) \times \mathbb{R}^n)$ is replaced by a local boundedness condition:

THEOREM 4.8 Feynman-Kac, viscosity sense
Assume that

$$u(t,x) = \mathbb{E}^{\mathbb{Q}}\left[g(X_T)e^{-\int_t^T r(s,X_s)ds} + \int_t^T e^{-\int_t^s r(u,X_u)du} f(s, X_s)\, ds \,\Big|\, X_t = x\right]$$

is locally bounded. Then u is a viscosity solution to

$$-\partial_t u(t,x) - \mathcal{L}u(t,x) - f(t,x) + r(t,x)u = 0, \qquad u(T,x) = g(x)$$

where $\mathcal{L} = \sum_{i=1}^n b_i(t,x)\partial_i + \frac{1}{2}\sum_{i,j=1}^n \sum_{k=1}^d \sigma_{i,k}(t,x)\sigma_{j,k}(t,x)\partial_{ij}$.

4.6.4 Numerical scheme: The Barles-Souganadis framework

Finally we give without proof a generalization of the Lax equivalence theorem for nonlinear PDEs.

THEOREM 4.9 Barles-Souganadis equivalence theorem [49]
If Scheme (4.4) satisfies the consistency, monotonicity, and stability proper-ties, its solution u_h converges locally uniformly to the unique viscosity solution to (4.1).

We recall that the scheme is said to be monotone if

$$\mathcal{S}_h(u_h(t,x),[u_h]_{t,x}) \leq \mathcal{S}_h(u_h(t,x),[v_h]_{t,x})$$

for all $[v_h]_{t,x} \geq [u_h]_{t,x}$ and for all $(t,x) \in \mathcal{G}_h\backslash\{t=T\}$.

Conclusion

In this chapter, we have presented important properties of nonlinear parabolic PDEs and their link with option pricing. In particular, the comparison principle was linked with the no-arbitrage condition. At this point, the reader should be able to formulate (possibly singular) stochastic control problems in terms of (possibly variational) HJB equations. In the next chapter, we present a collection of examples of nonlinear PDEs arising in quantitative finance. They arise for instance when one considers the pricing of American-style options, the uncertain volatility model, the uncertain lapse and mortality model, transaction costs, illiquid markets, different rates for borrowing and lending, counterparty risk valuation, or the pricing under delta and gamma constraints.

4.7   Exercises

4.7.1   The American timer put

In this exercise, the asset X_t follows local volatility dynamics:

$$dX_t = \sigma(t,X_t)X_t\,dW_t$$

An American timer put gives the holder the right, but not the obligation, to receive at any time a put option with strike K. Additionally, if the variance $V_t = \langle \ln X \rangle_t$ exceeds the variance budget $\bar{v}$ or if $t = T$, the option cancels and delivers a put option with strike K. The fair value can then be written as

$$u(t,x,v) = \sup_{\tau \in \mathcal{T}_{tT}} \mathbb{E}^{\mathbb{Q}}[(K - X_{\tau \wedge \tau_{\bar{v}}})^+ \,|\, X_t = x, V_t = v]$$

with the stopping time $\tau_{\bar{v}} = \inf\{t \geq 0 \,|\, V_t \geq \bar{v}\}$.

1. Prove that $u(t, x, v)$ is the solution to the variational inequality:

$$\max \left(\partial_t u(t, x, v) + \sigma(t, x)^2 \left(\frac{x^2}{2} \partial_x^2 u(t, x, v) + \partial_v u(t, x, v) \right), \right.$$
$$\left. (K - x)^+ - u(t, x, v) \right) = 0 \qquad \forall (t, x, v) \in [0, T) \times \mathbb{R}_+ \times [0, \bar{v}]$$

with the boundary conditions

$$u(t, x, \bar{v}) = (K - x)^+, \qquad \forall (t, x) \in [0, T) \times \mathbb{R}_+$$
$$u(T, x, v) = (K - x)^+, \qquad \forall (x, v) \in \mathbb{R}_+ \times [0, \bar{v}]$$

2. Prove that in the limit where T is much larger than $\mathbb{E}^{\mathbb{Q}}[\tau_{\bar{v}}]$ (i.e., $\mathbb{E}^{\mathbb{Q}}[\tau_{\bar{v}}] \ll T$), an American timer put is model-independent and its price is given by the fair value of an American put option in the Black-Scholes model with unit volatility and maturity $T = \bar{v}$.

3. Repeat this exercise but replace the put option by a call option.

4.7.2 Super-replication price in a stochastic volatility model

Let us introduce an SVM defined under a risk-neutral measure $\mathbb{Q}^0$ by

$$\frac{dX_t}{X_t} = a_t \, dW_t$$
$$da_t = b(a_t) \, dt + \sigma(a_t) \, dZ_t$$

with W_t and Z_t two correlated Brownian motions: $\langle W, Z \rangle_t = \rho t$.

1. Prove that the super-replication price of a vanilla option with payoff $g(X_T)$ can be written as a singular stochastic control problem

$$\mathcal{S}_0(g(X_T)) = \sup_{\lambda \in \Lambda_{0,T}} \mathbb{E}^{\mathbb{Q}_\lambda}[g(X_T)] \tag{4.29}$$

where $\Lambda_{0,T}$ is the set of all adapted processes λ_t such that for all $t \in [0, T]$, $\lambda_t \in [0, \infty)$. The dynamics of X_t reads under the ELMM $\mathbb{Q}_\lambda$:

$$\frac{dX_t}{X_t} = a_t \, dW_t$$
$$da_t = \left(b(a_t) + \sqrt{1 - \rho^2} \sigma(a_t) \lambda_t \right) dt + \sigma(a_t) \, dZ_t'$$

2. Write the HJB equation.

3. Deduce that the super-replication price coincides with the concave envelope of g: $\mathcal{S}_0(g(X_T)) = g^{\text{conc}}(X_0)$.

4.7.3   The robust utility indifference price

Following our discussion on utility indifference price in Section 3.2.1, we relax our assumption that the historical measure $\mathbb{P}^{\text{hist}}$ (and the volatility of the underlying) is known. We will assume that the volatility is uncertain and we define the value $u(x - C, F_T)$ for a claim $F_T = g(X_T)$ as

$$u(C, F_T) \equiv \sup_{\Delta \text{ adm.}} \inf_{\sigma \in \Sigma_{0,T}} \mathbb{E}^{\mathbb{Q}_\sigma} \left[U \left(C - F_T + \pi_T \right) \right]$$

where $\Sigma_{0,T}$ is the set of all adapted processes σ_t such that for all $t \in [0, T]$, $\sigma_t \in [0, \infty)$. Under the martingale measure $\mathbb{Q}_\sigma$, we have

$$dX_t = \sigma_t X_t \, dW_t, \quad d\pi_t = \Delta_t \, dX_t, \quad \pi_0 = 0$$

Here for the sake of simplicity, we take a zero interest rate. The (robust) utility indifference price is defined as the price $\bar{C}$ solution to

$$u \left(\bar{C}, g(X_T) \right) = u \left(0, 0 \right)$$

1. Set $v(t, x, \pi) \equiv \sup_\Delta \inf_\sigma \mathbb{E}^{\mathbb{Q}}[U \left(p - g(X_T) + \pi_T \right) | X_t = x, \pi_t = \pi]$ with p a constant. Prove that $v(t, x, \pi) = v(x, \pi)$ is independent of t and write the HJB variational inequality satisfied by $v(x, \pi)$.

2. We define the Legendre-Fenchel transform of v and U:

$$v^*(x, q) \equiv \sup_{\pi \geq 0} \{\pi q - v(x, \pi)\}, \qquad q \in \mathbb{R}$$

$$U^*(q) \equiv \sup_{\pi \geq 0} \{\pi q - U(\pi)\}, \qquad q \in \mathbb{R}$$

   Prove that v^* satisfies

$$\max \left(\partial_x^2 v^*(x, q), U^*(q) + (g(x) - p)q - v^*(x, q) \right) = 0 \qquad (4.30)$$

3. Deduce that v can be written as

$$v(x, \pi) = U^{\text{conv}} \left(\pi + p - \sup_{\sigma \in \Sigma_{0,T}} \mathbb{E}^{\mathbb{Q}}[g(X_T) | X_0 = x] \right)$$

   with U^{conv} the convex envelop of U.

4. Conclude that the robust utility indifference price coincides with the robust super-replication price

$$\bar{C} = \sup_{\sigma \in \Sigma_{0,T}} \mathbb{E}^{\mathbb{Q}}[g(X_T)] = g^{\text{conc}}(X_0)$$

Chapter 5

Examples of Nonlinear Problems in Finance

> In the fall of 1972 President Nixon announced that the rate of increase of inflation was decreasing. This was the first time a sitting president used the third derivative to advance his case for reelection.
>
> — Hugo Rossi, in *Mathematics Is an Edifice, Not a Toolbox*, Notices of the AMS 43, no. 10, October 1996.

Many recent issues in option pricing, ranging from parameter uncertainty to illiquidity, from optimal exercise to calibration of local stochastic volatility models or local correlation models to market smiles, from transaction costs to credit valuation adjustment, are described by nonlinear PDEs. In this chapter, as an appetizer for the rest of the book, we present various practical examples of such nonlinear PDEs. Finite difference schemes are described in a one-dimensional setup. Probabilistic methods are required in the multi-dimensional setup and will be extensively covered in the following chapters.

5.1 American options

A first basic example of nonlinear problem in finance is the pricing of American options. An American payoff can be described as an $\mathcal{F}_t$-adapted stochastic process $(F_t, t \in [0, T])$. An American option gives the holder the right but not the obligation to choose a stopping time $\tau(\omega)$ and exercise the option with payoff $F_{\tau(\omega)}$. As explained in Chapter 6 (see Theorem 6.1), once an ELMM $\mathbb{Q} \sim \mathbb{P}^{\text{hist}}$ has been picked out, an arbitrage-free price for such a claim is given by

$$u(t, x) = \sup_{\tau \in \mathcal{T}_{tT}} \mathbb{E}^{\mathbb{Q}}[F_\tau | X_t = x]$$

Here, for simplicity, we have taken zero interest rates. This is precisely an optimal stopping time problem and from Section 4.5 (see also Section 6.7), when $F_t = g(t, X_t)$, $u(t, x)$ satisfies the variational inequality:

$$\max\left(\partial_t u(t, x) + \mathcal{L}u(t, x), g(t, x) - u(t, x)\right) = 0$$

where $\mathcal{L}$ is the Itô generator of X.

Numerical scheme

This variational inequality can be discretized using a θ-scheme, leading to a so-called linear complementary problem:

$$\max\left(\frac{u_i^{n+1} - u_i^n}{\Delta t} + (1 - \theta)\left(\mathcal{L}u^n\right)_i + \theta\left(\mathcal{L}u^{n+1}\right)_i, g(t_n, x_i) - u_i^n\right) = 0$$

with the terminal condition $u_i^N = g(T, x_i)$. We split this scheme into two equations where Equation (5.1) requires only a classical (linear) PDE solver:

$$\frac{u_i^{n+1} - u_i^{n,1}}{\Delta t} + (1 - \theta)\left(\mathcal{L}u^{n,1}\right)_i + \theta\left(\mathcal{L}u^{n+1}\right)_i = 0 \qquad (5.1)$$

$$u_i^n = \max\left(u_i^{n,1}, g(t_n, x_i)\right)$$

American options will be extensively studied in Chapter 6, together with chooser options, which are a generalization of American options. The passport option (see Section 5.9) is an example of chooser option.

5.2   The uncertain volatility model

For American options, the nonlinearity comes from the possibility of early exercise that is given to the holder of the option, not from modeling assumptions. Other nonlinearities arise in finance as a result of modeling choices. This is the case, for instance, for the uncertain volatility model that we now describe.

5.2.1   The model

We model the asset X_t by a positive local Itô $(\mathcal{F}_t, \mathbb{Q})$-martingale:

$$dX_t = \sigma_t X_t \, dW_t$$

The volatility, which is an $\mathcal{F}_t$-adapted process, is unspecified for the moment. For the sake of simplicity, we have assumed zero interest rates, repos, and dividends. Let us then consider an option delivering a payoff F_T at maturity T which is a function of the asset path $(X_t, 0 \le t \le T)$. In the uncertain volatility model (in short UVM) introduced in [36, 159], the volatility is uncertain. As a minimal modeling hypothesis, we assume only that the volatility is valued in a compact interval $[\underline{\sigma}, \overline{\sigma}]$. We then define the time-t value u_t of the option as the solution to a stochastic control problem:

$$u_t = \sup_{[t,T]} \mathbb{E}^{\mathbb{Q}}[F_T | \mathcal{F}_t] \qquad (5.2)$$

Here, $\sup_{[t,T]}$ means that the supremum is taken over all $(\mathcal{F}_t)$-adapted processes $(\sigma_s)_{t \leq s \leq T}$ such that for all $s \in [t, T]$, σ_s belongs to the domain $[\underline{\sigma}, \overline{\sigma}]$. In Section 5.2.3, we justify the choice (5.2) by showing that u_t corresponds to the seller's super-replication price under the assumption of uncertain volatility.

5.2.2 Pricing vanilla options

For vanilla payoffs $F_T = g(X_T)$, the HJB PDE (4.17) reads

$$\partial_t u(t, x) + \frac{1}{2} \sup_{\sigma \in [\underline{\sigma}, \overline{\sigma}]} \sigma^2 x^2 \partial_x^2 u(t, x) = 0$$

Taking the supremum over σ, we get a fully nonlinear PDE called the Black-Scholes-Barenblatt equation (in short BSB)

$$\partial_t u(t, x) + \frac{1}{2} x^2 \Sigma \left(\partial_x^2 u(t, x) \right)^2 \partial_x^2 u(t, x) = 0, \quad (t, x) \in [0, T) \times \mathbb{R}_+^* \quad (5.3)$$

$$u(T, x) = g(x), \quad x \in \mathbb{R}_+^*$$

with $\Sigma(\Gamma) = \underline{\sigma} \mathbf{1}_{\Gamma < 0} + \overline{\sigma} \mathbf{1}_{\Gamma \geq 0}$. If we consider a convex payoff, u coincides with the Black-Scholes price with the upper volatility $\overline{\sigma}$. From a technical point of view, if $g \in C_b^3(\mathbb{R}_+)$ and $\overline{\sigma} < \infty$, then $u \in C^{1,2}([0, T) \times \mathbb{R}_+)$ as stated in Theorem 4.4. From the verification theorem (Theorem 4.3), we have that $u_t = u(t, X_t)$.

5.2.3 Robust super-replication

In Section 1.1.4, the buyer's and seller's super-replication prices have been defined with respect to a historical measure $\mathbb{P}^{\text{hist}}$ (more precisely the definition involves the space of ELMM $\sim \mathbb{P}^{\text{hist}}$), meaning that the volatility process of the asset is fixed. In practice, the volatility process is unknown. In the uncertain volatility framework, we consider instead the set $\mathcal{P}$ of all measures $\mathbb{Q}$ under which X_t is a local martingale such that

$$\underline{\sigma}^2 \leq \frac{d \langle \ln X \rangle_t}{dt} \equiv \sigma_t^2 \leq \overline{\sigma}^2$$

The measures $\mathbb{Q} \in \mathcal{P}$ are singular to each other as the process X admits different volatilities under different measures. The seller's super-replication price can be defined as (with zero interest rate, repo, and dividend for the sake of simplicity):

DEFINITION 5.1 Seller's super-replication price in the UVM   *The seller's super-replication price in the UVM is defined by*

$$S_t(F_T) = \inf \left\{ z \in \mathcal{F}_t \;\middle|\; \text{there exists an admissible portfolio } \Delta \text{ such that} \right.$$

$$\left. \pi_T^S \equiv z + \int_t^T \Delta_s \, dX_s - F_T \geq 0 \quad \mathbb{Q}-\text{a.s} \quad \forall \mathbb{Q} \in \mathcal{P} \right\} \quad (5.4)$$

Similarly as Theorem 1.3, we have (for a rigorous proof, see [92])

THEOREM 5.1

$$S_t(F_T) = \sup_{\mathbb{Q} \in \mathcal{P}} \mathbb{E}^{\mathbb{Q}}[F_T | \mathcal{F}_t] = \sup_{[t,T]} \mathbb{E}^{\mathbb{Q}}[F_T | \mathcal{F}_t]$$

This justifies our initial choice (5.2).

PROOF   We assume that $F_T = g(X_T)$ with $g \in C_b^3(\mathbb{R}_+)$ and $\overline{\sigma} < \infty$.
(i) We set $z = u(t, X_t) = \sup_{\mathbb{Q} \in \mathcal{P}} \mathbb{E}^{\mathbb{Q}}[g(X_T) | \mathcal{F}_t]$ and $\Delta_s = \partial_x u(s, X_s)$ with u the solution of PDE (5.3). As $u \in C^{1,2}([0, T) \times \mathbb{R}_+)$, Itô's lemma gives

$$z + \int_t^T \Delta_s \, dX_s - g(X_T) = \int_t^T \left(-\partial_s u(s, X_s) - \frac{1}{2} X_s^2 \sigma_s^2 \partial_x^2 u(s, X_s) \right) ds$$

Using the BSB PDE (5.3), we get

$$z + \int_t^T \Delta_s \, dX_s - g(X_T) = \frac{1}{2} \int_t^T X_s^2 \partial_x^2 u(s, X_s)(\Sigma(\partial_x^2 u(s, X_s))^2 - \sigma_s^2) \, ds \geq 0$$

$\mathbb{Q}$-a.s. for all $\mathbb{Q} \in \mathcal{P}$ as $\left(\Sigma \left(\partial_x^2 u(s, X_s) \right)^2 - \sigma_s^2 \right) \partial_x^2 u(s, X_s) \geq 0$ for all adapted σ_s such that for all $s \in [t, T]$, $\sigma_s \in [\underline{\sigma}, \overline{\sigma}]$. This implies that $S_t(g(X_T)) \leq u(t, X_t) \equiv \sup_{\mathbb{Q} \in \mathcal{P}} \mathbb{E}^{\mathbb{Q}}[g(X_T) | \mathcal{F}_t]$ as we have obtained that the delta-hedging strategy $(z = u(t, X_t), \Delta_s = \partial_x u(s, X_s))$ super-replicates the payoff.
(ii) By following the same arguments with the optimal stochastic control σ^*, we obtain

$$z + \int_t^T \Delta_s \, dX_s - g(X_T) = \frac{1}{2} \int_t^T X_s^2 \partial_x^2 u(s, X_s)(\Sigma(\partial_x^2 u(s, X_s))^2 - (\sigma_s^*)^2) \, ds = 0$$

By taking the conditional expectation w.r.t the measure $\mathbb{Q}^*$ under which $\sigma_t = \sigma_t^*$, we obtain $z \geq \mathbb{E}^{\mathbb{Q}^*}[g(X_T) | \mathcal{F}_t]$. We have used that the local martingale $\int_t^T \Delta_s \, dX_s$ bounded from below is a supermartingale. This implies that $S_t(g(X_T)) \geq \mathbb{E}^{\mathbb{Q}^*}[g(X_T) | \mathcal{F}_t] = \sup_{\mathbb{Q} \in \mathcal{P}} \mathbb{E}^{\mathbb{Q}}[g(X_T) | \mathcal{F}_t]$. □

5.2.4 A finite difference scheme

In practice, the nonlinear PDE (5.3) is not analytically solvable. In low dimension, we can implement a finite difference scheme. By setting $z = \ln \frac{x}{X_0}$, we consider the following θ-scheme:

1. Set $u_i^N = g(X_0 e^{z_i})$.

2. Predictor:

$$\frac{u_i^{n+1} - u_i^n}{\Delta t} + \frac{1}{2}\Sigma(\Gamma_i^{n+1})^2 \left(\theta\Gamma_i^{n+1} + (1-\theta)\Gamma_i^{n,(1)}\right) = 0$$

with $\Gamma_i^{n+1} = \frac{u_{i+1}^{n+1} + u_{i-1}^{n+1} - 2u_i^{n+1}}{\Delta z^2} - \frac{u_{i+1}^{n+1} - u_{i-1}^{n+1}}{2\Delta z}$ and $\Gamma_i^{n,(1)} = \frac{u_{i+1}^n + u_{i-1}^n - 2u_i^n}{\Delta z^2} - \frac{u_{i+1}^n - u_{i-1}^n}{2\Delta z}$. Note that this scheme is explicit in $\Sigma(\Gamma_i^{n+1})$.

3. Corrector: Set $\Gamma_i^n = \frac{1}{2}\left(\Gamma_i^{n,(1)} + \Gamma_i^{n+1}\right)$ and solve

$$\frac{u_i^{n+1} - u_i^n}{\Delta t} + \frac{1}{2}\Sigma(\Gamma_i^n)^2 \left(\theta\Gamma_i^{n+1} + (1-\theta)\Gamma_i^{n,(1)}\right) = 0$$

Monte Carlo methods for solving the UVM are unavoidable in high dimension and will be the subject of Chapter 9.

5.3 Transaction costs: Leland's model

As already noticed in [41, 108], the pricing equation resulting from the Leland transaction costs model [155] is highly similar to the nonlinear PDE describing the pricing of options in the UVM (see Section 5.2). In dimension one, it is given by PDE (5.3) with

$$\Sigma(\Gamma)^2 = \begin{cases} \underline{\sigma}^2 \equiv \sigma^2(1-A) & \text{if } \Gamma < 0 \\ \overline{\sigma}^2 \equiv \sigma^2(1+A) & \text{if } \Gamma \geq 0 \end{cases} \tag{5.5}$$

where the constant A is given below. Note that the parabolic condition (see Definition 4.1) is satisfied if and only if $A < 1$.

This nonlinear PDE can be obtained by a heuristic profit and loss analysis as follows: Imagine that we have sold an option and that we delta-hedge it. The corresponding (self-financing) portfolio π_t at time t is

$$\pi_t = -u(t, X_t) + \Delta_t X_t$$

For the sake of simplicity, we assume zero interest rates. The variation of this portfolio between t and $t + dt$ is

$$d\pi_t = -du(t, X_t) + \Delta_t \, dX_t - kX_t\mathbb{E}[|\Delta_{t+\delta t} - \Delta_t||X_t]$$

where we have included transaction costs proportional to a constant k depending on the individual investor; δt is the re-hedging frequency. Taking X_t to be a geometric Brownian motion with a constant volatility σ for the sake of simplicity, we get

$$d\pi_t = -\left(\partial_t u(t, X_t) + \frac{1}{2}\sigma^2 X_t^2 \partial_x^2 u(t, X_t)\right) dt + (\Delta_t - \partial_x u)\, \sigma X_t\, dW_t$$
$$- kX_t \mathbb{E}[|\Delta_{t+\delta t} - \Delta_t||X_t]$$

By assuming that the impact of the transaction cost is small, one can consider that the portfolio can still be delta-hedged with $\Delta_t = \partial_x u(t, X_t)$ as in the Black-Scholes framework. Then we get

$$d\pi_t = -\left(\partial_t u(t, X_t) + \frac{1}{2}\sigma^2 X_t^2 \partial_x^2 u(t, X_t)\right) dt$$
$$- k\sigma X_t^2 |\partial_x^2 u(t, X_t)| \mathbb{E}[|\Delta W_{\delta t}|] + O(k\delta t)$$

Finally, by using the identity $\mathbb{E}[|\Delta W_{\delta t}|] = \sqrt{\frac{2}{\pi}}\sqrt{\delta t}$, we get

$$d\pi_t = -\left(\partial_t u(t, X_t) + \frac{1}{2}\sigma^2 X_t^2 \partial_x^2 u(t, X_t)\right) dt$$
$$- \sqrt{\frac{2}{\pi}}\frac{k\sigma}{\sqrt{\delta t}} X_t^2 |\partial_x^2 u(t, X_t)| \delta t + O(k\delta t)$$

The no-arbitrage condition imposes that $d\pi_t = 0$ $\mathbb{Q}$-a.s, and by heuristically replacing δt by dt with the constant $\frac{k\sigma}{\sqrt{\delta t}}$ kept fixed, we get the pricing equation

$$\partial_t u(t, x) + \frac{1}{2}x^2 \Sigma \left(\partial_x^2 u(t, x)\right)^2 \partial_x^2 u(t, x) = 0 \tag{5.6}$$

where Σ is given by (5.5) with $A = 2\sqrt{\frac{2}{\pi}\frac{k}{\sigma \delta t}}$.

In the general case of d assets $\{X^\alpha\}_{1 \leq \alpha \leq d}$, each with a transaction costs constant k_α and a common re-hedging frequency δt, the pricing equation reads

$$\partial_t u + H(x, D^2 u) = 0$$

with the Hamiltonian

$$H(X, \Gamma) = \frac{1}{2}\sum_{\alpha,\beta=1}^{d} \rho^{\alpha\beta} X^\alpha X^\beta \sigma^\alpha \sigma^\beta \Gamma^{\alpha\beta}$$
$$+ \sqrt{\frac{2}{\pi \delta t}} \sum_{\alpha=1}^{d} k_\alpha X^\alpha \sqrt{\sum_{\beta,\gamma=1}^{d} X^\beta X^\gamma \rho^{\beta\gamma} \sigma^\beta \sigma^\gamma \Gamma^{\alpha\beta} \Gamma^{\alpha\gamma}} \tag{5.7}$$

By setting $\bar{\Gamma}^{\beta\gamma} \equiv X^\beta X^\gamma \sigma^\beta \sigma^\gamma \Gamma^{\beta\gamma}$ and $A_\alpha = \sqrt{\frac{2}{\pi\delta t}\frac{k_\alpha}{\sigma^\alpha}}$, the Hamiltonian becomes

$$H(\bar{\Gamma}) = \frac{1}{2}\sum_{\alpha,\beta=1}^{d} \rho^{\alpha\beta}\bar{\Gamma}^{\alpha\beta} + \sum_{\alpha=1}^{d} A_\alpha \sqrt{\sum_{\beta,\gamma=1}^{d} \rho^{\beta\gamma}\bar{\Gamma}^{\alpha\beta}\bar{\Gamma}^{\alpha\gamma}}$$

The parabolic condition (see Definition 4.1), which is not automatically satisfied as our PDE has been generated by a heuristic profit and loss argument, reads

$$0 \le A_\alpha < 1 \qquad \forall \alpha \in \{1,\dots,d\}$$
$$|\rho^{\alpha\beta}| \le -\frac{A_\alpha + A_\beta}{2} + \sqrt{(1-A_\alpha)(1-A_\beta)} \qquad \forall \alpha,\beta \in \{1,\dots,d\}$$

5.4 Illiquid markets

5.4.1 Feedback effects

By definition, a market is liquid when the *elasticity* parameter is small. The elasticity parameter ϵ is given by the ratio of relative change in price X_t to change in the net demand D_t:

$$\frac{dX_t}{X_t} = \epsilon\, dD_t \tag{5.8}$$

We observe empirically that when the demand increases (resp. decreases), the price rises (decreases). The parameter ϵ is therefore nonnegative and we will assume that it is also constant. Another interesting (because more realistic) assumption would be to define ϵ as a stochastic process.

The small traders who tend to apply the same hedging strategy can be considered as a large trader and the hedging feedback effects become very important. This can speed up a crash situation. Below, we use the simple relation (5.8) to analyze the influence of dynamic trading strategies on the prices of derivatives in financial markets.

Let us call $D(t, W_t, X_t)$ the demand of all the traders in the market. It depends on time t, a Brownian process W_t, and the price X_t. X_t is assumed to follow a local volatility model under $\mathbb{P}^{\text{hist}}$

$$\frac{dX_t}{X_t} = \mu(t, X_t)\, dt + \sigma(t, X_t)\, dW_t \tag{5.9}$$

with $\mu(t, X_t)$, the (historical) return, and $\sigma(t, X_t)$, the volatility. The process W_t models both the information the traders have on the demand and the fluctuation of the price X_t. Applying Itô's lemma to the function $D(t, X_t, W_t)$, we obtain

$$dD_t = (\partial_t D + \frac{1}{2}\sigma^2 X_t^2 \partial_x^2 D + \frac{1}{2}\partial_W^2 D + \sigma X_t \partial_{xW} D) \, dt + \partial_x D \, dX_t + \partial_W D \, dW_t$$

$$= (\partial_t D + \frac{1}{2}\sigma^2 X_t^2 \partial_x^2 D + \frac{1}{2}\partial_W^2 D + \sigma X_t \partial_{xW} D + \mu X_t \partial_x D) \, dt$$
$$+ (\sigma X_t \partial_x D + \partial_W D) \, dW_t \tag{5.10}$$

As we have also

$$dD_t = \frac{1}{\epsilon}\frac{dX_t}{X_t} = \frac{1}{\epsilon}(\mu(t, X_t) \, dt + \sigma(t, X_t) \, dW_t) \tag{5.11}$$

by identifying the coefficients for dt and dW_t in (5.10) and (5.11), we obtain μ and σ as functions of the derivatives of D:

$$\mu(t, x) = \epsilon \frac{\partial_t D + \frac{1}{2}\sigma^2 x^2 \partial_x^2 D + \sigma x \partial_{xW} D + \frac{1}{2}\partial_W^2 D}{1 - \epsilon x \partial_x D} \tag{5.12}$$

$$\sigma(t, x) = \frac{\epsilon \partial_W D}{1 - \epsilon x \partial_x D} \tag{5.13}$$

We will now assume that the market is composed of a group of *small traders* on the one hand who do not modify the prices of the market—they only sell or buy small amounts of assets—and a *large trader* on the other hand. One can consider a large trader as an aggregate of small traders following the same strategy given by their hedging position $\Delta_t = \Delta(t, X_t)$. We will choose the demand D_{small} of small traders in order to reproduce the Black-Scholes lognormal process for X_t (5.9) with a constant (historical) return $\mu = \mu_0$ and a constant volatility $\sigma = \sigma_0$. The solution is given by

$$D_{\text{small}} = \frac{1}{\epsilon}(\mu_0 t + \sigma_0 W_t) \tag{5.14}$$

Now, we include the effect of a large trader whose demand D_{large} is generated by his trading strategy

$$D_{\text{large}} = \Delta(t, X_t)$$

It is implicitly assumed that Δ only depends on t and X_t and not explicitly on W_t. The hedging position Δ is then added to the demand of the small traders D_{small} and the total net demand is given by

$$D_t = D_{\text{small}} + \Delta(t, X_t) \tag{5.15}$$

By inserting (5.15) in (5.13)–(5.12), one finds the volatility and return as a function of Δ:

$$\mu(t, x) = \frac{\mu_0 + \epsilon(\partial_t \Delta + \frac{1}{2}\sigma^2 x^2 \partial_x^2 \Delta)}{1 - \epsilon x \partial_x \Delta} \tag{5.16}$$

$$\sigma(t, x) = \frac{\sigma_0}{1 - \epsilon x \partial_x \Delta} \tag{5.17}$$

This relation describes a feedback effect of dynamic hedging on volatility and return. The Black-Scholes analysis can be trivially modified to incorporate the hedging feedback effect. As the model specified by (5.9), (5.16), and (5.17) is complete, we get that the value $u(t, X_t)$ of a vanilla option solves the *nonlinear Black-Scholes* PDE (assuming zero interest rates for simplicity)

$$\partial_t u + \frac{1}{2} x^2 \Sigma(x, \partial_x^2 u)^2 \partial_x^2 u = 0 \qquad (5.18)$$

with

$$\Sigma(x, \Gamma)^2 = \frac{\sigma_0^2}{(1 - \epsilon x \Gamma)^2}$$

Note the similarity of this nonlinear PDE with the UVM PDE (5.3) and the pricing equation (5.6) in Leland's model. The Hamiltonian reads

$$H(x, \Gamma) = \frac{\sigma_0^2 x^2 \Gamma}{2(1 - \epsilon x \Gamma)^2}$$

This PDE has three major flaws:

- First, the Hamiltonian blows up for finite gamma: when $\epsilon x \Gamma = 1$, the Hamiltonian is infinite.

- Second, due to the cost of illiquidity, we expect $H(x, \Gamma)$ to be greater than or equal to $\frac{1}{2} x^2 \sigma_0^2 \Gamma$, which is the Hamiltonian for liquid markets ($\epsilon = 0$). However, this is satified only if $\epsilon x \Gamma \leq 2$.

- Third, this PDE is *not* parabolic: the parabolic condition (see Definition 4.1) holds if and only if $\epsilon x |\Gamma| < 1$. This condition is not satisfied for a call option. Outside this domain, this PDE is not well defined and the comparison principle is not satisfied, leading to arbitrage opportunities. This means that the replication paradigm cannot be used.

Imagine a favorable case where at maturity T, for all x, $\epsilon x |\partial_x^2 u| < 1$. Solving the PDE backward, when $\epsilon x \partial_x^2 u$ approaches 1 from below, $-\partial_t u$ tends to $+\infty$ and the solution u blows up, so we can consider that the gamma cannot exit the domain $\{\epsilon x \Gamma < 1\}$. But even on this restricted domain, the parabolicity is not satisfied since the Hamiltonian $H(x, \Gamma)$ is decreasing in Γ for $\epsilon x \Gamma < -1$. This is problematic as, in general, the gamma $\partial_x^2 u$ can access this region. In [110], the well-posedness issue of this parabolic PDE is addressed by arbitrarily modifying the Hamiltonian for small and large gamma. More precisely, the nonlinear PDE (5.18) is replaced by

$$\partial_t u + \frac{\sigma_0^2 x^2}{2} \max\left(\alpha_0, \frac{\partial_x^2 u}{(1 - \min(\alpha_1, \epsilon x \partial_x^2 u))^2} \right) = 0 \qquad (5.19)$$

with $\alpha_0 = 0.02$ and $\alpha_1 = 0.85$. The interpretation of this modification (in particular the cutoffs α_0 and α_1) in terms of hedging and feedback effects is not obvious.

For the three reasons explained above, PDE (5.18) should be disregarded. The correct way to solve the lack of parabolicity is to consider the parabolic envelope of the Hamiltonian; see Section 5.4.6 and [76]. The second problem might *not* be solved by considering the parabolic envelope of H and should be tackled separately.

From a modeling point of view, we have incorporated a liquidity risk by a feedback effect on the volatility (and the drift). This is significantly different from the Leland model where the transaction costs induced by the delta hedging strategy have been modeled by including a term that is proportional to the absolute value of the gamma in the P&L at maturity (see Section 5.3). In the next section, we try to reconcile these two approaches in a unified manner. By considering a simple model of a limit order book, we characterize explicitly the cost of hedging and the feedback effect on the volatility. Then, we show that the super-replication price is linked to a well defined parabolic PDE.

5.4.2   A simple model of a limit order book

We follow the approach developed initially by Obizhaeva and Wang [167] (see also [30] for a generalization) for describing limit order books. We denote by A_t the ask price at time t, B_t the bid price, and $X_t = (A_t + B_t)/2$ the mid-price. The density at time t of our limit order book at the price y for $y > A_t$ (resp. $y < B_t$) is $f_+(y, A_t)$ (resp. $f_-(y, B_t)$). The term $f_+(y, A_t) \, dy$ (resp. $f_-(y, B_t) \, dy$) represents the volume on the sell (resp. buy) side available at the price $[y, y+dy]$. The cost of an order of size $\Delta_{t_i} - \Delta_{t_{i-1}} \equiv \delta \Delta_{t_i} = \delta \Delta_{t_i}^+ - \delta \Delta_{t_i}^-$ at t_i^- is by definition

$$\text{Cost}(\delta \Delta_{t_i}) = \int_{A_{t_i^-}}^{\overline{X}_{t_i}} y f_+(y, A_{t_i^-}) \, dy - \int_{\underline{X}_{t_i}}^{B_{t_i^-}} y f_-(y, B_{t_i^-}) \, dy \qquad (5.20)$$

where the new bid $\underline{X}_{t_i}$ and ask $\overline{X}_{t_i}$ prices just after the order are given by

$$\delta \Delta_{t_i}^+ = \int_{A_{t_i^-}}^{\overline{X}_{t_i}} f_+(y, A_{t_i^-}) \, dy \qquad (5.21)$$

$$\delta \Delta_{t_i}^- = \int_{\underline{X}_{t_i}}^{B_{t_i^-}} f_-(y, B_{t_i^-}) \, dy \qquad (5.22)$$

We denote

$$F_+(x, A_{t-}) \equiv \int_{A_{t-}}^{x} f_+(y, A_{t-})\, dy$$

$$F_-(x, B_{t-}) \equiv \int_{B_{t-}}^{x} f_-(y, B_{t-})\, dy$$

The new mid-price just after this order at t_i^+ is from (5.21)–(5.22):

$$X_{t_i}^* \equiv \frac{1}{2}\left(F_+^{-1}(\delta\Delta_{t_i}^+, A_{t_i^-}) + F_-^{-1}(-\delta\Delta_{t_i}^-, B_{t_i^-})\right) \tag{5.23}$$

Then the market reacts to this order by moving instantaneously the mid-price to

$$X_{t_i^+} \equiv X_{t_i^-} + k(X_{t_i}^* - X_{t_i^-}) \tag{5.24}$$

with $k \in [0, 1]$. For $k = 0$, the market moves back to the initial mid-price $X_{t_i^-}$. For $k = 1$, the market moves all the way to $X_{t_i}^*$. Between two orders (t_i, t_{i+1}), the mid-price fluctuates, due to small trades, as

$$X_{t_{i+1}^-} = X_{t_i^+} + \sigma(t_i, X_{t_i^-})\sqrt{t_{i+1} - t_i}\,\epsilon_i$$

with $\epsilon_i \sim \mathcal{N}(0, 1)$ and $\sigma(t_i, X_{t_i})$ the volatility of the mid-price. Finally, the cost is

$$\text{Cost}(\delta\Delta_{t_i}) = \int_0^{\delta\Delta_{t_i}^+} F_+^{-1}(y, A_{t_i^-})\, dy - \int_{-\delta\Delta_{t_i}^-}^{0} F_-^{-1}(y, B_{t_i^-})\, dy \tag{5.25}$$

For use below, we note that the cost function (5.25) and the mid-price (5.24) can be expanded as

$$\text{Cost}(\delta\Delta_{t_i}) = A_{t_i^-}\delta\Delta_{t_i}^+ - B_{t_i^-}\delta\Delta_{t_i}^-$$
$$+ \frac{1}{f_+(A_{t_i^-}, A_{t_i^-})}\frac{(\delta\Delta_{t_i}^+)^2}{2} + \frac{1}{f_-(B_{t_i^-}, B_{t_i^-})}\frac{(\delta\Delta_{t_i}^-)^2}{2} + O(\delta\Delta_{t_i}^3) \tag{5.26}$$

and,

$$X_{t_{i+1}^-} - X_{t_i^-} = \frac{k}{2}\Big(\frac{\delta\Delta_{t_i}^+}{f_+(A_{t_i^-}, A_{t_i^-})} - \frac{\delta\Delta_{t_i}^-}{f_-(B_{t_i^-}, B_{t_i^-})}$$
$$- \frac{f_+'(A_{t_i^-}, A_{t_i^-})}{f_+^3(A_{t_i^-}, A_{t_i^-})}\frac{(\delta\Delta_{t_i}^+)^2}{2} - \frac{f_-'(B_{t_i^-}, B_{t_i^-})}{f_-^3(B_{t_i^-}, B_{t_i^-})}\frac{(\delta\Delta_{t_i}^-)^2}{2}\Big)$$
$$+ \sigma(t_i, X_{t_i^-})\sqrt{t_{i+1} - t_i}\,\epsilon_i + O(\delta\Delta_{t_i}^3) \tag{5.27}$$

where f_-' denotes the derivative of f_- w.r.t. y (left derivative if $y = B_t$), and f_+' the derivative of f_+ w.r.t. y (right derivative if $y = A_t$).

5.4.3 Continuous-time limit: Some intuition

For the sake of simplicity, we assume that our limit order book is symmetric around the mid-price and the bid-ask spread is negligible, i.e., $X_t = A_t = B_t$. We set $f_+(X_t, X_t) = f_-(X_t, X_t) \equiv 1/(2\epsilon(X_t))$ with $\epsilon : \mathbb{R}_+ \to \mathbb{R}_+$ a nonnegative differentiable function on $\mathbb{R}_+$; ϵ will be interpreted as the market elasticity. The elasticity ϵ can also have an explicit time dependence. By symmetry we have $f'_-(X_t, X_t) = -f'_+(X_t, X_t)$. This quantity is typically positive and denoted by $\eta(X_t)$.

Throughout this section, we consider a one-dimensional Brownian motion W defined on a probability space $(\Omega, \mathcal{F}, \mathbb{P})$ and denote by $\mathcal{F}_t$ the filtration generated by W.

As in [76], we restrict the delta process Δ_t to be of the form

$$\Delta_t = \Delta_0 + \int_0^t \alpha_s \, ds + \int_0^t \Gamma_s \, dX_s + \sum_{j=0}^n y^j 1_{t \geq \tau_j} \tag{5.28}$$

so that it has finite quadratic variation. Here $\tau_0 \equiv 0 < \tau_1 < \cdots < \tau_n \equiv T$ is an increasing sequence of $[0, T]$-valued $(\mathcal{F}_t)$-stopping times; y^j is an $\mathbb{R}$-valued measurable random variable; and α and Γ are two adapted processes.

Disregarding for the moment the jump components, we obtain by taking the continuous-time limit of (5.26) that the cost induced by our time-continuous trading strategy is

$$\int_0^T X_t \, d\Delta_t + \int_0^T d\langle X, \Delta \rangle_t + \int_0^T \epsilon(X_t) \, d\langle \Delta \rangle_t$$

Integrating by parts, this can be replaced by

$$-\int_0^T \Delta_t \, dX_t + \int_0^T \epsilon(X_t) \, d\langle \Delta \rangle_t + X_T \Delta_T - X_0 \Delta_0 \tag{5.29}$$

Adding the jump part, we get

$$\sum_{j=0}^n L(X_{\tau_j}, y_j) - \int_0^T \Delta_t \, dX_t + \int_0^T \epsilon(X_t) \, d\langle \Delta \rangle_t + X_T \Delta_T - X_0 \Delta_0 \tag{5.30}$$

where $L(X_{\tau_j}, y_j)$ is the cost of an order of size y_j:

$$L(x, \Delta) \equiv \int_0^{\Delta^+} F_+^{-1}(y, x) \, dy - \int_{-\Delta^-}^0 F_-^{-1}(y, x) \, dy$$

Similarly, the market impact on the mid-price due to the delta hedging strategy is from (5.27)

$$dX_t = k \left(\epsilon(X_t) \, d\Delta_t + d\langle \epsilon(X.), \Delta \rangle_t \right) + \sigma(t, X_t) \, dW_t \tag{5.31}$$

Indeed, in the limit where the size of the time steps $t_{i+1} - t_i$ goes to zero,

$$\lim \sum_i \eta(X_{t_i^-})(\delta \Delta_{t_i})^2 \text{sgn}(\delta \Delta_{t_i}) = 0$$

The mid-price jumps at each date $(\tau_j)_{0 \leq j \leq n}$ according to (5.24):

$$X_{\tau_j^+} = X_{\tau_j^-} + k \left(\frac{F_+^{-1}\left(y_j^+, X_{\tau_j^-}\right) + F_-^{-1}\left(-y_j^-, X_{\tau_j^-}\right)}{2} - X_{\tau_j^-} \right) \quad (5.32)$$

The diffusion term $\sigma(t, X_t) \, dW_t$ models the fluctuation of the mid-price between t and $t+dt$ due to small trades. The volatility $\sigma(t, x)$ is a local volatility function, calibrated to the mid-prices of vanilla options.

REMARK 5.1   In Obizhaeva and Wang [167], the authors choose $f_{\pm}(y, x) = \frac{1}{2\epsilon(x)}$ independent of y. In this case we have $\eta(x) \equiv 0$ and $L(x, \Delta) = x\Delta + \epsilon(x)\Delta^2$.   ⬚

This simple limit order book model highlights that the cost of hedging depends on the quadratic variation of the delta as in [75, 76, 48], and the market impact on the mid-price involves the variation of the delta, leading to a non-trivial feedback effect on the drift and the volatility.

5.4.4   Admissible trading strategies

As in [76] and in the previous section, we restrict the process Δ_t to be of the form (5.28). We disregard the jump component as it can been shown that it is not optimal to have jumps in Δ (see Proposition 1 in [76]), except at $t = 0$ and $t = T$. The P&L then reads

$$\text{P\&L}_T^{\Delta} \equiv V_0 + X_0 \Delta_0 - g(X_T) - X_T \Delta_T - L(X_T, -\Delta_T) - L(X_0, \Delta_0)$$
$$+ \int_0^T \Delta_t \, dX_t - \int_0^T \epsilon(X_t) \, d\langle \Delta \rangle_t$$

REMARK 5.2 Liquid market   When Δ is small, we have

$$L(x, \Delta) = x\Delta + \epsilon(x)\Delta^2 + O(\Delta^3)$$

For $\epsilon \equiv 0$, the P&L reduces to the well known expression in a liquid market:

$$V_0 - g(X_T) + \int_0^T \Delta_t \, dX_t$$

⬚

5.4.5   Perfect replication paradigm

Let us assume that an (admissible) delta hedging strategy can be written as a function of t and X_t, $\Delta_t = \partial_x u(t, X_t)$, for some smooth function $u \in C^{1,2}([0,T] \times \mathbb{R}_+)$. By using Itô's lemma, we get

$$P\&L_T^\Delta = V_0 - L(X_0, \partial_x u(0, X_0)) + u(T, X_T) - u(0, X_0)$$
$$+ \int_0^T \left(-\partial_t u\, dt - \frac{1}{2}\partial_x^2 u \left(1 + 2\epsilon(X_t)\partial_x^2 u\right) d\langle X\rangle_t \right)$$
$$- g(X_T) - X_T \partial_x u(T, X_T) + X_0 \partial_x u(0, X_0) - L(X_T, -\partial_x u(T, X_T))$$

From (5.31), the dynamics on X_t has the form

$$dX_t = \frac{\sigma(t, X_t)}{1 - k\epsilon(X_t)\partial_x^2 u(t, X_t)}\, dW_t + \cdots dt$$

In particular,

$$d\langle X\rangle_t = \frac{\sigma(t, X_t)^2}{(1 - k\epsilon(X_t)\partial_x^2 u(t, X_t))^2}\, dt$$

If we now define $u(t, x)$ as the solution of the nonlinear PDE

$$\partial_t u(t, x) + \frac{\sigma^2(t, x)}{2}\frac{1 + 2\epsilon(x)\partial_x^2 u}{(1 - k\epsilon(x)\partial_x^2 u)^2}\partial_x^2 u = 0$$

with a terminal condition that satisfies the ODE $u(T, x) = g(x) + x\partial_x u(T, x) + L(x, -\partial_x u(T, x))$, the P&L at maturity cancels with

$$V_0 = u(0, X_0) + L(X_0, \partial_x u(0, X_0)) - X_0 \partial_x u(0, X_0)$$

meaning that the claim $g(X_T)$ can be perfectly replicated. We write this nonlinear PDE as

$$\partial_t u(t, x) + \frac{\sigma(t, x)^2}{2}H(x, \partial_x^2 u) = 0 \tag{5.33}$$

with the Hamiltonian H defined by

$$H(x, \Gamma) \equiv \frac{\Gamma(1 + 2\epsilon(x)\Gamma)}{(1 - k\epsilon(x)\Gamma)^2}$$

PDE (5.33) has the same flaws as PDE (5.18):

- The Hamiltonian blows up for finite gamma.

- Contrary to what we expect from a model including illiquidity costs, the Hamiltonian $H(x, \Gamma)$ is not always greater than or equal to Γ, which is the value of H in liquid markets ($\epsilon \equiv 0$). This happens only if $k\epsilon\Gamma \leq 2 + 2/k$.

- This PDE is *not* parabolic: as we have $\partial_\Gamma H(x,\Gamma) = \frac{1+(4+k)\epsilon(x)\Gamma}{(1-k\epsilon(x)\Gamma)^3}$, the operator H is parabolic only on the domain $k\epsilon(x)\partial_x^2 u \in (-k/(4+k), 1)$. Outside this domain, this PDE is not well defined and the comparison principle is not satisfied, leading to arbitrage opportunities. This means that the replication paradigm cannot be used here.

For all these reasons, PDE (5.33) should not be considered. Below we explain how to solve the first and third issues by considering the parabolic envelope of H. The second problem—the fact that the model tells that it may be less costly to hedge in an illiquid market in some regions of (x, Γ)—should be tackled separately. If the terminal condition is unchanged $(u(T,x) = g(x))$, the simplest, crudest way to solve it is to replace $H(x,\Gamma)$ by $\max(H(x,\Gamma),\Gamma)$.

5.4.6   The parabolic envelope of the Hamiltonian

Following closely [76], we deduce the following theorem:

THEOREM 5.2
The seller super-replication price $S_0(g(X_T))$ is equal to

$$u(0, X_0) + L(X_0, \partial_x u(0, X_0)) - \partial_x u(0, X_0)X_0$$

where u is the unique viscosity solution of the nonlinear PDE

$$\partial_t u(t,x) + \frac{\sigma^2(t,x)}{2}H^*(x, \partial_x^2 u) = 0 \tag{5.34}$$

with the terminal condition that solves the ODE

$$u(T,x) = g(x) + L(x, -\partial_x u(T,x)) + x\partial_x u(T,x)$$

and where $H^*(x,\Gamma) = \inf_{\beta \geq 0} H(x, \Gamma + \beta)$.

The Hamiltonian $H^*(x,\Gamma)$ is the largest lower bound of $H(x,\Gamma)$ that is nondecreasing in Γ, and therefore defines a parabolic operator. H^* is called the *parabolic envelope* of H. An explicit computation gives (see Figure 5.1)

$$H^*(x,\Gamma) = \begin{cases} -\frac{1}{4(2+k)\epsilon(x)} & \text{if } k\epsilon(x)\Gamma < -\frac{k}{4+k} \\ H(x,\Gamma) & \text{if } k\epsilon(x)\Gamma \in [-\frac{k}{4+k}, \frac{2}{4+k}) \\ \frac{2}{k^2\epsilon(x)} & \text{if } k\epsilon(x)\Gamma \geq \frac{2}{4+k} \end{cases}$$

Contrary to PDE (5.33) or (5.18), PDE (5.34) is well-posed and depends only on a local volatility function σ, calibrated to mid-prices of vanilla options, on the elasticity function $\epsilon(x)$, and on the parameter k. To prevent the Hamiltonian to be smaller than Γ, we can either compute the parabolic envelope of $\max(H(x,\Gamma),\Gamma)$, or replace $H^*(x,\Gamma)$ by $\max(H^*(x,\Gamma),\Gamma)$, which is not the

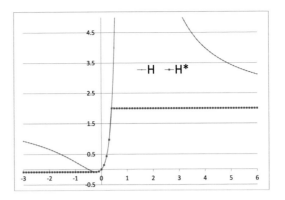

Figure 5.1: Hamiltonian $\Gamma \mapsto H(x,\Gamma)$ versus its parabolic envelope H^* for $k = 1$ and $\epsilon(x) = 1$.

same. Note that if H satisfies $H(x,\Gamma) \geq \Gamma$, then the parabolic envelope H^* will automatically satisfy $H^*(x,\Gamma) \geq \Gamma$. However, the terminal condition now depends on the elasticity ϵ, and the price at inception is not $u(0, X_0)$, but the sum of $u(0, X_0)$ and the liquidity cost at inception that depends on ϵ as well, so we cannot simply compare the prices for different choices of ϵ by only comparing the Hamiltonians.

Interpretation

How do we interpret β, H^* and the whole PDE (5.34)? The final and initial delta hedging illiquidity costs are respectively included in the terminal condition and added to the initial value of u so we can ignore them when we try to determine the Hamiltonian. Let us consider that we sell the option at time t_0 and start hedging it at the same time. The asset price at this time is denoted by x_0. The super-replication price is

$$u(t_0, x_0) = \inf \left\{ v \; \middle| \; \text{there exists an admissible } \Delta \text{ such that} \right.$$

$$\left. v + \int_{t_0}^T \Delta_r \, dX_r - \int_{t_0}^T \epsilon(X_r) \, d\langle \Delta \rangle_r \geq g(X_T), \quad \mathbb{Q} - \text{a.s} \right\}$$

We only consider continuous Δ's of the form $d\Delta_r = \alpha_r \, dr + \Gamma_r \, dX_r$; Δ_{t_0}, α_r and Γ_r are our controls. From the dynamic programming equation, the super-replication price satisfies the inequality

$$u(t_0, x_0) + \int_{t_0}^{t_0+h} \Delta_r \, dX_r - \int_{t_0}^{t_0+h} \epsilon(X_r) \, d\langle \Delta \rangle_r$$

$$- u(t_0 + h, x_0 + \Delta X) \geq 0, \quad \mathbb{Q} - \text{a.s} \quad (5.35)$$

for any h such that $t_0 + h \leq T$, with $\Delta X \equiv X_{t_0+h} - x_0$. By assuming that u is smooth and using Itô's lemma, we expand below the left-hand side at the first order in h. We first proceed with the stochastic integral:

$$
\begin{aligned}
\int_{t_0}^{t_0+h} \Delta_r \, dX_r &= \int_{t_0}^{t_0+h} \left(\Delta_{t_0} + \int_{t_0}^{r} d\Delta_u \right) dX_r \\
&= \Delta_{t_0} \Delta X + \int_{t_0}^{t_0+h} dX_r \int_{t_0}^{r} d\Delta_u \\
&= \Delta_{t_0} \Delta X + \int_{t_0}^{t_0+h} dX_r \int_{t_0}^{r} \Gamma_u \, dX_u + O(h^{\frac{3}{2}}) \\
&= \Delta_{t_0} \Delta X + \Gamma_{t_0} \frac{d\langle X \rangle_t}{dt}\Big|_{t_0} \int_{t_0}^{t_0+h} dW_r \int_{t_0}^{r} dW_u + O(h^{\frac{3}{2}}) \\
&= \Delta_{t_0} \Delta X + \Gamma_{t_0} \frac{d\langle X \rangle_t}{dt}\Big|_{t_0} \frac{(\Delta W^2 - h)}{2} + O(h^{\frac{3}{2}})
\end{aligned}
$$

with $\Delta W \equiv W_{t_0+h} - W_{t_0}$ and with $d\langle X \rangle_t = \frac{\sigma(t,X_t)^2}{(1-k\epsilon(X_t)\Gamma_t)^2} \, dt$. Proceeding similarly with the other terms, we obtain that, at the first order in h, (5.35) is equivalent to

$$
\left(-\partial_t u(t_0, x_0) - \frac{1}{2} \frac{\sigma^2(t_0, x_0)}{(1 - k\epsilon(x_0)\Gamma_{t_0})^2} \partial_x^2 u(t_0, x_0) \right) h
$$
$$
- \epsilon(x_0)\Gamma_{t_0}^2 \frac{\sigma^2(t_0, x_0)}{(1 - k\epsilon(x_0)\Gamma_{t_0})^2} h + (\Delta_{t_0} - \partial_x u(t_0, x_0)) \, \Delta X
$$
$$
+ \frac{1}{2} \frac{\sigma^2(t_0, x_0)}{(1 - k\epsilon(x_0)\Gamma_{t_0})^2} (\Gamma_{t_0} - \partial_x^2 u(t_0, x_0)) \, (\Delta W^2 - h) \geq 0
$$

$\mathbb{Q}$-a.s. with $\Delta W \equiv W_{t_0+h} - W_{t_0}$. For this inequality to be satisfied, we must choose $\Delta_{t_0} = \partial_x u(t_0, x_0)$ because ΔX is the dominating term in $O(\sqrt{h})$ and is a random variable with arbitrary sign. This does *not* mean that $\Delta_t = \partial_x u(t, X_t)$ for all $t \in [t_0, T]$. This only means that the control Δ_{t_0} must be equal to $\partial_x u(t_0, x_0)$. The delta strategy depends on the date t_0 when it was initiated. For instance, if u is concave at (t_0, x_0), one may decide to hold the same amount Δ_{t_0} of assets for a while, thus avoiding to pay the price for illiquidity. The term $\beta_{t_0} \equiv \Gamma_{t_0} - \partial_x^2 u(t_0, x_0)$, proportional to ΔW^2, must be nonnegative. From the definition of H, we get

$$
\left(-\partial_t u(t_0, x_0) - \frac{\sigma^2(t_0, x_0)}{2} H\left(x_0, \beta_{t_0} + \partial_x^2 u(t_0, x_0)\right) \right) h
$$
$$
+ \frac{1}{2} \frac{\sigma^2(t_0, x_0)}{(1 - k\epsilon(x_0)\Gamma_{t_0})^2} \beta_{t_0}^2 \, \Delta W^2 \geq 0, \quad \mathbb{Q}\text{-a.s} \quad (5.36)
$$

The pricing equation, which is parabolic and hence well-posed, reflects the fact that there exists a $\beta_{t_0} \geq 0$ such that the first term vanishes:

$$\partial_t u(t, x) + \frac{\sigma^2(t, x)}{2} \inf_{\beta \geq 0} H\left(x, \beta + \partial_x^2 u(t, x)\right) = 0$$

Then (5.36) becomes

$$\frac{1}{2} \frac{\sigma^2(t_0, x_0)}{(1 - k\epsilon(x_0)\Gamma_{t_0})^2} \beta^*(t_0, x_0)^2 \, \Delta W^2 \geq 0$$

which is nonnegative, proving that the inequality (5.35) is satisfied: we have super-replicated the payoff $g(X_T)$.

This analysis suggests that an ill defined PDE based on a replication argument can be turned into a well defined parabolic PDE by taking the parabolic envelope. The resulting PDE is then linked to a super-replication price. A careful proof is given in [186].

5.5 Super-replication under delta and gamma constraints

5.5.1 Super-replication under delta constraints

As a new example of friction, we impose that the self-financing strategy must satisfy delta constraints:

$$\underline{\Delta} \leq \Delta_t \leq \overline{\Delta} \tag{5.37}$$

where we denote by Δ_t the quantity of assets invested at time t in the risky asset (the amount in dollars would be $\Delta_t X_t$). The discounted self-financing wealth process satisfies

$$d\tilde{\pi}_t = \Delta_t d\tilde{X}_t$$

Taking $\underline{\Delta} = 0$ means that the financial market does not allow short selling of the asset. The buyer's (resp. seller's) super-replication price under delta constraints is defined as follows:

DEFINITION 5.2 Buyer's super-replication price under delta constraint

$$\mathcal{B}_t(F_T) = \sup \left\{ z \in \mathcal{F}_t \;\middle|\; \text{there exists an admissible portfolio } \Delta \text{ satisfying (5.37)} \right.$$

$$\left. \text{such that } \tilde{\pi}_T^B \equiv -D_{0t}z + \int_t^T \Delta_t \, d\tilde{X}_t + D_{0T} F_T \geq 0 \;\; \mathbb{P}^{\text{hist}} - a.s. \right\}$$

DEFINITION 5.3 Seller's super-replication price under delta constraint

$$\mathcal{S}_t(F_T) = \inf \left\{ z \in \mathcal{F}_t \;\middle|\; \text{there exists an admissible portfolio } \Delta \text{ satisfying (5.37)} \right.$$

$$\left. \text{such that } \tilde{\pi}_T^S \equiv D_{0t} z + \int_t^T \Delta_t \, d\tilde{X}_t - D_{0T} F_T \geq 0 \; \mathbb{P}^{\text{hist}} - a.s. \right\}$$

Let us assume that the asset X follows local volatility dynamics.[1] Then super-replication prices are given by:

THEOREM 5.3 Super-replication prices under delta constraint [87]

We have

$$\mathcal{B}_t(F_T) = \inf_{\lambda \in \Lambda_{t,T}} \mathbb{E}^{\mathbb{Q}} \left[D_{tT} F_T + \int_t^T D_{ts} \Delta \left(b(s, X_s) - r_s X_s - \sigma(s, X_s) \lambda_s \right) ds \middle| \mathcal{F}_t \right]$$

$$\mathcal{S}_t(F_T) = \sup_{\lambda \in \Lambda_{t,T}} \mathbb{E}^{\mathbb{Q}} \left[D_{tT} F_T - \int_t^T D_{ts} \Delta \left(b(s, X_s) - r_s X_s - \sigma(s, X_s) \lambda_s \right) ds \middle| \mathcal{F}_t \right]$$

where $\Lambda_{t,T}$ is the set of all $(\mathcal{F}_s)$-adapted processes $(\lambda_s, t \leq s \leq T)$,

$$\Delta(x) = \begin{cases} \overline{\Delta} x & \text{if } x \geq 0 \\ \underline{\Delta} x & \text{if } x < 0 \end{cases}$$

and X_t satisfies

$$dX_t^i = \left(b_i(t, X_t) - \sum_{j=1}^d \sigma_{i,j}(t, X_t) \lambda_t^j \right) dt + \sum_{j=1}^d \sigma_{i,j}(t, X_t) \, dW_t^j$$

with W_t a $\mathbb{Q}$-Brownian motion.

The proof is not reported here as it copycats the proof of Theorem 1.3. The presence of constraints does not affect the general methodology of the proof. In particular the minimization over Δ_t satisfying (5.37) of the functional $H(Y)$ appearing in this proof gives the source term

$$\int_t^T D_{ts} \Delta \left(b(s, X_s) - r_s X_s - \sigma(s, X_s) \lambda_s \right) ds$$

in the seller's super-replication price.

---

[1] If we assume that X follows a stochastic volatility model, we have to also take the infimum (resp. supremum) over the drift of the stochastic volatility process; see Exercise 4.7.2.

5.5.2   HJB and variational inequality

These super-replication problems read as singular stochastic control problems where the control λ_t is unbounded, i.e., $\underline{\lambda} = -\infty$ and $\overline{\lambda} = \infty$. Assuming, for the sake of simplicity, that $d = 1$ and that X_t is a complete model with zero interest rates, repos, and dividends, the HJB equation satisfied by the seller's super-replication price $u(t, X_t) \equiv \mathcal{S}_t(g(X_T))$ for a vanilla payoff $F_T = g(X_T)$ reads as (see Equation (4.18))

$$\partial_t u(t,x) + \mathcal{L}u(t,x) + \sup_{\alpha \in \mathbb{R}}\{-\Delta(\alpha) + \alpha\partial_x u(t,x)\} = 0$$

where, instead of $\lambda \in \mathbb{R}$, we use the control $\alpha = b(t,x) - \sigma(t,x)\lambda \in \mathbb{R}$ (α is the drift of X). Here, $\mathcal{L} = \frac{1}{2}\sigma(t,x)^2\partial_x^2$. The Hamiltonian is finite if and only if

$$\underline{\Delta} \leq \partial_x u(t,x) \leq \overline{\Delta}$$

Applying Theorems 4.6 and 4.7, u is a viscosity solution to the variation inequality

$$\min\left(-\partial_t u(t,x) - \mathcal{L}u(t,x), -\partial_x u(t,x) + \overline{\Delta}, \partial_x u(t,x) - \underline{\Delta}\right) = 0 \quad (5.38)$$

with the terminal condition $u(T,x) = \hat{g}(x)$ being itself a viscosity solution to the obstacle problem

$$\min\left(\hat{g}(x) - g(x), -\partial_x \hat{g}(x) + \overline{\Delta}, \partial_x \hat{g}(x) - \underline{\Delta}\right) = 0$$

This means that $\hat{g}$ is the smallest function greater than or equal to g that satisfies the delta constraints.

5.5.3   Adding gamma constraints

We add gamma constraints, which can be linked to liquidity constraints:

$$\underline{\Gamma} \leq X_t^2\Gamma_t \leq \overline{\Gamma} \quad (5.39)$$

The factor X_t^2 has only been introduced for convenience. The process Γ_t is defined as the variation of Δ_t:

$$d\Delta_t = \mu_t\, dt + \Gamma_t\, dX_t \quad (5.40)$$

This situation is more involved than in the previous section as the control Δ_t itself is assumed to follow the dynamics (5.40). The seller's super-replication price can then be obtained by introducing in the proof of Theorem 1.3 a Lagrange multiplier (a stochastic process playing the same role as the shadow price in the proof of Theorem 1.3) associated to the constraint (5.40). We do not follow this route. From the variational inequality (5.38), one can guess that the price u is a viscosity solution to the variation inequality

$$\min\left(-\partial_t u(t,x) - \mathcal{L}u(t,x), -\partial_x u(t,x) + \overline{\Delta}, \partial_x u(t,x) - \underline{\Delta},\right.$$
$$\left. -x^2\partial_x^2 u(t,x) + \overline{\Gamma}, -x^2\partial_x^2 u(t,x) + \overline{\Gamma}\right) = 0 \quad (5.41)$$

Unfortunately, except in the case where $\underline{\Gamma} = -\infty$, this variational inequality is not parabolic and therefore from the no-arbitrage condition u cannot be the solution of such an equation. The correct result is as follows.

Under gamma constraints with $\underline{\Gamma} = -\infty$, the seller's super-replication price u is proved [185, 79, 80] to be the viscosity solution to the following variational PDE

$$\min\left(-\partial_t u - \mathcal{L}u, \overline{\Gamma} - x^2\partial_x^2 u, \overline{\Delta} - \partial_x u, -\underline{\Delta} + \partial_x u\right) = 0 \quad (5.42)$$

where the terminal condition $\hat{g}(x)$ is itself the viscosity solution to the obstacle problem

$$\min\left(\hat{g}(x) - g(x), \overline{\Gamma} - x^2\partial_x^2 \hat{g}, \overline{\Delta} - \partial_x \hat{g}, -\underline{\Delta} + \partial_x \hat{g}\right) = 0 \quad (5.43)$$

Under general delta and gamma constraints, u is the viscosity solution to

$$H^*(\partial_t u, x^2\partial_x^2 u, \partial_x u) = 0 \quad (5.44)$$

with

$$H^*(p, A, B) \equiv \sup_{\beta \geq 0} H(p, A + \beta, B)$$

$$H(p, A, B) = \min\left(-p - \frac{1}{2}\sigma^2(t,x)A, \overline{\Gamma} - A, -\underline{\Gamma} + A, \overline{\Delta} - B, -\underline{\Delta} + B\right)$$

with the terminal condition $v(T, x) = \hat{g}(x)$ solution to the obstacle problem (5.43). The Hamiltonian H^* is the parabolic envelope of H (see Section 5.4.6 for an interpretation of H^*).

5.5.4 Numerical algorithm

5.5.4.1 The global procedure

Below, we focus on the case where $\underline{\Gamma} = -\infty$. In a one-asset framework, the nonlinear PDE (5.42)–(5.43) can be solved with a simple algorithm that can be described by the following steps:

1. Solve (5.43) for the payoff $\hat{g}(x)$. This can be achieved efficiently using the primal-dual algorithm as presented below.

2. Solve the pricing PDE $\partial_t u + \mathcal{L}u = 0$ by backward induction. Apply at each time step the primal-dual algorithm. As the primal-dual algorithm is fast (see below), this procedure is equivalent in computational time to the pricing of an American option using a successive over-relaxation (SOR) method. We should remark that for commonly used equity models such as the Black-Scholes model, Bergomi's variance swap model (see Section 3.1.2), or the uncertain volatility model (see Section 5.2), the operator $\mathcal{L}$ commutes with $\partial_{\ln x}$: $[\partial_{\ln x}, \mathcal{L}] = 0$. As a consequence, in these cases we do not need to apply the primal-dual algorithm at each time step in the backward induction as the constraints on the delta and the gamma are preserved by the pricing semigroup acting on $\hat{g}$. Note that this simplification is not valid for the local volatility model, or in case of nonlinear dividends $D(X)$. This simplification does not hold either in the presence of a discrete path-dependent variable Z: the price value at an observation date is modified as

$$u(t^-, x, Z) = u(t^+, x, f(x, Z))$$

As a consequence, the constraint is not preserved anymore and the smoothing method must be used again.

Below, we present the primal-dual algorithm as introduced in [63] to solve the obstacle problem (5.43).

5.5.4.2   The primal-dual algorithm

The discretization of (5.43) on a log-space grid ($z_i = \ln \frac{X_i}{X_0}, \hat{g}_i \equiv \hat{g}(X_0 e^{z_i}), i \in \{0, \ldots, N-1\}$) with a step Δz gives the following:

Find $\{\hat{g}_i\} \in \mathbb{R}^N$ such that $\min\left(\hat{g} - g, A^{\overline{\Gamma}}\hat{g} - b^{\overline{\Gamma}}, A^{\overline{\Delta}}\hat{g} - b^{\overline{\Delta}}, A^{\underline{\Delta}}\hat{g} - b^{\underline{\Delta}}\right) = 0$

$$(5.45)$$

where the A's are $N \times N$ matrices, and the b's and the g's belong to $\mathbb{R}^N$:

$$A^{\overline{\Gamma}}_{i,i-1} = -1 - \frac{\Delta z}{2}, \quad A^{\overline{\Gamma}}_{i,i} = 2, \quad A^{\overline{\Gamma}}_{i,i+1} = -1 + \frac{\Delta z}{2}, \quad b^{\overline{\Gamma}}_i = -\overline{\Gamma}\Delta z^2$$

$$A^{\overline{\Delta}}_{i,i} = 1, \quad A^{\overline{\Delta}}_{i,i+1} = -1, \quad b^{\overline{\Delta}}_i = -\overline{\Delta}X_i\Delta z$$

$$A^{\underline{\Delta}}_{i,i-1} = -1, \quad A^{\underline{\Delta}}_{i,i} = 1, \quad b^{\underline{\Delta}}_i = \underline{\Delta}X_i\Delta z$$

All other entries of the A's are zero. Note that $A^{\underline{\Delta}}$ (resp. $A^{\overline{\Delta}}$) has been discretized using a one-step backward (resp. forward) finite difference scheme. This ensures that $A^{\underline{\Delta}}$ (resp. $A^{\overline{\Delta}}$) is a monotone operator, i.e., all components of A^{-1} are nonnegative. We fix the boundary conditions at $i = N - 1$ as follows (Neumann)

$$\hat{g}'_{N-1} = \begin{cases} g'(\infty) & \text{if } \underline{\Delta} < g'(\infty) < \overline{\Delta} \\ \underline{\Delta} & \text{if } g'(\infty) \leq \underline{\Delta} \\ \overline{\Delta} & \text{if } \overline{\Delta} \leq g'(\infty) \end{cases}$$

and similarly at $i = 0$. Note that another choice could be $\hat{g}_{N-1} = g(\infty)$ and $\hat{g}_0 = g(-\infty)$ (Dirichlet).

Next, we reformulate the obstacle problem into a more convenient form: By setting $\mathcal{A} = \{0, 1, 2, 3\}$, $B(0) = I_N$, $B(1) = A^{\overline{\Gamma}}$, $B(2) = A^{\overline{\Delta}}$, $B(3) = A^{\underline{\Delta}}$, $c(0) = g$, $c(1) = b^{\overline{\Gamma}}$, $c(2) = b^{\overline{\Delta}}$, $c(3) = b^{\underline{\Delta}}$ (I_N being the $N \times N$ identity matrix), the above obstacle problem (5.45) can be cast into the following:

$$\text{Find } \{\hat{g}_i\} \in \mathbb{R}^N \text{ such that } \min_{\alpha \in \mathcal{A}} (B(\alpha)\hat{g} - c(\alpha)) = 0 \qquad (5.46)$$

The primal-dual meta-algorithm can then be described by the following steps [63]:

Primal-dual algorithm:

1. Initialize $\alpha^{(0)} \in \mathcal{A}^N = \{0, 1, 2, 3\}^N$. A convenient choice is $\alpha^{(0)} = (0, \ldots, 0)$.

2. Iterate for $k \geq 0$:

   (a) For $\alpha^{(k)} \in \mathcal{A}^N$, the $N \times N$ matrix $B(\alpha^{(k)})$ is defined as follows: The i-th row of matrix $B(\alpha^{(k)})$ is the i-th row of the matrix $B(\alpha_i^{(k)})$ where $\alpha_i^{(k)} \in \mathcal{A}$ denotes the i-th component of $\alpha^{(k)}$. Similarly, the i-th component of $c(\alpha^{(k)})$ is defined as $c(\alpha_i^{(k)})$. Then, we set $g^{(k)} = B(\alpha^{(k)})^{-1}c(\alpha^{(k)})$. This involves the inversion of a tridiagonal $N \times N$ matrix. If $k \geq 1$ and $\hat{g}^{(k)} = \hat{g}^{(k-1)}$, then stop. Otherwise, go to the next step.

   (b) For every $i \in \{0, \ldots, N-1\}$, take

   $$\alpha_i^{(k+1)} \equiv \text{argmin}_{\alpha \in \mathcal{A}^N} \left(B(\alpha)\hat{g}^{(k)} - c(\alpha) \right)$$

   (c) Set $k := k + 1$ and go to (a).

THEOREM 5.4 Primal-dual algorithm [63]
The primal-dual algorithm converges in at most $\text{Card}(\mathcal{A})^N$ *iterations, i.e.,* $\hat{g}^{(k)} = \hat{g}^*$ *for some* $k \leq \text{Card}(\mathcal{A})^N$.

The algorithm computes the solution via a sequence of trial values $\hat{g}^{(k)}$ and policies $\alpha^{(k)}$ under an alternating sequence of policy improvement and policy evaluation steps. In practical situations, one can observe that the primal-dual algorithm converges in a few iterations like the simplex method in linear programming optimization. Below, we present numerical examples that illustrate the algorithm.

5.5.5 Numerical experiments

5.5.5.1 Call option with gamma constraints

The analytical solution to (5.43) for a call option with strike K under gamma constraint is given by

$$\hat{g}(x) = \begin{cases} (x-K)^+ \equiv \sigma^2(1-A) & \text{if } x \geq x_1 \\ \overline{\Gamma}\left(\ln\frac{x_0}{x} + \frac{x}{x_0} - 1\right) & \text{if } x \in [x_0, x_1] \\ 0 & \text{if } x \leq x_0 \end{cases}$$

with $x_1 = \left(\frac{1}{x_0} - \frac{1}{\overline{\Gamma}}\right)^{-1}$ and x_0 is defined by

$$0 < x_0 < K \wedge \overline{\Gamma}, \qquad \ln\left(1 - \frac{x_0}{\overline{\Gamma}}\right) + \frac{K}{\overline{\Gamma}} = 0$$

We check this analytical solution against our numerical method with $\overline{\Gamma} = 0.5$ and find a perfect fit (see Figure 5.2).

5.5.5.2 Call spread option with delta and gamma constraints

We choose $\overline{\Delta} = 0.3$, $\underline{\Delta} = 0.01$, $\overline{\Gamma} = 0.5$, and $\underline{\Gamma} = -\infty$. We smooth a call spread payoff $95\% - 105\%$, i.e., $(\frac{X_T}{X_0} - 0.95)^+ - (\frac{X_T}{X_0} - 1.05)^+$, under the above delta and gamma constraints for which we have no analytical solution (see Figure 5.3).

5.6 The uncertain mortality model for reinsurance deals

Another example of nonlinear PDE arises when one considers pricing of reinsurance deals in the uncertain mortality model. In this case, the PDE is semilinear. Examples of reinsurance deals are the so-called GMxB deals. GMxB stands for GMIB (Guaranteed Minimum Investment Benefit) or GMDB (Guaranteed Minimum Death Benefit). For the sake of simplicity, let us ignore the usual lapse feature of GMxB deals. Those deals then consist of only two payoffs: at a maturity T, the seller of the option pays the payoff $g(X_T)$ with X_T the value of an asset; in the case where the insurance subscriber dies before maturity, the seller pays at time t the payoff $g_D(X_t)$. The index D stands for default. Like in credit modeling, we model the time to default τ (here, the time of death) by the first time to default of a Poisson process with an intensity λ_t (the mortality rate). In the uncertain mortality model, this rate λ_t is assumed to be uncertain and to range into the interval $[\underline{\lambda}(t), \overline{\lambda}(t)]$. The fair value of this option is given by (see Proposition 8.1 where u is interpreted

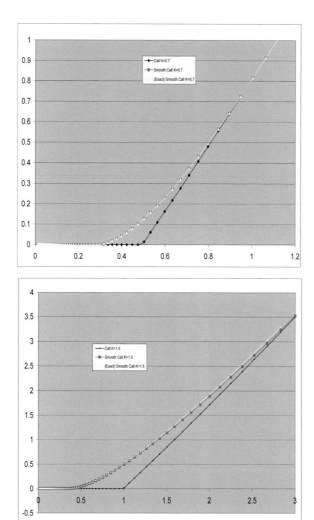

Figure 5.2: Call option with strikes $K = 0.5$ and $K = 1$ and with gamma constraint $\overline{\Gamma} = 0.5$.

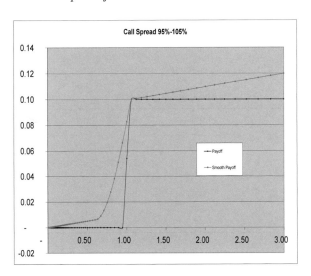

Figure 5.3: Call spread option $(\frac{X_T}{X_0} - 0.95)^+ - (\frac{X_T}{X_0} - 1.05)^+$ with delta and gamma constraints. $\underline{\Delta} = 0.01$, $\overline{\Delta} = 0.3$, $\underline{\Gamma} = -\infty$, and $\overline{\Gamma} = 0.5$.

as a super-replication price under uncertain intensity)

$$u(t, x) = \sup_{\lambda \in \Lambda_{t,T}} \mathbb{E}^{\mathbb{Q}}[g(X_T)\mathbf{1}_{\tau \geq T} + g_{\mathrm{D}}(X_\tau)\mathbf{1}_{\tau < T}|X_t = x]$$

where $\Lambda_{t,T}$ is the set of all adapted processes λ such that for all $s \in [t, T]$, $\lambda_s \in [\underline{\lambda}(s), \overline{\lambda}(s)]$. The HJB equation reads (see Example 1.1)

$$\partial_t u(t, x) + \mathcal{L}u(t, x) + \sup_{\lambda \in [\underline{\lambda}(t), \overline{\lambda}(t)]} \lambda\left(g_{\mathrm{D}}(x) - u\right) = 0$$

which is equivalent to

$$\partial_t u(t, x) + \mathcal{L}u(t, x) + \Lambda\left(t, g_{\mathrm{D}}(x) - u\right) = 0 \qquad (5.47)$$

with the terminal condition $u(T, x) = g(x)$ and Λ defined by

$$\Lambda(t, y) = \begin{cases} \overline{\lambda}(t)y & \text{if } y \geq 0 \\ \underline{\lambda}(t)y & \text{if } y < 0 \end{cases}$$

The uncertain lapse and mortality model for reinsurance deals will be the subject of Chapter 8. An example of PDE implementation of (5.47) is described in Section 8.6. In Sections 8.7 to 8.9, we will discuss Monte Carlo methods.

5.7 Different rates for borrowing and lending

Let us assume that one can lend at an interest rate $\underline{r}$, and borrow at another interest rate $\bar{r}$, with $\underline{r} \leq \bar{r}$. How does this impact the valuation of financial derivatives?

Let us assume for simplicity that the underlying asset follows local volatility dynamics. Then the usual pricing PDE, which assumes constant and equal interest rate r for borrowing and lending,

$$\partial_t u(t, x) + \frac{1}{2}\sigma(t, x)^2 x^2 \partial_x^2 u(t, x) + rx\partial_x u(t, x) - ru(t, x) = 0 \qquad (5.48)$$

$$u(T, x) = g(x)$$

must be modified. In PDE (5.48), the interest rate r accounts for the fact that, over intervals $[t, t + dt]$, (a) cash is either borrowed or invested, depending on the sign of the delta $\partial_x u(t, x)$, and (b) cash is either borrowed or invested, depending on the sign of the option value $u(t, x)$. For instance, the seller of a European call option must borrow money to buy shares for delta-hedging, which makes the option more expensive, hence the term $+rx\partial_x u(t, x)$. He or she can also invest the positive premium of the option, which makes the option cheaper, hence the term $-ru(t, x)$. After netting, he or she must borrow money if $x\partial_x u(t, x) - u(t, x) \geq 0$, or conversely he or she can invest $u(t, x) - x\partial_x u(t, x)$ if this quantity is positive.

As a consequence, when we assume different rates for borrowing and lending, PDE (5.48) must be replaced by

$$\partial_t u(t, x) + \frac{1}{2}\sigma(t, x)^2 x^2 \partial_x^2 u(t, x) + R(x\partial_x u(t, x) - u(t, x)) = 0 \qquad (5.49)$$

where

$$R(y) = \begin{cases} \bar{r}y & \text{if } y \geq 0 \\ \underline{r}y & \text{otherwise} \end{cases}$$

$$= \underline{r}y + (\bar{r} - \underline{r})y^+ \qquad (5.50)$$

The value $v(t, x) = -u(t, x)$ of being short the option is then a solution to

$$\partial_t v(t, x) + \frac{1}{2}\sigma(t, x)^2 x^2 \partial_x^2 v(t, x) - R(v(t.x) - x\partial_x v(t, x)) = 0 \qquad (5.51)$$

$$v(T, x) = -g(x)$$

Notice that, from (5.50), the nonlinearity is $(x\partial_x u - u)^+$. PDE (5.49) is thus semilinear, and involves the positive part function. In Section 5.8 below, we give another example of semilinear PDE involving the positive part function. Note also that we have

$$\sup_{r \in \{\underline{r}, \bar{r}\}} \left(\partial_t u(t, x) + \frac{1}{2}\sigma(t, x)^2 x^2 \partial_x^2 u(t, x) + rx\partial_x u(t, x) - ru(t, x) \right) = 0$$

which expresses the problem of option pricing under different rates for borrowing and lending as a stochastic control problem, or HJB PDE; see Section 4.4. The control set $\{\underline{r}, \overline{r}\}$ can be replaced by the whole interval $[\underline{r}, \overline{r}]$, since

$$\sup_{r \in \{\underline{r}, \overline{r}\}} (rx\partial_x u(t, x) - ru(t, x)) = \sup_{r \in [\underline{r}, \overline{r}]} (rx\partial_x u(t, x) - ru(t, x))$$

5.8 Credit valuation adjustment

The recent financial crisis has highlighted the importance of *credit valuation adjustment* (in short CVA) when pricing derivative contracts. Counterparty risk, also known as default risk, is the risk that a counterparty may default and fail to make future payments. For Markovian models, this market imperfection leads naturally to semilinear PDEs. More precisely, the nonlinearity in the pricing equation affects none of the differential terms and involves the positive part of the mark-to-market value of the derivative (or portfolio of derivatives) upon default. In short, depending on the (modeling) choice of the mark-to-market value of the derivative upon default, we will get two types of PDEs that can be schematically written as (see below for details)

$$\partial_t u + \mathcal{L}u + r_0 u + r_1 u^+ = 0, \quad u(T, x) = g(x) \tag{5.52}$$

and

$$\partial_t u + \mathcal{L}u + r_0 u + r_1 M + r_2 M^+ = 0, \quad u(T, x) = g(x) \tag{5.53}$$
$$\partial_t M + \mathcal{L}M + r_3 M = 0, \quad M(T, x) = g(x)$$

$\mathcal{L}$ is the Itô generator of a multi-dimensional diffusion process and the r_i's are functions of t and x. PDE (5.52) is semilinear while PDE (5.53) is linear, with a source term which is itself the solution to a linear PDE. Note the similarity between PDEs (5.47) and (5.52).

The numerical solution to Equations (5.52) and (5.53) is a computationally very intensive task. For multi-asset portfolios, these PDEs, which suffer from the curse of dimensionality, cannot be solved with finite difference schemes. We must rely on probabilistic methods that we will introduce in Chapter 13.

PDE derivation

We first consider unilateral counterparty risk: we assume that only one counterparty, denoted with C, may default. The other counterparty, B, cannot default. We assume that B is allowed to dynamically trade in the d underlying assets $X^1, \ldots, X^d$. Additionally, in order to hedge his credit risk on C, B can trade in a default risky bond, denoted P_t^C. Furthermore, the prices of the underlyings are not altered by the counterparty default, which is modeled by a Poisson jump process. For the sake of simplicity, we consider a constant intensity. This assumption can be easily relaxed, in particular, the intensity can follow an Itô diffusion. We denote by u the value of B's *long position* in a single derivative contracted with C, given that C has not defaulted so far. In practice, netting agreements apply to the global mark-to-market value of a pool of derivative positions; u would then denote the aggregate value of these derivatives. For simplicity we assume zero repo and dividends, and a constant interest rate r. Under the local volatility model, the market is complete, and under the unique risk-neutral measure $\mathbb{Q}$ the processes $X_t = \left(X_t^1, \ldots, X_t^d \right) \in \mathbb{R}_+^d$ and P_t^C satisfy

$$\frac{dX_t^i}{X_t^i} = r\,dt + \sum_{j=1}^d \sigma_{i,j}(t, X_t^i)\,dW_t^j$$

$$\frac{dP_t^C}{P_t^C} = (r + \lambda_C)\,dt - dJ_t^C$$

with W_t a d-dimensional Brownian motion, and J_t^C a Poisson jump process with intensity λ_C. The no-arbitrage condition gives that $e^{-rt}u(t, X_t)$ is a $\mathbb{Q}$-martingale, characterized by (see Example 1.1)

$$\partial_t u + \mathcal{L}u + \lambda_C\,(\tilde{u} - u) - ru = 0$$

where $\mathcal{L}$ denotes the Itô generator of X and $\tilde{u}$ the derivative value just after the counterparty has defaulted. At the default event, $\tilde{u}$ is given by

$$\tilde{u} = RM^+ - M^-$$

with M the mark-to-market value of the derivative to be used in the unwinding of the position upon default and R the recovery rate. Indeed, if the mark-to-market value M is positive, meaning that M should be received from the counterparty, only a fraction RM will be received in case of default of C (it is usual to take $R = 0.4$). If the mark-to-market value M is negative, meaning that M should be received by the counterparty, we will pay M in case of default of C.

In the case of bilateral counterparty risk, the above PDE is replaced by

$$\partial_t u + \mathcal{L}u + \lambda_C\,(\tilde{u}_C - u) + \lambda_B\,(\tilde{u}_B - u) - ru = 0$$

with

$$\tilde{u}_B = M^+ - RM^-$$

with λ_B the issuer's default intensity. For the sake of simplicity, we focus only on the unilateral counterparty risk below.

There is an ambiguity in the market about the convention for the mark-to-market value to be settled at default. There are two natural conventions (see [68] for discussions about the relevance of these conventions): the mark-to-market of the derivative is evaluated at the time of default with provision for counterparty risk or without.

1. Provision for counterparty risk, $M = u$:

$$\partial_t u + \mathcal{L}u - (1 - R)\lambda_C u^+ - ru = 0, \quad u(T, x) = g(x) \qquad (5.54)$$

In the particular case when the payoff $g(x)$ is negative (resp. positive), the solution is given by $e^{-r(T-t)}\mathbb{E}_{t,x}[g(X_T)]$ (resp. $e^{-(r+(1-R)\lambda_C)(T-t)}\mathbb{E}_{t,x}[g(X_T)]$).

2. No provision for counterparty risk ($M(t, x) = e^{-r(T-t)}\mathbb{E}_{t,x}[g(X_T)]$):

$$\partial_t u + \mathcal{L}u + \lambda_C\left(RM^+ - M^- - u\right) - ru = 0, \quad u(T, x) = g(x)$$
$$\partial_t M + \mathcal{L}M - rM = 0, \quad M(T, x) = g(x)$$

REMARK 5.3   In the case of collateralized positions, counterparty risk applies to the variation of the mark-to-market value of the corresponding positions experienced over the time it takes to qualify a failure to pay margin as a default event—typically a few days. The nonlinearity u_t^+ should be substituted with $(u_t - u_{t-\Delta})^+$ where $\Delta > 0$ is this delay.   ☐

By proper discounting and replacing u by $-u$ for the sake of the presentation, these two PDEs can be cast into normal forms

NLPDE : $\partial_t u + \mathcal{L}u + \beta\left(u^+ - u\right) = 0$ $\qquad\qquad\qquad$ (5.55)

LPDE : $\partial_t u + \mathcal{L}u + \dfrac{\beta}{1 - R}\left((1 - R)\mathbb{E}_{t,x}[g(X_T)]^+ + R\mathbb{E}_{t,x}[g(X_T)] - u\right) = 0$

$\qquad\qquad\qquad\qquad\qquad\qquad\qquad\qquad\qquad\qquad\qquad\qquad$ (5.56)

with the terminal condition $u(T, x) = g(x)$ and with $\beta \equiv \lambda_C(1 - R) \in \mathbb{R}_+$. In Chapter 13 we will use marked branching diffusions to solve PDEs (5.55) and (5.56) in high dimension.

The computation of CVA using marked branching diffusions will be studied in detail in Section 13.4. It is interesting to note that a semilinear PDE similar to (5.55) also arises in the pricing of American options as we explain in Section 6.2.

5.9   The passport option

To end this chapter on examples of nonlinear PDEs arising in option pricing, let us consider the pricing of passport options. Such options have been introduced by Bankers Trust in the mid-1990s and studied in several articles, see for instance [139, 32, 182, 126] and the references therein. A passport option is an option on the balance of a trading account: each day, or week, or month, the holder decides to be long ($\Delta_{t_i} = +1$) or short ($\Delta_{t_i} = -1$) some index X and receives at maturity $T = t_N$ some function $F_T = g(\pi_T)$ of the portfolio value π_T, where

$$\pi_T = \sum_{i=1}^{N} \Delta_{t_{i-1}}(X_{t_i} - X_{t_{i-1}})$$

By considering the continuous time limit, the fair value of this option can be written as a stochastic control problem:

$$u(t, x, \pi) = \sup_{\Delta \in \Delta_{t,T}} \mathbb{E}^{\mathbb{Q}}[g(\pi_T)|X_t = x, \pi_t = \pi]$$

with $d\pi_t = \Delta_t \, dX_t$, and where $\Delta_{t,T}$ is the set of all adapted processes $(\Delta_s)_{t \leq s \leq T}$ such that for all $s \in [t, T]$, $\Delta_s \in \{-1, +1\}$. Assuming for the sake of simplicity a Black-Scholes model for the stock process with a volatility σ and zero interest rate, repo, and dividend yield, the HJB PDE associated to this control problem writes ($\Delta^2 = 1$)

$$\partial_t u(t, x, \pi) + \sigma^2 x^2 \sup_{\Delta \in \{-1, +1\}} \left(\frac{1}{2}\partial_x^2 + \frac{1}{2}\partial_\pi^2 + \Delta\partial_{x\pi} \right) u = 0$$

$$u(T, x, \pi) = g(\pi)$$

i.e., u is solution to the nonlinear two-dimensional parabolic PDE

$$\partial_t u(t, x, \pi) + \sigma^2 x^2 \left(\frac{1}{2}\partial_x^2 u + \frac{1}{2}\partial_\pi^2 u + |\partial_{x\pi} u| \right) = 0 \qquad (5.57)$$

$$u(T, x, \pi) = g(\pi)$$

The optimal control is $\Delta^* = \text{sgn}(\partial_{x\pi} u)$.

The pricing of the passport option will be investigated in more detail when we study the pricing of chooser options, see Sections 6.10 and 6.13.2.

Conclusion

In the rest of the book, we present various Monte Carlo methods for solving some of the nonlinear PDEs that we presented in this chapter, in a high-dimensional setup. Some of these methods are original contributions to the field, namely, regression methods and dual methods for pricing chooser options (Chapter 6), the Monte Carlo approaches for pricing in the uncertain lapse and mortality model (Chapter 8) and in the uncertain volatility model (Chapter 9), the Markovian projection method and the particle method for calibrating local stochastic volatility models to market prices of vanilla options, with or without stochastic interest rates (Chapter 11), the $a+b\lambda$ technique for building local correlation models that calibrate to market prices of vanilla options on a basket (Chapter 12), and a new stochastic representation of solutions of some nonlinear PDEs based on marked branching diffusions (Chapter 13).

5.10   Exercises

5.10.1   Transaction costs

Prove Equation (5.7).

5.10.2   Super-replication under delta constraints

By following closely the proof of Theorem 1.3, prove Theorem 5.3.

Chapter 6

Early Exercise Problems

> If you want a happy ending, that depends, of course, on where you stop your story.
>
> — Orson Welles

This chapter is dedicated to the pricing and hedging of American options. This is a typical nonlinear problem that cannot be solved by the standard Monte Carlo method. American-style options were introduced in Section 5.1. We recall the theory of the valuation of American options, and then review numerical methods, with a particular focus on Monte Carlo methods, which are unavoidable when the number of variables exceeds three.[1] We also cover chooser options, a generalization of American options in which the stopping times can be multiple. Primal and dual methods are illustrated by concrete numerical examples of option pricing: a multi-asset convertible bond, the restrikable put, and the so-called passport option.

6.1 Super-replication of American options

American options were introduced in Section 5.1. We recall that $\mathcal{T}_{tT}$ denotes the set of all stopping times τ such that $\tau \in [t, T]$ $\mathbb{P}^{\text{hist}}$-a.s.. Let us assume that we buy (or sell) and delta-hedge an American option written on m assets, say $X_t^1, \ldots, X_t^m$. Proceeding similarly as we did for European options in Section 1.1.4, we define both the buyer's price and the seller's super-replication price of an American claim. We recall that $\cdot$ denotes the usual scalar product in $\mathbb{R}^m$: $\Delta_t \cdot X_t = \sum_{i=1}^m \Delta_t^i X_t^i$.

DEFINITION 6.1 Buyer's price   *The buyer's super-replication price of an American claim at time $t \in [0, T]$ (provided the option has not been*

---

[1] Quantization methods may actually be faster up to dimension four.

exercised before t) is defined by

$$\mathcal{B}_t(F) = \sup \left\{ z \in \mathcal{F}_t \, \bigg| \, \text{there exists a stopping time } \tau \in \mathcal{T}_{tT} \text{ and an admissible} \right.$$

$$\left. \text{portfolio } \Delta \text{ such that } -D_{0t}z + \int_t^\tau \Delta_s \cdot d\tilde{X}_s + D_{0\tau} F_\tau \geq 0 \quad \mathbb{P}^{\text{hist}} - \text{a.s.} \right\}$$

DEFINITION 6.2 Seller's price   *The seller's super-replication price of an American claim at time $t \in [0, T]$ (provided the option has not been exercised before t) is defined by*

$$\mathcal{S}_t(F) = \inf \left\{ z \in \mathcal{F}_t \, \bigg| \, \text{there exists an admissible portfolio } \Delta \text{ such that} \right.$$

$$\left. \forall s \in [t, T], \quad D_{0t}z + \int_t^s \Delta_s \cdot d\tilde{X}_s - D_{0s} F_s \geq 0 \quad \mathbb{P}^{\text{hist}} - \text{a.s.} \right\} \quad (6.1)$$

REMARK 6.1   The seller's price is also equal to

$$\mathcal{S}_t(F) = \inf \left\{ z \in \mathcal{F}_t \, \bigg| \, \text{there exists an admissible portfolio } \Delta \text{ such that for} \right.$$

$$\left. \text{any stopping time } \tau \in \mathcal{T}_{tT}, \quad D_{0t}z + \int_t^\tau \Delta_s \cdot d\tilde{X}_s - D_{0\tau} F_\tau \geq 0 \quad \mathbb{P}^{\text{hist}} - \text{a.s.} \right\}$$

☐

We also introduce the concept of Snell envelope.

DEFINITION 6.3 Snell envelope   *We define the Snell envelope associated to $\mathbb{Q}$ and to the American option with actualized payoff $D_{0t}F_t$ as*

$$\mathcal{S}_t = \text{ess sup}_{\tau \in \mathcal{T}_{tT}} \, \mathbb{E}^{\mathbb{Q}}[D_{0\tau} F_\tau | \mathcal{F}_t] \quad (6.2)$$

REMARK 6.2   Throughout this book, for the sake of simplicity, we will simply denote by sup the ess sup operator, and by inf the ess inf operator.   ☐

Intuitively, the Snell envelope quantifies the maximum discounted gain from exercising the option after time t, if the option has not already been exercised, for a given ELMM. Since $\tau = t$ is a possible exercise strategy, the Snell envelope is always greater than the option discounted payout:

$$\mathcal{S}_t \geq D_{0t} F_t$$

The following theorem, analogous to Theorems 1.1 and 1.2, is crucial. It states that the valuation of an American option naturally boils down to computing a Snell envelope.

THEOREM 6.1

Assume there exists an ELMM $\mathbb{Q} \sim \mathbb{P}^{\text{hist}}$. *Then*

$$\mathcal{B}_t(F) \leq \sup_{\tau \in \mathcal{T}_{tT}} \mathbb{E}^{\mathbb{Q}}[D_{t\tau} F_\tau | \mathcal{F}_t] = D_{0t}^{-1} \mathcal{S}_t \leq \mathcal{S}_t(F)$$

In a complete market, we have

$$\mathcal{B}_t(F) = \mathcal{S}_t(F) = \sup_{\tau \in \mathcal{T}_{tT}} \mathbb{E}^{\mathbb{Q}}[D_{t\tau} F_\tau | \mathcal{F}_t] = D_{0t}^{-1} \mathcal{S}_t$$

where $\mathbb{Q}$ *denotes the unique ELMM, and the value of the American option is naturally defined as* $V_t \equiv D_{0t}^{-1} \mathcal{S}_t$.

REMARK 6.3   In an incomplete market, the habit is to pick one particular ELMM $\mathbb{Q}$ and use $V_t = D_{0t}^{-1} \mathcal{S}_t$ as the value of the option (see Section 1.1.6).   ⧄

PROOF   (a) Let z be $\mathcal{F}_t$-measurable. Suppose there exists a stopping time $\tau \in \mathcal{T}_{tT}$ and an admissible portfolio Δ such that

$$-D_{0t}z + \int_t^\tau \Delta_s \cdot d\tilde{X}_s + D_{0\tau} F_\tau \geq 0 \qquad \mathbb{P}^{\text{hist}} - \text{a.s.}$$

Since $\int_t^s \Delta_r \cdot d\tilde{X}_r, s \in [t, T]$ is a $\mathbb{Q}$-supermartingale, by the boundedness of τ,

$$\mathbb{E}^{\mathbb{Q}} \left[\int_t^\tau \Delta_s \cdot d\tilde{X}_s \bigg| \mathcal{F}_t \right] \leq 0$$

This leads to $z \leq \mathbb{E}^{\mathbb{Q}} \left[D_{t\tau} F_\tau \big| \mathcal{F}_t \right]$. Taking the supremum over all stopping times τ, we get

$$z \leq \sup_{\tau \in \mathcal{T}_{tT}} \mathbb{E}^{\mathbb{Q}} \left[D_{t\tau} F_\tau \big| \mathcal{F}_t \right]$$

Since this holds for all such z, we obtain that

$$\mathcal{B}_t(F) \leq \sup_{\tau \in \mathcal{T}_{tT}} \mathbb{E}^{\mathbb{Q}}[D_{t\tau} F_\tau \big| \mathcal{F}_t]$$

Proceeding similarly, one can show that

$$\mathcal{S}_t(F) \geq \sup_{\tau \in \mathcal{T}_{tT}} \mathbb{E}^{\mathbb{Q}}[D_{t\tau} F_\tau | \mathcal{F}_t]$$

(b) Now, assume that the market is complete. For any stopping time $\tau \in \mathcal{T}_{tT}$, the European payoff $G_T \equiv F_\tau / D_{\tau T}$ is attainable, which means by definition that there exists $z \in \mathbb{R}$ and an admissible portfolio Δ such that $(\int_0^t \Delta_s \cdot d\tilde{X}_s, t \in [0,T])$ is a true $\mathbb{Q}$-martingale and

$$-z + \int_0^T \Delta_s \cdot d\tilde{X}_s + D_{0T} G_T = 0$$

i.e.,

$$-z + \int_0^t \Delta_s \cdot d\tilde{X}_s + \int_t^T \Delta_s \cdot d\tilde{X}_s + D_{0\tau} F_\tau = 0$$

Hence we have $z - \int_0^t \Delta_s \cdot d\tilde{X}_s = \mathbb{E}^{\mathbb{Q}}[D_{0\tau} F_\tau | \mathcal{F}_t]$. As a consequence, for any stopping time $\tau \in \mathcal{T}_{tT}$, $\mathcal{B}_t(F) \geq D_{0t}^{-1} \mathbb{E}^{\mathbb{Q}}[D_{0\tau} F_\tau | \mathcal{F}_t]$, i.e.,

$$\mathcal{B}_t(F) \geq \sup_{\tau \in \mathcal{T}_{tT}} \mathbb{E}^{\mathbb{Q}} \left[D_{t\tau} F_\tau \Big| \mathcal{F}_t \right]$$

and because the reverse inequality always holds (see (a)), we get that

$$\mathcal{B}_t(F) = \sup_{\tau \in \mathcal{T}_{tT}} \mathbb{E}^{\mathbb{Q}}[D_{t\tau} F_\tau | \mathcal{F}_t]$$

(c) To conclude, we need to show that, in a complete market, if we put $z = \sup_{\tau \in \mathcal{T}_{tT}} \mathbb{E}^{\mathbb{Q}}[D_{t\tau} F_\tau | \mathcal{F}_t] = D_{0t}^{-1} \mathcal{S}_t$, then there exists an admissible portfolio Δ which super-replicates F_t, i.e., such that

$$\forall s \in [t,T], \quad D_{0t} z + \int_t^s \Delta_r d\tilde{X}_r \geq D_{0s} F_s$$

As the Snell envelope $\mathcal{S}$ is a $(\mathcal{F}_s)$-supermartingale w.r.t. $\mathbb{Q}$, then by the Doob-Meyer decomposition, we can write $\mathcal{S}_s$ as

$$\mathcal{S}_s = M_s - A_s$$

where $(M_s, s \in [t,T])$ is a $(\mathcal{F}_s)$-local martingale under $\mathbb{Q}$ with $M_t = D_{0t} z$ and $(A_s, s \in [t,T])$ is a nondecreasing process with $A_t = 0$. From the Brownian martingale representation theorem, we can write M_t as

$$M_s = D_{0t} z + \int_t^s \delta_r \, dW_r^1$$

for some $\mathcal{F}_s$-adapted process δ. Since the market is complete, the volatility of $(X^1, \ldots, X^m)$ is invertible (see Section 1.1.5) and this can be written as

$$M_s = D_{0t} z + \int_t^s \Delta_r \cdot d\tilde{X}_r$$

from which we deduce that

$$D_{0t} z + \int_t^s \Delta_r \cdot d\tilde{X}_r = \mathcal{S}_s + A_s \geq \mathcal{S}_s \geq D_{0s} F_s$$

□

6.2   American options and semilinear PDEs

In this section, for simplicity, we take zero interest rates. Assuming that the market is complete, the replication price of an American option with exercise payoff $g(x)$ is

$$u(t, x) = \sup_{\tau \in \mathcal{T}_{tT}} \mathbb{E}^{\mathbb{Q}}[g(X_\tau)|X_t = x] \qquad (6.3)$$

As explained in Section 4.5 (see also Section 6.7), it satisfies:

$$\max(\partial_t u + \mathcal{L}u, g(x) - u) = 0, \qquad u(T, x) = g(x) \qquad (6.4)$$

where $\mathcal{L}$ is the Itô generator of the diffusion X_t. Surprisingly, this variational inequality can be converted into a semilinear PDE. A similar PDE is derived in [54, 55] for the particular case of call/put options, and in the general case in [149, 45].

PROPOSITION 6.1 Semilinear PDE for American options
A function $u \in C^{1,2}(\mathbb{R}_+ \times \mathbb{R}^d)$ is a solution to the variational inequality (6.4) if and only if

$$\partial_t u(t, x) + \mathcal{L}u(t, x) + (\mathcal{L}g(x))^- \mathbf{1}_{\{g(x) = u(t,x)\}} = 0, \qquad u(T, x) = g(x) \qquad (6.5)$$

PROOF   By applying Itô's lemma (this requires some regularity on u and could be relaxed if we work with viscosity solutions):

$$g(X_T) = u(t, X_t) + \int_t^T (\partial_t + \mathcal{L})u(s, X_s)\, ds + \int_t^T Du(s, X_s)\, dX_s$$

By separating the American value $u(t, X_t)$ into the exercise and continuation regions, we deduce

$$g(X_T) = u(t, X_t) + \int_t^T Du(s, X_s)\, dX_s +$$

$$\int_t^T \left((\partial_s u + \mathcal{L}u)(s, X_s)\mathbf{1}_{\{u(s,X_s)>g(X_s)\}} + (\partial_s u + \mathcal{L}u)(s, X_s)\mathbf{1}_{\{u(s,X_s)=g(X_s)\}}\right) ds$$

As from the variational inequality (6.4) $\partial_t u + \mathcal{L}u = 0$ in the continuation region $\{u > g\}$, we obtain

$$g(X_T) = u(t, X_t) + \int_t^T Du(s, X_s)\, dX_s$$

$$+ \int_t^T (\partial_s u + \mathcal{L}u)(s, X_s)\mathbf{1}_{\{u(s,X_s)=g(X_s)\}}\, ds$$

Finally, by taking the expectation, and assuming that $\int_0^t Du(s, X_s)\, dX_s$ is a true $\mathbb{Q}$-martingale, we get

$$u(t, X_t) = \mathbb{E}^{\mathbb{Q}}[g(X_T)|\mathcal{F}_t] - \int_t^T \mathbb{E}^{\mathbb{Q}}[(\partial_s u + \mathcal{L}u)(s, X_s)\mathbf{1}_{\{u(s,X_s)=g(X_s)\}}|\mathcal{F}_t]\, ds$$

As $\partial_t u + \mathcal{L}u \le 0$,

$$-(\partial_t u + \mathcal{L}u)\mathbf{1}_{\{u(t,x)=g(x)\}} = (\partial_t u + \mathcal{L}u)^- \mathbf{1}_{\{u(t,x)=g(x)\}}$$
$$= (\mathcal{L}g(x))^- \mathbf{1}_{\{u(t,x)=g(x)\}}$$

Here we have assumed that the exercise domain $\mathcal{D} = \{(t, x) \mid u(t, x) = g(x)\}$ is an open set so that at the points where $u = g$ we also have $\mathcal{L}u = \mathcal{L}g$, and we have disregarded the discontinuity of $\partial_x^2 u$ at the boundary $\partial\mathcal{D}$. Leaving apart these technical difficulties, we obtain

$$u(t, X_t) = \mathbb{E}^{\mathbb{Q}}[g(X_T)|\mathcal{F}_t] + \mathbb{E}^{\mathbb{Q}}\left[\int_t^T (\mathcal{L}g(X_s))^- \mathbf{1}_{\{u(s,X_s)=g(X_s)\}}\, ds\middle|\mathcal{F}_t\right]$$

We complete the proof using Feynman-Kac's formula.   □

In the case of non-zero interest rates, PDE (6.5) reads

$$\partial_t u(t, x) + \mathcal{L}u(t, x) - r(t, x)u(t, x) + (\mathcal{L}g(x) - r(t, x)g(x))^- \mathbf{1}_{\{g(x)=u(t,x)\}} = 0$$
$$u(T, x) = g(x) \quad (6.6)$$

This PDE leads to an early exercise premium formula for general American options. The reader should convince himself or herself that the above semi-linear PDE ensures that $u(t, x) \ge g(t, x)$. By definition, this is the case in the continuation region $u(t, x) > g(t, x)$. When the solution enters into the exercise region $u(t, x) = g(t, x)$, then a positive reaction term forces the solution to stay in this region or to go back to the continuation domain.

REMARK 6.4 Penalty approximation   PDE (6.5) differs from (and is in fact much simpler than) the semilinear PDE obtained by penalizing solutions u that go below g (see [102], Section 6)

$$\partial_t u^\omega + \mathcal{L}u^\omega + \omega(g - u^\omega)^+ = 0$$

with ω large (one can prove that $\lim_{\omega \to \infty} u^\omega = u$).   □

Example 6.1 American call option on a dividend-paying asset
We have $g(x) = (x - K)^+$, $\mathcal{L} = \frac{1}{2}\sigma^2(t, x)\partial_x^2 + (r - q)x\partial_x$; q corresponds to a repo or proportional dividend. We obtain from (6.6) the well known early

exercise premium formula [73] for call options:

$$u(t, X_t) \equiv \sup_{\tau \in [t,T]} \mathbb{E}^{\mathbb{Q}}[e^{-r(\tau-t)}(X_\tau - K)^+ | \mathcal{F}_t] = \mathbb{E}^{\mathbb{Q}}[e^{-r(T-t)}(X_T - K)^+ | \mathcal{F}_t]$$

$$+ \int_t^T e^{-r(s-t)} \mathbb{E}^{\mathbb{Q}}[(rK - qX_s)^- \mathbf{1}_{\{u(s,X_s)=g(X_s)\}} | \mathcal{F}_t] \, ds$$

$$\square$$

COROLLARY 6.1 Explicit Doob-Meyer decomposition
The process $u(t, X_t) \equiv \sup_{\tau \in [t,T]} \mathbb{E}^{\mathbb{Q}}[g(X_\tau)|\mathcal{F}_t]$ is a càdlàg supermartingale that admits the explicit Doob-Meyer decomposition:

$$u(t, X_t) = M_t - A_t$$

where the martingale M_t is given by

$$M_t = \mathbb{E}^{\mathbb{Q}}\left[g(X_T) + \int_0^T (\mathcal{L}g(X_s))^- \mathbf{1}_{\{u(s,X_s)=g(X_s)\}} \, ds \middle| \mathcal{F}_t \right]$$

and the nondecreasing process A_t, vanishing at time 0, is given by

$$A_t = \int_0^t (\mathcal{L}g(X_s))^- \mathbf{1}_{\{u(s,X_s)=g(X_s)\}} \, ds$$

6.3 The dual method for American options

In this section, we present a second way of valuing American options. It is called a dual method, because the maximization problem (6.2) is transformed into a minimization problem (see also Theorems 1.3 and 5.3). It was simultaneously derived by Rogers [178] and by Haugh and Kogan in the early 2000s. A multiplicative version of this result was derived by Jamshidian [143]; see also [78].

THEOREM 6.2 [178], [125]
We have

$$\sup_{\tau \in \mathcal{T}_{tT}} \mathbb{E}^{\mathbb{Q}}\left[D_{t\tau} F_\tau \middle| \mathcal{F}_t \right] = \inf_{M \in \mathcal{M}_{t,0}} \mathbb{E}^{\mathbb{Q}}\left[\sup_{t \le s \le T} (D_{ts} F_s - M_s) \middle| \mathcal{F}_t \right]$$

where $\mathcal{M}_{t,0}$ denotes the set of all right-continuous martingales $(M_s, s \in [t, T])$ with $M_t = 0$. Moreover, the optimal martingale M^* is the martingale part of *the Doob-Meyer decomposition of the supermartingale $(\mathcal{S}_s, s \in [t, T])$, minus its value at time t.*

PROOF   Let $M \in \mathcal{M}_{t,0}$. For any stopping time $\tau \in \mathcal{T}_{tT}$, τ is bounded, the optional sampling theorem applies and we have

$$\mathbb{E}^{\mathbb{Q}}[D_{t\tau}F_\tau|\mathcal{F}_t] = \mathbb{E}^{\mathbb{Q}}[D_{t\tau}F_\tau - M_\tau|\mathcal{F}_t] \leq \mathbb{E}^{\mathbb{Q}}\left[\sup_{t \leq s \leq T}(D_{ts}F_s - M_s)\Big|\mathcal{F}_t\right]$$

Therefore

$$\sup_{\tau \in \mathcal{T}_{tT}} \mathbb{E}^{\mathbb{Q}}[D_{t\tau}F_\tau|\mathcal{F}_t] \leq \inf_{M \in \mathcal{M}_{t,0}} \mathbb{E}^{\mathbb{Q}}\left[\sup_{t \leq s \leq T}(D_{ts}F_s - M_s)\Big|\mathcal{F}_t\right] \qquad (6.7)$$

On the other hand, the Snell envelope, defined by (6.2), is a supermartingale and admits a Doob-Meyer decomposition[2]

$$\mathcal{S}_s = M_s - A_s$$

where $(M_s, s \in [t, T])$ is a right-continuous $\mathbb{Q}$-martingale and $(A_s, s \in [t, T])$ a nondecreasing process, with $A_t = 0$. The process $M^* = M - M_t$ belongs to $\mathcal{M}_{t,0}$, and for all $s \in [t, T]$,

$$D_{ts}F_s - M_s^* = D_{ts}F_s + M_t - \mathcal{S}_s - A_s \leq M_t = \mathcal{S}_t$$

because A_s is nonnegative and $\mathcal{S}_s \geq D_{ts}F_s$. As a consequence,

$$\mathbb{E}^{\mathbb{Q}}\left[\sup_{t \leq s \leq T}(D_{ts}F_s - M_s^*)\Big|\mathcal{F}_t\right] \leq \mathcal{S}_t = \sup_{\tau \in \mathcal{T}_{tT}} \mathbb{E}^{\mathbb{Q}}[D_{t\tau}F_\tau|\mathcal{F}_t]$$

which proves that we have equality in (6.7) and that the infimum is achieved by taking $M = M^*$.   ☐

Approximating numerically the Snell envelope, i.e., the optimal stopping time, leads to a lower bound for the price of the American option, whereas approximating the optimal martingale $M^* - M_0$ provides an upper bound. Computing both bounds is essential: one single bound gives a much too incomplete piece of information. For this reason, dual methods are crucial in optimal stopping problems.

REMARK 6.5   As explained by Joshi [145], M can be seen as a hedging strategy, and Theorem 6.2 states that by investing the buyer's price, the seller can hedge in such a way as to cover the payoff even if the buyer exercises at the maximum.   ☐

---

[2]The process $D_{0t}F_t$ should be of class DL in order to apply the Doob-Meyer result; see e.g., [13].

6.4   On the ownership of the exercise right

So far, we have considered American options in which it is the holder of the option who owns the exercise right, i.e., the right of choosing the most convenient stopping time. The case also exists where it is the issuer of the option who owns the exercise right. We speak of *putable options* in the first case, and of *callable options* in the second one. The results of Sections 6.1 and 6.3 concerning putable options, are easily adapted to the case of callable options. The price of a callable American option in a complete market is given by

$$\inf_{\tau \in \mathcal{T}_{tT}} \mathbb{E}^{\mathbb{Q}} \left[D_{t\tau} F_{\tau} \Big| \mathcal{F}_t \right]$$

where $\mathbb{Q}$ is the unique ELMM. Since

$$\inf_{\theta} f(\theta) = -\sup_{\theta}(-f(\theta)) \tag{6.8}$$

it is enough to consider only the case of putable options: if the option is actually callable, reverse the sign of the payoff F, price this new American option as if it were putable, and eventually reverse the sign of the price. The general equality (6.8) also shows that, in the case of callable options, the dual method consists of computing

$$\sup_{M \in \mathcal{M}_{t,0}} \mathbb{E}^{\mathbb{Q}} \left[\inf_{t \leq s \leq T} (D_{ts} F_s - M_s) \Big| \mathcal{F}_t \right]$$

6.5   On the finiteness of exercise dates

Often, in practice, American options can be exercised only at discrete dates $t_1 < \cdots < t_N = T$. They are then described by the finite collection of random variables $F_{t_1}, \ldots, F_{t_N}$, with F_{t_i} being $\mathcal{F}_{t_i}$-measurable. Such options are sometimes called *Bermudan* because, as the exercise rights are limited, they look slightly more European than their "purely" American counterparts, and the Bermuda Islands are located between America and Europe. However, pricing methods are similar for American and Bermudan options, and strongly differ from the pricing methods for European options.[3] The finiteness of exercise dates does indeed not matter all that much, so we will make no semantic difference between American and Bermudan options, and will always use the term *American.*

---

[3]This is in line with the Bermuda Islands being much closer to the American seashore than to the European coast.

In the case where the number of exercise dates is finite, the results of Sections 6.1 and 6.3 are easily adapted as follows. For a set of dates D, we denote by $\mathcal{T}_D$ the set of all stopping times τ such that, $\mathbb{P}^{\text{hist}}$−a.s., $\tau \in D$. For instance, $\mathcal{T}_{\{t_1,t_2,\dots,t_N\}\cap[t,T]}$ is the set of all stopping times τ such that, $\mathbb{P}^{\text{hist}}$−a.s., $\tau \in \{t_1,t_2,\dots,t_N\} \cap [t,T]$. Define the buyer's price by replacing $\mathcal{T}_{tT}$ by $\mathcal{T}_{\{t_1,t_2,\dots,t_N\}\cap[t,T]}$ in (6.1), define the seller's price by replacing $\forall s \in [t,T]$ by $\forall s \in \{t_1,t_2,\dots,t_N\} \cap [t,T]$ in (6.1), and define the Snell envelope by replacing $\mathcal{T}_{tT}$ by $\mathcal{T}_{\{t_1,t_2,\dots,t_N\}\cap[t,T]}$ in (6.2). Assume there exists a local martingale measure $\mathbb{Q} \sim \mathbb{P}^{\text{hist}}$. Then

$$\mathcal{B}_t(F) \leq \sup_{\tau \in \mathcal{T}_{\{t_1,t_2,\dots,t_N\}\cap[t,T]}} \mathbb{E}^{\mathbb{Q}}[D_{t\tau}F_\tau|\mathcal{F}_t] \leq \mathcal{S}_t(F)$$

In a complete market, we have

$$\mathcal{B}_t(F) = \mathcal{S}_t(F) = \sup_{\tau \in \mathcal{T}_{\{t_1,t_2,\dots,t_N\}\cap[t,T]}} \mathbb{E}^{\mathbb{Q}}[D_{t\tau}F_\tau|\mathcal{F}_t] = D_{0t}^{-1}\mathcal{S}_t \equiv V_t \quad (6.9)$$

In particular, the value of the option at time t_i (if it has not been exercised before time t_i) can then be written as

$$V_{t_i} = \sup_{\tau \in \mathcal{T}_{\{t_i,\dots,t_N\}}} \mathbb{E}^{\mathbb{Q}}[D_{t_i\tau}F_\tau|\mathcal{F}_{t_i}], \quad (6.10)$$

In the dual manner,

$$V_t = \inf_{M \in \mathcal{M}_{0,t}} \mathbb{E}^{\mathbb{Q}}\left[\sup_{t \leq t_i \leq T}(D_{tt_i}F_{t_i} - M_{t_i})\Big|\mathcal{F}_t\right] \quad (6.11)$$

where $\mathcal{M}_{0,t}$ now stands for the set of all martingales defined at the discrete times t and $t_i \in (t,T]$, with $M_t = 0$. In this case, the optimal martingale M^* simply reads

$$M_t^* = 0$$
$$M_{t_i}^* - M_{t_{i-1}\vee t}^* = \mathcal{S}_{t_i} - \mathbb{E}^{\mathbb{Q}}[\mathcal{S}_{t_i}|\mathcal{F}_{t_{i-1}\vee t}], \quad t_i \in (t,T]$$

Note that $t_{i-1} \vee t = \max(t_{i-1},t)$ is worth t_{i-1}, except for the first date t_i such that $t_i > t$. Since $\mathcal{S}_t = D_{0t}V_t$, the second equation also reads

$$M_{t_i}^* - M_{t_{i-1}\vee t}^* = D_{0t_i}V_{t_i} - \mathbb{E}^{\mathbb{Q}}[D_{0t_i}V_{t_i}|\mathcal{F}_{t_{i-1}\vee t}], \quad t_i \in (t,T] \quad (6.12)$$

In the case of finite exercise dates, given an ELMM $\mathbb{Q}$, the value V_{t_i} of the American option at time t_i (if it has not been exercised before time t_i) is solution to the following dynamic programming problem

$$\begin{cases} V_{t_N} = F_{t_N} \\ V_{t_i} = \max\left(F_{t_i}, \mathbb{E}^{\mathbb{Q}}[D_{t_i t_{i+1}}V_{t_{i+1}}|\mathcal{F}_{t_i}]\right) \end{cases} \quad (6.13)$$

The continuation value $C_{t_i} = \mathbb{E}^{\mathbb{Q}}[D_{t_i t_{i+1}} V_{t_{i+1}} | \mathcal{F}_{t_i}]$ satisfies:

$$\begin{cases} C_{t_N} = -\infty \\ C_{t_i} = \mathbb{E}^{\mathbb{Q}}[D_{t_i t_{i+1}} \max(F_{t_{i+1}}, C_{t_{i+1}}) | \mathcal{F}_{t_i}] \end{cases} \tag{6.14}$$

Indeed, at date t_i, the holder chooses between exercising, which pays him F_{t_i}, and holding the American option after t_i, i.e., being rich of the continuation value $C_{t_i} = \mathbb{E}^{\mathbb{Q}}[D_{t_i t_{i+1}} V_{t_{i+1}} | \mathcal{F}_{t_i}]$. This recursive procedure provides a constructive way to build the optimal stopping time in (6.10): if the option has not been exercised before t_i, an optimal stopping time τ_i^* is

$$\tau_i^* = \inf\{t_j \geq t_i \,|\, V_{t_j} = F_{t_j}\} \tag{6.15}$$

In particular, at inception, an optimal stopping time is

$$\tau^* = \tau_1^* = \inf\{t_i \,|\, V_{t_i} = F_{t_i}\} \tag{6.16}$$

6.6 On the accounting of multiple coupons

Often, in practice, the set of exercise dates is finite and the option's holder receives some continuation coupon $C_{t_i}^c$ at time t_i if he decides not to exercise at that time, and the exercise coupon $C_{t_i}^e$ if otherwise. This situation can be cast into our general framework by defining

$$F_{t_i} = \sum_{j<i} D_{t_i t_j} C_{t_j}^c + C_{t_i}^e$$

That is, one must capitalize the past coupons $C_{t_j}^c$ and add them to the exercise coupon $C_{t_i}^e$ to get the equivalent F_{t_i}. As $t_j < t_i$,

$$D_{t_i t_j} = \exp\left(\int_{t_j}^{t_i} r_s \, ds\right) \geq 1$$

is a capitalization factor. In this case, V_{t_i} represents the value of holding the option from inception at time t_i, which differs from the resale value at time t_i. The resale value $V_{t_i}^r$ does not take the coupons paid before t_i into account:

$$V_{t_i}^r = V_{t_i} - \sum_{j<i} D_{t_i t_j} C_{t_j}^c$$

In the case where no coupon is served if the option's holder does not exercise, all $C_{t_i}^c$'s are zero and $V = V^r$: there is no need to distinguish between the value of holding the option and the resale value of the option.

Equations (6.13)–(6.16) are modified as follows: V_{t_i}, the value of holding the option from inception at time t_i, satisfies the dynamic programming problem

$$\begin{cases} V_{t_N} = \sum_{j<N} D_{t_N t_j} C^c_{t_j} + C^e_{t_N} \\ V_{t_i} = \max \left(\sum_{j<i} D_{t_i t_j} C^c_{t_j} + C^e_{t_i}, \mathbb{E}^{\mathbb{Q}}[D_{t_i t_{i+1}} V_{t_{i+1}} | \mathcal{F}_{t_i}] \right) \end{cases} \quad (6.17)$$

In terms of the resale value $V^r_{t_i}$, this reads

$$\begin{cases} V^r_{t_N} = C^e_{t_N} \\ V^r_{t_i} = \max \left(C^e_{t_i}, C^c_{t_i} + \mathbb{E}^{\mathbb{Q}}[D_{t_i t_{i+1}} V^r_{t_{i+1}} | \mathcal{F}_{t_i}] \right) \end{cases} \quad (6.18)$$

The corresponding continuation value $C_{t_i} = \mathbb{E}^{\mathbb{Q}}[D_{t_i t_{i+1}} V_{t_{i+1}} | \mathcal{F}_{t_i}]$ satisfies

$$\begin{cases} C_{t_N} = -\infty \\ C_{t_i} = \mathbb{E}^{\mathbb{Q}} \left[D_{t_i t_{i+1}} \max \left(\sum_{j<i+1} D_{t_{i+1} t_j} C^c_{t_j} + C^e_{t_{i+1}}, C_{t_{i+1}} \right) \Big| \mathcal{F}_{t_i} \right] \end{cases} \quad (6.19)$$

The resale continuation value is the continuation value less the capitalized past coupons:

$$C^r_{t_i} = C_{t_i} - \sum_{j<i} D_{t_i t_j} C^c_{t_j} = C^c_{t_i} + \mathbb{E}^{\mathbb{Q}}[D_{t_i t_{i+1}} V^r_{t_{i+1}} | \mathcal{F}_{t_i}]$$

In terms of the resale continuation value, (6.19) reads

$$\begin{cases} C^r_{t_N} = -\infty \\ C^r_{t_i} = C^c_{t_i} + \mathbb{E}^{\mathbb{Q}}[D_{t_i t_{i+1}} \max(C^e_{t_{i+1}}, C^r_{t_{i+1}}) | \mathcal{F}_{t_i}] \end{cases} \quad (6.20)$$

If the option has not been exercised before t_i, an optimal stopping time τ^*_i is

$$\tau^*_i = \inf \left\{ t_j \geq t_i \, \Big| \, V_{t_j} = \sum_{k<j} D_{t_j t_k} C^c_{t_k} + C^e_{t_j} \right\}$$

$$= \inf \left\{ t_j \geq t_i \, \Big| \, V^r_{t_j} = C^e_{t_j} \right\}$$

At inception, an optimal stopping time is

$$\tau^* = \tau^*_1 = \inf \left\{ t_i \, \Big| \, V_{t_i} = \sum_{j<i} D_{t_i t_j} C^c_{t_j} + C^e_{t_i} \right\}$$

$$= \inf \left\{ t_i \, | \, V^r_{t_i} = C^e_{t_i} \right\}$$

REMARK 6.6 When using the dual method (6.12), one must not forget to capitalize the past coupons when defining the optimal martingale M^*. To be precise, if, in order to get an estimate of the optimal martingale M^*, one approximates $V_{t_i} \simeq \mathbb{E}^{\mathbb{Q}}[D_{t_i \bar{\tau}} F_{\bar{\tau}} | \mathcal{F}_{t_i}]$ using some nearly optimal exercise strategy $\bar{\tau}$, then the approximate V_{t_i} must include all (capitalized) coupons paid before $\bar{\tau}$, *even those delivered before t_i*. This is because V_{t_i} represents the value of holding the option until time t_i, and not the resale value at time t_i. $\qquad \square$

6.7 Finite difference methods for American options

To introduce finite difference methods for American options, it is convenient to first consider the case where the number of exercise dates is finite, say $t_1 < \cdots < t_N = T$. In the case where the American payout is vanilla, i.e., $F_{t_i} = g(X_{t_i})$, the price of the American option at time t is a function of (t, X_t), say $u(t, X_t)$. Obviously, $u(T, x) = g(x)$. We now proceed backward. For $t \in (t_i, t_{i+1})$, assuming zero interest rates for the sake of simplicity,

$$\partial_t u(t, x) + \mathcal{L}u(t, x) = 0 \qquad (6.21)$$

where $\mathcal{L}$ denotes the infinitesimal generator of the process X, because no exercise is allowed between t_i and t_{i+1}. At date t_i, the holder chooses between exercising, which pays him $g(X_{t_i})$, and holding the American option after t_i, i.e., being rich of $u(t_i^+, X_{t_i})$, where $u(t_i^+, x)$ is the solution at time t_i of PDE (6.21) on (t_i, t_{i+1}). Hence, the American option at time t_i is worth

$$u(t_i, x) = \max(u(t_i^+, x), g(x))$$

and $u(t_i, x)$ is the terminal condition for PDE (6.21) to be solved on the next interval (t_{i-1}, t_i). Note that $u(t_i, x)$ is always greater than or equal to $g(x)$. Moreover, the PDE solver provides an optimal strategy given by

$$\tau^* = \inf\{s \in \{t_1, \ldots, t_N\} \cap [t, T] \,|\, u(s, X_s) = g(X_s)\}$$

Indeed, for all $s \in \{t_1, \ldots, t_N\}$ such that $s < \tau^*$, $u(s, X_s) > g(X_s)$, which means that the option's holder should not exercise.

In the continuous time limit, this sequence of PDEs naturally becomes a variational inequality, namely

$$\max\left(\partial_t u(t, x) + \mathcal{L}u(t, x), g(x) - u(t, x)\right) = 0$$

with terminal condition $u(T, x) = g(x)$. This equation simply means that, when we solve PDE (6.21) backward between t and $t - dt$, for each x only two cases occur:

- if we get a result $u(t - dt, x)$ which is smaller than $g(x)$, we replace $u(t - dt, x)$ by $g(x)$, and in this situation $\partial_t u(t, x) + \mathcal{L}u(t, x) \leq 0$ because the new $u(t - dt, x)$ is greater than the old one;

- otherwise, we keep the result $u(t - dt, x)$, and $\partial_t u(t, x) + \mathcal{L}u(t, x) = 0$.

In the case of non-zero interest rates $r(t, X_t)$, the variational inequality reads

$$\max\left(\partial_t u(t, x) + \mathcal{L}u(t, x) - r(t, x)u(t, x), g(x) - u(t, x)\right) = 0$$

The optimal stopping time is given by

$$\tau^* = \inf\{s \geq t \,|\, u(s, X_s) = g(X_s)\}$$

The case of path-dependent options requires that one adds the relevant path-dependent variables to X (see Remark 1.2).

6.8 Monte Carlo methods for American options

Finite difference methods look just perfect. Not only do they provide the price of the American option at inception $(t = 0)$, but they also give the price $u(t, x)$ at any future date $t > 0$, as well as the delta $\partial_x u(t, x)$ and the gamma $\partial_x^2 u(t, x)$, i.e., they provide the price and the hedge at any date t and in any market condition x.

Unfortunately, they suffer from the curse of dimensionality, which means that, when the number of variables, be they assets or path-dependent variables, exceeds three, the computational time becomes prohibitive. This is the usual situation in financial markets, where multi-asset options are common, and where one must often model stochastic volatility and/or stochastic interest rates, in addition to the underlying itself, or to the underlyings themselves. One must then turn to simulation-based methods, which only depend minimally on dimensionality. Such methods are also known as Monte Carlo methods (see Chapter 2 for a reminder on those methods). Their application to the valuation of American contracts has given rise to a huge literature, and we review below a short selection of what we consider the most efficient Monte Carlo algorithms.

6.8.1 Why is it difficult?

Monte Carlo methods for valuation of American options are most easily explained in the context of a finite number of exercise dates $t_1 < \cdots < t_N$. For the case of continuous exercise dates, see Section 6.8.9. The value V_0 of the option at time $t = 0$ can then be written as (6.9), i.e., as a supremum over stopping times, or as (6.11), i.e., as an infimum over martingales. In a naive approach, one can always try to pick a stopping time (resp. a martingale), estimate the objective function to maximize (resp. minimize) by simulating paths and computing empirical averages, and then maximize over the set of stopping times (resp. over the set of martingales). Unfortunately, these sets are very large, the supremum and the infimum of the primal and dual formulations are thus very hard to evaluate through classical optimization procedures, and such an approach would consume a huge amount of computational time.[4]

Equations (6.16) and (6.12) provide expressions of an optimal stopping time and an optimal martingale. However, these expressions depend on the futures values V_{t_i} of the option. Of course, these futures values are unknown: we are precisely looking for the price of the option! The Monte Carlo valuation of American contracts, which necessitates the estimation of the optimal stopping strategy, or dually, the estimation of the optimal martingale, hence requires that one be able to estimate all the future values V_{t_i} in all possible model scenarios. This statement does not look encouraging at all! Note that the classical Monte Carlo method, used in the pricing of European options, does not provide information on the values of the option at future dates and in all possible future model scenarios. It only gives the value of the option at the current date $t = 0$. This is what makes the pricing of American options by simulation a difficult problem.

However, we can look at Equations (6.16) and (6.12) in a positive way: they tell us that if we are able to build estimates of the futures values V_{t_j}, $j > i$, then we will obtain estimates of the optimal stopping time and of the optimal martingale. A crucial observation, exploited by Longstaff and Schwartz [157] for the primal problem (maximizing over all stopping times), and by Andersen and Broadie [34] for the dual problem (minimizing over the martingales), is that even inaccurate estimates of the future values V_{t_i} can lead to accurate estimates of the optimal stopping time τ^* and of the optimal martingale M^*. Before looking at the corresponding algorithms, let us first describe the Tsitsiklis-Van Roy algorithm. Here we do not account for multiple coupons (see Remark 6.10)

6.8.2 Estimating V_{t_i}: The Tsitsiklis-Van Roy algorithm

A very naive way to estimate V_{t_i} is to launch nested Monte Carlo procedures. Namely, at date t_i, for each path, one simulates subpaths until t_{i+1}. For each subpath, one must compare $F_{t_{i+1}}$ and $V_{t_{i+1}}$. To get an estimate of $V_{t_{i+1}}$, one then simulates subsubpaths until t_{i+2}, and so on and so forth. This naive procedure is explosive and too time consuming.

A much more clever approach due to Tsitsiklis and Van Roy [191] consists of exploiting the dynamic programming Equations (6.13). In this approach, one simulates paths until maturity $T = t_N$, and then approximates on each path ω the value $V_{t_i}(\omega)$ by $\hat{V}_{t_i}(\omega)$, where $\hat{V}_{t_N} = F_{t_N}$ and, for $i = N-1, N-2, \ldots, 1$,

$$\hat{V}_{t_i} = \max(F_{t_i}, \hat{C}_{t_i})$$
$$\hat{C}_{t_i} = \mathbb{E}^{\mathbb{Q}}[D_{t_i t_{i+1}} \hat{V}_{t_{i+1}} | \mathcal{F}_{t_i}] \tag{6.22}$$

---

[4]However, it is always possible to optimize over a set of suitably parameterized stopping times or martingales; see Section 6.8.6.

Of course, the conditional expectations cannot be computed exactly, and are replaced by suitable estimates. There are many ways to approximate the conditional expectations, some of which we will review in Section 6.8.5.

Tsitsiklis and Van Roy do not use their estimation of the values V_{t_i} to get an estimate of the optimal strategy τ^*, nor to get an estimate of the optimal martingale M^*.[5] They directly estimate V_0 by

$$\hat{V}_0 = \mathbb{E}^{\mathbb{Q}}[D_{0t_1}\hat{V}_{t_1}]$$

where $\mathbb{E}^{\mathbb{Q}}$ is replaced by the empirical average over all simulated paths.

It is very hard to get a precise estimate using this algorithm. Indeed, it is very difficult to build an estimate $\hat{C}_{t_i}$ of the value of conditional expectation $C_{t_i} = \mathbb{E}^{\mathbb{Q}}[D_{t_i t_{i+1}} V_{t_{i+1}}|\mathcal{F}_{t_i}]$ that is accurate for all ω, i.e., in all possible model scenarios at time t_i. The error on $C_{t_i}(\omega)$ propagates because the next conditional expectation to compute is $\hat{C}_{t_{i-1}} = \mathbb{E}^{\mathbb{Q}}[D_{t_{i-1}t_i} \max(F_{t_i}, \hat{C}_{t_i})|\mathcal{F}_{t_{i-1}}]$.

REMARK 6.7   It often happens that $\hat{C}_{t_i}(\omega)$ does not approximate accurately $C_{t_i}(\omega)$ for all ω, but that $\{\omega \,|\, F_{t_i}(\omega) \geq C_{t_i}(\omega)\}$ is well approximated by $\{\omega \,|\, F_{t_i}(\omega) \geq \hat{C}_{t_i}(\omega)\}$. Stated otherwise, $\hat{C}_{t_i}(\omega)$ may be far from $C_{t_i}(\omega)$ on vast zones of ω, but the exercise region at time t_i may be very well captured. Now, as far as the pricing of American options is concerned, one does not really care how much $C_{t_i}(\omega)$ precisely is for each model scenario ω: one only needs to know whether $C_{t_i}(\omega)$ is smaller or greater than $F_{t_i}(\omega)$. That is the basic idea behind the Longstaff-Schwartz algorithm, which we now describe. □

6.8.3   Estimating τ_i^*: The Longstaff-Schwartz algorithm

Longstaff and Schwartz [157] concentrate on the estimation of the optimal stopping times τ_i^*; see (6.15). Since

$$\tau_i^* = \inf\{t_j \geq t_i \,|\, C_{t_j} \leq F_{t_j}\} \tag{6.23}$$

they build estimates $\hat{\tau}_i$ and $\hat{C}_{t_i}$ in the following way: $\hat{\tau}_N = t_N$ and, for $i = N-1, \ldots, 1$,

$$\hat{C}_{t_i} = \mathbb{E}^{\mathbb{Q}}[D_{t_i \hat{\tau}_{i+1}} F_{\hat{\tau}_{i+1}}|\mathcal{F}_{t_i}] \tag{6.24}$$

$$\hat{\tau}_i = \inf\{t_j \geq t_i \,|\, \hat{C}_{t_j} \leq F_{t_j}\} \tag{6.25}$$

Here again, conditional expectations are replaced by suitable approximations (see Section 6.8.5). Note that

$$\hat{\tau}_i = \begin{cases} t_i & \text{if } \hat{C}_{t_i} \leq F_{t_i} \\ \hat{\tau}_{i+1} & \text{otherwise} \end{cases} \tag{6.26}$$

---

[5]The authors had probably no idea of the dual formulation, which was obtained years after they published their work.

Eventually, the authors estimate the optimal strategy by $\hat{\tau} = \hat{\tau}_1$ and the price V_0 by

$$\hat{V}_0 = \mathbb{E}^{\mathbb{Q}}[D_{0\hat{\tau}} F_{\hat{\tau}}] \qquad (6.27)$$

where $\mathbb{E}^{\mathbb{Q}}$ is replaced by the empirical average over all simulated paths.

The estimated continuation value (6.24) differs from its counterpart (6.22) in the Tsitsiklis-Van Roy algorithm: the random variable inside the conditional expectation is the actualized future (nearly) optimal cash flow itself, rather than the actualized approximate value of the option at the next date t_{i+1}. Stated otherwise, Longstaff and Schwartz approximate the continuation value C_{t_i} only to get an estimate of the optimal strategy τ_i^*. They are not interested in the absolute level of the continuation value, but only in its relative level compared to the exercise value. As explained in Remark 6.7, most of the time, the Longstaff-Schwartz estimate is more accurate than the Tsitsiklis-Van Roy estimate.

REMARK 6.8 Low-biased pricing requires a two-step procedure
A subtle point, which the authors seem to have missed in their original paper [157], is the following. Since they use values of asset paths at dates greater than t_i when they estimate $\hat{C}_{t_i} = \mathbb{E}^{\mathbb{Q}}[D_{t_i\hat{\tau}_{i+1}} F_{\hat{\tau}_{i+1}} | \mathcal{F}_{t_i}]$, the resulting $\hat{\tau}_i$ is not a $(\mathcal{F}_{t_i})$-stopping time; for a path ω, $\hat{\tau}_i(\omega)$ contains some information on the future of ω. For this reason, $\hat{V}_0$ may not be a low-biased estimate of V_0. For instance, implementing the Longstaff-Schwartz algorithm to price an American call on an asset that pays no dividend, leads to a figure which is greater than the true price, which is known to be the value of the European call in this case. To get a low-biased estimate, one may keep in memory the (nearly optimal) strategy $\hat{\tau}$ resulting from the above procedure with $\#p_1$ paths, and then, in a second step, draw $\#p_2$ new *independent* paths that are stopped according to $\hat{\tau}$. Usually $\#p_1$ can be much smaller than $\#p_2$. ⬚

REMARK 6.9 On the convergence of the Longstaff-Schwartz algorithm In their original paper [157], Longstaff and Schwartz give few details of the convergence of their algorithm. Clement *et al.* [83] carry out a detailed convergence analysis. They show convergence of the regression approximation to the true price and convergence of the Monte Carlo procedure for a fixed number of basis functions. Glasserman and Yu [115] and Stentoft [188] have studied situations in which the number of basis functions and the number of simulation paths increase together. More recent papers include [114] and [193]. The former extends the results by Glasserman and Yu to several Lévy models, while the latter analyzes the convergence properties of the Longstaff-Schwartz algorithm for bounded approximating sets. ⬚

REMARK 6.10 On the accounting of multiple coupons Consider the usual case where the option's holder receives some continuation coupon $C_{t_i}^c$ at time t_i if he decides not to exercise at that time, and the exercise coupon $C_{t_i}^e$ otherwise. In this case, following Section 6.6, Equations (6.23)–(6.27) become

$$\tau_i^* = \inf\{t_j \geq t_i \,|\, C_{t_j}^r \leq C_{t_j}^e\} \tag{6.28}$$

$$\hat{C}_{t_i}^r = C_{t_i}^c + \mathbb{E}^\mathbb{Q}\left[\sum_{i<j<\hat{\tau}_{i+1}} D_{t_i t_j} C_{t_j}^c + D_{t_i \hat{\tau}_{i+1}} C_{\hat{\tau}_{i+1}}^e \,\middle|\, \mathcal{F}_{t_i}\right] \tag{6.29}$$

$$\hat{\tau}_i = \inf\{t_j \geq t_i \,|\, \hat{C}_{t_j}^r \leq C_{t_j}^e\} \tag{6.30}$$

$$\hat{\tau}_i = \begin{cases} t_i & \text{if } \hat{C}_{t_i}^r \leq C_{t_i}^e \\ \hat{\tau}_{i+1} & \text{otherwise} \end{cases} \tag{6.31}$$

$$V_0 = \mathbb{E}^\mathbb{Q}\left[\sum_{i<\hat{\tau}} D_{0t_i} C_{t_i}^c + D_{0\hat{\tau}} C_{\hat{\tau}}^e\right] \tag{6.32}$$

Equation (6.29) says that, in the Longstaff-Schwartz algorithm, one must estimate the expected sum of discounted future optimal coupons given the current state of the market. In practice, this is often done using regression methods, i.e., one regresses the sum of discounted future optimal coupons against a set of regressors. Regressors can be any $\mathcal{F}_{t_i}$-measurable random variables, like asset values, running maximums or minimums, running average of asset values, etc. □

6.8.4 Estimating M^*: The Andersen-Broadie algorithm

The Longstaff-Schwartz algorithm provides a lower bound for the value of the American option. Andersen and Broadie [34] built an algorithm which produces an upper bound, based on the dual representation (6.11)–(6.12). They first run a primal algorithm (e.g., the Longstaff-Schwartz algorithm), which provides a sequence of (nearly) optimal stopping times $\hat{\tau}_i$. Then they estimate the optimal martingale M^* by replacing V_{t_i} by $V_{t_i}^{\hat{\tau}_i}$ in (6.12); $V_{t_i}^{\hat{\tau}_i}$ is the value at time t_i of the option that pays $F_{\hat{\tau}_i}$ at time $\hat{\tau}_i$. Stated otherwise, it is the value of the triggerable option (or autocallable option) defined by the (almost) optimal strategy $\hat{\tau}_i$. Eventually they plug their estimate $\hat{M}$ into (6.11) and estimate V_0 by

$$\hat{V}_0 = \mathbb{E}^\mathbb{Q}\left[\sup_{0 \leq t_i \leq T} (D_{0t_i} F_{t_i} - \hat{M}_{t_i})\right] \tag{6.33}$$

Note that

$$\hat{M}_{t_i} - \hat{M}_{t_{i-1}} = D_{0t_i} V_{t_i}^{\hat{\tau}_i} - \mathbb{E}^\mathbb{Q}[D_{0t_i} V_{t_i}^{\hat{\tau}_i} | \mathcal{F}_{t_{i-1}}]$$
$$= D_{0t_i} V_{t_i}^{\hat{\tau}_i} - D_{0t_{i-1}} V_{t_{i-1}}^{\hat{\tau}_{i-1}} - \mathbf{1}_{\hat{\tau}_{i-1}=t_{i-1}}\left(\mathbb{E}^\mathbb{Q}[D_{0t_i} V_{t_i}^{\hat{\tau}_i} | \mathcal{F}_{t_{i-1}}] - D_{0t_{i-1}} V_{t_{i-1}}^{\hat{\tau}_{i-1}}\right) \tag{6.34}$$

because

$$\mathbf{1}_{\hat{\tau}_{i-1}>t_{i-1}}\mathbb{E}^{\mathbb{Q}}[D_{0t_i}V_{t_i}^{\hat{\tau}_i}|\mathcal{F}_{t_{i-1}}]=\mathbf{1}_{\hat{\tau}_{i-1}>t_{i-1}}D_{0t_{i-1}}V_{t_{i-1}}^{\hat{\tau}_{i-1}}$$

Hence, the conditional expectation $\mathbb{E}^{\mathbb{Q}}[D_{0t_i}V_{t_i}^{\hat{\tau}_i}|\mathcal{F}_{t_{i-1}}]$ needs to be computed only when $\hat{\tau}_{i-1}$ indicates immediate exercise.

Practically, after running the primal algorithm, one simulates new *independent* paths, estimates $V_{t_i}^{\hat{\tau}_i}$ by launching subpaths, and is then able to estimate the martingale increment (6.34), which is eventually plugged into (6.33). To be precise, the algorithm can be decomposed in the following steps:

1. Simulate #p_1 paths from 0 to T to get the primal strategy ($\hat{\tau}_i, 1 \le i \le N$).

2. Simulate #p_3 paths[6] from 0 to T. For each of these paths, and for each exercise date t_i:

   - If $\hat{\tau}_i$ indicates continuation, i.e., $\hat{\tau}_i > t_i$, then simulate #p_{3a} independent subpaths starting from the state of the model at date t_i and estimate $D_{0t_i}\hat{V}_{t_i}^{\hat{\tau}_i}$ by

   $$\frac{1}{\#p_{3a}}\sum_{p_{3a}=1}^{\#p_{3a}}D_{0\hat{\tau}_i}^{(p_{3a})}F_{\hat{\tau}_i}^{(p_{3a})} \tag{6.35}$$

   - If $\hat{\tau}_i$ indicates exercise, i.e., $\hat{\tau}_i = t_i$, then estimate $D_{0t_i}\hat{V}_{t_i}^{\hat{\tau}_i}$ by

   $$D_{0t_i}F_{t_i} \tag{6.36}$$

   and, if $t_i < T$, simulate #p_{3b} independent subpaths starting from the state of the market at date t_i and estimate $\mathbb{E}^{\mathbb{Q}}[D_{0t_{i+1}}V_{t_{i+1}}^{\hat{\tau}_{i+1}}|\mathcal{F}_{t_i}]$ by

   $$\frac{1}{\#p_{3b}}\sum_{p_{3b}=1}^{\#p_{3b}}D_{0\hat{\tau}_{i+1}}^{(p_{3b})}F_{\hat{\tau}_{i+1}}^{(p_{3b})} \tag{6.37}$$

Then compute the approximate martingale $\hat{M}$ by plugging the estimates (6.35), (6.36), and (6.37) into (6.34). Eventually, compute

$$\hat{V}_0 = \frac{1}{\#p_3}\sum_{p_3=1}^{\#p_3}\sup_{0\le t_i\le T}\left(D_{0t_i}^{(p_3)}F_{t_i}^{(p_3)}-\hat{M}_{t_i}^{(p_3)}\right)$$

---

[6]We use the notation #p_3 because we reserve #p_2 for the number of paths used in the second step of the Longstaff-Schwartz algorithm (see Remark 6.8).

REMARK 6.11   Note that the nested Monte Carlo is *not* recursive: no subsubpath is ever drawn. Hence the number of simulated paths is at most

$$\#p_1 + \#p_3 \times (N-1) \times \max\left(\#p_{3a}, \#p_{3b}\right)$$

Remarkably, and fortunately, in most cases, even small values of $\#p_3$, $\#p_{3a}$ and $\#p_{3b}$, of order 10^2, give accurate results. ☐

REMARK 6.12   As explained in [34], $\hat{V}_0$ is a high-biased estimator of the upper bound

$$\mathbb{E}^{\mathbb{Q}}\left[\sup_{0 \le t_i \le T}\left(D_{0t_i}F_{t_i} - \hat{M}_{t_i}\right)\right]$$

Indeed, the nested Monte Carlo sampling error generates a zero mean error $\varepsilon_{t_i}^{(p_3)}$ for each term $\hat{M}_{t_i}^{(p_3)}$, independent of $D_{0t_i}^{(p_3)}F_{t_i}^{(p_3)} - \hat{M}_{t_i}^{(p_3)}$, so in fact $\hat{V}_0$ reads

$$\hat{V}_0 = \frac{1}{\#p_3}\sum_{p_3=1}^{\#p_3}\sup_{0 \le t_i \le T}\left(D_{0t_i}^{(p_3)}F_{t_i}^{(p_3)} - \hat{M}_{t_i}^{(p_3)} - \varepsilon_{t_i}^{(p_3)}\right)$$

As a consequence, denoting by η the random time at which $D_{0t_i}F_{t_i} - \hat{M}_{t_i}$ reaches its maximum, we have

$$\begin{aligned}
\mathbb{E}^{\mathbb{Q}}[\hat{V}_0] &= \mathbb{E}^{\mathbb{Q}}\left[\sup_{0 \le t_i \le T}\left(D_{0t_i}F_{t_i} - \hat{M}_{t_i} - \varepsilon_{t_i}\right)\right] \\
&\ge \mathbb{E}^{\mathbb{Q}}\left[D_{0\eta}F_{\eta} - \hat{M}_{\eta} - \varepsilon_{\eta}\right] \\
&= \mathbb{E}^{\mathbb{Q}}\left[D_{0\eta}F_{\eta} - \hat{M}_{\eta}\right] \\
&= \mathbb{E}^{\mathbb{Q}}\left[\sup_{0 \le t_i \le T}\left(D_{0t_i}F_{t_i} - \hat{M}_{t_i}\right)\right]
\end{aligned}$$

because the fact that η is independent of (ε_{t_i}) yields $\mathbb{E}^{\mathbb{Q}}\left[\varepsilon_{\eta}\right] = 0$. ☐

6.8.5   Conditional expectation approximation

The Tsitsiklis-Van Roy and Longstaff-Schwartz algorithms require computing conditional expectations at each exercise date. For this purpose, one can use

- Parametric regression:

$$\mathbb{E}^{\mathbb{Q}}[Y_{i+1}|X_i = x] \approx \sum_{k=1}^{N} c_k \phi_k(x)$$

- Nonparametric regression:

$$\mathbb{E}^{\mathbb{Q}}[Y_{i+1}|X_i = x] \approx \frac{\mathbb{E}^{\mathbb{Q}}[Y_{i+1}\delta_N(X_i - x)]}{\mathbb{E}^{\mathbb{Q}}[\delta_N(X_i - x)]}$$

with $\delta_N(\cdot)$ a kernel approximating a Dirac mass at zero.

- Likelihood ratio weight:

$$\mathbb{E}^{\mathbb{Q}}[Y_{i+1}|X_i = x] = \mathbb{E}^{\mathbb{Q}}\left[Y_{i+1}\frac{p(t_i, t_{i+1}, x, X_{i+1})}{p(0, t_i, X_0, x)}\right]$$

where $p(s, t, x, y)$ denotes the transition probability density function of X, i.e., the density of X_t at point y given that $X_s = x$.

- Malliavin's representation of conditional expectations: Malliavin calculus provides a way to compute conditional expectations (see [65]):

$$\mathbb{E}[G|F = x] = \frac{\mathbb{E}\left[\mathbf{1}_{[x,\infty)}(F)H(F;G)\right]}{\mathbb{E}\left[\mathbf{1}_{[x,\infty)}(F)H(F;1)\right]} \tag{6.38}$$

$H(F;G)$ is a weight that arises in the Malliavin integration by parts $\mathbb{E}[\phi'(F)G] = \mathbb{E}[\phi(F)H(F;G)]$. For instance, in dimension one, $H(F;1) = \delta\left(\gamma_F^{-1}DF\right)$, where D stands for the Malliavin derivative, γ for the Malliavin covariance matrix $\gamma_{ij} = \langle DF^i, DF^j\rangle$, and δ for the Skorokhod integral.

6.8.6   Parametric methods

We have seen in Section 6.8.1 that the maximum and the minimum arising in the primal and dual formulations of the American option pricing problem are very hard to evaluate through classical optimization procedures, because in both cases the optimization set—the set of all stopping times for the primal formulation, and the set of all martingales (starting from 0) in the dual one—is huge. However, it is always possible to maximize over a set of suitably parameterized stopping times or martingales. Let us give two examples, one concerning the primal formulation, and the other concerning the dual formulation.

Andersen, in a paper that deals with the valuation of Bermudan swaptions [33], searches for an early exercise boundary parameterized in intrinsic value and the values of still-alive swaptions. Loosely speaking, such an algorithm generates an accurate lower bound if and only if the optimal stopping time τ^* is close to the parameterized set of stopping times considered. A good parameterization can be hard to guess, especially in high-dimensional problems. Andersen suggests three different strategies, all of which are parameterized by a time-dependent barrier H_{t_i}:

- In the first strategy, exercise takes place at date t_i when the intrinsic value of the underlying swap exceeds H_{t_i}.

- The second strategy is a refinement that, in addition, checks whether the value of one or more of the remaining European swaptions exceeds the intrinsic swap value (if so, the exercise cannot be optimal).

- In the third strategy, the holder exercises at date t_i when the intrinsic swap value is greater than H_{t_i} plus the most expensive component swaption.

Then one optimizes on the H_{t_i}'s.

Building over the dual approach, Joshi and Theis [144] use a parameterized set of martingales M, and then optimize on the set of parameters to obtain the lowest upper bound (see Theorem 6.2). Here again they are interested in the pricing of Bermudan swaptions. The martingale M can be seen as a dynamic trading strategy. They consider weighted sums of European swaptions, each one associated to an exercise date of the Bermudan swaption, together with a short position in zero coupon bonds to ensure that the initial value is zero.

6.8.7 Quantization methods

Quantization is about discretization of a random variable. For instance, a Gaussian distribution is approached by weighted points $(x_i, w_i, 1 \leq i \leq N_q)$, i.e., by the discrete law $\sum_{1 \leq i \leq N_q} w_i \delta_{x_i}$. Expectations are thus replaced by finite sums. The error, which depends on the law μ of the random variable and on $x = (x_i)$, is controlled by the so-called distortion

$$D_{N_q}^{\mu,p}(x) = \int_{\mathbb{R}^d} \min_{1 \leq i \leq N_q} |x_i - y|^p \, \mu(dy)$$

which is, up to a power $1/p$, the minimum L^p quantization error. The distortion is comparable to $N_q^{-p/d}$ as the number of quantizers N_q tends to infinity. If ϕ is a Lipschitz-continuous function with Lipschitz coefficient $[\phi]_{\text{lip}}$, then

$$\left| \int_{\mathbb{R}^d} \phi \, d\mu - \sum_{i=1}^{N_q} \phi(x_i) \mu\left(C_i(x)\right) \right| \leq [\phi]_{\text{lip}} \, D_{N_q}^{\mu,p}(x)^{1/p} \qquad (6.39)$$

where the $C_i(x)$'s are the Voronoi cells attached to x:

$$C_i(x) = \{ y \mid \forall j \in \{1, \ldots, N_q\}, |y - x_i| \leq |y - x_j| \}$$

If the quantizer is optimal, we can gain one order in the speed of convergence: the r.h.s of (6.39) becomes $[\phi]_{\text{lip}} \, D_{N_q}^{\mu,p}(x)^{2/p}$.

As explained in [46], quantization may be used to approximate the price of a multi-asset American option. Every X_{t_i} $(1 \leq i \leq N)$ is first approximated by a discretization scheme, such as the Euler scheme. This defines a Markov chain, which is in turn replaced by a quantized approximation taking its values in a grid of size N_{t_i}. The N grids and their transition probability matrices

$$\pi_{kl}^i = \frac{\mathbb{Q}\left(X_{t_i} \in C_k(x^i), X_{t_{i+1}} \in C_l(x^{i+1})\right)}{\mathbb{Q}\left(X_{t_i} \in C_k(x^i)\right)}$$

form a discrete tree on which a pseudo-Snell envelope is devised by mimicking the regular dynamic programming formula (6.13). Using the quantization theory of random vectors, Bally and Pagès show the existence of, and how to build, a set of optimal grids, given the total number of elementary quantizers. Numerical examples in dimension up to 10 can also be found in [47]. Reference [71] is a more recent paper that explains how to speed up the quantization tree algorithm, with application to swing options.

REMARK 6.13   The first quantization method in the pricing of American options is due to Barraquand and Martineau [51]. They use partitions of the type $P_k(t_i) = \{x \mid F_{t_i}(x) \in Q_k(t_i)\}$ for partitions $(Q_k(t_i))$ of $\mathbb{R}_+$, and make the assumption that the process I_t defined by $X_t \in P_{I_t}(t)$ is approximately Markov.   ⬜

6.8.8   Mesh methods

Broadie and Glasserman [69] have proposed a stochastic mesh method, together with accompanying mesh density functions, to estimate the conditional expectations involved in the dynamic programming problem (6.13). This leads to a high-biased estimator, which they complement with a low-biased one. The method is prone to variance explosion, so its practical success requires using control variates adapted to the specific pricing problem.

The main limitation of this approach is that it requires the knowledge (and the existence!) of the transition density of the underlying process of asset prices and other state variables, which is very restrictive. To overcome this problem, Broadie *et al.* [70] have suggested a variation of the stochastic mesh method where the weights w involved in the estimation of the conditional expectations do not depend on the transition density anymore, but are computed via optimization, namely, via a least squares or entropy maximization under a moment matching constraint.

6.8.9   The case of continuous exercise dates

So far, we have described Monte Carlo methods for valuation of American options in the case when exercise dates are discrete. In the case where exercise dates are continuous, one can first discretize the exercise dates, say $t_i = i\Delta t$, then apply one of the above methods, and eventually make Δt go to zero.

6.9   Case study: Pricing and hedging of a multi-asset convertible bond

6.9.1   Introduction

To illustrate the Longstaff-Schwartz algorithm, as well as the dual method, we consider the following convertible bond:

- $d = 10$ stocks: Deutsche Telekom, E-On, ENL, France Telecom, HSBC, KPN, Nippon Steel, Nissan, Nokia, and Sanofi. Two stocks are Japanese, one is denominated in GBP, and seven are denominated in euros. All payments are made in euros.

- The maturity is $T = 5$ years.

- We define the time-t performance of stock i by $r_t^i = 100 \times S_t^i / S_0^i$. A basket performance B_t^i is defined by removing the best $d_h = 2$ performers and the worst $d_l = 2$ performers, and taking the arithmetical average of the remaining $d - d_h - d_l = 6$ performances.

- If the option is still alive, the issuer pays an annual coupon $C = 1.5$. C is paid at the end of the first and second years, then $C/4$ is paid quarterly.

- The holder may exercise every quarter, but not before the end of the second year. He or she then receives the above coupon plus $\beta = 1$ times the basket performance.

- If the holder never exercises, he or she receives 100 at maturity, on top of coupon $C/4$.

The product is called a convertible bond, because the holder owns a coupon-bearing bond that he or she can convert into equity, by exercising the option. The holder may have an interest in exercising early because the basket performance has negative drift. The negative drift results from the payment of dividends, and also from the removal of the best 2 and worst 2 performers. We aim at:

- Pricing this option. Obviously, this product cannot be priced using PDE or tree methods, due to the high number of stocks involved. Using the Longstaff-Schwartz (see Section 6.8.3) and Andersen-Broadie (see Section 6.8.4) algorithms, we will build close lower and upper bounds for the value of this option.

- Computing hedge ratios. In particular, we will compare "reoptimized exercise policy" hedge ratios and "fixed policy" hedge ratios on the one hand, and lower bound hedge ratios and upper bound hedge ratios on the other hand.

6.9.2 Modeling assumptions

Each stock is equipped with its local volatility and its Brownian motion. The 10 Brownian motions are correlated. As a pricing example, we pick $\rho_{ij} = 0.5$ for each $i \neq j$. Local volatility is calibrated from market data as of June 7th, 2007, using the Dupire formula. We use the Black-Scholes model for foreign exchange (FX) rates, choosing JPY/EUR and GBP/EUR volatilities to be 10% and the FX/stock correlations to be 0.2. Interest rate curves are deterministic, as are the repo curves and the (cash) dividends.

6.9.3 Pricing parameters and regressors

We use the pricing parameters in Figure 6.1. In particular, we discretize the stock processes at least every 0.1 year, in addition to the ex-dividend dates. We use $\#p_1 = 1,000$ paths to determine a (nearly optimal) exercise strategy, $\#p_2 = 40,000$ paths to get the lower bound using the Longstaff-Schwartz algorithm, and at most $\#p_3 \times \max(\#p_{3a}, \#p_{3b}) = 300 \times 300 = 90,000$ paths to get the upper bound using the Andersen-Broadie algorithm.

Table 6.1: Pricing parameters.

$\#p_1$	$\#p_2$	$\#p_3$	$\#p_{3a}$	$\#p_{3b}$	Δt
1,000	40,000	300	300	300	0.1

In the Longstaff-Schwartz method, the determination of a suboptimal but close-to-optimal strategy relies on an estimation of the continuation value (6.24). Using parametric regression (see Section 6.8.5), we will describe the continuation value as a function of relevant market variables, the so-called regressors. We choose to regress the actualized future optimal coupons (if continuation) to the constant 1.0 and the Black-Scholes value of the call on the "basket performance," assumed to be lognormal, struck at $100/\beta$. The Black-Scholes parameters are arbitrarily taken to be $\sigma = 15\%$, $r = 4\%$, $q = 0$. They define the shape of the nonconstant regressor, seen as a function of the basket performance.

The reason for this choice is the following. At the last exercise date before maturity, the continuation value is precisely the value of a call option on the basket performance, plus 100 times the zero coupon bond. At earlier dates, we expect the continuation value to behave approximately like a constant plus a call option on the basket performance. At the exercise date t_i, the algorithm automatically selects parameters a_i and b_i such that $a_i + b_i R_i$ achieves best fit of the sum of actualized future optimal coupons scatter plot in the least square sense, where

$$R_i = C_{\mathrm{BS}}(B_{t_i}, \sigma, r, q, T - t_i, K = 100/\beta)$$

The exercise strategy is the following: compare what you get from exercise (coupon plus basket performance) to what you own if you continue (coupon plus $a_i + b_i R_i$) and exercise if the former is greater than the latter. This may happen because the basket performance has a negative drift.

The fact that the real dynamics of the basket performance is not lognormal has little impact: the only thing that matters is whether the exercise region is correctly approximated or not. One doesn't need to estimate the continuation value everywhere with high accuracy to get a precise estimate of the exercise region (see Remark 6.7). Of course, one may test other values of σ, r, and q, and add as many regressors as one wants.

No other ingredients are needed to perform the dual algorithm, from which one gets an upper bound of the value of the option in the model.

6.9.4   Price and exercise strategy

The estimators of the price of the convertible bond are given in Table 6.2. The upper bound is only 65 bps above the lower bound, and 5% above from the price of the European version of this option, i.e., when early exercise is not allowed. This shows that the strategy we determine through backward regressions is accurate enough. Note that, had we only performed the primal algorithm, we would have only been able to conclude that the option's price is greater than 104.25—not a very precise statement! One quant may think that his or her choice of regressors makes perfect sense, and so this 104.25 *should* be close to the true price, but another quant may argue that regression is not precise enough. This can lead to endless discussions. Only the dual algorithm (or any algorithm that provides an upper bound) can put an end to those discussions. As a consequence, running the dual algorithm is not only important when pricing putable options (for which the primal algorithm only provides a lower bound, as is the case here), but also when pricing callable options, because the dual algorithm quantifies the quality of the approximation of the optimal strategy.

Table 6.2: Price of the multi-asset convertible bond.

	Lower bound	Upper bound	European option
Price	104.25	104.90	99.43
Std dev	0.09	0.10	0.12

Table 6.3: Strategy. The probability of never exercising is 4%.

t_i	2	2.25	2.5	2.75	3	3.25
a_i	94.11	94.97	94.92	95.44	95.69	96.51
b_i	0.84	0.81	0.87	0.85	0.93	0.93
$\mathbb{E}^{\mathbb{Q}}[R_i]$	19.59	19.45	18.92	19.00	18.03	17.64
Prob. of exercise	41%	11%	5%	7%	3%	3%
t_i	3.5	3.75	4	4.25	4.5	4.75
a_i	97.24	97.13	97.87	98.36	98.69	99.18
b_i	0.92	0.93	0.93	0.96	0.99	0.96
$\mathbb{E}^{\mathbb{Q}}[R_i]$	17.60	17.86	17.20	16.65	16.51	16.63
Prob. of exercise	3%	6%	4%	3%	2%	7%

Table 6.3 describes our proxy for the optimal exercise strategy. For instance, on June 7, 2009 (the first exercise date, corresponding to $t_i = 2$), the holder of the option should compare

- the exercise value: $C + 100 + (B_{t_i} - 100)_+$

- and the continuation value estimated by: $C + 94.11 + 0.84R_i$.

It also shows the average value of R_i, as computed from the sample, here 19.59, and the probability of exercising at this date, here 41%. Note that this product lives until maturity with probability of only 4%.

In Figure 6.1 we have plotted the exercise and continuation values at the first date of early redemption ($t_i = 2$), after removing common coupon C. Exercise is useful when the basket performance B is either low ($B \leq 84$) or high ($B \geq 128$). When the basket performance is low, the holder has an interest in exercising in order to receive 100 now and put this money in a bank account (coupon C was less than the Euro rates in 2007). There is little chance that the basket performance goes above 100 in the future. When the basket performance is high, the holder has an interest in exercising because the negative drift of the basket performance has a greater impact on the continuation value than the time value of the call and the future payment of coupons. See further tests in Section 6.9.7.

6.9.5   Hedge ratios

We compute hedge ratios of the multi-asset convertible bond by bumping parameters, while keeping the same Brownian paths. The delta in the direction of stock i is computed by bumping S_0^i (using the multiplicative factor 1.05) without recalibrating the local volatility. As for the vega in the direction of stock i, it is computed by bumping the whole implied volatility surface of stock i (using a 1% increment) and recalibrating the local volatility.

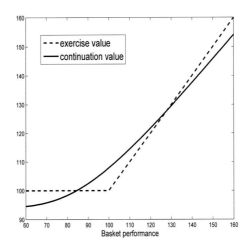

Figure 6.1: Exercise value and continuation value of the multi-asset convertible bond at the first date of early redemption (common coupon C has been removed).

After bumping the parameters, we can choose to reoptimize the exercise policy or not. We show further in Section 6.9.6 that for first order ratios, reoptimizing the strategy is unnecessary. In this section we only perform numerical checks. Moreover, we can compute the hedge ratios either from the lower bound price or from the upper bound price. We shall compare the hedge ratios for the four resulting combinations.

Vegas. Table 6.4 shows the numerical vegas. For instance, if the volatility of stock #10 experiences a one-point increase, the price goes approximately 0.42 up, i.e., 42 bps up. We observe that the four methods give indistinguishable price bumps (recall that the standard deviation on the price is about 10 bps).

Deltas. Table 6.5 shows the numerical deltas. The initial values of the ten stocks are, respectively, 13.75, 113, 26.23, 22.64, 9.265, 12.32, 879, 1337, 21.03 and 69.58, so these deltas correspond to the price bumps shown in Table 6.6. Again, the four methods give indistinguishable price bumps (recall again that the standard deviation on the price is about 10 bps).

Conclusion: practical computation of hedge ratios. Since the dual method is time consuming, hedge ratios should be computed using the Longstaff-Schwartz price only. Besides, as far as deltas and vegas are concerned, there is no need to reoptimize the exercise policy. Actually, this is a general result for first order ratios, as we now explain.

Table 6.4: Numerical vegas of the multi-asset convertible bond.

Reoptimized strategy		Fixed strategy	
Primal	Dual	Primal	Dual
3.52	2.76	3.96	4.18
7.39	6.74	7.55	7.31
14.86	14.48	14.32	13.64
17.96	18.36	18.98	18.68
25.42	23.57	26.14	22.63
30.26	28.35	29.96	27.70
32.31	29.56	32.04	29.31
37.29	36.39	37.37	33.31
39.36	39.83	40.19	36.88
42.76	44.51	44.46	42.02

6.9.6   First order ratios need no reoptimization of exercise strategy

Let us take the following example. We wish to compute the vega of a callable option in the Black-Scholes framework. We first compute a price when $\sigma = \sigma_0$. This price may be understood as the true price of the option in the Black-Scholes framework, or as the price obtained by running the Longstaff-Schwartz algorithm with a given set of regressors (but infinitely many paths). Attached to it is the optimal (or "Longstaff-Schwartz optimal") strategy "0" corresponding to $\sigma = \sigma_0$. This defines a pricing function V_0. $V_0(\sigma)$ is the price of the option when we adopt the exercise policy "0" and when the vol is σ. We can do the same starting from $\sigma = \sigma_1$ and get a pricing function $V_1(\sigma)$. We must have $V_1(\sigma_1) \leq V_0(\sigma_1)$ by construction (when $\sigma = \sigma_1$, the strategy "0" is not as good as strategy "1"). Symmetrically, $V_0(\sigma_0) \leq V_1(\sigma_0)$.

The true price is a function $V(\sigma)$, where the optimal strategy is reoptimized for each value of σ: $V(\sigma_0) = V_0(\sigma_0), V(\sigma_1) = V_1(\sigma_1)$, etc. V is thus the concave envelope of all the V_i's; see Figure 6.2. In particular, it is less than or equal to V_0. Now, both functions are equal when $\sigma = \sigma_0$. Hence, provided they are smooth, they must have the same first derivative at point σ_0. This means that $V'(\sigma_0) = V_0'(\sigma_0)$: to compute the vega $V'(\sigma_0)$, we may fix the strategy "0" and compute the finite difference $(V_0(\sigma_0 + \Delta\sigma_0) - V_0(\sigma_0)))/\Delta\sigma_0$. However, this is not true for higher order derivatives: $V''(\sigma_0)$ may be strictly smaller than $V_0''(\sigma_0)$.

Note that the same reasoning is not valid when we consider the price obtained using the dual method. Nevertheless, we have checked in the previous section that up to the Monte Carlo standard deviation, in our numerical experiment, we cannot distinguish between dual hedge ratios and primal ones.

Table 6.5: Numerical deltas of the multi-asset convertible bond.

Reoptimized strategy		Fixed strategy	
Primal	Dual	Primal	Dual
0.3348	0.3464	0.3387	0.3796
0.0440	0.0447	0.0415	0.0445
0.2204	0.2392	0.2248	0.2340
0.2179	0.2060	0.2325	0.2158
0.6648	0.7207	0.6937	0.7311
0.4131	0.4038	0.3929	0.3824
0.0043	0.0041	0.0041	0.0042
0.0036	0.0038	0.0036	0.0036
0.2018	0.2117	0.2105	0.2148
0.0653	0.0709	0.0708	0.0714

6.9.7   Why exercise the convertible bond?

We know that an American call on a stock S paying no dividend is worth the European call: the holder should never exercise. Within a probabilistic model, this is reflected by Jensen's inequality: the time-t value of the European call,[7] $P_{tT}\mathbb{E}^{\mathbb{Q}}[(S_T - K)_+|\mathcal{F}_t]$ is greater than or equal to $P_{tT}(\mathbb{E}^{\mathbb{Q}}[S_T|\mathcal{F}_t] - K)_+ = (S_t - KP_{tT})_+$, which is itself greater than or equal to the intrinsic value $(S_t - K)_+$. Hence, the time-t value of holding the American call, which is greater than or equal to the time-t value of the European call, is greater than or equal to the intrinsic value $(S_t - K)_+$, which is the value of exercising the American call.

In our case, as already mentioned, if the holder has an interest in exercising and getting the intrinsic value of the call option, despite the nonpayment of future coupons, it is because the forward of the basket performance $\mathbb{E}^{\mathbb{Q}}[B_T|\mathcal{F}_t]$ is low, smaller than B_t, so the above reasoning does not hold anymore. The low forward value results from removing the best 2 and worst 2 performers and from the payment of dividends. In the following, we check this by performing a few extra tests.

Pricing the discounted forward basket performance gives $P_{tT}\mathbb{E}^{\mathbb{Q}}[B_T] \approx 78.14$. Loosely speaking, this means that the discounted basket performance has a deterministic negative trend, at a linear rate which is a bit more than 4% a year. The value of holding the American call may then lie under the intrinsic value $(\beta B_t - K)_+$.

When we keep all 10 stocks to compute the basket performance, the discounted forward basket performance goes up to 81.41. This is still much lower than

---

[7] P_{tT} denotes the (deterministic) discount factor.

Table 6.6: Bumps in the price of the multi-asset convertible bond when the initial value of stock is multiplied by 1.05.

Reoptimized strategy		Fixed strategy	
Primal	Dual	Primal	Dual
0.2302	0.2382	0.2329	0.2610
0.2488	0.2527	0.2345	0.2514
0.2891	0.3137	0.2948	0.3068
0.2466	0.2332	0.2632	0.2443
0.3080	0.3339	0.3213	0.3387
0.2545	0.2488	0.2420	0.2356
0.1887	0.1810	0.1815	0.1866
0.2412	0.2553	0.2394	0.2413
0.2122	0.2226	0.2213	0.2259
0.2272	0.2468	0.2463	0.2483

100, so the main reason for the low forward basket performance is the payment of dividends. When dividends are removed, while keeping the 10 stocks in the basket, the discounted forward basket performance increases to 97.33. The fact that it still lies below 100 is due to the very low Japanese interest rates (there are two Japanese stocks in the basket), and also because we picked a positive correlation (0.2) between the stocks that are not expressed in euros and the corresponding FX. When we take this correlation to be −0.2, the discounted forward becomes 98.64. Besides, if we arbitrarily take the Japanese rate curve to be equal to the Euro rate curve, the discounted forward basket performance is worth 100.03: the positive correlation between the stocks that are not expressed in euros and the corresponding FX then compensates the positive difference between GBP and Euro rate curves (there is one GBP-denominated stock in the basket).

Table 6.7: Prices of the discounted forward of the basket performance when varying the number of stocks removed.

$d_l \setminus d_h$	0	1	2	3
0	97.33	89.90	84.40	79.50
1	102.81	95.14	89.60	84.75
2	107.92	99.88	94.21	89.32
3	113.04	104.52	98.65	93.64

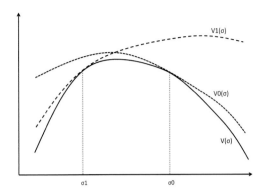

Figure 6.2: First order hedge ratios need no reoptimization of exercise strategy.

Table 6.8: Impact of removing worst and best stock performances on the price of the convertible bond and its European version.

$d_h = d_l$	0	1	2	3
European	110.50	108.83	108.17	107.79
European	100.92	99.46	98.94	98.68

We give in Table 6.7 a few prices of the discounted forward of the basket performance when the number of stocks removed varies. We removed the dividends but kept the true rate curves. Removing the same increasing number of worst and best perfs decreases the price of the discounted forward of the basket performance, from 97.33 to 93.64.

Some of the corresponding (lower bound) prices of the convertible option are given in Table 6.8, which also considers the case where dividends are taken into account. Here we used $\#p_1 = 1,000$ and $\#p_2 = 4,000$, so the standard deviation of the price estimator is much larger. We observe that, as said before, the payment of dividends has more impact on the price of the multi-asset convertible bond than removing worst and best stock performances.

6.10 Introduction to chooser options

Chooser options look like American options but differ in several ways:

- The holder of exercise rights may exercise several times.

- When he or she exercises, he or she potentially modifies all future coupons. Stated otherwise, at each exercise date, he or she can choose between various products. This is the reason why these options are called "chooser" options.

These characteristics imply that, from a pricing perspective,

- we must evaluate several continuation values, and

- each continuation value depends not only on path-dependent variables but also on past decisions of the holder.

The holder of exercise rights can exercise only at pre-specified dates, and no more than, say, n_T times; n_T hence denotes an initial number of tokens, and each time he or she exercises, the holder of exercise rights looses one token.

Like American options, chooser options exist in their callable and putable versions. In the callable version, the issuer of the option is the holder of the exercise rights. In the putable version, it is the buyer who owns the exercise rights. As examples, one can think of the following two putable products.

- **The multi-times restrikable put**: The holder is "long" a put, i.e., owns a put, with strike K, say 90%, and maturity T, say 1 year. He or she can restrike this put, at predefined dates, say every month, but no more than n_T times. In case of restrike, the new strike is some proportion α, say 90%, of the current underlying value. Some small restrike fee ϕ may apply. In such a case, the holder must pay ϕ to the issuer when he or she restrikes the put.

- **The "passport option:"** Each day, or week, or month, the holder decides to be long ($\varepsilon_{t_i} = +1$) or short ($\varepsilon_{t_i} = -1$) some index X and receives the payout $g(\pi_T)$ at maturity T, where $\pi_T = \sum_i \varepsilon_{t_i} \left(X_{t_{i+1}} - X_{t_i} \right)$. In this simple version, n_T is the number of days, or weeks, or months. We can think of versions where we limit the number of tokens.

In this book, we will mainly consider a small class of options within the huge world of chooser options, namely the *multi-token binary chooser options*:

DEFINITION 6.4 Multi-token binary chooser options *A multi-token binary chooser option is a chooser option where the holder of exercise rights*

- *initially owns a finite number of tokens, say n_T,*

- *makes a binary choice ("exercise" or not) at each exercise date t_i under the sole condition that he or she still has tokens, and*

- *looses one token each time he or she decides to exercise.*

The multi-times restrikable put and the passport option are examples of putable multi-token binary chooser options.

When he or she exercises, the holder of exercise rights modifies the present and future coupons. Hence, the coupon served at some cash-flow date t_i may depend on his or her past decisions. For instance, in the two-token case where $n_T = 2$, we must distinguish:

- $C_0(t_i)$: the coupon served at t_i if the holder of exercise rights has not exercised at dates $\leq t_i$,

- from $C_1(\tau_1, t_i)$: the coupon served at t_i if the holder of exercise rights has exercised once at dates $\leq t_i$, and if the exercise date was τ_1,

- and from $C_2(\tau_1, \tau_2, t_i)$: the coupon served at t_i if the holder of exercise rights has exercised twice at dates $\leq t_i$, and if the exercise dates were τ_1 and τ_2.

In the general case of n_T tokens, the past decisions of the issuer are described by $(n, \tau_1, \ldots, \tau_n)$ where n denotes the number of already played tokens, and $\tau_1, \ldots, \tau_n$ the exercise dates at which these tokens have been played.

6.11   Regression methods for chooser options

6.11.1   The "naive" primal algorithm

Here we present a naive adaptation of the Longstaff-Schwartz (LS) algorithm to chooser options. The principle of the LS algorithm is conserved: we launch $\#p_1$ paths and perform backward regressions (first step), and then price the triggerable product defined by the strategy resulting from the first step, using $\#p_2$ independent paths (second step), see Remark 6.8. The difference with the classical Bermudan case is that, because of the possibility of multiple exercises, the price at some exercise date t_i depends not only on path-dependent variables[8] *but also on the past decisions $(n, \tau_1, \ldots, \tau_n)$ of the issuer,* which cannot be inferred from the generated paths.[9]

---

[8]in the general sense, that is, mathematically, $\mathcal{F}_{t_i}$-measurable random variables, where $\mathcal{F}$ is the filtration generated by the Brownian motions we used to generate the $\#p_1$ copies of the assets. For instance, the spot at time t_i is a path-dependent variable.

[9]i.e., which are not $\mathcal{F}_{t_i}$-measurable.

As a consequence, in the first step of the naive primal algorithm for multi-token binary chooser options, at each exercise date t_i, we perform two regressions by possible value of $(n, \tau_1, \ldots, \tau_n)$, $n < n_T$:[10] one to estimate the "no exercise" value, or "continuation" value; one to estimate the exercise value. In both cases, we regress the actualized future optimal cash flows on some path-dependent variables. Note that the optimal cash flows at dates greater than or equal to t_{i+1} were found at the previous (backward) regression step, the one for exercise date t_{i+1}, under the past-decisions hypothesis $(n, \tau_1, \ldots, \tau_n)$ for the continuation value, and under the past-decisions hypothesis $(n + 1, \tau_1, \ldots, \tau_n, t_i)$ for the exercise value.

REMARK 6.14 It may happen that the option value, just before exercise date t_i, depends on the past decisions $(n, \tau_1, \ldots, \tau_n)$ only through a subset of these variables. For instance, in the case of the multi-times restrikable put, the option value only depends on (n, τ_n). Indeed, only the current strike and the number of remaining tokens matter. Hence, it is enough to perform two regressions by possible value of (n, τ_n). There are at most $1 + i(i - 1)/2$ of them at date t_i (fewer if $n_T < i - 1$), hence at most $O(N^3)$ in total, where N is the total number of exercise dates. This saves a lot of memory space and computation time. But this is not the case for the passport option, for which we need (in the naive perspective) to perform 2^{i-1} regressions at date t_i, hence $O(2^N)$ in total. ⬜

REMARK 6.15 The naive primal algorithm makes no use of the peculiar structure of multi-token binary chooser options. For instance, if the choice is not binary but m-ary, the naive primal algorithm consists in performing, at each exercise date t_i, m regressions by possible value of the past choices $(c_1, \ldots, c_{i-1})$, where each $c_i \in \{1, \ldots, m\}$ denotes the decision taken at date t_i. It involves at most $O(m^N)$ regressions in total, maybe less if the set of past choices $(c_1, \ldots, c_{i-1})$ has cardinal $< m^{j-1}$, depending on the option priced (think of tokens for instance). An example of a multi-token 3-ary chooser option is the extension of the "passport option" where each day, or week, or month, the issuer (or the holder) decides to be long ($\varepsilon_{t_i} = +1$) or neutral ($\varepsilon_{t_i} = 0$) or short ($\varepsilon_{t_i} = -1$) some index X. ⬜

We call this method the naive one because:

- It is natural and simple and mimics at low cost the well known LS algorithm.

- It cannot price options with many exercise dates in a reasonable amount of time, because the number of regressions to perform at each exercise

---

[10]When $n = n_T$, no regression is needed, because the option has become of European type.

date t_i is equal to the number of possible past decisions of the issuer, which may be exponentially explosive; see Remarks 6.14 and 6.15. For instance, think of the passport option with daily or weekly choices and a (reasonable) maturity of one year. For weekly exercise dates and a maturity of 1 year, $2^N = 2^{52} \simeq 10^{15}$, a huge number! Not to mention the case of a daily exercise, or the case of a m-ary choice, where we need to perform $O\left(m^N\right)$ regressions.

To get rid of this combinatorial issue, we have designed another primal algorithm in which the past decisions are summed up in a (small) number of decision-dependent variables. For instance, the value of the passport option at exercise date t_i depends on past decisions only through the value of $\pi_{t_i} = \sum_{\{j|t_{j+1}\leq t_i\}} \varepsilon_{t_j}\left(X_{t_{j+1}} - X_{t_j}\right)$. We now describe this "efficient" algorithm.

6.11.2   Efficient Monte Carlo valuation of chooser options

In this section, we suggest an *efficient* Monte Carlo algorithm for the valuation of general chooser options. The naive Monte Carlo algorithm described in Section 6.11.1 appears to work well in the case where the number of possible decisions is small, but in the other case, it consumes a very large amount of time and memory and is thus *inefficient*. The efficient Monte Carlo algorithm works for a large class of chooser options, namely the ones whose price at time t is a function of *only a few continuous decision-dependent variables*, plus possible discrete decision-dependent variables, provided they cannot take too many values.

6.11.2.1   The chooser options considered

The price at time t of a general chooser option can be written

$$P(t, \mathrm{pdv}_t, \mathrm{ddv}_t)$$

where

- pdv_t denotes some path-dependent variables in the general meaning, that is, mathematically, $\mathcal{F}_t$-measurable random variables, where $\mathcal{F}$ is the filtration generated by the Brownian motions we used to generate the assets paths. For instance, some asset value at time t is here considered a path-dependent variable,

- ddv_t denotes some decision-dependent variables, that is, variables depending on the past decisions of the issuer. In particular, these variables *are not* $\mathcal{F}_t$-measurable: we cannot deduce them from the knowledge of the sole past asset prices.

Here we only consider chooser options for which ddv_t *can reduce to*

- 1 or 2 continuous variables: cddv_t^1 and cddv_t^2

- A few possible discrete variables, dddv_t^i, taking only a small number of values

As an example, one can think of the passport option; see description in Section 6.10. In this case, the price at some date t reads $P_{\mathrm{eff}}(t, \mathrm{pdv}_t, \mathrm{ddv}_t)$ with

$$\mathrm{pdv}_t = (X_t, X_{t_i})$$
$$\mathrm{ddv}_t = (\mathrm{cddv}_t, \mathrm{dddv}_t)$$
$$\mathrm{cddv}_t = \pi_{t_i}$$
$$\mathrm{dddv}_t = \varepsilon_{t_i}$$

where t_i denotes the last exercise date $\leq t$, i.e., $t_i \leq t < t_{i+1}$. π_{t_i} cannot be inferred from the path $(X_s, 0 \leq s \leq t)$. It requires the knowledge of past decisions $\varepsilon_{t_j} \in \{-1, +1\}$, hence is a decision-dependent variable. In particular, at an exercise date t_i, under the assumption that we make the choice c_{t_i} $(= \varepsilon_{t_i})$ at that time, the price reads

$$P_{\mathrm{eff}}^{\mathrm{ex}}(t_i, X_{t_i}, \pi_{t_i}; c_{t_i})$$

and thus depends on only one path-dependent variable X_{t_i} and only one continuous decision-dependent variable π_{t_i}. In the naive perspective behind the naive algorithm, we *can* also write down the price at some date t as $P_{\mathrm{naive}}(t, \mathrm{pdv}_t, \mathrm{ddv}_t)$ with

$$\mathrm{pdv}_t = (X_{t_0}, X_{t_1}, \ldots, X_{t_i}, X_t)$$
$$\mathrm{ddv}_t = (\varepsilon_{t_1}, \varepsilon_{t_2}, \ldots, \varepsilon_{t_i})$$

where again t_i denotes the last exercise date $\leq t$, i.e., $t_i \leq t < t_{i+1}$. In particular, at an exercise date t_i, under the assumption that we make the choice c_{t_i} $(= \varepsilon_{t_i})$ at that time, the naive price function reads

$$P_{\mathrm{naive}}^{\mathrm{ex}}(t_i, X_{t_0}, X_{t_1}, \ldots, X_{t_i}, \varepsilon_{t_1}, \varepsilon_{t_2}, \ldots, \varepsilon_{t_{i-1}}; c_{t_i})$$

depends on the past of the path of X and $i - 1$ discrete decision-dependent variables, each one taking two values.

6.11.2.2 The efficient algorithm

The efficient algorithm will make use of the representation

$$P_{\mathrm{eff}}^{\mathrm{ex}}(t_i, X_{t_i}, \pi_{t_i}; c_{t_i})$$

rather than the naive representation

$$P_{\mathrm{naive}}^{\mathrm{ex}}(t_i, X_{t_0}, X_{t_1}, \ldots, X_{t_i}, \varepsilon_{t_1}, \varepsilon_{t_2}, \ldots, \varepsilon_{t_{i-1}}; c_{t_i})$$

Why is the first representation more efficient in the LS perspective? We must find what the optimal strategy is at date t_i given $\mathcal{F}_{t_i}$ (the past of the assets paths) and given ddv_{t_i}, the decision-dependent variables. Whatever representation we choose, ddv_{t_i} can take 2^{i-1} values given $\mathcal{F}_{t_i}$. In the naive representation,

$$\mathrm{ddv}_{t_i} = (\varepsilon_{t_1}, \varepsilon_{t_2}, \dots, \varepsilon_{t_{i-1}})$$

and in the efficient representation,

$$\mathrm{ddv}_{t_i} = \pi_{t_i} = \sum_{j<i} \varepsilon_{t_j} \left(X_{t_{j+1}} - X_{t_j} \right).$$

In the naive LS algorithm, we consider each and every possible past decision $(\varepsilon_{t_1}, \varepsilon_{t_2}, \dots, \varepsilon_{t_{i-1}})$. But in the effective LS method, we consider only some values of π_{t_i}. It is enough to define a grid of discrete values $v_{t_i}(k)$ for π_{t_i}, with reasonable size (much less than the 2^{i-1} we needed in the naive method!), and find what the optimal strategy is at date t_i given $\mathcal{F}_{t_i}$ (the past of the assets paths) and given that $\pi_{t_i} = v_{t_i}(k)$.

When we look for this optimal strategy, we pick one possible choice c_{t_i} at date t_i, and compute the sum of actualized future optimal cash flows on path p at date t_i given $(c_{t_i}, v_{t_i}(k))$. These optimal cash flows are

$$\mathrm{OCF}_{t_i}^p(t; c_{t_i}, v_{t_i}(k)) = \begin{cases} \mathrm{CF}^p(t; c_{t_i}, v_{t_i}(k)) & \text{if } t_i < t < t_{i+1} \\ \mathrm{OCF}_{t_{i+1}}^p\left(t; v_{t_i}(k) + c_{t_i}\left(X_{t_{i+1}} - X_{t_i}\right)\right) & \text{if } t_{i+1} \leq t \end{cases}$$

where $\mathrm{OCF}_{t_{i+1}}^p\left(t; v_{t_i}(k) + c_{t_i}\left(X_{t_{i+1}} - X_{t_i}\right)\right)$ is the optimal cash flow on path p at date t given that $\mathrm{cddv}_{t_{i+1}} = v_{t_i}(k) + c_{t_i}\left(X_{t_{i+1}} - X_{t_i}\right)$. The latter does not generally match a grid value $v_{t_{i+1}}(k)$, so that we must either

- project $v_{t_i}(k) + c_{t_i}\left(X_{t_{i+1}} - X_{t_i}\right)$ on the nearest point in the grid,

- or use finer interpolation methods, like cubic splines

to get the optimal strategy from t_{i+1} onward on path p, hence the optimal cash flows.

REMARK 6.16   There may appear inconsistencies between path-dependent variables and decision-dependent variables. For instance, if the running maximum $M_t = \max_{0 \leq s \leq t} X_s$ is part of pdv_t, and if $\mathrm{cddv}_t = X_\tau$ for some $\tau \leq t$, we must have $\mathrm{cddv}_{t_i}(k) \leq M_{t_i}$. The left-hand side is path-independent, whereas the right-hand side is not. This means that in the regression for $\mathrm{cddv}_{t_i}(k)$, we should ignore some paths—here, the paths p with $\mathrm{cddv}_{t_i}(k) > M_{t_i}^p$. $\qquad\Box$

6.12 The dual algorithm for chooser options

In the spirit of Rogers [178] and Haugh and Kogan [125], we now derive a dual algorithm to approximate an upper bound of putable chooser options.

6.12.1 The case of two tokens

In the case where the holder owns two tokens, we can write the initial value of the option as[11]

$$
V_0 = \inf_{\tau_1} \mathbb{E} \left[\sum_{t_j < \tau_1} D_{0t_j} C_0(t_j) + D_{0\tau_1} V_{\tau_1}^1 \right]
$$

where

$$
V_{\tau_1}^1 = \inf_{\tau_2 > \tau_1} \mathbb{E} \left[\sum_{\tau_1 \le t_j < \tau_2} D_{\tau_1 t_j} C_1(\tau_1, t_j) + \sum_{\tau_2 \le t_j \le T} D_{\tau_1 t_j} C_2(\tau_1, \tau_2, t_j) \Big| \mathcal{F}_{\tau_1} \right]
$$

We are thus back to the situation of a classical American option that pays at exercise date τ_1 the coupon $V_{\tau_1}^1$. This coupon is itself the value of the same chooser option, but with only one token, given $\mathcal{F}_{\tau_1}$. Unlike the classical Bermudan case, this coupon is model-dependent and must be evaluated during the pricing process. As a consequence, we cannot guarantee that any martingale $\hat{M}$ and estimate $\hat{V}_1$ inputted in Equation (6.40) will generate an upper bound of the price. However, if the optimal strategy is accurately estimated, the output of the dual algorithm is likely to be an upper bound of the option price.

In Section 6.8.4, we have seen that for a classical American option, a (good) upper bound is defined by

$$
\hat{V}_0 = \mathbb{E} \left[\sup_{t_i} \left(\sum_{t_j < t_i} D_{0t_i} C_0(t_i) + D_{0t_i} V_{t_i}^1 - \hat{M}_{t_i} \right) \right] \tag{6.40}
$$

where $\hat{M}_0 = 0$,

$$
\hat{M}_{t_i} - \hat{M}_{t_{i-1}} = D_{0t_i} V_{t_i}^{\hat{\tau}^i} - \mathbb{E}^{\mathbb{Q}} [D_{0t_i} V_{t_i}^{\hat{\tau}^i} | \mathcal{F}_{t_{i-1}}]
$$

$$
= D_{0t_i} V_{t_i}^{\hat{\tau}^i} - D_{0t_{i-1}} V_{t_{i-1}}^{\hat{\tau}^{i-1}} - \mathbf{1}_{\hat{\tau}_1^{i-1} = t_{i-1}} \left(\mathbb{E}^{\mathbb{Q}} [D_{0t_i} V_{t_i}^{\hat{\tau}^i} | \mathcal{F}_{t_{i-1}}] - D_{0t_{i-1}} V_{t_{i-1}}^{\hat{\tau}^{i-1}} \right) \tag{6.41}
$$

and $V_{t_i}^{\hat{\tau}^i}$ is the time-t_i value of the triggerable option defined by the (nearly optimal) strategy $\hat{\tau}^i$ (see Section 6.6, and in particular Remark 6.6, for how

---

[11]The coupons C_0, C_1, C_2 have been defined in Section 6.10.

to deal with multiple coupons). The strategy $\hat{\tau}^i = (\hat{\tau}_1^i, \dots, \hat{\tau}_{n_T}^i)$ denotes the strategy conditional on the fact that we have not used any token before t_i, i.e., given that $\tau_1 \geq t_i$. The naive primal algorithm for chooser options provides such a sequence of nearly optimal strategies $(\hat{\tau}^i)$. Following [34], the quantities $V_{t_i}^{\hat{\tau}^i}$ and $\mathbb{E}^{\mathbb{Q}}[D_{0t_i} V_{t_i}^{\hat{\tau}^i} | \mathcal{F}_{t_{i-1}}]$ can be estimated thanks to a nested Monte Carlo procedure. This allows us to compute $\hat{M}_{t_i}$ in (6.40). In the case of the two-token chooser option, unlike the classical American option case, in (6.40), we must also estimate $V_{t_i}^1$. This quantity can be estimated in the same nested Monte Carlo procedure, using the (nearly) optimal strategy, say $\bar{\tau}^i$, conditional on the fact that the first token is played at time t_i, i.e., given that $\tau_1 = t_i$. Again, this strategy is easy to retrieve using all the conditional (i.e., depending on the past decisions) nearly optimal strategies computed during the first step of the primal algorithm. Hence we suggest the following algorithm:

Dual algorithm for chooser options:

After simulating $\#p_3$ paths from 0 to T, for each path and each exercise date t_i (the maturity T being excluded):

- We simulate $\#p_{3a}$ subpaths starting from the state of the market at date t_i.[12]

- We estimate
$$V_{t_i}^1 = \inf_{\tau_2 > t_i} \mathbb{E}\Big[\sum_{t_i \leq t_j < \tau_2} D_{t_i t_j} C_1(t_i, t_j) + \sum_{\tau_2 \leq t_j \leq T} D_{t_i t_j} C_2(t_i, \tau_2, t_j) \Big| \mathcal{F}_{t_i} \Big]$$
by
$$\hat{V}_{t_i}^1 = \frac{1}{\#p_{3a}} \sum_{p=1}^{\#p_{3a}} \Big(\sum_{t_i \leq t_j < \bar{\tau}_2^i} D_{t_i, t_j}^p C_1^p(t_i, t_j) + \sum_{\bar{\tau}_2^i \leq t_j \leq T} D_{t_i, t_j}^p C_2^p(t_i, \bar{\tau}_2^i, t_j) \Big)$$

If $t_i = T$, we set $\hat{V}_T^1 = C_1(T, T)$.

- If we continue, i.e., if $\hat{\tau}_1^i > t_i$, we estimate $D_{0t_i} V_{t_i}^{\hat{\tau}^i}$ by
$$\sum_{t_j < t_i} D_{0t_j} C_0(t_j)$$

$$+ \frac{1}{\#p_{3a}} \sum_{p=1}^{\#p_{3a}} \Big(\sum_{t_i \leq t_j < \hat{\tau}_1^i} D_{0t_j}^p C_0^p(t_j) + \sum_{\hat{\tau}_1^i \leq t_j < \hat{\tau}_2^i} D_{0, t_j}^p C_1^p(\hat{\tau}_1^i, t_j)$$

$$+ \sum_{\hat{\tau}_2^i \leq t_j \leq T} D_{0t_j}^p C_2^p(\hat{\tau}_1^i, \hat{\tau}_2^i, t_j) \Big)$$

---

[12]Here, for the sake of simplicity, we use the same number $\#p_{3a}$ of subpaths for both the continuation and the exercise cases, and for the estimation of $V_{t_i}^1$.

When $t_i = T$, we take $D_{0T} V_T^{\hat{\tau}^i} = \sum_{t_j \leq T} D_{0t_j} C_0(t_j)$,

- If we exercise, i.e., if $\hat{\tau}_1^i = t_i$, then we estimate $D_{0t_i} V_{t_i}^{\hat{\tau}^i}$ by

$$\sum_{t_j < t_i} D_{0t_j} C_0(t_i) + D_{0t_i} \hat{V}_{t_i}^1$$

(in particular, by $\sum_{t_j < T} D_{0t_j} C_0(t_j) + D_{0T} C_1(T, T)$ if $t_i = T$) and we estimate $\mathbb{E}^{\mathbb{Q}}[D_{0t_i} V_{t_i}^{\hat{\tau}^i} | \mathcal{F}_{t_{i-1}}]$ by

$$\sum_{t_j < t_{i-1}} D_{0t_j} C_0(t_j)$$

$$+ \frac{1}{\#p_{3a}} \sum_{p=1}^{\#p_{3a}} \left(\sum_{t_{i-1} \leq t_j < \hat{\tau}_1^i} D_{0t_j}^p C_0^p(t_j) + \sum_{\hat{\tau}_1^i \leq t_j < \hat{\tau}_2^i} D_{0t_j}^p C_1^p(\hat{\tau}_1^i, t_j) \right.$$

$$\left. + \sum_{\hat{\tau}_2^i \leq t_j \leq T} D_{0t_j}^p C_2^p(\hat{\tau}_1^i, \hat{\tau}_2^i, t_j) \right)$$

Using (6.41), we are thus able to estimate $\hat{M}_{t_i}$ and $V_{t_i}^1$ for all exercise date t_i on each path. Averaging over the $\#p_3$ paths gives a Monte Carlo estimate of the upper bound (6.40). Note that $\hat{V}_1$ is a low-biased estimator of V_1, so we cannot guarantee that (6.40) will generate an upper bound of the price.

6.12.2   The general case

In the general case, we can still write

$$V_0 = \inf_{\tau_1} \mathbb{E} \left[\sum_{t_j < \tau_1} D_{0t_j} C_0(t_j) + D_{0\tau_1} V_{\tau_1}^1 \right]$$

but now $V_{\tau_1}^1$ is the value just after time τ_1 of the chooser option with $n_T - 1$ tokens, given $\mathcal{F}_{\tau_1}$. Hence we can very easily adapt the above algorithm. We only need to store the strategies

- $\hat{\tau}^i$: the (nearly) optimal strategy from time t_i onward given that $\tau_1 \geq t_i$

- $\bar{\tau}^i$: the (nearly) optimal strategy after time t_i given that $\tau_1 = t_i$,

which result from the first step of the primal algorithm; $\hat{\tau}^i$ allows us to estimate the quantities $V_{t_i}^{\hat{\tau}^i}$ and $\mathbb{E}^{\mathbb{Q}}[D_{0t_i} V_{t_i}^{\hat{\tau}^i} | \mathcal{F}_{t_{i-1}}]$, and hence to compute M_{t_i} in (6.40); $\bar{\tau}^i$ allows us to estimate the quantities $V_{t_i}^1$ in (6.40).

It is remarkable that the suggested algorithm involves no more than a nested Monte Carlo simulation, despite the possibility of multiple exercises. As in the American case, to get a lower bound algorithm, it is enough to simulate $\#p_3 \times \#p_{3a} \times (N-1)$ paths, and globally, to build a lower and an upper bound, it is enough to simulate $\#p_1 + \#p_2 + \#p_3 \times \#p_{3a} \times (N-1)$ paths.

6.13   Numerical examples of pricing of chooser options

6.13.1   The multi-times restrikable put

In this section we test our naive primal and dual algorithms by pricing the multi-times restrikable put. This derivative has been introduced in Section 6.10. It can be seen as the sum of:

- the European (K, T)-put and

- the chooser option, which pays at maturity the put spread $(\alpha S_{t_i} - S_T)_+ - (K - S_T)_+$ when exercised for the last time at time t_i, and zero if never exercised.

Hence the price at time $t = 0$ of the restrikable put is the sum of $P_0(K, T)$, the price of the European (K, T)-put, which is read from the market, and V_0, the time 0 value of the chooser option, and it is enough to price the latter, hereafter "the option."

Here we consider a version where the issuer can restrike only when the underlying is greater than some $\underline{S}$, say $\underline{S} = S_0 = 100\%$. To take this additional conditional feature into account, one can simply add a huge negative amount in the coupon paid at exercise date t_i if the spot at that date is less than $\underline{S}$. The product parameters are given in Table 6.9. We allow the number of tokens, n_T, to vary from 1 to 4.

Table 6.9: Parameters of the restrikable put.

K	$\underline{S}$	α	ϕ	T
90%	100%	90%	0	1

Table 6.10: Numbers of paths used for pricing the restrikable put.

$\#p_1$	$\#p_2$	$\#p_3$	$\#p_{3a} = \#p_{3b}$
2,000	2,000	100	100

The numbers of paths are given in Table 6.10. For these (small) values, we get standard deviations between 10 and 14 bps. The effective lower (resp.

upper) bound is the one given in the tables below minus (resp. plus), say, two standard deviations.

To test the naive and dual algorithms, we price the option in the Black-Scholes model with parameters given in Table 6.11, assuming no dividends. This is of course not a good model to price such an option, which bears much forward smile risk. Here we just want to test the efficiency of our algorithm. To get the (nearly) optimal strategy, in the first step of the primal naive algorithm, we perform regressions. We shall compare two cases: linear regressions and nonparametric regressions.

Table 6.11: Pricing parameters.

σ	r	q
30%	0	0

6.13.1.1 Nonparametric regressions

Recall that (n, τ_n) denotes the number of tokens already used, and the date at which the last token was used (see Remark 6.14). In this section, we estimate the value of the option at exercise date t_i, for a spot value S, given a choice $c \in \{0, 1\}$ (say, 0 for continuation, 1 for restrike) and a decision-dependent variable ddv $= (n, \tau_n)$, by computing

$$\frac{\sum_{p=1}^{\#P_1} Y_{t_i}^p(c, \text{ddv}) \varphi_h(S, S_{t_i}^p)}{\sum_{p=1}^{\#P_1} \varphi_h(S, S_{t_i}^p)}$$

where $Y_{t_i}^p(c, \text{ddv})$ stands for the sum of actualized future optimal cash flows on path p at date t_i given (c, ddv), and φ_h is a kernel with bandwidth h, i.e.,

$$\varphi_h(x, y) = \frac{1}{h} \varphi \left(\frac{x - y}{h} \right)$$

with $\varphi \geq 0$. We pick the Gaussian kernel $\varphi(x) = \exp(-x^2)$. We allow the bandwidth h to vary from 0.5 to 8—to be compared with $S_0 = 100$. Too small a bandwidth leads to an estimate which is very sensitive to the sample paths; too large a bandwidth leads to an estimate which badly captures the dependence of the option value on the spot value S. Typically, the user must fine-tune this bandwidth, by trying different values of h.

The results are shown in Table 6.12. We indeed observe that there is a value of h, around 3 or 4, which maximizes the lower bound value. This optimal value of h seems to also minimize the spread between upper and lower bounds.

This is not surprising because this spread is the best quantitative indicator of the accuracy of the regression, and the more accurate the regression, the higher the lower bound.

Table 6.12: Lower and upper bounds (LB–UB) using nonparametric regressions.

$n_T \setminus h$	0.5	1	2	3	4	6	8
1	2.14–2.47	2.09–2.33	2.17–2.36	2.22–2.36	2.16–2.33	2.11–2.28	2.08–2.31
2	2.84–3.07	2.89–3.02	3.01–3.08	3.04–3.11	3.01–3.11	2.96–3.12	2.86–3.10
3	3.20–3.27	3.32–3.42	3.34–3.38	3.37–3.41	3.52–3.57	3.44–3.57	3.34–3.54
4	3.32–3.38	3.46–3.50	3.48–3.50	3.65–3.67	3.70–3.73	3.65–3.71	3.54–3.71

It seems that the bigger the number of tokens, n_T, the smaller the spread between upper and lower bounds. This spread is only a few bps, when we ignore the standard deviation. When we take it into account, it is worth 40 or 50 bps. But we stress again that here, we picked very small pricing parameters $\#p_1, \#p_2$, and larger values would lead to a much smaller standard deviation. This big standard error also materializes in the fact that in Table 6.12 the upper bound for some bandwidth value can be smaller than the lower bound for another bandwidth value. To correct this, we repeat the same experiment with more paths; see Table 6.13. For $n_T = 4$, the results are shown in Table 6.14, with a standard deviation of around 7 bps. As one can see, the highest lower bound, $3.62 - 2 \times 0.07 = 3.48$, is lower than the lowest upper bound, $3.53 + 2 \times 0.07 = 3.67$.

Table 6.13: Second set of numbers of paths used for pricing the restrikable put with nonparametric regressions.

$\#p_1$	$\#p_2$	$\#p_3$	$\#p_{3a} = \#p_{3b}$
10,000	10,000	100	100

6.13.1.2　Linear regressions

Here we estimate the value of the option at exercise date t_i, given a choice $c \in \{0, 1\}$ and a decision-dependent variable $\mathrm{ddv} = (n, \tau_n)$, by regressing the $\#p_1$ values of $Y_{t_i}^p(c, \mathrm{ddv})$ linearly against some regressors, where again

Table 6.14: Lower bound (LB) and upper bound (UB) using nonparametric regressions, $\#p_1 = \#p_2 = 10,000$, $n_T = 4$.

h	0.5	2	4	8
LB–UB	3.51–3.53	3.57–3.61	3.62–3.67	3.54–3.71

$Y_{t_i}^p(c, \text{ddv})$ stands for the sum of actualized future optimal cash flows on path p at date t_i given (c, ddv). We allow the set of regressors to vary:

- Set 1: $1, S_{t_i}$

- Set 2: $1, S_{t_i}, S_{t_i}^2$

- Set 3: $1, S_{t_i}, S_{t_i}^2, S_{t_i}^3$

The results are shown in Table 6.15. Set 3 gives the higher lower bounds and the tighter spread between upper and lower bounds. Again, we observe that the higher the number of tokens, n_T, the smaller this spread. With larger numbers of paths (see Table 6.16), we get the results shown in Table 6.17, with a standard deviation of 8 bps. There is good agreement between the linear regression prices and the nonparametric regression prices.

Table 6.15: Lower and upper bounds using linear regressions.

$n_T \setminus$ Set	1	2	3
1	1.86–2.40	2.10–2.42	2.18–2.41
2	2.71–3.14	3.00–3.20	3.09–3.19
3	3.23–3.65	3.47–3.59	3.47–3.53
4	3.34–3.76	3.68–3.82	3.68–3.70

Table 6.16: Second set of numbers of paths used for pricing the restrikable put with linear regressions.

$\#p_1$	$\#p_2$	$\#p_3$	$\#p_{3a}$
6,000	6,000	100	100

Table 6.17: Lower bound (LB) and upper bound (UB) using linear regressions, $\#p_1 = \#p_2 = 6,000$, $n_T = 4$.

Set	1	2	3
LB–UB	3.35–3.81	3.62–3.68	3.64–3.66

6.13.2 The passport option

In this section we illustrate the use of our efficient algorithm for chooser options. We consider the passport option, which has been introduced in Sections 5.9 and 6.10. We consider the passport option with monthly choices which pays

$$g(\pi_T) = \max(0, \pi_T)$$

at maturity $T = 1$. We consider the Black-Scholes model with zero interest rates, repos, and dividends. We pick $\sigma_{BS} = 30\%$. In this model and for this g, the optimal strategy is known as being short the index S when $\pi \geq 0$, and long the index otherwise. Hence we can compare the price given by the efficient algorithm presented in Section 6.11.2.2 with the optimal price. We also compare to two specific strategies: those where we decide to be always short or always long the index; see Table 6.18.

Table 6.18: Price of the passport option.

	Efficient strat.	Optimal strat.	Always short	Always long
Price	12.62	12.86	11.96	12.00

Here, we get the efficient strategy by performing nonparametric regression on the index value, with a large value of the window size. Indeed, we know that here the optimal strategy depends only on π, not on S. We have used nearest point projection, instead of cubic splines (see the above section). We have used the parameters in Table 6.19. The efficient algorithm is able to capture most of the difference between the optimal strategy and the two benchmark strategies.

Table 6.19: Pricing parameters.

$\#p_1$	$\#p_2$	nb std dev for grid	nb pts in grid
1,000	16,000	5	100

Conclusion

The pricing of American options is the basic example of a nonlinear problem arising in finance: because of the embedded optimal stopping problem, the American price of the sum of two payoffs is not the sum of the two American prices. The nonlinearity materializes in the pricing equation through the max operator: the price is the solution to a variational inequality. We have described several probabilistic methods to solve this particular type of nonlinear PDEs, as well as their extensions to chooser options. We will elaborate on the Longstaff-Schwartz and Tsitsiklis-Van Roy algorithms to build Monte Carlo methods for the valuation of options in the uncertain lapse and mortality model in Chapter 8.

6.14 Exercise: Bounds on prices of American call options

Prove that the price $\mathcal{A}(T, K)$ of an American call option with maturity T and strike K on a stock paying a dividend yield q admits the following bounds:

$$\mathcal{C}(T, K) \leq \mathcal{A}(T, K) \leq \mathcal{C}(T, K) + q \int_0^T \mathcal{C}\left(s, \frac{rK}{q}\right) ds$$

with $\mathcal{C}(s, x)$ the price of a European call option with maturity s and strike x.

Chapter 7

Backward Stochastic Differential Equations

I find it hard to focus looking forward. So I look backward.

— Iggy Pop

In this chapter, we introduce backward stochastic differential equations (in short BSDEs). Recently, first order BSDEs (1-BSDEs) have attracted attention in the mathematical finance community, see El Karoui *et al.* [101] for references and a review of applications. One of the key features of 1-BSDEs is that they provide a probabilistic representation of solutions of nonlinear parabolic PDEs, generalizing the Feynman-Kac formula. However, the corresponding PDEs cannot be nonlinear in the second order derivative and are therefore connected to HJB equations with no control on the diffusion term. In [81], Cheridito *et al.* provide a stochastic representation for solutions of fully nonlinear parabolic PDEs by introducing a new class of BSDEs, the so-called second order BSDEs (in short 2-BSDEs). We first present a short accessible introduction that gives the main definitions and results, explains why 1-BSDEs provide probabilistic representations of nonlinear PDEs, and addresses the question of numerical simulation. Then we deal with reflected 1-BSDEs and their application to the valuation of American-style options. Eventually we introduce 2-BSDEs.

7.1 First order BSDEs

7.1.1 Introduction

We consider a d-dimensional Itô process defined by the (forward) SDE

$$dX_t = b(t, X_t)\, dt + \sigma(t, X_t)\, dW_t$$

with the initial condition $X_0 = x \in \mathbb{R}^d$. W is an r-dimensional standard Brownian motion. Under a classical Lipschitz condition on the drift $b \in \mathbb{R}^d$ and volatility $\sigma \in \mathbb{R}^{d \times r}$, this SDE admits a unique strong solution (see e.g.,

[13]). First order BSDEs (in short 1-BSDEs) differ from (forward) SDEs in that we impose the terminal value.

DEFINITION 7.1 1-BSDE [18]   *A solution to a (Markovian) 1-BSDE is a couple* (Y, Z) *of* $(\mathcal{F}_t)$*-adapted Itô processes taking values in* $\mathbb{R} \times \mathbb{R}^r$ *and satisfying*

$$dY_t = -f(t, X_t, Y_t, Z_t) \, dt + Z_t. \, dW_t \tag{7.1}$$

with the terminal *condition* $Y_T = g(X_T)$. *The above equation means that*

$$Y_t = g(X_T) + \int_t^T f(s, X_s, Y_s, Z_s) \, ds - \int_t^T Z_s. \, dW_s \tag{7.2}$$

The deterministic function f *is called the driver or generator. We emphasize that the diffusion term* Z *is a part of the solution.*

Below, we give conditions ensuring the uniqueness and existence of 1-BSDEs. The proof depends strongly on the martingale representation theorem.

REMARK 7.1   In the simple (linear) case where $f = 0$, (7.2) reads

$$g(X_T) = Y_t + \int_t^T Z_s. \, dW_s$$

This simply means that the price at time t of the option with payout $g(X_T)$ is Y_t (a $\mathcal{F}_t$-measurable r.v.), and the corresponding delta strategy is given by the adapted process Z. This interpretation of Y as the price and Z as the delta (up to the gradient of the volatility) will be made systematic in Proposition 7.1. In the general case where $f \neq 0$, the payoff $g(X_T)$ can be decomposed as the sum of the price Y_t at time t, the delta strategy from t to T, and a source term that depends on the price and the delta:

$$g(X_T) = Y_t - \int_t^T f(s, X_s, Y_s, Z_s) \, ds + \int_t^T Z_s. \, dW_s$$

◻

Example 7.1 Martingale representation theorem
We consider the case where f does not depend on Y and Z. By taking the conditional expectation on both sides of (7.2), we get

$$Y_t + \int_0^t f(s, X_s) \, ds = \mathbb{E}\left[g(X_T) + \int_0^T f(s, X_s) \, ds \middle| \mathcal{F}_t\right] \equiv M_t$$

M_t is an $(\mathcal{F}_t)$-martingale with terminal value $g(X_T)+\int_0^T f(s, X_s)\, ds$ at $t = T$. By using the martingale representation theorem, there exists a unique adapted process Z such that

$$M_t = M_0 + \int_0^t Z_s.\, dW_s$$

Define $Y_t \equiv M_t - \int_0^t f(s, X_s)\, ds$. Then Y is adapted and

$$Y_t = g(X_T) + \int_t^T f(s, X_s)\, ds - \int_t^T Z_s.\, dW_s$$

The pair (Y, Z) gives the solution to the BSDE. $\qquad\qquad$ ⬜

Existence and uniqueness of 1-BSDEs are stated in the following theorem:

THEOREM 7.1 Pardoux-Peng [170]
Assume that $f : [0, T] \times \mathbb{R}^d \times \mathbb{R} \times \mathbb{R}^r \to \mathbb{R}$ is a deterministic function satisfying the Lipschitz condition

$$|f(t, x, y_1, z_1) - f(t, x, y_2, z_2)| \le K\left(|y_1 - y_2| + |z_1 - z_2|\right) \tag{7.3}$$

for some constant K independent of $(y_1, y_2) \in \mathbb{R}$ and $(z_1, z_2) \in \mathbb{R}^r$, that

$$\mathbb{E}\left[\int_0^T |f(t, X_t, 0, 0)|^2\, dt\right] < \infty \tag{7.4}$$

and that $g \in \mathrm{L}^2(\Omega, \mathcal{F}, \mathbb{R}^d)$. Then there is a unique adapted solution (Y, Z) to (7.2) satisfying

$$\sup_{0 \le t \le T} \mathbb{E}\left[Y_t^2\right] + \mathbb{E}\left[\int_0^T |Z_t|^2\, dt\right] < \infty \tag{7.5}$$

PROOF (sketch) We consider the function $\Phi(U, V) = (Y, Z)$, which maps the processes U, V to the processes Y, Z defined by

$$Y_t = g(X_T) + \int_t^T f(s, X_s, U_s, V_s)\, ds - \int_t^T Z_s.\, dW_s$$

The existence and uniqueness of the couple (Y, Z) is ensured by the martingale representation theorem (see Example 7.1). This actually requires the assumptions (7.3)–(7.4). Under these assumptions, (Y, Z) satisfy the integrability condition (7.5) provided it also holds for (U, V). If we can prove that

Φ is a strictly contractive map under a proper norm, then we can complete the proof thanks to the Picard fixed point theorem. We set $(Y, Z) = \Phi(U, V)$, $(Y', Z') = \Phi(U', V')$, $(\bar{Y}, \bar{Z}) = (Y' - Y, Z' - Z)$, $(\bar{U}, \bar{V}) = (U' - U, V' - V)$ and $\bar{f}_s = f(s, X_s, U'_s, V'_s) - f(s, X_s, U_s, V_s)$. By applying Itô's lemma to $e^{\beta s} \bar{Y}_s^2$ with β a positive number left unspecified for the moment, we obtain

$$-\bar{Y}_0^2 = \int_0^T e^{\beta s} \left(\beta \bar{Y}_s^2 - 2\bar{Y}_s \bar{f}_s + \bar{Z}_s^2 \right) ds + 2 \int_0^T e^{\beta s} \bar{Y}_s \bar{Z}_s . dW_s$$

By taking the expectation (after proving that $\int_0^t e^{\beta s} \bar{Y}_s \bar{Z}_s . dW_s$ is a true martingale), we obtain

$$\mathbb{E}[\bar{Y}_0^2] + \int_0^T e^{\beta s} \mathbb{E}[\beta \bar{Y}_s^2 + \bar{Z}_s^2] ds = 2 \int_0^T e^{\beta s} \mathbb{E}[\bar{Y}_s \bar{f}_s] ds$$

$$\leq 2K \int_0^T e^{\beta s} \mathbb{E}[|\bar{Y}_s| (|\bar{U}_s| + |\bar{V}_s|)] ds$$

$$\leq 4K^2 \int_0^T e^{\beta s} \mathbb{E}[\bar{Y}_s^2] ds + \frac{1}{2} \int_0^T e^{\beta s} \mathbb{E}[\bar{U}_s^2 + |\bar{V}_s|^2] ds$$

In the second line, we have used the Lipschitz property of f and in the last line, the identity $|ab| \leq \frac{1}{2} \left(a^2 + b^2 \right)$ with $a = 2\sqrt{2}K|\bar{Y}_s|$ and $b = \frac{1}{\sqrt{2}}(|\bar{U}_s| + |\bar{V}_s|)$. By choosing $\beta = 1 + 4K^2$, we get

$$\mathbb{E}[\bar{Y}_0^2] + \int_0^T e^{\beta s} \mathbb{E}[(\bar{Y}_s^2 + \bar{Z}_s^2)] ds \leq \frac{1}{2} \int_0^T e^{\beta s} \mathbb{E}[\bar{U}_s^2 + |\bar{V}_s|^2] ds$$

which reads

$$||\Phi(U', V') - \Phi(U, V)||_\beta^2 = ||(\bar{Y}, \bar{Z})||_\beta^2 \leq \frac{1}{2}||(\bar{U}, \bar{V})||_\beta^2 = \frac{1}{2}||(U', V') - (U, V)||_\beta^2$$

with the norm $||(U, V)||_\beta^2 = \int_0^T e^{\beta s} \mathbb{E}[U_s^2 + |V_s|^2] ds$. This implies that the map Φ is strictly contractive on the Banach space of all (U, V) such that $||(U, V)||_\beta^2 < \infty$, and therefore admits a unique fixed point in this space. □

Example 7.2 Linear 1-BSDE

Let us consider the linear 1-BSDE (Y, Z)

$$dY_t = - (a_t Y_t + Z_t . b_t + c_t) dt + Z_t . dW_t$$

with $Y_T = g(X_T) \in L^2(\Omega)$ and a_t, b_t, and c_t three deterministic functions of time. The above theorem ensures that this BSDE admits a unique solution that we specify below. Let us introduce the Itô process

$$d\Gamma_t = \Gamma_t(a_t \, dt + b_t . dW_t), \qquad \Gamma_0 = 1$$

the solution to which is given by

$$\Gamma_t = \exp\left(\int_0^t b_r . \, dW_r - \frac{1}{2}\int_0^t |b_r|^2 \, dr + \int_0^t a_r \, dr\right)$$

By applying Itô's lemma to $\Gamma_t Y_t$, we obtain that $M_t \equiv \Gamma_t Y_t + \int_0^t c_r \Gamma_r \, dr$ is a martingale. We deduce that the unique solution to the BSDE is given by (Y, Z) where

$$Y_t = \Gamma_t^{-1}\mathbb{E}\left[g(X_T)\Gamma_T + \int_t^T c_r \Gamma_r \, dr \Big| \mathcal{F}_t\right] \tag{7.6}$$

and Z_t is the integrand whose existence and uniqueness is guaranteed from the martingale representation theorem applied to M_t. □

7.1.2 First order BSDEs provide an extension of Feynman-Kac's formula for semilinear PDEs

Let us consider the semilinear PDE

$$\partial_t u + \mathcal{L}u + f(t, x, u(t, x), \sigma(t, x)' D_x u(t, x)) = 0, \quad (t, x) \in [0, T) \times \mathbb{R}^d \tag{7.7}$$

with the terminal condition $u(T, x) = g(x)$ and $\mathcal{L}$ the Itô generator of X. Such an equation appears when we consider a stochastic control problem with no control on the diffusion coefficient. Examples of such HJB equations include the uncertain lapse and mortality model equation for the pricing of reinsurance deals (see Section 5.6 and Chapter 8). It turns out that $(Y_t = u(t, X_t), Z_t = \sigma(t, X_t)' D_x u(t, X_t))$ is a solution to the BSDE (7.2). To be precise, a straightforward application of Itô's lemma gives the following:

PROPOSITION 7.1 Generalization of Feynman-Kac's formula
Let u be a function of $C^{1,2}$ satisfying (7.7) and suppose that there exists a constant C such that, for each $(t, x) \in [0, T] \times \mathbb{R}^d$,

$$|\sigma(t, x)' D_x u(t, x)| \leq C(1 + |x|)$$

Then $(Y_t = u(t, X_t), Z_t = \sigma(t, X_t)' D_x u(t, X_t))$ is the unique solution to 1-BSDE (7.2).

PROOF The processes Y and Z are adapted, $Y_T = g(X_T)$, and by applying Itô's lemma

$$\begin{aligned}
dY_t &= (\partial_t u + \mathcal{L}u) \, dt + \sigma(t, X_t)' D_x u(t, X_t) \, dW_t \\
&= -f(t, X_t, u_t, \sigma(t, X_t)' D_x u(t, X_t)) + \sigma(t, X_t)' D_x u(t, X_t) \, dW_t \\
&= -f(t, X_t, Y_t, Z_t) + Z_t . \, dW_t
\end{aligned}$$

⬜

Conversely, under additional regularity conditions, we can show [18] that the solution to BSDE (7.2) is a viscosity solution to (7.7). We state the precise theorem for $d = r = 1$ (see [101] for $d, r > 1$).

THEOREM 7.2 Pardoux-Peng
We suppose $d = r = 1$ and that f and g are uniformly continuous with respect to x. Then the function $u(t, x) \equiv Y_t^x$ is a viscosity solution to PDE (7.7), where the superscript x means that $X_t = x$. Furthermore, if we suppose that for each $R > 0$ there exists a continuous function $m_R : \mathbb{R}_+ \to \mathbb{R}$ such $m_R(0) = 0$ and

$$|f(t, x, y, z) - f(t, x', y, z)| \le m_R(|x - x'|(1 + |z|))$$

for all $t \in [0, T]$ and $|x|, |x'|, |z| \le R$, then u is the unique viscosity solution to PDE (7.7).

REMARK 7.2 Classical Feynman-Kac   Taking a linear driver $f(t, X_t, Y_t) = a(t, X_t)Y_t + c(t, X_t)$, we reproduce the classical Feynman-Kac formula

$$u(t, x) = \mathbb{E}\left[g(X_T)e^{\int_t^T a(r, X_r)dr} + \int_t^T c(r, X_r)e^{\int_t^r a(s, X_s)ds}\, dr \,\bigg|\, X_t = x\right]$$

from our solution to the linear 1-BSDE (see Equation (7.6)).   ⬜

7.1.3   Numerical simulation

Imagine one wants to solve the semilinear PDE (7.7) and $d \ge 3$. Since $d \ge 3$, finite difference methods are very slow. This is known as the curse of dimensionality. Since BSDE (7.2) is a viscosity solution to (7.7), one can instead solve numerically BSDE (7.2), i.e., look for an approximation of $Y_0^x \equiv u(0, x)$. This is the reason why the numerical simulation of 1-BSDEs is important in practice. Recall that for BSDEs, Y_T is known (as a function of the time-T value of the forward process X_T) but Y_0 is unknown so we will proceed backward, from $t = T$ to $t = 0$.

Let us divide $(0, T)$ into subintervals (t_{i-1}, t_i), $1 \le i \le n$, and set $\Delta t_i = t_i - t_{i-1}$, $\Delta W_{t_i} = W_{t_i} - W_{t_{i-1}}$, and $\Delta = \max_i \Delta t_i$. Simulating the forward process X gives us values of $X_T = X_{t_n}$, hence values of $Y_T = Y_{t_n} = g(X_{t_n})$. Then we proceed backward to compute the values of Y_{t_i}, $i = n - 1, n - 2, \ldots, 2, 1, 0$. We have

$$Y_{t_i} - Y_{t_{i-1}} = -\int_{t_{i-1}}^{t_i} f(s, X_s, Y_s, Z_s)\, ds + \int_{t_{i-1}}^{t_i} Z_s.\, dW_s$$

We consider the naive Euler scheme

$$Y_{t_i}^\Delta - Y_{t_{i-1}}^\Delta = -f(t_{i-1}, X_{t_{i-1}}^\Delta, Y_{t_{i-1}}^\Delta, Z_{t_{i-1}}^\Delta)\Delta t_i + Z_{t_{i-1}}^\Delta \cdot \Delta W_{t_i} \qquad (7.8)$$

together with the terminal condition $Y_{t_n}^\Delta = g(X_{t_n}^\Delta)$. In general, there are no $\mathcal{F}_{t_{i-1}}$-measurable random variables $Y_{t_{i-1}}^\Delta, Z_{t_{i-1}}^\Delta$ satisfying the equation above. Therefore, we consider the (*implicit*) scheme obtained by taking conditional expectations $\mathbb{E}_{i-1} \equiv \mathbb{E}[\cdot|\mathcal{F}_{t_{i-1}}]$ on both sides of (7.8):

$$Y_{t_n}^\Delta = g(X_{t_n}^\Delta) \qquad (7.9)$$
$$Y_{t_{i-1}}^\Delta = \mathbb{E}_{i-1}[Y_{t_i}^\Delta] + f(t_{i-1}, X_{t_{i-1}}^\Delta, Y_{t_{i-1}}^\Delta, Z_{t_{i-1}}^\Delta)\Delta t_i \qquad (7.10)$$
$$Z_{t_{i-1}}^\Delta = \frac{1}{\Delta t_i}\mathbb{E}_{i-1}[Y_{t_i}^\Delta \Delta W_{t_i}] \qquad (7.11)$$

The last equation is obtained by multiplying both sides of (7.8) with ΔW_{t_i} and taking the conditional expectation $\mathbb{E}_{i-1}[\cdot]$. Note that for Bachelier's model, the Malliavin weight for the delta is precisely $\frac{\Delta W_T}{\Delta T}$. Equation (7.11) is the Malliavin estimate of the delta $Z_{t_{i-1}}^\Delta$ based on the price $Y_{t_i}^\Delta$ and on Bachelier's model Malliavin weight. We need to use regression approximations or Malliavin's integration by part formula to compute the conditional expectations $\mathbb{E}_{i-1}[Y_{t_i}^\Delta \Delta W_{t_i}]$ and $\mathbb{E}_{i-1}[Y_{t_i}^\Delta]$ (see Section 6.8.5). This scheme is *implicit* as $Y_{t_{i-1}}^\Delta$ appears on both sides of (7.10). As a consequence, we need to use a Picard fixed point method. In order to overpass this difficulty, we can consider *explicit* schemes where (7.10) is replaced by

$$Y_{t_{i-1}}^\Delta = \mathbb{E}_{i-1}[Y_{t_i}^\Delta + f(t_i, X_{t_i}^\Delta, Y_{t_i}^\Delta, Z_{t_i}^\Delta)\Delta t_i] \qquad (7.12)$$

or by

$$Y_{t_{i-1}}^\Delta = \mathbb{E}_{i-1}[Y_{t_i}^\Delta] + f(t_{i-1}, X_{t_{i-1}}^\Delta, \mathbb{E}_{i-1}[Y_{t_i}^\Delta], Z_{t_{i-1}}^\Delta)\Delta t_i \qquad (7.13)$$

Similarly, one can introduce θ-schemes like

$$Y_{t_{i-1}}^\Delta = \mathbb{E}_{i-1}[Y_{t_i}^\Delta] + \theta\mathbb{E}_{i-1}[f(t_i, X_{t_i}^\Delta, Y_{t_i}^\Delta, Z_{t_i}^\Delta)]\Delta t_i$$
$$+ (1-\theta)f(t_{i-1}, X_{t_{i-1}}^\Delta, Y_{t_{i-1}}^\Delta, Z_{t_{i-1}}^\Delta)\Delta t_i$$

or

$$Y_{t_{i-1}}^\Delta = \mathbb{E}_{i-1}[Y_{t_i}^\Delta] + \theta f(t_{i-1}, X_{t_{i-1}}^\Delta, \mathbb{E}_{i-1}[Y_{t_i}^\Delta], Z_{t_{i-1}}^\Delta)\Delta t_i$$
$$+ (1-\theta)f(t_{i-1}, X_{t_{i-1}}^\Delta, Y_{t_{i-1}}^\Delta, Z_{t_{i-1}}^\Delta)\Delta t_i$$

Scheme (7.9)–(7.11) converges as stated below.

THEOREM 7.3 Convergence [66]
Consider the implicit scheme (7.9)–(7.11). Define $Z_t^\Delta = Z_{t_i}^\Delta$, $Y_t^\Delta = Y_{t_i}^\Delta$ for $t \in [t_i, t_{i+1})$. Then

$$\limsup_{\Delta \to 0} \Delta^{-1}\left(\sup_{0 \le t \le T}\mathbb{E}[|Y_t^\Delta - Y_t|^2] + \mathbb{E}\left[\int_0^T |Z_t^\Delta - Z_t|^2\, dt\right]\right) < \infty$$

Note, however, that for a fixed number of Monte Carlo paths, the above schemes tend to diverge numerically when $\Delta \to 0$; see for instance Table 9.1 in the case of a second order BSDE. This comes probably from the fact that errors in the estimations of the successive Y_{t_i}'s, i.e., errors in the estimations of the conditional expectations, propagate and blow up from $i = n$ to $i = 0$. To avoid estimating conditional expectations, a new forward Monte Carlo scheme using branching diffusions will be presented in Chapter 13.

Example 7.3 CVA pricing
The semilinear PDE arising in the pricing of counterparty risk in case of risky close-out is (see Equation (5.55) and Chapter 13)

$$\partial_t u + \mathcal{L}u + \beta(u^+ - u) = 0, \qquad u(T, x) = g(x)$$

with $\mathcal{L}$ the Itô generator of a diffusion X_t. The corresponding BSDE is

$$dY_t = -\beta(Y_t^+ - Y_t)\,dt + Z_t.\,dW_t, \qquad Y_T = g(X_T)$$

This can be solved numerically using either of the following schemes:

$$Y_{t_{i-1}}^{\Delta} = \frac{1}{1 + \beta\Delta t_i \mathbf{1}_{\mathbb{E}_{i-1}[Y_{t_i}^{\Delta}] \le 0}} \mathbb{E}_{i-1}[Y_{t_i}^{\Delta}]$$

or $\quad Y_{t_{i-1}}^{\Delta} = \mathbb{E}_{i-1}[Y_{t_i}^{\Delta} + \beta\Delta t_i((Y_{t_i}^{\Delta})^+) - Y_{t_i}^{\Delta}]$

or $\quad Y_{t_{i-1}}^{\Delta} = \mathbb{E}_{i-1}[Y_{t_i}^{\Delta}] + \beta\Delta t_i(\mathbb{E}_{i-1}[Y_{t_i}^{\Delta}]^+ - \mathbb{E}_{i-1}[Y_{t_i}^{\Delta}])$

or θ-schemes. Note that these schemes do not depend on Z^{Δ}. ☐

7.2 Reflected first order BSDEs

Reflected BSDEs provide a stochastic representation of semilinear partial differential variational inequalities (PDI).

DEFINITION 7.2 Reflected 1-BSDE   *A solution to a (Markovian) reflected 1-BSDE is a triplet $(Y, Z, L) \in \mathbb{R} \times \mathbb{R}^r \times \mathbb{R}$ of $(\mathcal{F}_t)$-adapted Itô processes satisfying*

$$dY_t = -f(t, X_t, Y_t, Z_t)\,dt + Z_t.\,dW_t - dL_t$$

with the terminal condition $Y_T = g(X_T)$ and L_t a nondecreasing càdlàg process satisfying the minimum action condition

$$\int_0^T (Y_t - g(X_t))dL_t = 0$$

The measure L_t is minimal in the sense that it is supported by the set of times t where Y_t touches the boundary defined by g. The above equation means that

$$Y_t = g(X_T) + \int_t^T f(s, X_s, Y_s, Z_s)\, ds - \int_t^T Z_s. dW_s + (L_T - L_t) \quad (7.14)$$

Let us consider the semilinear PDI

$$\max\left(\partial_t u + \mathcal{L}u + f(t, x, u(t, x), \sigma(t, x)' D_x u(t, x)), g - u\right) = 0 \quad (7.15)$$

with the terminal condition $u(T, x) = g(x)$ and $\mathcal{L}$ the Itô generator of X. Such an equation appears when one prices American-style options, see Section 6.7. As in Section 7.1.2, one can show that (see Proposition 6.1 for a similar computation)

$$Y_t = u(t, X_t)$$
$$Z_t = \sigma(t, X_t)' D_x u(t, X_t)$$
$$L_t = (f + \mathcal{L}g)^- \mathbf{1}_{Y_t = g(X_t)}$$

is the unique solution to BSDE (7.14). We recall that $(f + \mathcal{L}g)^-$ stands for the negative part of $f + \mathcal{L}g$. As a consequence, the numerical simulation of reflected BSDEs is a way to approximate solutions of Equation (7.15). This is of interest in a high-dimensional setting. Following the discussion in Section 7.1.3, we can build a backward scheme for the pricing of American options by setting:

$$Y_{t_{i-1}}^\Delta = \max\left(\mathbb{E}_{i-1}[Y_{t_i}^\Delta] + f(t_{i-1}, X_{t_{i-1}}^\Delta, Y_{t_{i-1}}^\Delta, Z_{t_{i-1}}^\Delta)\Delta t_i, g(X_{t_{i-1}})\right) \quad (7.16)$$
$$Z_{t_{i-1}}^\Delta = \frac{1}{\Delta t_i}\mathbb{E}_{i-1}[Y_{t_i}^\Delta \Delta W_{t_i}]$$

In the particular case where $f = 0$, the above scheme degenerates into the programming equation for American options:

$$Y_{t_{i-1}}^\Delta = \max\left(\mathbb{E}_{i-1}[Y_{t_i}^\Delta], g(X_{t_{i-1}})\right) \quad (7.17)$$

7.3 Second order BSDEs

7.3.1 Introduction

In 1-BSDEs, second order derivatives only arise linearly through Itô's formula from the quadratic variation of the state process. We now introduce second order BSDEs for which the corresponding PDE can be nonlinear in the second order derivatives and are therefore connected to HJB equations with a control on the diffusion coefficient. Examples of such HJB equations include the Black-Scholes-Barenblatt equation (5.3), the Leland equation (5.7), Equation (5.18) for illiquid markets or the pricing equation (5.57) for the passport option.

We consider a forward process X satisfying

$$dX_t = b(t, X_t)\, dt + \sigma(t, X_t)\, dW_t, \qquad X_0 = x \in \mathbb{R}^d$$

where W is an r-dimensional standard Brownian motion.

DEFINITION 7.3 Second order BSDE [81]   *Let $(Y_t, Z_t, \Gamma_t, \alpha_t)_{t \in [0,T]}$ be a quadruple of $(\mathcal{F}_t)$-adapted processes taking values in $\mathbb{R}$, $\mathbb{R}^d$, $\mathcal{S}^d$, and $\mathbb{R}^d$ respectively.*[1] *We call (Y, Z, Γ, α) a solution to a (Markovian) 2-BSDE if*

$$dY_t = -f(t, X_t, Y_t, Z_t, \Gamma_t)\, dt + Z_t \diamond \sigma(t, X_t)\, dW_t \qquad (7.18)$$
$$dZ_t = \alpha_t\, dt + \Gamma_t \sigma(t, X_t)\, dW_t$$
$$Y_T = g(X_T)$$

where $\diamond$ denotes the Stratonovich integral. The use of the Stratonovich product is only for convenience and can be replaced by an Itô integral

$$dY_t = \left(-f(t, X_t, Y_t, Z_t, \Gamma_t) + \frac{1}{2}\mathrm{tr}\left(\sigma(t, X_t)\sigma(t, X_t)'\Gamma_t\right)\right) dt + Z_t.\sigma(t, X_t)\, dW_t$$

7.3.2   Second order BSDEs provide an extension of Feynman-Kac's formula for fully nonlinear PDEs

Let us consider the fully nonlinear parabolic PDE

$$\partial_t u(t, x) + f(t, x, u, D_x u, D_x^2 u) = 0, \qquad (t, x) \in [0, T) \times \mathbb{R}^d \qquad (7.19)$$
$$u(T, x) = g(x)$$

As in Proposition 7.1, a straightforward application of Itô's formula gives the following:

PROPOSITION 7.2 [81]
Let u be a smooth function (smooth enough to apply Itô's formula) satisfying PDE (7.19). Then $(Y_t = u(t, X_t), Z_t = D_x u(t, X_t), \Gamma_t = D_x^2 u(t, X_t), \alpha_t = (\partial_t + \mathcal{L})D_x u(t, X_t))$ is a solution to 2-BSDE (7.18), where $\mathcal{L}$ is the Itô generator of X.

Recently, in [187], Touzi, Soner, and Zhang proved existence for 2-BSDEs under uniform Lipschitz conditions. Below we give a uniqueness result. The precise result is based on two assumptions:

---

[1] $\mathcal{S}^d$ stands for the space of d-dimensional symmetric matrices.

A(f): $f : [0, T) \times \mathbb{R}^d \times \mathbb{R} \times \mathbb{R}^d \times \mathcal{S}^d \to \mathbb{R}$ is continuous, Lipschitz-continuous in y uniformly in (t, x, z, Γ) and for some $C, p > 0$

$$|f(t, x, y, z, \Gamma)| \leq C \left(1 + |y| + |x|^p + |z|^p + |\Gamma|^p\right)$$

A(comp): if $\bar{u} : [0, T] \times \mathbb{R}^d \to \mathbb{R}$ (resp. $\underline{u}$) is a lower semi-continuous (resp. u.s.c) viscosity supersolution (resp. subsolution) of (7.19) with $\bar{u}(t, x) \geq -C(1 + |x|^p)$ and $\underline{u}(t, x) \leq C(1 + |x|^p)$, then $\bar{u}(T, \cdot) \geq \underline{u}(T, \cdot)$ implies that $\bar{u} \geq \underline{u}$ on $[0, T] \times \mathbb{R}^d$.

THEOREM 7.4 Uniqueness [81]
Under **A(f)** *and* **A(comp)**, *for every g having polynomial growth, there is at most one solution to (7.18).*

7.3.3 Numerical simulation

The numerical simulation of 2-BSDEs is a way to build approximations of solutions to PDE (7.19). Proceeding by analogy with 1-BSDEs, we obtain the following (implicit) discretization scheme for 2-BSDEs as reported in [81]:

Scheme 2-BSDE:

$$Y_{t_n}^{\Delta} = g(X_{t_n}^{\Delta})$$
$$Z_{t_n} = Dg(X_{t_n})$$
$$Y_{t_{i-1}}^{\Delta} = \mathbb{E}_{i-1}[Y_{t_i}^{\Delta}] + \Big(f(t_{i-1}, X_{t_{i-1}}^{\Delta}, Y_{t_{i-1}}^{\Delta}, Z_{t_{i-1}}^{\Delta}, \Gamma_{t_{i-1}}^{\Delta}) \quad\quad (7.20)$$
$$\qquad\qquad - \frac{1}{2}\mathrm{tr}[\sigma(t_{i-1}, X_{t_{i-1}}^{\Delta})\sigma(t_{i-1}, X_{t_{i-1}}^{\Delta})'\Gamma_{t_{i-1}}^{\Delta}]\Big)\Delta t_i$$
$$Z_{t_{i-1}}^{\Delta} = \frac{1}{\Delta t_i}\sigma(t_{i-1}, X_{t_{i-1}}^{\Delta})'^{-1}\mathbb{E}_{i-1}[Y_{t_i}^{\Delta}\Delta W_{t_i}]$$
$$\Gamma_{t_{i-1}}^{\Delta} = \frac{1}{\Delta t_i}\mathbb{E}_{i-1}[Z_{t_i}^{\Delta}\Delta W_{t_i}']\sigma(t_{i-1}, X_{t_{i-1}}^{\Delta})^{-1}$$

The quantity P&L $\times \Delta t_i$ where

$$\text{P\&L} \equiv f(t_{i-1}, X_{t_{i-1}}^{\Delta}, Y_{t_{i-1}}^{\Delta}, Z_{t_{i-1}}^{\Delta}, \Gamma_{t_{i-1}}^{\Delta})$$
$$\qquad\qquad - \frac{1}{2}\mathrm{tr}[\sigma(t_{i-1}, X_{t_{i-1}}^{\Delta})\sigma(t_{i-1}, X_{t_{i-1}}^{\Delta})'\Gamma_{t_{i-1}}^{\Delta}] \quad (7.21)$$

is the so-called gamma-theta P&L between t_{i-1} and t_i. This scheme requires a final condition for $Z_{t_n} = Dg(X_{t_n})$. Since the payoffs usually considered in finance are not smooth, this scheme may perform poorly. We suggest a new scheme in Section 9.4.1.

Using an induction argument, one can show that the random variables $Y_{t_i}^{\Delta}$, $Z_{t_i}^{\Delta}$, and $\Gamma_{t_i}^{\Delta}$ are deterministic functions of $X_{t_i}^{\Delta}$. It then follows that the conditional expectations above can be replaced by

$$\mathbb{E}_{i-1}[Y_{t_i}^{\Delta} \Delta W_{t_i}] = \mathbb{E}[Y_{t_i}^{\Delta} \Delta W_{t_i} | X_{t_{i-1}}^{\Delta}]$$
$$\mathbb{E}_{i-1}[Z_{t_i}^{\Delta} \Delta W_{t_i}'] = \mathbb{E}[Z_{t_i}^{\Delta} \Delta W_{t_i}' | X_{t_{i-1}}^{\Delta}]$$

REMARK 7.3   Note that PDE (7.19) does not depend on the function σ, which can be chosen arbitrarily.   □

The convergence of this scheme has been obtained recently in [105] in the case where the diffusion coefficient dominates the partial gradient of the Theta-Gamma P&L with respect to its Hessian component.

Note that in Section 9.4.1 we will suggest a new numerical scheme for simulating BSDEs, which basically uses the Malliavin weight directly for the Black-Scholes gamma instead of using twice the Malliavin weight for the Bachelier delta. As mentioned earlier in Section 7.1.3, for a fixed number of Monte Carlo paths, BSDE schemes may diverge numerically when $\Delta \to 0$, as shown for instance in Table 9.1. Nevertheless, in Section 9.5, we report prices of various payoffs valued in the uncertain volatility model using numerical simulation of BSDEs that are quite accurate.

Of course, one can also consider explicit schemes like

$$Y_{t_{i-1}}^{\Delta} = \mathbb{E}_{i-1}\left[Y_{t_i}^{\Delta} + \left(f(t_i, X_{t_i}^{\Delta}, Y_{t_i}^{\Delta}, Z_{t_i}^{\Delta}, \Gamma_{t_i}^{\Delta})\right.\right.$$
$$\left.\left. - \frac{1}{2}\mathrm{tr}[\sigma(t_i, X_{t_i}^{\Delta})\sigma(t_i, X_{t_i}^{\Delta})'\Gamma_{t_i}^{\Delta}]\right)\Delta t_i\right] \quad (7.22)$$

or

$$Y_{t_{i-1}}^{\Delta} = \mathbb{E}_{i-1}[Y_{t_i}^{\Delta}] + \left(f(t_{i-1}, X_{t_{i-1}}^{\Delta}, \mathbb{E}_{i-1}[Y_{t_i}^{\Delta}], Z_{t_{i-1}}^{\Delta}, \Gamma_{t_{i-1}}^{\Delta})\right.$$
$$\left. - \frac{1}{2}\mathrm{tr}[\sigma(t_{i-1}, X_{t_{i-1}}^{\Delta})\sigma(t_{i-1}, X_{t_{i-1}}^{\Delta})'\Gamma_{t_{i-1}}^{\Delta}]\right)\Delta t_i \quad (7.23)$$

Example 7.4 Passport options

The two-dimensional nonlinear parabolic PDE for the pricing of passport options (see Section 5.9)

$$\partial_t u(t, S, \pi) + \sigma^2 S^2 \left(\frac{1}{2}\partial_S^2 u + \frac{1}{2}\partial_\pi^2 u + |\partial_{S\pi} u|\right) = 0$$

can be solved by simulating the following 2-BSDE where $\overline{\Delta} \in \{-1, +1\}$ is arbitrarily fixed:

$$dS_t = \sigma S_t \, dW_t$$
$$d\pi_t = \sigma S_t \overline{\Delta} \, dW_t$$
$$dY_t = \sigma^2 S_t^2 \left(\overline{\Delta} \Gamma_t^{S\pi} - |\Gamma_t^{S\pi}| \right) dt + \sigma S_t \left(Z_t^S + Z_t^\pi \overline{\Delta} \right) dW_t$$
$$dZ_t^S = \alpha^S \, dt + \sigma S_t \left(\Gamma_t^{SS} + \Gamma_t^{S\pi} \overline{\Delta} \right) dW_t$$
$$dZ_t^\pi = \alpha^\pi \, dt + \sigma S_t \left(\Gamma_t^{\pi S} + \Gamma_t^{\pi\pi} \overline{\Delta} \right) dW_t$$
$$Y_T = g(\pi_T), \qquad Z_T^S = 0, \qquad Z_T^\pi = g'(\pi_T)$$

Conclusion

BSDEs provide stochastic representations of solutions of nonlinear parabolic PDEs, generalizing the Feynman-Kac formula. As a consequence, they are a very useful tool when one needs to solve such a PDE in high dimension. Numerical schemes for BSDEs lead to approximations of solutions of such PDEs.

We will use BSDEs in the next two chapters to solve nonlinear pricing equations:

- First order BSDEs will allow us to price reinsurance deals in the uncertain lapse and mortality model in Chapter 8.

- Second order BSDEs will allow us to price options in the uncertain volatility model in Chapter 9.

7.4   Exercise: An example of semilinear PDE

We consider the semilinear PDE

$$\partial_t u + \mathcal{L}u + f(u) = 0, \qquad u(T, x) = g(x)$$

with $\mathcal{L}$ the Itô generator of a process X_t.

1. Write the BSDE associated to this PDE.

2. Write a numerical scheme for solving this BSDE.

3. Implement your scheme for $f(u) = \beta \left(u^2 - u \right)$, $\mathcal{L} = \frac{1}{2} x^2 \sigma_{\text{BS}}^2 \partial_x^2 u$, and $g(x) = \mathbf{1}_{x>1}$, with $\beta = 0.05$, $\sigma_{\text{BS}} = 0.2$, and $T = 10$ years. Compare with Table 13.3.

Chapter 8

The Uncertain Lapse and Mortality Model

> If man were immortal, do you realize what his meat bills would be?
>
> — Woody Allen

Reinsurance deals have been briefly introduced in Section 5.6. In this chapter, we present the uncertain lapse and mortality model (ULMM) for the valuation of such contracts. The price of those long-term deals depends on mortality rates and lapse rates. One cannot hedge against the future movements of these non-financial variables. The ULMM provides a worst-case price, under the assumption that the mortality and lapse rates stay within a certain range before the maturity of the deal.

8.1 Reinsurance deals

Reinsurance deals are sold by banks to insurance companies that need to hedge their book of variable annuities. Variable annuities are individual insurance policies sold to retail buyers. Typically, the payout of a variable annuity depends on the performance of a fund, or a fund of funds, and has three characteristics: (a) a capital is guaranteed at maturity, (b) a capital is guaranteed in case of death, and (c) the subscriber can lapse at any time. To lapse means to definitively cancel the product. A typical reinsurance deal then consists of three payoffs:

- At maturity, in the case where the insurance subscriber is still alive, the issuer delivers a put on the net asset value (NAV) X of the underlying fund, whose strike is denoted K_{mat}. Typically, $K_{\mathrm{mat}} \approx 90\% X_0$, where X_0 is the NAV of the underlying fund at inception:

$$u^{\mathrm{mat}}(x) = (K_{\mathrm{mat}} - x)_+$$

- In the case where the insurance subscriber dies before maturity, the issuer delivers a put on X whose strike is denoted K_D. Typically, $K_D = 100\% X_0$:

$$u^D(t, x) = (K_D - x)_+$$

- Each month, the issuer receives a constant fee.

All three payoffs are canceled as soon as the insurance subscriber lapses, which he or she can do whenever he or she wants:

$$u^L(t, x) = 0$$

Moreover, often, all three payoffs become up-and-out after some date and until maturity. The up-and-out barrier can be changed every day by the insurance subscriber. It must belong to a list of discrete possible barriers, or to a continuum of possible barriers, and must be higher than the current NAV of the fund.

Reinsurance deals are often called GMxB. For instance, GMIB stands for Guaranteed Minimum Investment Benefit and corresponds to the payoff u^{mat}. GMDB stands for Guaranteed Minimum Death Benefit and corresponds to the payoff u^D. Typical maturities are on the order of ten years.

This chapter is organized as follows. In Section 8.2, we first look at the case where the lapse and mortality rates are known deterministic functions of time. Then we introduce the uncertain lapse and mortality model (ULMM) in Section 8.3. The case of path-dependent payoffs is dealt with in Section 8.4, and the pricing of the option on the choice of the up-and-out barrier in Section 8.5. After describing examples of finite difference scheme implementations in Section 8.6, we introduce original Monte Carlo methods in Sections 8.7 and 8.8. In Section 8.9, we compare with the Monte Carlo method derived from the numerical approximation of first order BSDEs. Eventually, we show and analyze the results of numerical experiments for both PDE methods and Monte Carlo methods in Sections 8.10 and 8.11.

8.2   The deterministic lapse and mortality model

We model death (D) and lapse (L) as independent default events, jointly independent of the path ($X_t, 0 \leq t \leq T$) of the NAV of the underlying fund. They are supposed to be first jump times of two independent Poisson processes, τ^D and τ^L, with respective deterministic intensities λ_t^D and λ_t^L. Once one of these defaults occurs, the product is canceled, and the NAV is unchanged (no delta-hedging loss is incurred).

Due to the random default events, perfect hedging is impossible: defaults are unpredictable and we cannot hedge them out. In Section 1.2 we derived a pricing equation for hedging in average, given the path of the underlyings, in the case of one random event; see Example 1.1. In the case of reinsurance deals, we must consider two random events, the fair value of this contract is

$$u(t,x) = \mathbb{E}_{t,x}^{\mathbb{Q}}\Big[u^{\mathrm{mat}}(X_T)\mathbf{1}_{\tau^D \geq T}\mathbf{1}_{\tau^L \geq T} - \int_t^T \alpha\, dt$$

$$+ u^D(\tau^D, X_{\tau^D})\mathbf{1}_{\tau^D < T}\mathbf{1}_{\tau^D \leq \tau^L} + u^L(\tau^L, X_{\tau^L})\mathbf{1}_{\tau^L < T}\mathbf{1}_{\tau^L < \tau^D}\Big] \quad (8.1)$$

and, slightly adapting the derivation of PDE (1.22) in Example 1.1, we get that it is the solution to the PDE

$$\partial_t u + \frac{1}{2}\sigma^2 x^2 \partial_x^2 u - \alpha + \lambda_t^D\left(u^D - u\right) + \lambda_t^L\left(u^L - u\right) = 0 \quad (8.2)$$

with terminal condition $u(T,x) = u^{\mathrm{mat}}(x)$. Here we have assumed that the fund has constant volatility σ, and we have modeled the fee the issuer receives monthly through the rate α. We have also assumed zero interest and repo rates. In order to justify our choice (8.1) (and thus its pricing equation (8.2)), below we give a financial interpretation in terms of the P&L of a delta-hedged strategy, averaged over all random default events.

Interpretation of Equation (8.2)

Let us assume that the price of the reinsurance deal is a deterministic function u of current time t and current underlying X_t. With probability

$$1 - p_t \equiv 1 - \exp\left(-\int_0^t \left(\lambda_u^D + \lambda_u^L\right) du\right) \quad (8.3)$$

a default has occurred before t and the P&L of the issuer's delta-hedged position incurred between t and $t + dt$ is zero. Given that no default has occurred before t (this event has probability p_t):

- with probability $1 - \lambda_t^D\, dt - \lambda_t^L\, dt$, no default occurs before $t + dt$ and the P&L of the issuer's delta-hedged position incurred between t and $t + dt$ is, at first order in dt,

$$- (u(t + dt, X_t + dX_t) - u(t, X_t)) + \Delta\, dX_t + \alpha\, dt$$
$$= -\partial_t u\, dt - \frac{1}{2}\partial_x^2 u\, d\langle X \rangle_t + (\Delta - \partial_x u)\, dX_t + \alpha\, dt,$$

- with probability $\lambda_t^D\, dt$, death occurs between t and $t + dt$ and the P&L of the issuer's delta-hedged position incurred between t and $t + dt$ is $-\left(u^D - u\right)$;

- with probability $\lambda_t^L\, dt$, lapse occurs between t and $t + dt$ and the P&L of the issuer's delta-hedged position incurred between t and $t + dt$ is $-\left(u^L - u\right)$;

- both defaults occur between t and $t + dt$ with probability $O(dt^2)$.

As a consequence, at first order in dt, the P&L of the issuer's delta-hedged position incurred between t and $t + dt$ is

$$d\text{P\&L}_t = \mathbf{1}_{\text{death or lapse} < t} \times 0 - \mathbf{1}_{\text{death} \in [t, t+dt]} \left(u^D - u \right) - \mathbf{1}_{\text{lapse} \in [t, t+dt]} \left(u^L - u \right)$$
$$+ \mathbf{1}_{\text{death\&lapse} > t + dt} \left(- \partial_t u \, dt - \frac{1}{2} \partial_x^2 u \, d\langle X \rangle_t + (\Delta - \partial_x u) \, dX_t + \alpha \, dt \right)$$

Due to the default events, this random variable cannot be canceled out by a suitable choice of (u, Δ). However, conditionally on the underlying path, by independence of the couple of default times with the underlying path, its average value, at first order in dt, is worth

$$\mathbb{E}^{\mathbb{Q}} \left[d\text{P\&L}_t \middle| (X_u, 0 \le u \le T) \right] = p_t \bigg(- \partial_t u \, dt - \frac{1}{2} \partial_x^2 u \, d\langle X \rangle_t$$
$$+ (\Delta - \partial_x u) \, dX_t + \alpha \, dt - \lambda_t^D \, dt \left(u^D - u \right) - \lambda_t^L \, dt \left(u^L - u \right) \bigg)$$

This is the value of the P&L of the issuer's delta-hedged position incurred between t and $t + dt$ averaged over default events, and only over default events: no average is taken over underlying values.

In the Black-Scholes model with volatility σ, $d\langle X \rangle_t = \sigma^2 X_t^2 \, dt$ and by choosing $\Delta = \partial_x u$ and solving PDE (8.2) with terminal condition $u(T, x) = u^{\text{mat}}(x)$, we build a pricing function u such that the default risk is hedged in average. Indeed, if defaults are mutually independent and independent of the underlying path, pricing M such products on the same underlying but with M independent default risks with function u yields the total P&L per contract:

$$\frac{1}{M} \sum_{m=1}^{M} \int_0^T \left(\mathbf{1}_{\text{death}_m \& \text{lapse}_m > t + dt} \left(- \partial_t u - \frac{1}{2} \sigma^2 X_t^2 \partial_x^2 u + \alpha \right) dt \right.$$
$$\left. - \mathbf{1}_{\text{death}_m \in [t, t+dt]} \left(u^D - u \right) - \mathbf{1}_{\text{lapse}_m \in [t, t+dt]} \left(u^L - u \right) \right)$$
$$= \int_0^T \left(\left(\frac{1}{M} \sum_{m=1}^{M} \mathbf{1}_{\text{death}_m \& \text{lapse}_m > t + dt} \right) \left(- \partial_t u - \frac{1}{2} \sigma^2 X_t^2 \partial_x^2 u + \alpha \right) dt \right.$$
$$\left. - \left(\frac{1}{M} \sum_{m=1}^{M} \mathbf{1}_{\text{death}_m \in [t, t+dt]} \right) \left(u^D - u \right) - \left(\frac{1}{M} \sum_{m=1}^{M} \mathbf{1}_{\text{lapse}_m \in [t, t+dt]} \right) \left(u^L - u \right) \right)$$
$$\xrightarrow[M \to \infty]{} \int_0^T p_t \left(- \partial_t u - \frac{1}{2} \sigma^2 X_t^2 \partial_x^2 u + \alpha - \lambda_t^D \left(u^D - u \right) - \lambda_t^L \left(u^L - u \right) \right) dt$$

which is worth zero because u is a solution to (8.2). The limit results from the strong law of large numbers. Hence, the (unaveraged) total P&L per contract vanishes, in the limit when the number of contracts tends to infinity.

8.3 The uncertain lapse and mortality model

We now consider that the mortality rate and the lapse rate are uncertain. In Section 5.6, we have derived the Hamilton-Jacobi-Bellman equation for the valuation of a reinsurance deal in the case where there is only one type of default event, namely, the death of the insurance subscriber, when the mortality rate is uncertain, see Equation (5.47). One can easily extend this result to the case where two types of default can occur (death and lapse), with both rates of default being uncertain. We then get the following nonlinear pricing PDE:

$$\partial_t u + \frac{1}{2}\sigma^2 x^2 \partial_x^2 u - \alpha$$
$$+ \lambda^D \left(t, u^D - u\right)\left(u^D - u\right) + \lambda^L \left(t, u^L - u\right)\left(u^L - u\right) = 0 \quad (8.4)$$

with terminal condition $u(T,x) = u^{\mathrm{mat}}(x)$, where

$$\lambda^D(t,y) = \begin{cases} \overline{\lambda}^D(t) & \text{if } y \geq 0 \\ \underline{\lambda}^D(t) & \text{otherwise} \end{cases} \qquad (8.5)$$

$$\lambda^L(t,y) = \begin{cases} \overline{\lambda}^L(t) & \text{if } y \geq 0 \\ \underline{\lambda}^L(t) & \text{otherwise} \end{cases}$$

for some given functions of time $\overline{\lambda}^D$, $\underline{\lambda}^D$, $\overline{\lambda}^L$ and $\underline{\lambda}^L$.

REMARK 8.1 More generally, we can consider functions $\lambda^D\left(t, x, u^D - u\right)$ and $\lambda^L\left(t, x, u^L - u\right)$ defined using given functions $\overline{\lambda}^D$, $\underline{\lambda}^D$, $\overline{\lambda}^L$, and $\underline{\lambda}^L$ that all depend on (t, x). For instance, one can consider

$$\lambda^L(t,x,y) = \begin{cases} \overline{\lambda}^L(t)m(x) & \text{if } y \geq 0 \\ \underline{\lambda}^L(t)m(x) & \text{otherwise} \end{cases}$$

where m is some lapse multiplier function; m is normalized to be 1 around the money, greater than 1 for large NAVs and less than 1 for small NAVs. This way, we can model that subscribers are more likely to lapse when the NAV of the underlying fund is large. □

PROPOSITION 8.1

Assume zero interest rates, repos, and dividends. Assume that X follows a Black-Scholes diffusion with constant volatility σ. Then the solution u to the ULMM equation (8.4) is the lowest price guaranteeing that the conditional expectation of the global P&L of the issuer's delta-hedged position incurred between inception and maturity, given the underlying path, is nonnegative in the following context of uncertainty on the rates:

(i) *the death rate process (λ_t^D) is unknown but is adapted and belongs to the moving corridor $[\underline{\lambda}^D(t), \overline{\lambda}^D(t)]$,*

(ii) *the lapse rate process (λ_t^L) is unknown but is adapted and belongs to the moving corridor $[\underline{\lambda}^L(t), \overline{\lambda}^L(t)]$.*

PROOF   This is clear from the P&L analysis of previous section. Indeed, conditionally on the whole underlying path $(X_u, 0 \le u \le T)$ and on the values of the processes λ^D and λ^L until time t, the average value of the P&L of the issuer's delta-hedged position incurred between t and $t + dt$ is

$$p_t \left\{ -\partial_t u \, dt - \frac{1}{2}\partial_x^2 u \, d\langle X \rangle_t + \alpha \, dt - \lambda_t^D \, dt \left(u^D - u \right) - \lambda_t^L \, dt \left(u^L - u \right) \right\}$$

where p_t denotes the probability that no default has occurred before t and is given by (8.3). Now,

$$\lambda_t^D \left(u^D - u \right) \le \lambda^D \left(t, u^D - u \right) \left(u^D - u \right)$$
$$\lambda_t^L \left(u^L - u \right) \le \lambda^L \left(t, u^D - u \right) \left(u^L - u \right)$$

because of the corridor hypotheses. As a consequence, the above average P&L is greater than $p_t \, dt$ times

$$-\partial_t u - \sigma^2 X_t^2 \partial_x^2 u + \alpha - \lambda^D \left(t, u^D - u \right) \left(t, u^D - u \right) - \lambda^L \left(t, u^D - u \right) \left(t, u^L - u \right)$$

which equals zero because u is a solution to (8.4). This ensures that the global P&L of the issuer's delta-hedged position incurred between inception and maturity, given the underlying path, is nonnegative.
Besides, the scenario where $\lambda_t^D = \lambda^D \left(t, u^D(t, X_t) - u(t, X_t) \right)$ and $\lambda_t^L = \lambda^L \left(t, u^L(t, X_t) - u(t, X_t) \right)$ satisfies the adaptation and corridor hypotheses and cancels out the global P&L of the issuer's delta-hedged position incurred between inception and maturity, given the underlying path, which completes the proof.   ☐

REMARK 8.2   The insurance subscriber can always lapse. Assume that the fee α matches exactly the managing fee that the subscriber pays to the insurance company. A rational policyholder should lapse as soon as $u^L > u$. A GMxB deal is therefore an American option. Why not price it as such? It is always possible to price reinsurance deals as American options, but the resulting price is often far from the market quotes. In fact, it has been observed that, so far, policyholders do not exercise optimally, which is taken into account through the $\overline{\lambda}^L$ function. The $\overline{\lambda}^L$ function accounts for the fact that policyholders may not lapse when they should. The $\underline{\lambda}^L$ function accounts for the fact that policyholders may lapse when they should not. Choosing $\overline{\lambda}^L = +\infty$ and $\underline{\lambda}^L = 0$ boils down to pricing the GMxB as an American option:

$$\max\left(\partial_t u + \frac{1}{2}\sigma^2 x^2 \partial_x^2 u - \alpha + \lambda^D\left(t, u^D - u\right)\left(u^D - u\right), u^L - u\right) = 0 \quad (8.6)$$

with terminal condition $u(T, x) = u^{\mathrm{mat}}(x)$. One can imagine that, in the future, a new business of financial advice will help individual policyholders exercise optimally. When such a business develops, banks will tend to increase $\overline{\lambda}^L$ and decrease $\underline{\lambda}^L$. ⬜

REMARK 8.3 Uncertain Lapse, Mortality, and Volatility Model
In Equation (8.4), we considered uncertain lapse rates and uncertain mortality rates. We can also take uncertain volatility into account. This is natural as, often, the volatility of the underlying fund cannot be traded. We then speak of uncertain lapse, mortality, and volatility model (ULMVM). From Section 5.2, the ULMVM is simply described by the following nonlinear PDE

$$\partial_t u + \frac{1}{2}\sigma^2(\partial_x^2 u)x^2\partial_x^2 u - \alpha$$
$$+ \lambda^D\left(t, u^D - u\right)\left(u^D - u\right) + \lambda^L\left(t, u^L - u\right)\left(u^L - u\right) = 0 \quad (8.7)$$

with terminal condition $u(T, x) = u^{\mathrm{mat}}(x)$, where

$$\sigma(\Gamma) = \begin{cases} \overline{\sigma} & \text{if } \Gamma \geq 0 \\ \underline{\sigma} & \text{otherwise} \end{cases} \quad (8.8)$$

for some given scalars $\overline{\sigma}$ and $\underline{\sigma}$. The functions λ^D and λ^L are given by (8.5). As we already mentioned in the case of the (pure) ULMM, we can more generally consider a function $\sigma\left(t, x, \partial_x^2 u\right)$ defined using given functions $\overline{\sigma}$ and $\underline{\sigma}$ that all depend on (t, x). The ULMM is simply described by Equation (8.7), in which one picks $\underline{\sigma} = \overline{\sigma}$, or, more generally, $\underline{\sigma}(t, x) = \overline{\sigma}(t, x)$ (See exercise 9.6). One can easily extend Proposition 8.1: The solution u to the ULMVM equation (8.7) is the lowest price guaranteeing that the conditional expectation of the global P&L of the issuer's delta-hedged position incurred between inception and maturity, given the underlying path, is non negative in the following uncertain context:

(i) the volatility process (σ_t) is unknown but is adapted and belongs to the corridor $[\underline{\sigma}, \overline{\sigma}]$,

(ii) the death rate process (λ_t^D) is unknown but is adapted and belongs to the moving corridor $[\underline{\lambda}^D(t), \overline{\lambda}^D(t)]$,

(iii) the lapse rate process (λ_t^L) is unknown but is adapted and belongs to the moving corridor $[\underline{\lambda}^L(t), \overline{\lambda}^L(t)]$.

□

8.4 Path-dependent payoffs

So far, we have considered vanilla reinsurance deals. Exotic reinsurance deals also trade in the market. In those deals, all the payoffs u^{mat}, u^D, u^L and the fee α may depend on path-dependent variables. Below, we consider examples where some of those payoffs depend on realized volatility, or on the maximum past NAV of the underlying fund. In the case where the payoff depends on the (continuously compounded) realized volatility, it makes no sense to price in the (pure) ULMM with constant volatility, because this model does not take volatility risk into account. One may price these options in the ULMVM, where the volatility is uncertain as well (see Remark 8.3).

8.4.1 Payoffs depending on realized volatility

Let us first consider the case where all payoffs depend on the current underlying and the current realized variance $V_t = \int_0^t d\langle X \rangle_s / X_s^2$: the payoff at maturity $u^{\mathrm{mat}}(x, v)$, the payoff $u^D(t, x, v)$ delivered in case of death, the payoff $u^L(t, x, v)$ delivered in case of lapse, and the fee $\alpha(t, x, v)$. In such a case, the ULMVM pricing equation (8.7) becomes

$$
\partial_t u + \sigma^2 \left(\frac{x^2}{2} \partial_x^2 u + \partial_v u \right) \left(\frac{x^2}{2} \partial_x^2 u + \partial_v u \right) - \alpha
$$
$$
+ \lambda^D \left(t, u^D - u \right) \left(u^D - u \right) + \lambda^L \left(t, u^L - u \right) \left(u^L - u \right) = 0 \quad (8.9)
$$

with terminal condition $u(T, x, v) = u^{\mathrm{mat}}(x, v)$, where the functions $\lambda^D(\cdot, \cdot)$, $\lambda^L(\cdot, \cdot)$, and $\sigma(\cdot)$ are given in (8.5) and (8.8). Such a framework embraces the cases of

- **Variable fees**: In this exotic version of the GMxB deal, $u(T, x, v) = (K_{\mathrm{mat}} - x)_+$, $u^D(t, x, v) = (K_D - x)_+$ and $u^L(t, x, v) = 0$ as before, but

$$
\alpha(t, x, v) = \alpha_0 \varphi(v/t) \quad (8.10)
$$

  where v/t is the annualized realized variance and φ is a nondecreasing function, e.g.,

$$
\varphi(v) = \max \left(\varphi_{\min}, \min \left(\varphi_{\max}, (\varphi_{\max} - \varphi_{\min}) \frac{v - \sigma_{\mathrm{lo}}^2}{\sigma_{\mathrm{hi}}^2 - \sigma_{\mathrm{lo}}^2} + \varphi_{\min} \right) \right)
$$

- **Variable nominal put**: In this version, $u^D(t, x, v) = (K_D - x)_+$, $u^L(t, x, v) = 0$ and $\alpha(t, x, v) = \alpha_0$ is constant as before, but

$$u^{\mathrm{mat}}(x, v) = \psi(v/T)\,(K_{\mathrm{mat}} - x)_+, \qquad (8.11)$$

where v/T is the annualized realized variance and ψ is a nonincreasing function, e.g.,

$$\psi(v) = \max\left(\psi_{\min}, \min\left(1, 1 - (1 - \psi_{\min})\frac{v - \sigma_{\mathrm{lo}}^2}{\sigma_{\mathrm{hi}}^2 - \sigma_{\mathrm{lo}}^2}\right)\right) \qquad (8.12)$$

where $\psi_{\min} < 1$.

In those cases, the buyer accepts a part of the volatility risk, which makes the product cheaper.

8.4.2 Payoffs depending on the running maximum

We may also consider the case where all payoffs depend on X_t and on the maximum

$$M_t = \max_{T_0 \leq r \leq t} X_r$$

of the underlying levels over $[T_0, t]$: the payoff at maturity $u^{\mathrm{mat}}(x, M)$, the payoff $u^D(t, x, M)$ delivered in case of death, the payoff $u^L(t, x, M)$ delivered in case of lapse and the fee $\alpha(t, x, M)$. This embraces the case of cliquet strikes: pick some $\nu > 1$, consider the plain vanilla GMxB described in Section 8.1, assume that the GMIB strike K_{mat} and the GMDB strike K_D are locked during some period $[0, T_0]$ (say $T_0 = 1$ year), but after date T_0:

- when the underlying X_t exceeds νX_0, the GMIB strike becomes νK_{mat} and the GMDB strike becomes νK_D;

- when the underlying X_t exceeds $\nu^2 X_0$, the GMIB strike becomes $\nu^2 K_{\mathrm{mat}}$ and the GMDB strike becomes $\nu^2 K_D$;

- when the underlying X_t exceeds $\nu^3 X_0$, the GMIB strike becomes $\nu^3 K_{\mathrm{mat}}$ and the GMDB strike becomes $\nu^3 K_D$, etc.

Stated otherwise, $\alpha(t, x, M) = \alpha_0$, $u^L(t, x, M) = 0$, and

$$u^D(t, x, M) = (\mathrm{Strike}_D(M) - x)_+ \qquad \forall t \in [T_0, T]$$
$$u^{\mathrm{mat}}(x, M) = (\mathrm{Strike}_{\mathrm{mat}}(M) - x)_+$$

where

$$\mathrm{Strike}_D(M) = \left(\mathbf{1}_{\{M < \nu X_0\}} + \sum_{n=1}^{\infty} \mathbf{1}_{\{\nu^n X_0 \leq M < \nu^{n+1} X_0\}} \nu^n\right) K_D$$

$$\mathrm{Strike}_{\mathrm{mat}}(M) = \left(\mathbf{1}_{\{M < \nu X_0\}} + \sum_{n=1}^{\infty} \mathbf{1}_{\{\nu^n X_0 \leq M < \nu^{n+1} X_0\}} \nu^n\right) K_{\mathrm{mat}}$$

Since the strikes of the puts ratchet when the underlying goes up, this product is more expensive than the plain vanilla GMxB, for a given spread α_0. Stated otherwise, the value of α_0 that cancels out the price at inception is higher.

8.5　Pricing the option on the up-and-out barrier

So far we have ignored the trigger feature of the deal. When there is no option on the choice of the up-and-out barrier, i.e., when the up-and-out barrier is predetermined and left unchanged during the life of the option, the ULMM price of the option is the solution to PDE (8.4) with null boundary conditions at the barrier. When the up-and-out barrier is allowed to change, let us assume that it may take any value within a discrete list of barriers, provided it is greater than the current underlying level. Hence, in a finite difference scheme, once one has a price $u(t + \Delta t, x)$ for all x in a grid, one builds n_B prices $u_{b_i}(t, x)$, $i \in \{1, \ldots, n_B\}$, one for each possible value b_i of the barrier, by putting the relevant Dirichlet condition on the high x boundary. For each x in the grid, one then sets

$$u(t, x) = \max_{b_i > x} u_{b_i}(t, x)$$

This way, one builds up the worst-case price for the issuer. Of course, this procedure also applies in the case when the price u depends on a path-dependent variable.

8.6　An example of PDE implementation

Assume we want to solve the ULMVM PDE (8.9) for payoffs depending on realized volatility. To compute $u(t, x, v)$ from $u(t + \Delta t, x, v)$, we may do the following:

1. **Predictor:**

   (a) We fix v and compute the finite difference estimates of $\frac{1}{2}x^2\partial_x^2 u + \partial_v u$, $u^D - u$ and $u^L - u$ at time $t + \Delta t$ to determine the values of the coefficients $\sigma\left(\frac{1}{2}x^2\partial_x^2 u + \partial_v u\right)$, $\lambda^D\left(t, u^D - u\right)$, and $\lambda^L\left(t, u^L - u\right)$.

   (b) Solving the fully implicit one-dimensional difference scheme on the variable x, with null boundary conditions at $x = b_i$, yields $v_{b_i}^+(t, x, v)$. For the low spot boundary conditions, we may impose zero gamma. We then set

$$v_{b_i}(t, x, v) = v_{b_i}^+ \left(t, x, v + \sigma \left(\frac{1}{2} x^2 \partial_x^2 u + \partial_v u \right)^2 \Delta t \right)$$

To this end, we may use cubic splines on the variable v for $v_{b_i}^+$.

2. **Corrector**:

   (a) We then recompute finite difference estimates of $\frac{1}{2} x^2 \partial_x^2 u + \partial_v u$, $u^D - u$, $u^L - u$ using $1 - \theta$ times the estimates based on v_{b_i}, plus θ times the first estimates (those based on u at time $t + \Delta t$), for some $\theta \in [0, 1]$. This allows us to determine new values of the coefficients $\sigma \left(\frac{1}{2} x^2 \partial_x^2 u + \partial_v u \right)$, $\lambda^D \left(t, u^D - u \right)$, $\lambda^D \left(t, u^D - u \right)$, and $\lambda^L \left(t, u^L - u \right)$.

   (b) Solving again the fully implicit one-dimensional difference scheme on the variable x, but with these new coefficients, yields $u_{b_i}^+(t, x, v)$. We then set

   $$u_{b_i}(t, x, v) = u_{b_i}^+ \left(t, x, v + \sigma \left(\frac{1}{2} x^2 \partial_x^2 u + \partial_v u \right)^2 \Delta t \right)$$

3. **Option on the choice of the up-and-out barrier**: Eventually, we set

   $$u(t, x, v) = \max_{b_i > x} u_{b_i}(t, x, v).$$

The first two steps are called *predictor steps*. The next two steps are called *corrector steps*. We speak of *predictor-corrector schemes*.
When we reach (backward) the up-and-out start date T_{UO}, we set

$$u(T_{\mathrm{UO}}^-, x, v) = \begin{cases} u(T_{\mathrm{UO}}, x, v) & \text{if } x \le b_{n_B} \\ 0 & \text{otherwise} \end{cases}$$

and proceed as above, except that there is now a single solution to compute (no multiple b_i's) and that we now impose the zero gamma condition for both high and low x boundaries.

REMARK 8.4   Actually we would rather work with the variable $z = \log(x/X_0)$ than with x itself. We may also use the non-time-monotonic variable $V_t + \sigma_0^2(T - t)$, rather than V_t itself, where σ_0 is a fixed constant that we may choose between $\underline{\sigma}$ and $\overline{\sigma}$. This allows us to use a grid of values of $V_t + \sigma_0^2(T - t)$ that does not depend on time. ⧠

The above numerical scheme easily adapts to the case where the payoffs depend on the maximum past spot value, described in Section 8.4.2. The path-dependent variable is now M. When we solve the PDE backward from $t + \Delta t$

to t, we think of M as the maximum just after date t. The maximum just after t (the value of M in $u(t + \Delta t, x, M)$, whence the value of M in $v_{b_i}^+(t, x, M)$) is the maximum of (i) the maximum just after $t - \Delta t$ (the value of M in $v_{b_i}(t, x, M)$) and (ii) the value of x at time t. As a consequence, the new equations are simply

$$v_{b_i}(t, x, M) = v_{b_i}^+(t, x, \max(M, x))$$
$$u_{b_i}(t, x, M) = u_{b_i}^+(t, x, \max(M, x))$$

Assume that $T_0 < T_{\mathrm{UO}}$. When we reach (backward) the up-and-out start date T_{UO}, we set

$$u(T_{\mathrm{UO}}^-, x, M) = \begin{cases} u(T_{\mathrm{UO}}, x, M) & \text{if } x \leq b_{n_B} \\ 0 & \text{otherwise} \end{cases}$$

and proceed as above, except that there is now a single solution to compute (no multiple b_i's) and that we now impose the zero gamma condition for both high and low x boundaries.

Eventually, when we reach (backward) the date T_0, at which the strikes begin to ratchet, we set

$$u(T_0^-, x) = u(T_0, x, x)$$

and solve the standard PDE on x backward until today (the price does not depend on M anymore).

8.7   Monte Carlo pricing

We now explain how to price options in the ULMM using Monte Carlo methods. The nonlinearity in Equation (8.4) only affects the pricing function u itself, and not its first order derivative $\partial_x u$ nor its second order derivative $\partial_x^2 u$. This means that, in order to build a Monte Carlo method for pricing in the ULMM, we need to be able to estimate the value of the option in the future on all sample paths, but we need no future delta or gamma estimation. As a consequence, pricing in the ULMM with Monte Carlo can be achieved in a way which is very similar to pricing an American option with Monte Carlo, and we can price an option in the ULMM by simply adapting the Longstaff-Schwartz algorithm, as we now show.

For the sake of clarity, let us ignore the option on the up-and-out barrier for the moment. Section 8.8 deals specifically with this issue. One way to adapt the Longstaff-Schwartz algorithm is as follows.

Step 1: Backward induction

First we simulate N_1 random future market paths. To estimate the value $u_{t_k}^p$ of a reinsurance deal at some future date t_k on the path p, we use a backward induction to estimate $u(t_k, \cdot)$, $k = N, N-1, \ldots, 2, 1$. The $\cdot$ stands for x, or (x, v), or (x, M), etc. At date t_k, we simply estimate the sum of future discounted coupons $\sum_{t_j > t_k} D_{t_k t_j} C_{t_j}^{t_k, p}$ as a function of market variables that are known at date t_k. Mathematically, the explaining factors are $\mathcal{F}_{t_k}$-measurable random variables. In practice, this is usually achieved through (parametric or non-parametric) regression. The future coupons $C_{t_j}^{t_k, p}$ must be averaged on all default events (lapse and mortality):

$$
\begin{aligned}
C_{t_j}^{t_k, p} = {} & N_{t_j}^{t_k, p} C_{t_j}^{c, p} \\
& + N_{t_{j-1}}^{t_k, p} \left(1 - \exp\left(-\lambda_{t_j}^{L, p} \Delta t_j \right) \right) \exp\left(-\lambda_{t_j}^{D, p} \Delta t_j \right) C_{t_j}^{L, p} \\
& + N_{t_{j-1}}^{t_k, p} \left(1 - \exp\left(\lambda_{t_j}^{D, p} \Delta t_j \right) \right) C_{t_j}^{D, p}
\end{aligned}
\tag{8.13}
$$

where we have set

$$
\begin{cases}
N_{t_k}^{t_k, p} = 1 \\
N_{t_j}^{t_k, p} = N_{t_{j-1}}^{t_k, p} \exp\left(-\lambda_{t_j}^{L, p} \Delta t_j \right) \exp\left(-\lambda_{t_j}^{D, p} \Delta t_j \right),
\end{cases}
\qquad t_j > t_k
\tag{8.14}
$$

and $\Delta t_j = t_j - t_{j-1}$. Let us comment on Equations (8.13)–(8.14):

- The future coupon $C_{t_j}^{t_k, p}$ served at date t_j, averaged on all default events, is a sum of three terms: (a) continuation (no lapse-no death), (b) lapse and no death, and (c) death.

- $N_{t_j}^{t_k, p}$ is the proportion of people who are still bound by the contract at time t_j among those who were bound by the contract at time t_k on the path p.

- $C_{t_j}^{c, p}$ is the coupon served in case of continuation, i.e., when no default occurs. For instance:

  - For the vanilla reinsurance deal described in Section 8.1, when $t_j < T$, $C_{t_j}^{c, p} = -\alpha \Delta t_j$, and, at maturity, $C_T^{c, p} = (K_{\mathrm{mat}} - X_T^p)_+ - \alpha \Delta t_N$.
  - For the payoffs depending on realized volatility, described in Section 8.4.1, when $t_j < T$, $C_{t_j}^{c, p} = -\alpha(t_j, X_{t_j}, V_{t_j}) \Delta t_j$, and, at maturity, $C_T^{c, p} = u_{\mathrm{mat}}(X_T, V_T) - \alpha(T, X_T, V_T) \Delta t_N$.

- If both defaults occur simultaneously, we assume that death comes first:

  - $N_{t_{j-1}}^{t_k, p} \left(1 - \exp\left(-\lambda_{t_j}^{L, p} \Delta t_j \right) \right) \exp\left(-\lambda_{t_j}^{D, p} \Delta t_j \right)$ is the proportion of people who lapse at time t_j and do not die at this date among those who were bound by the contract at time t_k on the path p.
  - $N_{t_{j-1}}^{t_k, p} \left(1 - \exp\left(-\lambda_{t_j}^{D, p} \Delta t_j \right) \right)$ is the proportion of people who die at time t_j among those who were bound by the contract at time t_k on the path p.

- We assume that the fee is paid even on the date of lapse or on the date of death. For instance, for the vanilla reinsurance deal described in Section 8.1, $C_{t_j}^{L,p} = -\alpha \Delta t_j$ and $C_{t_j}^{D,p} = \left(K_D - X_{t_j}^p \right)_+ - \alpha \Delta t_j$.

- As we proceed by backward induction, the $\lambda_{t_j}^{L,p}$ and $\lambda_{t_j}^{D,p}$ are known at date $t_k < t_j$.

Once we have the estimate u_{t_k} of the function $u(t_k, \cdot)$ at time t_k, for instance, as the result of a regression procedure, we simply define

$$\lambda_{t_k}^{L,p} = \begin{cases} \overline{\lambda}^{L,p}(t_k) & \text{if } u_{t_k}^{L,p} - u_{t_k}^p \geq 0 \\ \underline{\lambda}^{L,p}(t_k) & \text{otherwise} \end{cases} \tag{8.15}$$

$$\lambda_{t_k}^{D,p} = \begin{cases} \overline{\lambda}^{D,p}(t_k) & \text{if } u_{t_k}^{D,p} - u_{t_k}^p \geq 0 \\ \underline{\lambda}^{D,p}(t_k) & \text{otherwise} \end{cases} \tag{8.16}$$

where $u_{t_k}^{L,p}$ (resp. $u_{t_k}^{D,p}$) is the value of the lapse payoff $u^L(t_k, \cdot)$ (resp. the death payoff $u^D(t_k, \cdot)$) on path p at date t_k. With these definitions of $\lambda_{t_k}^{L,p}$ and $\lambda_{t_k}^{D,p}$, we can now estimate the price $u(t_{k-1}, \cdot)$ at date t_{k-1}, etc. This is how the backward induction works.

Step 2: Pricing

We simulate N_2 new independent paths and compute the price of the "auto-callable" product

$$\frac{1}{N_2} \sum_{p=1}^{N_2} \sum_{t_j > 0} D_{0 t_j} C_{t_j}^p$$

with

$$C_{t_j}^p = N_{t_j}^p C_{t_j}^{c,p} \tag{8.17}$$
$$+ N_{t_{j-1}}^p \left(1 - \exp\left(-\lambda_{t_j}^{L,p} \Delta t_j \right) \right) \exp\left(-\lambda_{t_j}^{D,p} \Delta t_j \right) C_{t_j}^{L,p}$$
$$+ N_{t_{j-1}}^p \left(1 - \exp\left(\lambda_{t_j}^{D,p} \Delta t_j \right) \right) C_{t_j}^{D,p}$$

and

$$\begin{cases} N_0^p = 1 \\ N_{t_j}^p = N_{t_{j-1}}^p \exp\left(-\lambda_{t_j}^{L,p} \Delta t_j \right) \exp\left(-\lambda_{t_j}^{D,p} \Delta t_j \right), & t_j > 0 \end{cases} \tag{8.18}$$

During this second step, the $\lambda_{t_k}^{L,p}$ and $\lambda_{t_k}^{D,p}$ ($1 \leq p \leq N_2$) are determined by the rule (8.15)–(8.16) where we use for u_{t_k} the functional estimate $u(t_k, \cdot)$ computed during the first step.

REMARK 8.5 Note that for $t_j > t_k$

$$C_{t_j}^{t_{k-1},p} = C_{t_j}^{t_k,p} \exp\left(-\lambda_{t_k}^{L,p} \Delta t_k \right) \exp\left(-\lambda_{t_k}^{D,p} \Delta t_k \right)$$

i.e., we can compute the cash flows iteratively. ⬜

8.8 Monte Carlo pricing of the option on the up-and-out barrier

In Section 8.7, we have ignored the up-and-out feature. When $t \geq T_{\text{UO}}$, the buyer holds an option on the up-and-out barrier. As explained in Section 8.5, in order to price the worst case for the issuer, we compute the American price of this option:

$$u\left(t_k, \cdot\right) = \max_{b_i > x} u_{b_i}\left(t_k, \cdot\right), \qquad t_k \geq T_{\text{UO}}$$

where the b_i's are the authorized barriers, and $u_{b_i}\left(t_k, \cdot\right)$ is the price of the product at time t_k given that the buyer chooses the barrier b_i on the interval $[t_k, t_{k+1}]$. Again, the $\cdot$ stands for x, or (x, V), or (x, M), etc. We denote by $b^*\left(t_k, \cdot\right)$ the optimal barrier:

$$u\left(t_k, \cdot\right) = \max_{b_i > x} u_{b_i}\left(t_k, \cdot\right) = u_{b^*\left(t_k, \cdot\right)}\left(t_k, \cdot\right).$$

When $t_k < T_{\text{UO}}$, there is no up-and-out barrier and we simply compute $u\left(t_k, \cdot\right) = u_{b=\infty}\left(t_k, \cdot\right)$.

Let $u^p_{b_i, t_k}$ be the value of a reinsurance deal at some future date t_k on the path p, given that the insurance subscriber has chosen b_i as the up-and-out barrier level over $[t_k, t_{k+1}]$ (with $b_i = \infty$ if $t_k < T_{\text{UO}}$). We estimate $u^p_{b_i, t_k}$ along the lines of Section 8.7, i.e., by performing recursive backward functional estimations of $u_{b_i}(t_k, \cdot)$, $k = N, N-1, \ldots, 2, 1$. At date t_k, we estimate the sum of future discounted coupons $\sum_{t_j > t_k} D_{t_k t_j} C^{t_k, p}_{t_j}$ as a function of explaining market variables that are known at date t_k. The future coupons $C^{t_k, p}_{t_j}$, averaged on default events, are paid only if the up-and-out barrier has not been reached:

$$C^{t_k, p}_{t_j} = \mathbf{1}_{E^{t_k, p, b_i}_{t_j}} \bar{C}^{t_k, p}_{t_j}$$

$$\bar{C}^{t_k, p}_{t_j} = N^{t_k, p}_{t_j} C^{c, p}_{t_j}$$
$$+ N^{t_k, p}_{t_{j-1}} \left(1 - \exp\left(-\lambda^{L, p}_{t_j} \Delta t_j\right)\right) \exp\left(-\lambda^{L, p}_{t_j} \Delta t_j\right) C^{L, p}_{t_j}$$
$$+ N^{t_k, p}_{t_{j-1}} \left(1 - \exp\left(\lambda^{D, p}_{t_j} \Delta t_j\right)\right) C^{D, p}_{t_j}$$

$$E^{t_k, p, b_i}_{t_j} = \left\{\forall t_l \in [t_{k+1}, t_{j-1}] \cap [T_{\text{UO}}, \infty), \ X^p_{t_{l+1}} \leq b^*\left(t_l, \cdot\right)\right\} \cap \left\{X^p_{t_{k+1}} \leq b_i\right\}$$

where the $N^{t_k, p}_{t_j}$ have been defined in (8.14).

As $t_j > t_k$, the $\lambda^{L, p}_{t_j}$, $\lambda^{D, p}_{t_j}$ and $b^*\left(t_l, \cdot\right)$ have already been computed in the previous steps of the backward induction. Once we have the estimates $u_{b_i}(t_k, \cdot)$, we define $u(t_k, \cdot) = \max_{b_i > x} u_{b_i}(t_k, \cdot)$, and

$$\lambda_{t_k}^{L,p} = \begin{cases} \overline{\lambda}^{L,p}(t_k) & \text{if } u_{t_k}^{L,p} - u_{t_k}^{p} \geq 0 \\ \underline{\lambda}^{L,p}(t_k) & \text{otherwise} \end{cases}$$

$$\lambda_{t_k}^{D,p} = \begin{cases} \overline{\lambda}^{D,p}(t_k) & \text{if } u_{t_k}^{D,p} - u_{t_k}^{p} \geq 0 \\ \underline{\lambda}^{D,p}(t_k) & \text{otherwise} \end{cases}$$

$$b^*(t_k, \cdot) = \arg\max_{b_i > x} u_{b_i}(t_k, \cdot)$$

so that we are now able to estimate $u(t_{k-1}, \cdot)$, etc. At the end of the backward induction, all $\lambda_{t_k}^{L}$, $\lambda_{t_k}^{D}$, and $b^*(t_k, \cdot)$ have been estimated as functions of the regressors, we simulate N_2 new independent paths and compute

$$\frac{1}{N_2} \sum_{p=1}^{N_2} \sum_{t_j > 0} D_{0t_j} C_{t_j}^{p}$$

with

$$C_{t_j}^{p} = \mathbf{1}_{E_{t_j}^{p,b_i}} \bar{C}_{t_j}^{p}$$

$$\bar{C}_{t_j}^{p} = N_{t_j}^{p} C_{t_j}^{c,p}$$
$$+ N_{t_{j-1}}^{p} \left(1 - \exp\left(-\lambda_{t_j}^{L,p} \Delta t_j\right)\right) \exp\left(-\lambda_{t_j}^{L,p} \Delta t_j\right) C_{t_j}^{L,p}$$
$$+ N_{t_{j-1}}^{p} \left(1 - \exp\left(\lambda_{t_j}^{D,p} \Delta t_j\right)\right) C_{t_j}^{D,p}$$

$$E_{t_j}^{p,b_i} = \left\{\forall t_l \in [t_1, t_{j-1}] \cap [T_{UO}, \infty), X_{t_{l+1}}^{p} \leq b^*(t_l, \cdot)\right\} \cap \left\{X_{t_1}^{p} \leq b_i\right\}$$

where the N_{t_j} have been defined in (8.18).

8.9 Link with first order BSDEs

In Chapter 7, we have seen that solutions of semilinear PDEs can be represented as solutions of 1-BSDEs. Hence, numerical schemes for solving 1-BSDEs allow us to approximate solutions of semilinear PDEs. Recall the ULMM pricing equation:

$$\partial_t u + \frac{1}{2}\sigma^2 x^2 \partial_x^2 u - \alpha$$
$$+ \lambda^D\left(t, u^D - u\right)\left(u^D - u\right) + \lambda^L\left(t, u^L - u\right)\left(u^L - u\right) = 0 \quad (8.19)$$

where the functions $\lambda^D(\cdot, \cdot)$ and $\lambda^L(\cdot, \cdot)$ are defined in (8.5). From Proposition 7.1, $u(0, x)$ can be represented by the solution Y_0^x to the 1-BSDE

$$dY_t = -f(t, X_t, Y_t, Z_t)\, dt + Z_t\, dW_t$$

with the terminal condition $Y_T = u^{\mathrm{mat}}(X_T)$, where

$$dX_t = \sigma X_t \, dW_t, \qquad X_0 = x$$

and

$$f(t, x, y, z) = -\alpha + \lambda^D \left(t, u^D(t, x) - y\right)\left(u^D(t, x) - y\right)$$
$$+ \lambda^L \left(t, u^L(t, x) - y\right)\left(u^L(t, x) - y\right)$$

Note that in this case, f does not depend on z. As f is Lipschitz-continuous with respect to (t, x, y), we deduce from Theorem 7.1 that this 1-BSDE admits a unique solution given by $Y_t = u(t, X_t)$ and $Z_t = \partial_x u(t, X_t)$.

Following Equation (7.9), this BSDE can be discretized by an implicit Euler-like scheme:

$$Y^\Delta_{t_{i-1}} = \mathbb{E}^{\mathbb{Q}}_{i-1}[Y^\Delta_{t_i}] - \alpha \Delta t_i$$
$$+ \lambda^D \left(t_{i-1}, u^D(t_{i-1}, X^\Delta_{t_{i-1}}) - Y^\Delta_{t_{i-1}}\right)\left(u^D(t_{i-1}, X^\Delta_{t_{i-1}}) - Y^\Delta_{t_{i-1}}\right)\Delta t_i$$
$$+ \lambda^L \left(t_{i-1}, u^L(t_{i-1}, x^\Delta_{t_{i-1}}) - Y^\Delta_{t_{i-1}}\right)\left(u^L(t_{i-1}, x^\Delta_{t_{i-1}}) - Y^\Delta_{t_{i-1}}\right)\Delta t_i \quad (8.20)$$

with $Y^\Delta_{t_n} = u^{\mathrm{mat}}(X^\Delta_{t_n})$, where $t_n = T$, $\Delta t_i = t_i - t_{i-1}$ and $\mathbb{E}^{\mathbb{Q}}_{i-1} = \mathbb{E}^{\mathbb{Q}}[\cdot|\mathcal{F}_{t_{i-1}}]$. In this case we do not need to compute $Z^\Delta_{t_{i-1}}$ as f does not depend on z. Explicit alternatives, corresponding to schemes (7.12) and (7.13), are

$$Y^\Delta_{t_{i-1}} = \mathbb{E}^{\mathbb{Q}}_{i-1}\left[Y^\Delta_{t_i} - \alpha \Delta t_i \right.$$
$$+ \lambda^D \left(t_i, u^D(t_i, X^\Delta_{t_i}) - Y^\Delta_{t_i}\right)\left(u^D(t_i, X^\Delta_{t_i}) - Y^\Delta_{t_i}\right)\Delta t_i$$
$$\left. + \lambda^L \left(t_i, u^L(t_i, X^\Delta_{t_i}) - Y^\Delta_{t_i}\right)\left(u^L(t_i, X^\Delta_{t_i}) - Y^\Delta_{t_i}\right)\Delta t_i\right] \quad (8.21)$$

and

$$Y^\Delta_{t_{i-1}} = \mathbb{E}^{\mathbb{Q}}_{i-1}[Y^\Delta_{t_i}] - \alpha \Delta t_i$$
$$+ \lambda^D \left(t_{i-1}, u^D(t_{i-1}, X^\Delta_{t_{i-1}}) - \mathbb{E}^{\mathbb{Q}}_{i-1}[Y^\Delta_{t_i}]\right)\left(u^D(t_{i-1}, X^\Delta_{t_{i-1}}) - \mathbb{E}^{\mathbb{Q}}_{i-1}[Y^\Delta_{t_i}]\right)\Delta t_i$$
$$+ \lambda^L \left(t_{i-1}, u^L(t_{i-1}, X^\Delta_{t_{i-1}}) - \mathbb{E}^{\mathbb{Q}}_{i-1}[Y^\Delta_{t_i}]\right)\left(u^L(t_{i-1}, X^\Delta_{t_{i-1}}) - \mathbb{E}^{\mathbb{Q}}_{i-1}[Y^\Delta_{t_i}]\right)\Delta t_i$$
$$(8.22)$$

Denoting

$$\lambda^D_{t_i} = \lambda^D\left(t_i, u^D(t_i, X^\Delta_{t_i}) - Y^\Delta_{t_i}\right)$$
$$\lambda^L_{t_i} = \lambda^L\left(t_i, u^L(t_i, X^\Delta_{t_i}) - Y^\Delta_{t_i}\right)$$

Equation (8.21) can be rewritten as

$$Y^\Delta_{t_{i-1}} = \mathbb{E}^{\mathbb{Q}}_{i-1}\left[\left(1 - \lambda^D_{t_i}\Delta t_i - \lambda^L_{t_i}\Delta t_i\right)Y^\Delta_{t_i} + \lambda^D_{t_i}\Delta t_i u^D(t_i, X^\Delta_{t_i})\right.$$
$$\left. + \lambda^L_{t_i}\Delta t_i u^L(t_i, X^\Delta_{t_i}) - \alpha \Delta t_i\right] \quad (8.23)$$

Though it may not be clear at first sight, this is analogous to Equations (8.13)–(8.14). Referring to the pricing of American options, Equations (8.13)–(8.14) can be seen as the "Longstaff-Schwartz" version of Equation (8.23), which is of the "Tsitsiklis-Van Roy" type. Indeed, in the scheme (8.13)–(8.14), we regress the sum of future discounted coupons, whereas in (8.23) we directly regress the price $Y_{t_i}^\Delta$, which is itself the result of a previous regression. See Sections 6.8.2 and 6.8.3 for a comparison between the Tsitsiklis-Van Roy and Longstaff-Schwartz approaches to the valuation of American options. Note that for small Δt_i

$$\exp\left(-\lambda_{t_i}^L \Delta t_i\right) \exp\left(-\lambda_{t_i}^D \Delta t_i\right) \approx 1 - \lambda_{t_i}^D \Delta t_i - \lambda_{t_i}^L \Delta t_i$$

$$1 - \exp\left(-\lambda_{t_i}^D \Delta t_i\right) \approx \lambda_{t_i}^D \Delta t_i$$

$$1 - \exp\left(-\lambda_{t_i}^L \Delta t_i\right) \approx \lambda_{t_i}^L \Delta t_i$$

and that, in Equations (8.13)–(8.14), the fee $\alpha \Delta t_i$ is included in the coupons $C_{t_i}^{c,p}$, $C_{t_i}^{D,p}$, and $C_{t_i}^{L,p}$.

REMARK 8.6   Assume a zero fee for simplicity, and that lapse is not allowed ($\lambda^L \equiv 0$). The implicit scheme (8.20) then reads

$$Y_{t_{i-1}}^\Delta = \mathbb{E}_{i-1}^\mathbb{Q}[Y_{t_i}^\Delta]$$
$$+ \lambda^D \left(t_{i-1}, u^D(t_{i-1}, X_{t_{i-1}}^\Delta) - Y_{t_{i-1}}^\Delta\right) \left(u^D(t_{i-1}, X_{t_{i-1}}^\Delta) - Y_{t_{i-1}}^\Delta\right) \Delta t_i \quad (8.24)$$

Since

$$u^D(t_{i-1}, X_{t_{i-1}}^\Delta) - Y_{t_{i-1}}^\Delta \geq 0 \iff u^D(t_{i-1}, X_{t_{i-1}}^\Delta) - \mathbb{E}_{i-1}^\mathbb{Q}[Y_{t_i}^\Delta] \geq 0$$

Equation (8.24) is equivalent to

$$Y_{t_{i-1}} =$$
$$\frac{1}{\overline{\lambda}^D \Delta t_i + 1} \left(\mathbb{E}_{i-1}^\mathbb{Q}[Y_{t_i}^\Delta] + u^D(t_{i-1}, X_{t_{i-1}}^\Delta)\overline{\lambda}^D \Delta t_i\right) \mathbf{1}_{u^D(t_{i-1}, X_{t_{i-1}}^\Delta) \geq \mathbb{E}_{i-1}^\mathbb{Q}[Y_{t_i}^\Delta]}$$
$$+ \frac{1}{\underline{\lambda}^D \Delta t_i + 1} \left(\mathbb{E}_{i-1}^\mathbb{Q}[Y_{t_i}^\Delta] + u^D(t_{i-1}, X_{t_{i-1}}^\Delta)\underline{\lambda}^D \Delta t_i\right) \mathbf{1}_{u^D(t_{i-1}, X_{t_{i-1}}^\Delta) < \mathbb{E}_{i-1}^\mathbb{Q}[Y_{t_i}^\Delta]}$$

In this case, the implicit scheme is actually explicit. It is very similar to the explicit scheme (8.22) which in this case reads

$$Y_{t_{i-1}} =$$
$$\left(\mathbb{E}_{i-1}^\mathbb{Q}[Y_{t_i}^\Delta]\left(1 - \overline{\lambda}^D \Delta t_i\right) + u^D(t_{i-1}, X_{t_{i-1}}^\Delta)\overline{\lambda}^D \Delta t_i\right) \mathbf{1}_{u^D(t_{i-1}, X_{t_{i-1}}^\Delta) \geq \mathbb{E}_{i-1}^\mathbb{Q}[Y_{t_i}^\Delta]}$$
$$+ \left(\mathbb{E}_{i-1}^\mathbb{Q}[Y_{t_i}^\Delta]\left(1 - \underline{\lambda}^D \Delta t_i\right) + u^D(t_{i-1}, X_{t_{i-1}}^\Delta)\underline{\lambda}^D \Delta t_i\right) \mathbf{1}_{u^D(t_{i-1}, X_{t_{i-1}}^\Delta) < \mathbb{E}_{i-1}^\mathbb{Q}[Y_{t_i}^\Delta]}$$

□

8.10   Numerical results using PDEs

An example of PDE implementation was described in Section 8.6. We analyze the ULMM/ULMVM price $u(t, x)$, first for the vanilla GMxB deal, and then for GMxB deals with path-dependent payoffs.

8.10.1   Vanilla GMxB deal

The vanilla GMxB deal was described in Section 8.1. The payouts delivered at maturity, $u^{\text{mat}}(x)$, and in case of death of the insurance subscriber, $u^D(t, x)$, are puts on the NAV X of the underlying fund. The payout served in the case where the insurance subscriber lapses is zero. As a consequence, the price of the vanilla GMxB in the ULMM has a put profile, with a move south due to the monthly fee paid to the issuer. The price is positive for low values of X, and negative for high values of X. By definition, the fair fee is the one that makes the product worth zero at inception when $x = (1 - \phi)X_0$, where ϕ denotes the fees (usually around 4%) taken at inception by the insurer. Since the product always has positive gamma, the ULMVM would always select maximum volatility (see Remark 8.3). Hence, we price in the pure ULMM with constant volatility.

In Figure 8.1, we graph the price of the vanilla GMxB deal in the ULMM as a function of X, the current NAV of the underlying fund, computed by solving PDE (8.4) along the lines of Section 8.6. We can check that, if we increase the curve $\overline{\lambda}^L(t)$, i.e., if we increase the maximum rate of lapse, the price rises. As expected, the impact of increasing $\overline{\lambda}^L(t)$ is greater for large values of X. Indeed, for these large values of X, the lapse payoff $u^L = 0$ is greater than the option price, and hence the ULMM automatically selects the curve $\overline{\lambda}^L(t)$, which produces the maximum price. Increasing the maximum rate of lapse infinitely yields the American price of the lapse option. The American price of the lapse option must always be greater than the lapse payoff itself (zero), which plays the role of the exercise payoff in the American pricing formulation. We check on Figure 8.1 that this is indeed the case.

We can easily degenerate the ULMM to price with a single lapse curve $\lambda^L(t)$: it is enough to set $\underline{\lambda}^L(t) = \overline{\lambda}^L(t) = \lambda^L(t)$. For instance, the "minimum lapse price" is obtained by requiring that the single lapse curve equals the minimum lapse curve $\underline{\lambda}^L(t)$. Similarly, the "maximum lapse price" is obtained by using as the single lapse curve, the maximum lapse curve $\overline{\lambda}^L(t)$. It is instructive to compare the ULMM price, the minimum lapse price, and the maximum lapse price in Figure 8.2. We check that the ULMM price is indeed above the other two prices. It is close to the minimum lapse price when X is low, because in such a case the price of the option is greater than the lapse payoff (zero), so that the ULMM automatically selects the minimum lapse curve.

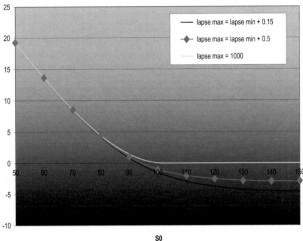

Figure 8.1: Price of the vanilla GMxB deal for various lapse max curves. The minimum lapse rate is flat at 2%, except for year 1 (0%) and year 2 (0.5%). $X_0 = 100$, $K_{\text{mat}} = 91.2\%X_0$, $K_D = 100\%X_0$, $\alpha = 2.20\%$, $T = 10$, $\sigma = 13\%$.

Symmetrically, when X is large, it is the maximum lapse curve which is mostly used so that the ULMM price is close to the maximum lapse price.

The same comparison of prices for death rate curves shows that the pricer uses almost exclusively the maximum death rate curve: the ULMM price is indistinguishable from the "maximum death price." Indeed the ULMM price curve is (almost) always smaller than the death payoff curve $u^D(t,x) = (K_D - x)_+$, at all dates, because $K_D = 100\%X_0$ is quite large. This is reported in the optimal death and lapse rate graphs as we now explain.

Figure 8.3 shows the mortality rate to be chosen at some time t (on the x-axis) as a function of the NAV of the underlying fund (on the y-axis). It shows that from the inception date (March 2008) to September 2012, the issuer should always use the maximum death rate, whatever the level of the NAV of the underlying fund, as well as from September 2016 to the maturity (March 2018). It also shows that from September 2012 to September 2016, there is a (narrow) interval $[\underline{x}^*(t), \overline{x}^*(t)]$ within which the issuer should use the minimum death rate. Stated otherwise, there is a small "$\underline{\lambda}^D$ bubble" in the (t,x)-plane. This bubble represents the few values of (t,x) for which $u(t,x)$ is greater than the death payoff $u^D(t,x) = (K_D - x)_+$.

Whereas $u(t,x)$ is almost always smaller than $u^D(t,x)$, it is not always smaller than $u^L(t,x) = 0$, since for low underlying levels u is positive. Figure 8.4

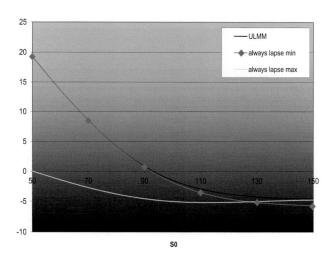

Figure 8.2: Price of the vanilla GMxB using (a) the optimal lapse rate (ULMM), (b) only the minimum lapse rate, and (c) only the maximum lapse rate. The minimum lapse rate is flat at 2%, except for year 1 (0%) and year 2 (0.5%). The maximum lapse rate equals the minimum lapse rate plus 15%. The minimum mortality rate increases linearly over $[0, T]$, from 0.5% to 1.5%. The maximum mortality rate equals the minimum mortality rate. $X_0 = 100$, $K_{\mathrm{mat}} = 91.2\% X_0$, $K_D = 100\% X_0$, $\alpha = 2.20\%$, $T_{\mathrm{UO}} = 3$, one trigger only at $120\% X_0$, $T = 10$, $\sigma = 13\%$.

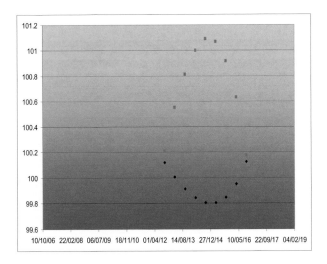

Figure 8.3: Optimal mortality frontier in the time–price plane. The parameters are those of Figure 8.2.

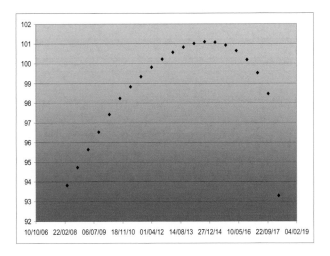

Figure 8.4: Optimal lapse frontier in the time–price plane. The parameters are those of Figure 8.2.

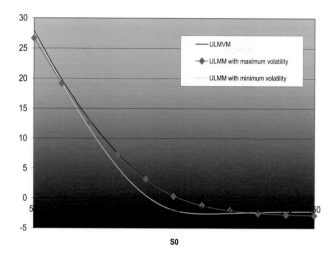

Price of the GMxB with variable fees

Figure 8.5: Price of the GMxB with variable fees using (a) the optimal volatility, (b) always the maximum volatility, and (c) always the minimum volatility. The minimum lapse rate is flat at 2%. The maximum lapse rate is flat at 6%. The minimum mortality increases linearly over $[0, T]$, from 1% to 3%. The maximum mortality rate increases linearly over $[0, T]$, from 5% to 15%. $X_0 = 100$, $K_{mat} = 91.2\%X_0$, $K_D = 100\%X_0$, $T_{UO} = 3$, one trigger only at $120\%X_0$, $T = 10$, $\underline{\sigma} = 0.065$, $\overline{\sigma} = 0.13$, $\varphi_{min} = 0.9\%$, $\varphi_{max} = 1.2\%$, $\sigma_{lo} = 0.07$, $\sigma_{hi} = 0.13$.

shows that, for each time slice t, the mark-to-market $(u^L - u)(t, x)$ changes sign once and only once, when x reaches some critical level $x^*(t)$. Hence the issuer should use the minimum lapse rate when $X_t \leq x^*(t)$ and the maximum lapse rate when $X_t > x^*(t)$.

8.10.2 GMxB with variable fees

The GMxB deal with variable fees was described in Section 8.4.1. Since the fees depend on realized volatility, the option's gamma is not positive everywhere and it makes sense to use the ULMVM in which not only the lapse and death rates are uncertain but the volatility is also (see Remark 8.3). Figure 8.5 shows the ULMVM price, the low-volatility price (the ULMM price with constant volatility $\underline{\sigma}$), and the high-volatility price (the ULMM price with constant volatility $\overline{\sigma}$).

As expected, the ULMVM price, in which the volatility is uncertain, is always above both ULMM prices, in which the volatility is certain. For x around

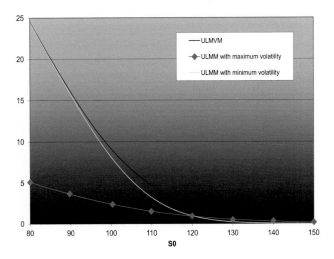

Figure 8.6: Price of the GMxB with variable put nominal using (a) the optimal volatility, (b) always the maximum volatility, and (c) always the minimum volatility, with $q = 0.03365$, $T = 3$, no up-and-out feature, no default (all λ's are zero), zero fee, $\psi_{\min} = 0.2$, $K_{\mathrm{mat}} = 100$, $\underline{\sigma} = 0.065$, $\overline{\sigma} = 0.13$, $\sigma_{\mathrm{lo}} = 0.07$, $\sigma_{\mathrm{hi}} = 0.13$.

80–110, it is close to the ULMM price with constant volatility $\overline{\sigma}$. This is because the effect of high volatility on the values of both puts is higher than its counterpart, namely the higher monthly fee. In the wings, i.e., when the underlying is either very low or very high, the ULMVM price is close to the ULMM price with constant volatility $\underline{\sigma}$. In this region indeed, the vegas of the puts are small, so it is the effect of the volatility on the fee that is prominent.

8.10.3 GMxB with variable put nominal

The GMxB deal with variable put nominal was described in Section 8.4.1. Here also, because the nominal of the put delivered at maturity depends on realized volatility, the option's gamma is not positive everywhere and it makes sense to price in the ULMVM. Figure 8.6 shows the ULMVM price, the low-volatility price (the ULMM price with constant volatility $\underline{\sigma}$), and the high-volatility price (the ULMM price with constant volatility $\overline{\sigma}$).

The curves in Figure 8.6 are not surprising, given that we have picked $\underline{\sigma} \leq \sigma_{\mathrm{lo}} \leq \sigma_{\mathrm{hi}} \leq \overline{\sigma}$. Deep in the money, the ULMVM price is close to the ULMM price with constant volatility $\sigma = \underline{\sigma}$. In this region, the gamma $\partial_x^2 u$ is close to zero, whereas $\partial_v u$ is negative, since the nominal function ψ is decreasing (see Equation (8.12)). Hence, $\frac{x^2}{2}\partial_x^2 u + \partial_v u$ is negative and, from (8.9), the ULMVM automatically selects the minimum volatility. Picking $\sigma = \underline{\sigma}$ at all times yields the lowest V_T possible, and hence the highest nominal. It also yields the minimum put value, but deep in the money, this effect is secondary, compared to the nominal effect.

At the critical underlying value $x \approx 120$, both ULMM prices, the one using $\underline{\sigma}$, and the other using $\overline{\sigma}$, coincide. Each price is the result of an equilibrium between two effects: when volatility rises, the put value also rises but the nominal decreases. One effect perfectly compensates the other for $x \approx 120$. At this underlying value, the ULMVM price is significantly higher than both ULMM prices, since it incorporates dynamic switches from $\underline{\sigma}$ to $\overline{\sigma}$ according to the current dominating effect.

8.10.4 GMxB with cliqueting strikes

The GMxB deal with cliquet strikes was described in Section 8.4.2. The product always has positive gamma so that the ULMVM pricer always selects the maximum volatility $\overline{\sigma}$. In Figure 8.7, we compare the ULMM price with the minimum lapse and maximum lapse prices as defined in Section 8.10.1. There are only tiny differences between the ULMM price and the minimum lapse price. Sometimes the maximum lapse rate curve is used but when it is, i.e., when the ULMM price u is negative, this maximum lapse rate multiplies $u^L - u = -u$, which is small. Unlike the vanilla GMxB price, the price of the GMxB with cliquet strikes is positive for high underlying values, due to the ratchet feature and to the repo being higher than the interest rate. As a consequence, only at medium underlying levels is the maximum lapse rate curve used, but for those underlying levels the lapse rate has little impact because $u^L - u$ is small.

In Figure 8.8, we compare the ULMM price with the minimum mortality and maximum mortality prices. Not surprisingly, the maximum mortality rate is privileged by the Avellaneda pricer. Indeed $u^D - u$ is almost always positive, with the noticeable exception of high underlying levels.

8.11 Numerical results using Monte Carlo

We have described a Monte Carlo algorithm for the valuation of reinsurance deals in the ULMM in Section 8.7. In this section, we run this algorithm and check that the Monte Carlo price and the Monte Carlo lapse and mortality frontiers agree with the ones we get when we use the PDE solver.

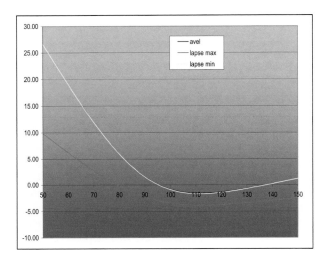

Figure 8.7: Price of the GMxB with cliqueting strikes using (a) the optimal lapse rate ("avel"), (b) only the maximum lapse rate, and (c) only the minimum lapse rate. Curves (a) and (c) almost coincide. Curve (a) is slightly above Curve (c) around the money. The minimum lapse rate is flat at 2%, except for year 1 (0%) and year 2 (0.5%). The maximum lapse rate equals the minimum lapse rate plus 15%. The minimum mortality rate increases linearly over $[0, T]$, from 0.5% to 1%. The maximum mortality rate equals the minimum mortality rate. No up-and-out feature. $X_0 = 100$, $K_{\text{mat}} = 90\% X_0$, $K_D = 100\% X_0$, $\nu = 10\%$, $\alpha = 2.80\%$, $T_0 = 1$, $T = 5$, $\sigma = 13\%$.

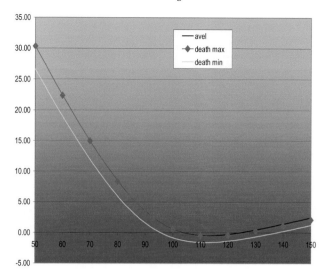

Figure 8.8: Price of the GMxB with cliqueting strikes using (a) the optimal mortality rate ("avel"), (b) only the maximum mortality rate, and (c) only the minimum mortality rate. The parameters are as in Figure 8.7.

8.11.1 Vanilla GMxB deal

We first consider the vanilla GMxB described in Section 8.1. In the Monte Carlo pricing, we use time steps of length one month: $\Delta t_j = 1/12$; the fee is paid every month. As regressors we use the constant 1.0 and the Black-Scholes value of the put on x struck at K_{mat}, with the constant volatility $\sigma = 13\%$ (the one used in the ULMM), and with the residual maturity as duration. We use $N_1 = 10,000$ paths to obtain the frontiers, and $N_2 = 10,000$ independent paths to get the price. Note that the frontiers are well estimated even if we use a much smaller value of N_1, for instance 2,000.

The Monte Carlo and PDE lapse frontiers are shown in Figure 8.9. Both frontiers agree very well. In particular, the critical underlying level above which we should use the maximum lapse value is a function of time which is first increasing, then capped at 110 (the lowest up-and-out barrier value) at the up-and-out start date (December 2008), then decreasing. As for mortality frontiers, Figure 8.10 shows that both frontiers are very close too. Except for the last dates, there is a small interval of underlying levels where the ULMM uses the minimum death rate. The maximum death rate is selected outside.

The prices are shown in Table 8.1. The Monte Carlo prices are just a few bps from the PDE prices. They should lie a bit below the PDE prices, because of the suboptimality of the frontier, but the up-and-out feature introduces some numerical distortion in prices when the underlying is high. With no up-and-

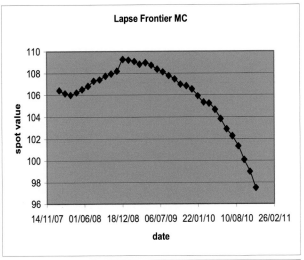

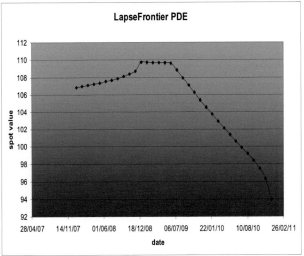

Figure 8.9: Optimal lapse frontier in the time–price plane. Up: using Monte Carlo. Bottom: using PDE. The minimum lapse rate is flat at 2%. The maximum lapse increases linearly over $[0, T]$ from 10% to 13%. The minimum mortality rate is flat at 1%. The maximum mortality rate is flat at 2%. $X_0 = 100$, $K_{\mathrm{mat}} = 91.2\% X_0$, $K_D = 100\% X_0$, $\alpha = 2\%$, $T_{\mathrm{UO}} = 1$, five triggers at $110, 120, 130, 140, 150\% X_0$, $T = 3$, $\sigma = 13\%$. We price using a flat rate curve at 1.5%, and a flat repo curve at 3.5%. A lapse multiplier function is used as in Remark 8.1, increasing from 0.125 for $x = 0.5X_0$ to 3 for $x = 1.5X_0$.

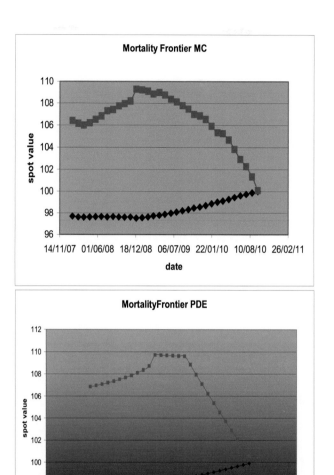

Figure 8.10: Optimal mortality frontier in the time–price plane. Up: using Monte Carlo. Bottom: using PDE. The parameters are as in Figure 8.9.

out feature, the Monte Carlo is extremely accurate, as shown in Figures 8.11 and 8.12, and in Table 8.2.

Table 8.1: Prices of the vanilla GMxB deal. Up-and-out barriers: 110, 120, 130, 140, 150.

Spot value	80	100	120
PDE	11.35	1.61	−1.43
Monte Carlo	11.37	1.66	−1.29

Table 8.2: Prices of the vanilla GMxB deal. No Up-and-out barrier.

Spot value	80	100	120
PDE	11.26	0.95	−2.88
Monte Carlo	11.27	0.95	−2.90

8.11.2   GMxB with variable fee and variable put nominal

In this second numerical test of the Monte Carlo method for reinsurance deals, we consider a GMxB deal with variable fee and variable put nominal. This product is the combination of the two payoffs described in Section 8.4.1: both the monthly fee and the put nominal depend on realized variance; see Equations (8.10) and (8.11).

In the Monte Carlo simulation, since the realized variance is computed on a monthly basis, the quantity V_{t_k}/t_k is random, even in the ULMM model in which volatility is constant. As a consequence, the lapse and mortality frontiers depend not only on x, but also on v. We ignore this dependency on V (it is small in the constant volatility model) and only use the constant 1.0 and the Black-Scholes value of the put struck at K_{mat} as regressors, thus introducing some suboptimality.

Figures 8.13 and 8.14 show the lapse and mortality frontiers when there is no up-and-out feature. The Monte Carlo and PDE frontiers (projected on the X space) are very close. In Table 8.3, the spread between the PDE price and the Monte Carlo price is a manifestation of the suboptimality described above— the PDE frontiers actually depend on the realized variance V as well. When we add up-and-out barriers at 110, 120, 130, 140, 150, we get Figures 8.15 and

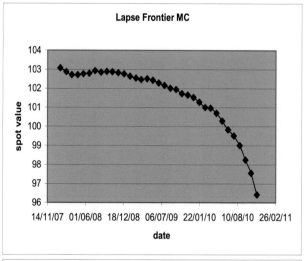

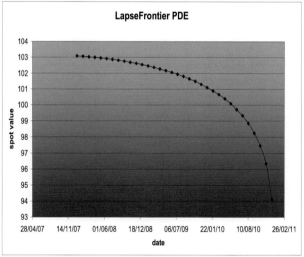

Figure 8.11: Optimal lapse frontier in the time–price plane. No up-and-out barrier. Up: using Monte Carlo. Bottom: using PDE. The parameters are as in Figure 8.9, except for the up-and-out barriers.

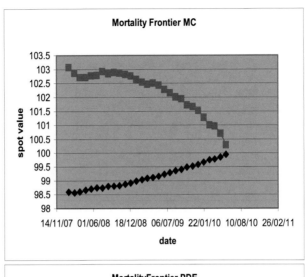

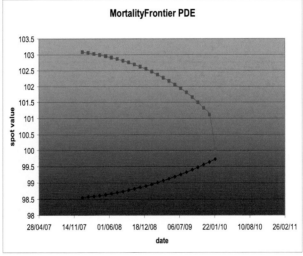

Figure 8.12: Optimal mortality frontier in the time–price plane. No up-and-out barrier. Up: using Monte Carlo. Bottom: using PDE. The parameters are as in Figure 8.9, except for the up-and-out barriers.

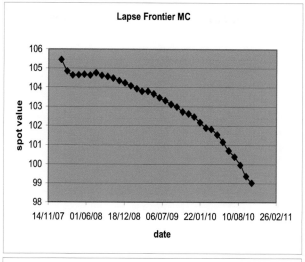

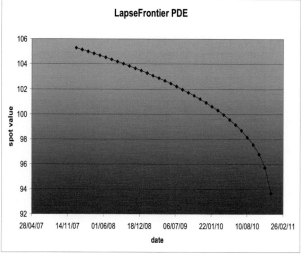

Figure 8.13: Optimal lapse frontier in the time–price plane for the GMxB with variable fee and variable put nominal. No up-and-out barrier. Up: using Monte Carlo. Bottom: using PDE. $\psi_{min} = 0.5$, $\varphi_{min} = 0.5$, $\varphi_{max} = 1.5$, $\sigma_{lo} = 7\%$, $\sigma_{hi} = 13\%$. The constant volatility is 10%. The other parameters are as in Figure 8.9.

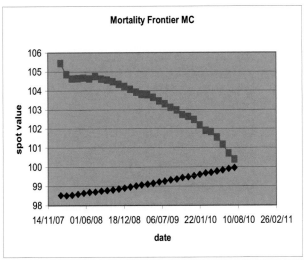

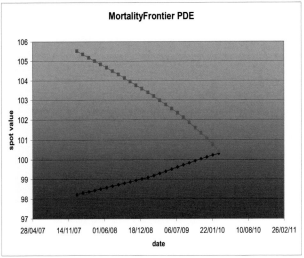

Figure 8.14: Optimal mortality frontier in the time–price plane for the GMxB with variable fee and variable put nominal. No up-and-out barrier. Up: using Monte Carlo. Bottom: using PDE. The parameters are as in Figure 8.13.

Table 8.3: Prices of the GMxB deal with variable fee and variable put nominal. No Up-and-out barrier. The parameters are as in Figure 8.13.

Spot value	80	100	120
PDE	10.32	1.22	−1.47
Monte Carlo	10.07	1.14	−1.54

Table 8.4: Prices of the GMxB deal with variable fee and variable put nominal. Up-and-out barriers: 110, 120, 130, 140, 150. The parameters are as in Figure 8.13.

Spot value	80	100	120
PDE	10.19	1.49	−0.76
Monte Carlo	10.09	1.45	−0.65

8.16. In both PDE frontiers, we observe a peak whose summit is the lowest up-and-out barrier (110) at the up-and-out start date. The regressions cannot reveal this peak very clearly, but the frontiers are globally well estimated by the Monte Carlo algorithm. Moreover, the Monte Carlo prices are just a few bps below the PDE ones (see Table 8.4).

Conclusion

In this chapter, we introduced the uncertain lapse and mortality model. Such a model is useful when one prices a reinsurance deal because it provides a worst-case scenario, under the assumption that mortality and lapse rates, whose future movements cannot be hedged, stay within a certain range.

Finite difference schemes are useful in low dimension only. We have exhibited original Monte Carlo alternatives. The first method extends the Longstaff-Schwartz algorithm. The second one uses the numerical approximation of first order BSDEs. Both methods are actually very similar: using an analogy with the valuation of American options, the second one can be seen as the "Tsitsiklis-Van Roy" version of the first one. Numerical tests prove that the probabilistic algorithms are accurate. One can rely on them to determine the price and hedge, as well as the impact of lapse and mortality assumptions.

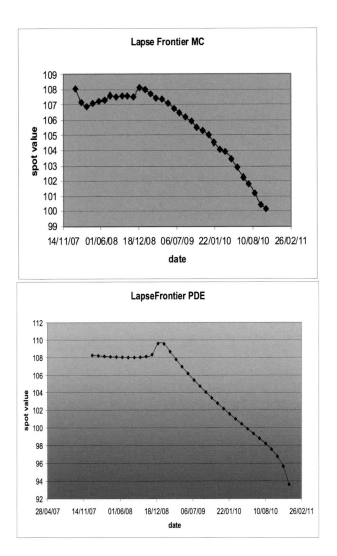

Figure 8.15: Optimal lapse frontier in the time–price plane for the GMxB with variable fee and variable put nominal. Up-and-out barriers: 110, 120, 130, 140, 150. Up: using Monte Carlo. Bottom: using PDE. The parameters are as in Figure 8.13.

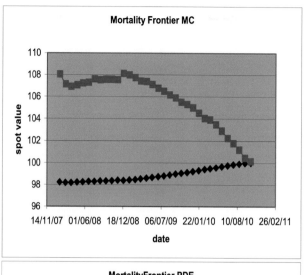

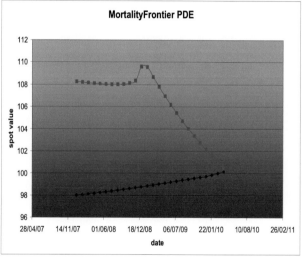

Figure 8.16: Optimal mortality frontier in the time–price plane for the GMxB with variable fee and variable put nominal. Up-and-out barriers: 110, 120, 130, 140, 150. Up: using Monte Carlo. Bottom: using PDE. The parameters are as in Figure 8.13.

Chapter 9

The Uncertain Volatility Model

Il n'est pas certain que tout soit incertain.[1]

— Blaise Pascal

We have introduced the uncertain volatility model in Section 5.2 in the one-dimensional setting. This model has long attracted the attention of practitioners as it provides a worst-case pricing scenario for the sell-side. As seen in Section 5.2, the valuation of financial derivatives based on this model requires solving a nonlinear PDE. One can rely on finite difference schemes only when the number of variables (that is, underlyings and path-dependent variables) is small—in practice no more than three. In all other cases, the numerical valuation seems out of reach. In this chapter, we suggest two new, accurate, easy-implementable Monte Carlo methods which hardly suffer from the curse of dimensionality. The first method requires a parameterization of the optimal covariance matrix and consists in a series of backward low-dimensional optimizations. The second method relies heavily on second order backward stochastic differential equations. Parts of this research have been published in [120].

9.1   Introduction

The uncertain volatility model (in short UVM), developed simultaneously by Avellaneda *et al.* [36] and Lyons [159] in the mid-1990s, assumes that a risky asset follows a controlled diffusion under a risk-neutral measure

$$dX_t = \sigma_t X_t \, dW_t$$

The control (σ_t), representing the uncertain volatility process of the asset X, is valued in the interval $[\underline{\sigma}, \overline{\sigma}]$. It is also supposed to be adapted to the Brownian filtration, that is, it is not allowed to look into the future. The UVM selects one volatility at each time such that the value of the option under

---

[1] "It is not certain that everything is uncertain."

consideration is maximized. In this way, it provides a worst-case scenario for the sell-side. The valuation of an option can be written as the solution (in the viscosity sense) of an Hamilton-Jacobi-Bellman (HJB) equation with a control on the diffusion coefficient. This leads to a *nonlinear* second order PDE, the so-called Black-Scholes-Barenblatt PDE (in short BSB):

$$\partial_t u(t, x) + \frac{1}{2} x^2 \Sigma \left(\partial_x^2 u(t, x) \right)^2 \partial_x^2 u(t, x) = 0, \qquad (t, x) \in [0, T) \times \mathbb{R}_+^* \quad (9.1)$$

with some terminal condition $u(T, x) = g(x)$, and

$$\Sigma(\Gamma) = \underline{\sigma} \mathbf{1}_{\Gamma < 0} + \overline{\sigma} \mathbf{1}_{\Gamma \geq 0}$$

(see Section 5.2). This PDE is called *fully* nonlinear because the nonlinearity affects the second order space derivative $\partial_x^2 u$, the so-called "gamma." In practice, this highly nonlinear PDE is not solvable and we must rely on a finite difference scheme. This method suffers from the curse of dimensionality when the number of variables—assets or path-dependent variables[2]—is large. A probabilistic representation of the BSB equation leading to a Monte Carlo valuation would solve this problem. Unfortunately, the classical link between Monte Carlo and finite difference methods as stated by the Feynman-Kac formula is only valid for *linear* second order parabolic PDEs (see Section 1.2). In this chapter, to get rid of this curse of dimensionality, we suggest two Monte Carlo approaches to valuate options in the UVM. Both approaches are original. Both consist of (a) estimating the optimal covariance matrix, and (b) computing a lower bound of the price using this (necessarily suboptimal) estimate. By the BSB PDE, the optimal covariance matrix at date t is a function of the gammas of the product at that time. Note, however, that the gammas are often very poorly estimated by the naive method consisting of double differentiating a continuation value proxy with respect to the asset values. For instance, the Longstaff-Schwartz (LS) algorithm [157] is designed to produce good estimates of the value of the product in the future, not to build good proxies for the gammas in the future.

The first approach requires that the user specifies a (relevant) parameterization of the optimal covariance matrix and consists of a series of backward optimizations on the parameter. It is instructive to compare such an approach with the Monte Carlo valuation of American options suggested by Longstaff and Schwartz and explained in Section 6.8.3: the parameterization of the optimal covariance matrix is analogous to the choice of regressors; the backward optimizations are similar to the successive least-squares procedures. Note that the parameterization step, as well as the choice of regressors in the LS method, require human intelligence, namely an accurate analysis of the

---

[2]In this chapter, we consider path-dependent variables whose values can change only at discrete dates. For instance, we exclude continuously computed realized variance, but we allow realized variance computed on a monthly basis. However, see Remark 9.3.

payoff. Nevertheless, we will explain how to build tools to help us input good parameterizations.

The second method makes use of *backward* stochastic differential equations (in short BSDEs) as described in Chapter 7. We will apply this tool to obtain our second Monte Carlo implementation of the UVM.

As far as the authors know, it is the first time that Monte Carlo implementations of the UVM, which are unavoidable in high dimension, are suggested in literature. Such methods dramatically enlarge the range of options that can be priced—and hedged—under uncertain volatility. After the seminal works by Avellaneda *et al.* [36] and Lyons [159], Avellaneda and Buff [39] and Buff [2] have considered the pricing in the UVM of a basket of options, including barrier and American options, written on a single asset. In the single-asset case, Smith [184] has also studied the pricing of American options; Martini [160] has established a link between the UVM and American options; Leblanc and Martini [154] have focused on the case where $\overline{\sigma} = +\infty$; Forsyth *et al.* [108] have developed a fully implicit PDE method to price discretely observed barrier options; and Pooley *et al.* [173] have looked at the convergence properties of some numerical schemes. Pooley *et al.* [174] also looked at some numerical details peculiar to the two-asset case. Meyer [165] has applied the BSB PDE to static hedging. Zhang and Wang [194] have recently developed a fitted finite volume method to numerically solve the BSB PDE. As for Monte Carlo methods, Fahim *et al.* [105] used BSDEs for the numerical computation of the solution to the mean curvature flow in dimensions two and three and for two- and five-dimensional HJB equations arising in the theory of portfolio optimization.

The chapter is organized as follows. In Section 9.2, we recall the main results about the UVM and we set our notations. Section 9.3 describes the parametric method, first in the general multi-asset case, then with a special focus on the single-asset and two-asset cases. In Section 9.4, we use 2-BSDEs to build our second Monte Carlo approach. Eventually, our two Monte Carlo techniques are illustrated with numerical examples in the last section.

9.2 The model

We have introduced the uncertain volatility model in Section 5.2 in the one-dimensional setting. We now present the general multi-dimensional setting. Let $(\Omega, \mathcal{F}, \mathbb{Q})$ be a probability space equipped with a d-dimensional Brownian motion W. We denote $(\mathcal{F}_t)$ the natural filtration of W. The market is assumed to be made of d assets, the values of which are represented by a d-dimensional positive local Itô $(\mathcal{F}_t, \mathbb{Q})$-martingale X. Let us consider an option delivering some payoff F_T at maturity T. F_T is some function of $(X_t, 0 \le t \le T)$, the path followed by the assets. The time-t value u_t of the option in the UVM is the solution to a maximization problem (see Section 5.2.3 for an interpretation of u_t as a super-replication price under uncertain covariances):

$$u_t = \sup_{[t,T]} \mathbb{E}^{\mathbb{Q}}[F_T | \mathcal{F}_t] \qquad (9.2)$$

$$dX_t^\alpha = \sigma_t^\alpha X_t^\alpha \, dW_t^\alpha$$

$$d\langle W^\alpha, W^\beta \rangle_t = \rho_t^{\alpha\beta} \, dt \qquad 1 \le \alpha < \beta \le d$$

where $\sup_{[t,T]}$ means that the supremum is taken over all $(\mathcal{F}_s)$-adapted processes $(\xi_s)_{t \le s \le T} \equiv ((\sigma_s^\alpha, \rho_s^{\alpha\beta})_{1 \le \alpha < \beta \le d})_{t \le s \le T}$ such that for all $s \in [t, T]$, ξ_s belongs to some compact domain D. The domain D must be such that for all $\xi = (\sigma^\alpha, \rho^{\alpha\beta})_{1 \le \alpha < \beta \le d} \in D$, the covariance matrix $(\rho^{\alpha\beta} \sigma^\alpha \sigma^\beta)_{1 \le \alpha, \beta \le d}$, with $\rho^{\beta\alpha} = \rho^{\alpha\beta}$ and $\rho^{\alpha\alpha} = 1$, is positive semi-definite. We will consider domains D of the form $D = [\underline{\sigma}, \overline{\sigma}]$ when $d = 1$, and $D = [\underline{\sigma}^1, \overline{\sigma}^1] \times [\underline{\sigma}^2, \overline{\sigma}^2] \times [\underline{\rho}, \overline{\rho}]$ when $d = 2$.

Pricing vanilla options

Let us first consider vanilla payoffs $F_T = g(X_T)$. The payoff function g is assumed to be continuous with quadratic growth. From the HJB principle, i.e., applying Theorems 4.6 and 4.7, we obtain

THEOREM 9.1 Black-Scholes-Barenblatt PDE [36]
We have $u_t = u(t, X_t)$ where u is the unique (viscosity) solution with quadratic growth of the nonlinear PDE

$$\partial_t u(t, x) + H(x, D_x^2 u(t, x)) = 0, \qquad (t, x) \in [0, T) \times (\mathbb{R}_+^*)^d \qquad (9.3)$$

with the terminal condition $u(T, x) = g(x)$ and the Hamiltonian

$$H(X, \Gamma) = \frac{1}{2} \max_{(\sigma^\alpha, \rho^{\alpha\beta})_{1 \le \alpha < \beta \le d} \in D} \sum_{\alpha, \beta = 1}^d \rho^{\alpha\beta} \sigma^\alpha \sigma^\beta X^\alpha X^\beta \Gamma^{\alpha\beta} \qquad (9.4)$$

Note that if $g \in C^3((\mathbb{R}_+^*)^d)$, then $u \in C^{1,2}([0, T) \times (\mathbb{R}_+^*)^d)$; see Theorem 4.4.

REMARK 9.1 $d = 1$ For $d = 1$, the above PDE reduces to (9.1). If we consider a convex (resp. concave) payoff, u coincides with the Black-Scholes price with the upper volatility $\overline{\sigma}$ (resp. lower volatility $\underline{\sigma}$) as $\sigma(\partial_x^2 u) = \overline{\sigma}$ (resp. $\sigma(\partial_x^2 u) = \underline{\sigma}$).

REMARK 9.2 Transaction costs It is interesting to note the similarity of the Black-Scholes-Barenblatt PDE with the pricing equation resulting from the Leland transaction costs model [155] sketched in Section 5.3. In the general case of d assets, the Hamiltonian reads as (5.7). The fact that H is only a function of the gamma can be seen as a consequence of the invariance under axioms I1, I2 set in Section 4.2. □

Pricing with path-dependent variables

When the price of an option depends on path-dependent variables whose values can change only at discrete dates, one solves PDE (9.3) between two such dates t_l and t_{l-1} for fixed values of the path-dependent variables A, and defines

$$u(t_{l-1}^-, X, A) = u(t_{l-1}^+, X, \varphi(A, X))$$

with the function φ linking the past and new values of the path-dependent variables. For instance, if the option value depends on a monthly computed realized variance,

$$A_t^1 = \sum_{\{l \mid t_l \leq t\}} \left(\ln \frac{X_{t_l}}{X_{t_{l-1}}} \right)^2$$

$$A_t^2 = X_{\sup_{\{l \mid t_l \leq t\}} t_l}$$

$$\varphi(X, A) = A^1 + \left(\ln \frac{X}{A^2} \right)^2$$

REMARK 9.3 On path-dependent variables whose values can change continuously   It is only for the sake of simplicity that, in this chapter, we consider path-dependent variables whose values can change only at discrete dates. What happens in the case where the price of an option depends on path-dependent variables whose values can change continuously, is that the Hamiltonian H may not involve only the gammas. For instance, in the single-asset case, if the price $u(t, x, v)$ of an option depends on the continuously compounded realized variance v, the Hamiltonian reads

$$H(x, \partial_x^2 u, \partial_v u) = \max_{\underline{\sigma} \leq \sigma \leq \overline{\sigma}} \sigma^2 \left(\frac{1}{2} x^2 \partial_x^2 u + \partial_v u \right)$$

$$= \Sigma^2 \left(\frac{1}{2} x^2 \partial_x^2 u + \partial_v u \right)$$

i.e., the optimal volatility is either $\underline{\sigma}$ or $\overline{\sigma}$, depending on the sign, not of the gamma $\partial_x^2 u$, but of $\frac{1}{2} x^2 \partial_x^2 u + \partial_v u$. Our Monte Carlo approaches are easily adapted to such a case; see Remarks 9.6 and 9.9. □

Computation of the Hamiltonian

Equation (9.4) is a constrained programming problem. In the particular case where $d = 2$, which we will consider in our numerical experiments, the Hamiltonian reads

$$H(X, \Gamma) =$$
$$\max_{(\sigma^1, \sigma^2, \rho) \in D} \left(\frac{1}{2}(\sigma^1)^2 (X^1)^2 \Gamma^{11} + \frac{1}{2}(\sigma^2)^2 (X^2)^2 \Gamma^{22} + \rho \sigma^1 \sigma^2 X^1 X^2 \Gamma^{12} \right)$$

and has a closed-form solution when $D = [\underline{\sigma}^1, \overline{\sigma}^1] \times [\underline{\sigma}^2, \overline{\sigma}^2] \times [\underline{\rho}, \overline{\rho}]$. Indeed, the optimal correlation is bang-bang: it is either $\underline{\rho}$ or $\overline{\rho}$, depending on the sign of the cross-gamma Γ^{12}: $R(\Gamma^{12}) = \underline{\rho} \mathbf{1}_{\Gamma^{12} < 0} + \overline{\rho} \mathbf{1}_{\Gamma^{12} \geq 0}$. Then, the problem

$$\max_{\sigma^1, \sigma^2} \phi(\sigma^1, \sigma^2)$$
$$\phi(\sigma^1, \sigma^2) = \left(\frac{1}{2}(\sigma^1)^2 (X^1)^2 \Gamma^{11} + \frac{1}{2}(\sigma^2)^2 (X^2)^2 \Gamma^{22} + R(\Gamma^{12}) \sigma^1 \sigma^2 X^1 X^2 \Gamma^{12} \right)$$

is a two-dimensional quadratic form maximization under a double inequality constraint, which has an explicit solution because one can first freeze σ^2 and maximize in the σ^1 variable to get

$$\sigma^{*1}(\sigma^2) = \mathbf{1}_{\Gamma^{11} < 0} \left(\left(\left(-\frac{R(\Gamma^{12}) \sigma^2 X^2 \Gamma^{12}}{X^1 \Gamma^{11}} \right) \vee \underline{\sigma}^1 \right) \wedge \overline{\sigma}^1 \right)$$
$$+ \mathbf{1}_{\Gamma^{11} \geq 0} \begin{cases} \underline{\sigma}^1 & \text{if } \phi(\underline{\sigma}^1, \sigma^2) > \phi(\overline{\sigma}^1, \sigma^2) \\ \overline{\sigma}^1 & \text{otherwise} \end{cases}$$

and then compute the maximum μ of $\psi(\sigma^2) = \phi(\sigma^{*1}(\sigma^2), \sigma^2)$. Now, $\mu = \mathbf{1}_{\Gamma^{11} \geq 0}(\mu_1 \vee \mu_2) + \mathbf{1}_{\Gamma^{11} < 0}(\mu_1 \vee \mu_2 \vee \mu_3)$ where

$$\mu_1 = \max_{\underline{\sigma}^2 \leq \sigma^2 \leq \overline{\sigma}^2} \phi(\underline{\sigma}^1, \sigma^2)$$

$$\mu_2 = \max_{\underline{\sigma}^2 \leq \sigma^2 \leq \overline{\sigma}^2} \phi(\overline{\sigma}^1, \sigma^2)$$

$$\mu_3 = \max_{\underline{\sigma}^2 \leq \sigma^2 \leq \overline{\sigma}^2, \ \underline{\sigma}^1 \leq -\frac{R(\Gamma^{12}) \sigma^2 X^2 \Gamma^{12}}{X^1 \Gamma^{11}} \leq \overline{\sigma}^1} \phi\left(-\frac{R(\Gamma^{12}) \sigma^2 X^2 \Gamma^{12}}{X^1 \Gamma^{11}}, \sigma^2 \right)$$

and each μ_i is itself the maximum of a polynomial of degree 2 under a double inequality constraint, the solution to which has a closed form. Note that, contrary to the single-asset case, the optimal volatility $(\Sigma^1(X, \Gamma), \Sigma^2(X, \Gamma))$ may take values out of the boundary. However, if $\underline{\rho} = \overline{\rho} = 0$, it necessarily hits the boundary. In the general case of d assets, we can first maximize in ρ_{ij} as above. Equation (9.4) then becomes a constrained quadratic programming problem that can be solved efficiently using Lemke's algorithm; see [10].

In practice, PDE (9.3) is not solvable and we must rely on a finite difference scheme. But standard finite difference schemes can only be implemented when the number of variables—assets or path-dependent variables—is no more than three.[3] To get rid of this curse of dimensionality, we suggest two Monte Carlo approaches. In the next section, we first focus on the parametric one.

9.3 The parametric approach

9.3.1 The ideas behind the algorithm

The two main ingredients in the parametric approach are the following.

First, we cut the optimization problem (9.2) into a series of smaller and simpler ones. To this end, let us discretize the time interval $[0, T]$, say $[0, T] = \cup_{i=1}^{n} [t_{i-1}, t_i]$. By the dynamic programming principle (see Section 4.4.3), we can proceed backward and split the original maximization problem, written for $t = 0$, into n consecutive smaller ones:

$$V_{t_i} = \sup_{[t_i, t_{i+1}]} \mathbb{E}^{\mathbb{Q}}[F_T | \mathcal{F}_{t_i}]$$

$$dX_t^\alpha = \sigma_t^\alpha X_t^\alpha \, dW_t^\alpha$$

$$d\langle W^\alpha, W^\beta \rangle_t = \rho_t^{\alpha\beta} \, dt, \qquad 1 \leq \alpha < \beta \leq d$$

where, for $t \in [t_{i+1}, T]$, $\xi_t \equiv (\sigma_t^\alpha, \rho_t^{\alpha\beta})_{1 \leq \alpha < \beta \leq d}$ is the solution to the previously solved maximization problem. We now approximate each small problem by the following simpler one, say (P_i): find a D-valued deterministic function a_{t_i} of some (relevant) $\mathcal{F}_t$-measurable random variables (say, X_t and some path-dependent variables that we sum up under the notation A_t) so as to maximize $\mathbb{E}^{\mathbb{Q}}[F_T | \mathcal{F}_{t_i}]$, with $\xi_t = a_{t_i}(X_t, A_t)$ for $t \in [t_i, t_{i+1})$, and with the already optimized $\xi_t = a_{t_j}^*(X_t, A_t)$ for $t \in [t_j, t_{j+1})$, $j \geq i + 1$. For instance, a component of the vector A_t may represent some realized variance, some realized maximum, some mean value, or some past value of an asset.

The second ingredient concerns the resolution to problem (P_i). Even after the simplifications above, we are left with an optimization problem in which the maximization set is huge: each point in this set is a deterministic function a_{t_i} of (X, A) taking values in D. In particular, a_{t_i} lives in an infinite-dimensional space. Even if we discretize the domain D (say, q possible values) and the domain of possible values for (X, A) (say, p possible values), enumerating the maximization set takes too much time. This is essentially because the set of functions of $\{1, \ldots, p\}$ onto $\{1, \ldots, q\}$ has cardinal q^p, a quantity which grows exponentially with p.

---

[3] However, see von Petersdorff *et al.* [192] for sparse matrix methods.

To overcome this issue, we *decide* to restrict the maximization domain to a parameterized set of *relevant* functions we want to test in the optimization procedure. To be precise, we will only test a_{t_i}'s of the form $\lambda_{t_i}(\cdot; \theta)$, for θ in some parameter set $\Theta \subset \mathbb{R}^l$. The $\cdot$ stands for (X, A). A typical example is $\lambda_{t_i}(X, A; \theta) = \xi_1$ if $\gamma_{t_i}(X, A; \theta) \geq 0$, $= \xi_2$ otherwise, where ξ_1 and ξ_2 are two given points in the domain D. At each date t_i, we are hence left with a maximization over the low-dimensional parameter θ, a problem that can be solved much more quickly than the initial optimization over all possible functions a_{t_i}. A similar approach was used in Andersen [33] to compute the fair value of Bermudan swaptions (see Section 6.8.6).

9.3.2   The algorithm

We can now build a Monte Carlo algorithm as follows:

- Simulate N_1 replications of X with some diffusion, for instance, the lognormal diffusion with some arbitrary volatilities and correlations $\hat{\xi} = (\hat{\sigma}^\alpha, \hat{\rho}^{\alpha\beta})_{1 \leq \alpha < \beta \leq d}$.

- For $i = n - 1, \ldots, 0$, find a numerical solution θ_i^* of the maximization problem (Q_i):

$$\sup_{\theta_i \in \Theta} h(\theta_i), \qquad h(\theta_i) = \frac{1}{N_1} \sum_{p=1}^{N_1} F_T^{(p)} \qquad (9.5)$$

$$dX_t^\alpha = \sigma_t^\alpha X_t^\alpha \, dW_t^\alpha$$

$$d\langle W^\alpha, W^\beta \rangle_t = \rho_t^{\alpha\beta} \, dt$$

where

$$\xi_t \equiv (\sigma_t^\alpha, \rho_t^{\alpha\beta})_{1 \leq \alpha < \beta \leq d} = \begin{cases} \hat{\xi} & \text{if } t \in [0, t_i), \\ \lambda_{t_i}(X_t, A_t; \theta_i) & \text{if } t \in [t_i, t_{i+1}), \\ \lambda_{t_j}(X_t, A_t; \theta_j^*) & \text{if } t \in [t_j, t_{j+1}), j \geq i + 1 \end{cases} \qquad (9.6)$$

- Independently, simulate N_2 replications of X using $\xi_t = \lambda_{t_i}(X_t, A_t; \theta_i^*)$ for $t \in [t_i, t_{i+1})$ and compute $\frac{1}{N_2} \sum_{p=1}^{N_2} H_T^{(p)}$.

In Step 1, we must pick some numerical volatilities and correlations $\hat{\xi} = (\hat{\sigma}^\alpha, \hat{\rho}^{\alpha\beta})_{1 \leq \alpha < \beta \leq d}$. Section 9.4.4 deals with this issue. In Step 2, solving (Q_i) requires that we compute $h(\theta_i)$ for many values of θ_i. Each $F_T^{(p)}$ depends on θ_i, because it is a function of path p whose volatility at date $t \in [t_i, t_{i+1})$ depends on θ_i. As a consequence, to compute each $h(\theta_i)$, we must resimulate the N_1 paths from t_i to T. No Brownian increments are drawn at Step 2: the resimulations only consist of multiplying the Brownian increments drawn at Step 1 by new volatilities.

REMARK 9.4 Solving (Q_i) requires picking a guess θ_i^0. For $i \leq n - 2$, we choose $\theta_i^0 = \theta_{i+1}^*$. ⬚

REMARK 9.5 An elegant way to avoid resimulations involves multiplying F_T by the relevant likelihood ratio. Indeed, if we denote $\mathbb{Q}^{\theta_i}$ as the probability under which (9.6) holds and $p_{\hat{\xi}}(s, t, x, y)$ as the lognormal density, we have

$$
\mathbb{E}^{\theta_i}[F_T] = \mathbb{E}^{\hat{\xi}} \left[F_T \prod_{t_i \leq t_k < T} \frac{p_{\xi_{t_k}}}{p_{\hat{\xi}}}(t_k, t_{k+1}, X_{t_k}, X_{t_{k+1}}) \right]
$$

where the t_k's are the discretization times in the simulation of process X, and include the t_i's. A similar technique is used in Broadie and Glasserman [69] to price American options in high dimension. Unfortunately, the likelihood ratio has great variance, unless D is "small," so that empirical averages very poorly estimate the right-hand side expectation above. ⬚

REMARK 9.6 In Remark 9.3, we pointed out that in the case where the price of an option depends on path-dependent variables whose values can change continuously, the Hamiltonian H may depend not only on the gammas, and hence may differ from (9.4). In the parametric approach, one directly parameterizes the optimal covariance matrix, regardless of the form of the Hamiltonian. As a consequence, this method works whether the path-dependent variables change continuously or only at discrete dates. ⬚

9.3.3 Choice of the parameterization

The parameterization of the maximization set, i.e., the choice of relevant functions λ_{t_i}'s, is a crucial step in our procedure. A wrong choice would lead to a bad estimate of the optimal volatilities and correlations, and hence to a bad lower bound price. To build a good parameterization, one can proceed as follows:

- Choose some relevant path-dependent variables $A = (A^1, \ldots, A^q)$.

- For a grid of dates, asset values, and path-dependent values (t, X, A), compute Monte Carlo gammas $\Gamma(t, X, A)$ in the Black-Scholes model with some covariance matrix $\hat{\xi}$.

- For each point (t, X, A) in the grid, build the solution $(\sigma^{*\alpha}(t, X, A), \rho^{*\alpha\beta}(t, X, A))_{1 \leq \alpha < \beta \leq d}$ to the problem

$$
H(X, \Gamma(t, X, A)) = \frac{1}{2} \max_{(\sigma^\alpha, \rho^{\alpha\beta})_{1 \leq \alpha < \beta \leq d}} \sum_{\alpha, \beta = 1}^{d} \rho^{\alpha\beta} \sigma^\alpha \sigma^\beta X^\alpha X^\beta \Gamma^{\alpha\beta}(t, X, A)
$$

- For each date t in the time grid, graph the optimal solutions $(X, A) \mapsto \sigma^{*\alpha}(t, X, A)$ and $(X, A) \mapsto \rho^{*\alpha\beta}(t, X, A)$ and guess a parameterization for them.

Since this preliminary process involves many computations of Monte Carlo prices, it is very time-consuming *but* it is done once and for all for each product. An alternative tool using our second Monte Carlo approach is sketched in Remark 9.8.

9.3.4   The single-asset case

To illustrate the general algorithm stated above, let us focus on the single-asset case. In practice, a simulation-based valuation method is needed even in this case, when the value of an option depends on three or more path-dependent variables. We know that the price of an option in the UVM is the solution to the BSB PDE, and that this PDE is of the bang-bang type: at each date, the optimal volatility is either $\underline{\sigma}$ or $\overline{\sigma}$. This means that we can restrict the maximization set to functions λ_{t_i} taking values in $\{\underline{\sigma}, \overline{\sigma}\}$. Stated otherwise, it is enough to specify functions $\gamma_{t_i}(\cdot; \theta)$ such that $\lambda_{t_i}(X, A; \theta_i) = \overline{\sigma}$ if $\gamma_{t_i}(X, A; \theta_i) \geq 0$, $= \underline{\sigma}$ otherwise. The sign of $\gamma_{t_i}(\cdot; \theta)$ hence represents the sign of the gamma at date $t \in [t_i, t_{i+1})$.

This particular form for λ_{t_i} speeds up evaluations of $h(\theta_i)$ in Step 2 (see (9.6)) because only some of the N_1 paths must be resimulated. Imagine that we resimulated paths and computed the value $h(\theta_i^m)$ for some value θ_i^m. The next step in our solver for (Q_i) requires that we do the same for some next value θ_i^{m+1}. It often happens that for many paths p, we have

$$\forall t_k \in [t_i, t_{i+1}), \quad \lambda_{t_i}(X_{t_k}^{(p)}, A_{t_k}^{(p)}; \theta_i^{m+1}) = \lambda_{t_i}(X_{t_k}^{(p)}, A_{t_k}^{(p)}; \theta_i^m)$$

so these paths need not be resimulated. This happens when $\gamma_{t_i}(X_{t_k}^{(p)}, A_{t_k}^{(p)}; \theta_i^{m+1})$ and $\gamma_{t_i}(X_{t_k}^{(p)}, A_{t_k}^{(p)}; \theta_i^m)$ have same sign for all discretization dates t_k's in the (small) interval $[t_i, t_{i+1})$. In particular, if γ_{t_i} is continuous in θ_i, this happens for large enough m's, provided the sequence $(\theta_i^m)_{m \in \mathbb{N}}$ converges as expected to some maximum θ_i^* (such that $\gamma_{t_i}(X_{t_k}^{(p)}, A_{t_k}^{(p)}; \theta_i^*) \neq 0$ for all $t_k \in [t_i, t_{i+1})$).

Also, provided that γ_{t_i} is continuous in θ_i, h is stepwise constant as a function of θ_i. As a consequence, optimization routines based on the computation of gradients and Hessians are of no use. We use the so-called downhill simplex method as reported in [4].

9.3.5   The two-asset case

In the two-asset case with $D = [\underline{\sigma}^1, \overline{\sigma}^1] \times [\underline{\sigma}^2, \overline{\sigma}^2] \times [\underline{\rho}, \overline{\rho}]$, our parametric approach requires that we provide:

- a function $\kappa_{t_i}(\cdot; \theta_i)$ such that $\rho_{t_i}(X, A; \theta_i) = \rho$ if $\kappa_{t_i}(X, A; \theta_i) \geq 0$, $= \underline{\rho}$ otherwise, i.e., κ_{t_i} represents the sign of the cross-gamma;

- a function $\sigma_{t_i}^1(\cdot; \theta_i)$ taking values in $[\underline{\sigma}^1, \overline{\sigma}^1]$;

- a function $\sigma_{t_i}^2(\cdot; \theta_i)$ taking values in $[\underline{\sigma}^2, \overline{\sigma}^2]$.

Although we know that the solution is not of the bang-bang type, we can pick two functions $\sigma_{t_i}^1(\cdot; \theta_i)$ and $\sigma_{t_i}^2(\cdot; \theta_i)$ taking values resp. in $\{\underline{\sigma}^1, \overline{\sigma}^1\}$ and $\{\underline{\sigma}^2, \overline{\sigma}^2\}$. This speeds up the resolution to (Q_i) as explained in the single-asset case section.

Numerical examples with one or two assets will be given in Section 9.5. Before, we present our second approach for valuing options by simulation in the UVM. This second approach uses a link between nonlinear PDEs and BSDEs that we introduced in Chapter 7.

9.4 Solving the UVM with BSDEs

BSB PDE (9.3) is a particular case of (7.19) with the driver H depending only on x and $D_x^2 u$:

$$H(X, \Gamma) = \frac{1}{2} \max_{(\sigma^\alpha, \rho^{\alpha\beta})_{1 \leq \alpha < \beta \leq d} \in D} \sum_{\alpha, \beta=1}^{d} \rho^{\alpha\beta} \sigma^\alpha \sigma^\beta X^\alpha X^\beta \Gamma^{\alpha\beta}$$

From (7.18), the 2-BSDE associated to the BSB equation is

$$dX_t^\alpha = \hat{\sigma}^\alpha X_t^\alpha \, dW_t^\alpha, \quad d\langle W^\alpha, W^\beta \rangle_t = \hat{\rho}^{\alpha\beta} \, dt, \quad 1 \leq \alpha < \beta < d$$

$$dY_t = -H(X_t, \Gamma_t) \, dt + \sum_{\alpha=1}^{d} Z_t^\alpha \diamond \hat{\sigma}^\alpha X_t^\alpha \, dW_t^\alpha \tag{9.7}$$

$$dZ_t^\alpha = A_t^\alpha \, dt + \sum_{\beta=1}^{d} \Gamma_t^{\alpha\beta} \hat{\sigma}^\beta X_t^\beta \, dW_t^\beta$$

$$Y_T = g(X_T)$$

We are free to choose the diffusion $\sigma(\cdot, \cdot)$; we pick a lognormal dynamics for X with some constant volatility $\hat{\sigma}^\alpha$ and some constant correlation $\hat{\rho}^{\alpha\beta}$.

From Theorem 9.1, under regularity and growth assumptions on the payoff g, the solution u to PDE (9.3) is smooth. By Proposition 7.2, $Y_t = u(t, X_t)$ is a solution to (9.7). Since D is compact, $\mathbf{A(f)}$ and $\mathbf{A(comp)}$ hold (see Section 7.3.2) so that Theorem 7.4 applies and Y is the unique solution to (9.7).

9.4.1 A new numerical scheme

Numerical schemes for 2-BSDEs were introduced in Section 7.3.3. Here, we suggest a new numerical scheme that is particularly well suited to the UVM. This new numerical scheme for the BSB 2-BSDE reads, after dropping the Δ superscript:

Scheme UVM:

$$X_{t_i}^\alpha = X_0^\alpha e^{-(\hat\sigma^\alpha)^2 \frac{t_i}{2} + \hat\sigma^\alpha W_{t_i}^\alpha}, \qquad \mathbb{E}^{\mathbb{Q}}[\Delta W_{t_i}^\alpha \Delta W_{t_i}^\beta] = \hat\rho^{\alpha\beta} \Delta t_i$$

$$Y_{t_n} = g(X_{t_n})$$

$$Y_{t_{i-1}} = \mathbb{E}^{\mathbb{Q}}[Y_{t_i} | X_{t_{i-1}}]$$

$$+ \left(H(X_{t_{i-1}}, \Gamma_{t_{i-1}}) - \frac{1}{2} \sum_{\alpha,\beta=1}^d \hat\rho^{\alpha\beta} \hat\sigma^\alpha \hat\sigma^\beta X_{t_{i-1}}^\alpha X_{t_{i-1}}^\beta \Gamma_{t_{i-1}}^{\alpha\beta} \right) \Delta t_i \tag{9.8}$$

$$(\Delta t_i)^2 \hat\sigma^\alpha \hat\sigma^\beta X_{t_{i-1}}^\alpha X_{t_{i-1}}^\beta \Gamma_{t_{i-1}}^{\alpha\beta}$$

$$= \mathbb{E}^{\mathbb{Q}}\left[Y_{t_i} \left(U_{t_i}^\alpha U_{t_i}^\beta - \Delta t_i \hat\rho_{\alpha\beta}^{-1} - \Delta t_i \hat\sigma^\alpha U_{t_i}^\alpha \delta_{\alpha\beta} \right) \Big| X_{t_{i-1}} \right]$$

with $U_{t_i}^\alpha \equiv \sum_{\beta=1}^d \hat\rho_{\alpha\beta}^{-1} \Delta W_{t_i}^\beta$. Compared to Scheme (7.20), we have changed the discretization for the gamma Γ by explicitly introducing the Malliavin weight for a lognormal diffusion with volatility $\hat\sigma$ and correlation $\hat\rho$. This scheme performs better in our numerical experiments. A different conclusion was reached in [105] where Scheme (7.20) was shown to perform better. Note that in [105] the payoffs tested are smooth $(g(x) = x^\eta)$.

In the particular case where P&L = 0 (see (7.21) and (9.9)), the nonlinear PDE reduces to a Black-Scholes PDE and unlike (7.20), this scheme is exact. Considering the case where $i = n$ helps us understand what Scheme (9.8) does: since we simulated lognormal X's, the price $Y_{t_{n-1}}$ is the sum of the Black-Scholes price $\mathbb{E}^{\mathbb{Q}}[Y_{t_n} | X_{t_{n-1}}]$ and the gamma-theta P&L correction

$$\left(H(X_{t_{n-1}}, \Gamma_{t_{n-1}}) - \frac{1}{2} \sum_{\alpha,\beta=1}^d \hat\rho^{\alpha\beta} \hat\sigma^\alpha \hat\sigma^\beta X_{t_{n-1}}^\alpha X_{t_{n-1}}^\beta \Gamma_{t_{n-1}}^{\alpha\beta} \right) \Delta t_n \tag{9.9}$$

This last term requires that we estimate the gamma at time t_{n-1}. We use the Black-Scholes gamma, as given by the last equation in (9.8), which uses the Malliavin weight for the lognormal diffusion.

Scheme (9.8) requires computing $\frac{d(d+1)}{2} + 1$ conditional expectations at each discretization date. For this purpose, as for the valuation of American options, one can use different choices (see Section 6.8.5). Parametric regressions are used in [116] and Malliavin's weights in [66] in the case of 1-BSDEs.

REMARK 9.7 Transaction costs For transaction costs, Scheme (9.8) is replaced by

$$Y_{t_{i-1}} = \mathbb{E}^{\mathbb{Q}}[Y_{t_i} | X_{t_{i-1}}]$$

$$+ \sqrt{\frac{2}{\pi \delta t}} \sum_{\alpha=1}^{d} k_\alpha X_{t_{i-1}}^\alpha \Delta t_i \sqrt{\sum_{\beta,\gamma=1}^{d} \rho^{\beta\gamma} \sigma^\beta \sigma^\gamma X_{t_{i-1}}^\beta X_{t_{i-1}}^\gamma \Gamma_{t_{i-1}}^{\alpha\beta} \Gamma_{t_{i-1}}^{\alpha\gamma}}$$

where X follows an Itô diffusion process. ☐

In our numerical experiments for the second approach, we have used non-parametric regressions in one dimension and parametric regressions in two dimensions with suitable basis functions. Likelihood ratio weights have too much variance. Before presenting our results, we test the algorithm in the case of an at-the-money call option with payoff $(X_T - X_0)^+$. This enables us to state the details of our algorithm.

9.4.2 First example: At-the-money call option

We take $T = 1$, $X_0 = 100$, $\underline{\sigma} = 0.1$, $\overline{\sigma} = 0.2$; see Table 9.1. The true price, $\mathcal{C} = 7.97$, is the Black-Scholes price with the upper volatility as stated in Remark 9.1. We denote by $N_1 = 2^{M_1}$ the number of paths used to compute the conditional expectations.

As Δ goes small, we need more and more simulations to obtain an accurate price. Gobet *et al.* [116], for 1-BSDEs, and Fahim *et al.* [105], for 2-BSDEs, also noticed that the numerical scheme diverges when Δ goes to zero, M_1 being fixed. For instance, for $\Delta = 1/8$, the price has not converged yet with $M_1 = 17$. Alanko and Avellaneda [29] recently suggested a nice and simple trick to overcome this issue.

Furthermore, as observed in this simple example, the algorithm has an unpredictable bias. In order to build a low-biased estimate, we simulate N_2 replications of

$$dX_t^\alpha = \sigma_t^{*\alpha} X_t^\alpha \, dW_t^\alpha, \qquad d\langle W^\alpha, W^\beta \rangle_t = \rho_t^{*\alpha\beta} \, dt, \qquad 1 \le \alpha < \beta \le d$$

in an independent second Monte Carlo procedure, where $\sigma_t^{*\alpha}$ and $\rho_t^{*\alpha\beta}$ are the solutions to

$$\max_{(\sigma^\alpha, \rho^{\alpha\beta})_{1 \le \alpha < \beta \le d} \in D} \sum_{\alpha,\beta=1}^{d} \rho^{\alpha\beta} \sigma^\alpha \sigma^\beta X_t^\alpha X_t^\beta \Gamma_t^{\alpha\beta} \qquad (9.10)$$

and $\Gamma_t^{\alpha\beta} = \varphi(t, X_t)$, with φ the result of the regression step. Since the co-variance matrix is suboptimal, we obtain a low-biased estimator. This is a commonly used technique for the pricing of American options in Monte Carlo (see for instance the description of the Longstaff-Schwartz algorithm in Section 6.8.3). Below, we choose $N_2 = 2^{15}$ paths and a time step of $\Delta_2 = 1/400$ for the forward discretization of X (see Table 9.2). In Table 9.2 and in the following tables of this chapter, we highlight the maximum numerical price using bold characters.

Table 9.1: At-the-money call option valued using the BSDE approach with volatility $\hat{\sigma} = 0.15$. The true price is $C = 7.97$.

Δ	M_1	12	13	14	15	16	17
1/2	Price	8.04	8.06	8.00	8.00	8.00	8.00
1/4	Price	8.53	8.29	8.00	7.87	7.89	7.86
1/8	Price	9.28	8.72	8.02	7.79	7.78	7.68

Table 9.2: At-the-money call option valued using the BSDE approach with volatility $\hat{\sigma} = 0.15$ and an independent second Monte Carlo run. The true price is $C = 7.97$.

Δ	M_1	12	13	14	15	16	17
1/2	Price	7.93	7.94	7.95	7.95	**7.96**	**7.96**
1/4	Price	7.88	7.90	7.92	7.93	7.95	7.94
1/8	Price	7.53	7.93	7.58	7.60	**7.96**	**7.96**

9.4.3 The algorithm

The final meta-algorithm for pricing can be summarized by the following steps:

1. Simulate N_1 replications of X with a lognormal diffusion.

2. Apply the backward algorithm (9.8) using a regression approximation. In a high-dimensional problem, the parametric regression is the most appropriate.

3. Simulate N_2 independent replications of X using the gamma functions computed at the previous step.

Note that the payoffs g that we will use in our numerical experiments below do not satisfy the regularity condition $g \in C^3$ under which we stated the existence and uniqueness of 2-BSDE (9.7). However, even for these non-smooth payoffs, our discretization scheme seems numerically convergent.

REMARK 9.8 Another tool to choose a relevant parameterization: Combination of the BSDE and parametric approaches   Here we present an alternative to Section 9.3.3. We may use the BSDE algorithm to choose an efficient parameterization for the parametric approach. For each date t in the time grid, we graph the optimal $\sigma_t^{*\alpha}$ and $\rho_t^{*\alpha\beta}$ solutions to (9.10) as functions of X_t and guess a parameterization for them. Note that contrary to our original determination of a parameterization, this process involves only one Monte Carlo computation.   ▯

9.4.4   About the generation of the first N_1 paths

In both algorithms, the volatility function $\sigma(\cdot, \cdot)$ is arbitrary. We can choose different volatility functions $\sigma(\cdot, \cdot)$'s to generate the first N_1 replications of X. Different $\sigma(\cdot, \cdot)$'s lead to different sets $(X^{(p)}, A^{(p)})_{1 \leq p \leq N_1}$, which serve in the optimization or regression procedures, hence to different optimal covariance matrix estimates. Of course, the exact matrix does not depend on $\sigma(\cdot, \cdot)$. Here, we have chosen a lognormal diffusion with a mid-volatility $\hat{\sigma}$, but other choices are possible. For instance:

- Before proceeding to Step 3, we may repeat Steps 1 and 2, replacing $(\hat{\rho}^{\alpha\beta}\hat{\sigma}^\alpha\hat{\sigma}^\beta)$ by the optimal covariance matrix estimate as computed at Step 2. This should improve the estimate and result in a better lower bound for the price in Step 3. In the BSDE approach, this should reduce the contribution of the gamma-theta P&L term (9.9) to $Y_{t_{i-1}}$ in (9.8) and therefore the impact of the error in the computation of $\Gamma_{t_{i-1}}$ through regression.

- Allowing the $(X^{(p)}, A^{(p)})_{1 \leq p \leq N_1}$ points to completely cover the diffusion support of the UVM should result in more precise optimization or regression results. This can be achieved by allowing the volatility function $\sigma(\cdot, \cdot)$ to depend on path number p. For instance, in the single-asset case, for payoffs depending on some realized variance, we may choose

$$\sigma^{(p)} = \begin{cases} \underline{\sigma} & \text{if } p < N_1/3 \\ (\underline{\sigma} + \overline{\sigma})/2 & \text{if } N_1/3 \leq p < 2N_1/3 \\ \overline{\sigma} & \text{if } 2N_1/3 \leq p \end{cases}$$

9.5 Numerical experiments

Here we present numerical results for both approaches. In our experiments, we take $T = 1$, and for each asset α, $X_0^\alpha = 100$, $\underline{\sigma}^\alpha = 0.1$, $\overline{\sigma}^\alpha = 0.2$, and we use the constant mid-volatility $\hat{\sigma}^\alpha = 0.15$ to generate the first N_1 replications of X. We also pick $t_i = i/n$, so that $\Delta = 1/n$. In the pricing stage, the $N_2 = 2^{15}$ replications of X use a time step $\Delta_2 = 1/400$. **Param:** In the gamma calibration stage, we pick $N_1 = 2^{M_1}$ with $M_1 = 12$, and the N_1 replications of X use a time step $t_{k+1} - t_k = 1/100$. **BSDE:** We allow M_1 to vary from 12 to 17.

9.5.1 Options with one underlying

Let us first consider three payoffs that depend on a single asset: a call spread, a digital option, and a call Sharpe.

Call spread. (See Table 9.3) Let us test our two algorithms in the case of a call spread option with payoff $(X_T - K_1)^+ - (X_T - K_2)^+$. We pick $K_1 = 90$ and $K_2 = 110$. The true price (PDE) is $\mathcal{C}_{\mathrm{PDE}} = 11.20$ and the Black-Scholes price with the mid-volatility is $C_{\mathrm{BS}} = 9.52$. **Param:** Due to the shape of the call spread payout, we pick the following parameterization of the volatility: $\theta_i \in \mathbb{R}$ and $\lambda_{t_i}(X; \theta_i) = \overline{\sigma}$ if $\theta_i - \ln(X/X_0) \geq 0$, $= \underline{\sigma}$ otherwise, i.e., $\gamma_{t_i}(X; \theta_i) = \theta_i - \ln(X/X_0)$. For instance, for $\Delta = 1/4$, the numerical optimal gamma frontier is given by $(\theta_0^*, \theta_1^*, \theta_2^*, \theta_3^*) = (0.02, 0.02, 0.02, 0.02)$. **BSDE:** We use non-parametric regressions.

Table 9.3: Call spread valued using the parametric approach (column 2) and the BSDE approach (columns 4 to 9). The true price (PDE) is $\mathcal{C}_{\mathrm{PDE}} = 11.20$.

Δ	Param	M_1	12	13	14	15	16	17
$1/2$	**11.19**	Price	11.08	11.07	11.06	11.06	11.06	11.06
$1/4$	**11.19**	Price	11.01	**11.12**	11.06	11.07	11.11	11.11
$1/8$	11.18	Price	10.74	10.55	10.73	11.01	11.04	11.11

Since our parameterization of the optimal gamma frontier is exact in this case, the parametric approach gives very accurate results. Besides, as $\theta_{t_i}^*$ varies very slowly with t_i, this method proves to be efficient even with $\Delta = 1/2$, i.e., even when the gamma frontier is updated twice a year. The BSDE approach captures the right magnitude of the price but is not able to produce a lower bound greater than 11.12.

In Table 9.4, we illustrate the fact that although the diffusion coefficient $\hat{\sigma}$ can be chosen arbitrarily in both algorithms, a convenient choice can lead to better numerical results (see Section 9.4.4). We have computed the price of the above call spread option as a function of the volatility $\hat{\sigma}$. We pick $\Delta = 1/4$ and $M_1 = 16$. We also report the Black-Scholes price with volatility $\hat{\sigma}$. **Param:** The one-dimensional downhill simplex method, with initial guess $\theta_3^0 = 0$ and a first simplex side of 0.02, gives the same optimal $(\theta_0^*, \theta_1^*, \theta_2^*, \theta_3^*) = (0.02, 0.02, 0.02, 0.02)$ for all $\hat{\sigma}$ from 2% to 30%. For $\hat{\sigma} = 50\%$, it finds $(\theta_0^*, \theta_1^*, \theta_2^*, \theta_3^*) = (0, 0, 0, 0)$. **BSDE:** The numerical volatility $\hat{\sigma}$ that gives the most accurate result is $\hat{\sigma} = 15\%$.

Table 9.4: Call spread valued using the parametric and BSDE approaches with different volatilities $\hat{\sigma}$. The true price (PDE) is $\mathcal{C}_{\text{PDE}} = 11.20$. $\Delta = 1/4$ and $M_1 = 16$. We have included the Black-Scholes price with volatility $\hat{\sigma}$.

Algo/$\hat{\sigma}$	2%	5%	10%	15%	20%	30%	50%
Param	**11.19**	**11.19**	**11.19**	**11.19**	**11.19**	**11.19**	11.14
BSDE	9.36	10.72	11.01	**11.12**	11.07	11.00	10.81
Black-Scholes	10.00	9.97	9.76	9.52	9.30	8.87	8.06

Digital option. (See Table 9.5) Let us now test the algorithms with the digital option delivering $100 \times \mathbf{1}_{X_T \geq K}$. We pick $K = 100$. The true price (PDE) is $\mathcal{C}_{\text{PDE}} = 63.33$ and the Black-Scholes price with the mid-volatility is $C_{\text{BS}} = 46.54$. **Param:** Given that the payoffs are similar, we pick the same parameterization as for the call spread. **BSDE:** We use non-parametric regressions.

Table 9.5: Digital option valued using the parametric approach (column 2) and the BSDE approach (columns 4 to 9). The true price (PDE) is $\mathcal{C}_{\text{PDE}} = 63.33$ and the Black-Scholes price with the mid-volatility is $C_{\text{BS}} = 46.54$.

Δ	Param	M_1	12	13	14	15	16	17
1/2	63.13	Price	62.83	62.83	62.74	62.75	62.75	62.74
1/4	**63.14**	Price	62.53	**62.86**	62.77	62.35	62.45	62.43
1/8	62.68	Price	60.06	59.16	60.56	60.59	60.94	60.53

Again the parameterization of the optimal gamma frontier is exact in this case, so the parametric method gives very accurate results. The BSDE method performs well, except for too small Δ's. This example shows that discontinuous payoffs can be well priced by our algorithms.

Call Sharpe. (See Table 9.6) To finish with the single-asset examples, let us test the algorithms with a call Sharpe option delivering $(X_T - 100)^+ / \sqrt{V_T}$ where $V_T = \frac{1}{T} \sum_{l=1}^{12} \left(\ln \frac{X_{t_l}}{X_{t_{l-1}}} \right)^2$ is the monthly realized volatility. At time t, the option value depends on X_t and on the two path-dependent variables

$$A_t^1 = \sum_{\{l|t_l \leq t\}} \left(\ln \frac{X_{t_l}}{X_{t_{l-1}}} \right)^2$$

$$A_t^2 = X_{\sup_{\{l|t_l \leq t\}} t_l}$$

Param: We take $\theta_i = (\theta_i^1, \theta_i^2) \in \mathbb{R}^2$ and pick $\gamma_{t_i}(X, A; \theta_i) = \theta_i^1 \sqrt{A^1} + \theta_i^2 - \ln(X/X_0)$. **BSDE:** It is notably difficult to find a convenient basis to compute the conditional expectations and we assume as a first approximation that $\mathbb{E}_{i-1}^{\mathbb{Q}}[\cdot] \equiv \mathbb{E}^{\mathbb{Q}}[\cdot|X_{t_{i-1}}, A_{t_{i-1}}^1, A_{t_{i-1}}^2] \simeq \mathbb{E}^{\mathbb{Q}}[\cdot|X_{t_{i-1}}]$. The latter is computed using a one-dimensional non-parametric regression.

Table 9.6: Call Sharpe valued using the parametric approach (column 2) and the BSDE approach (columns 4 to 9). The true price (PDE) is $\mathcal{C}_{\text{PDE}} = 58.4$. The Black-Scholes price with mid-volatility is $\mathcal{C}_{\text{BS}} = 40.71$.

Δ	Param	M_1	12	13	14	15	16	17	18
1/2	54.98	Price	47.73	47.18	48.82	48.09	48.10	48.01	48.09
1/4	**55.55**	Price	46.93	47.34	48.01	48.92	48.67	49.38	49.44
1/12	54.32	Price	48.03	49.26	49.78	50.87	51.11	51.66	**52.12**

In this case, the optimal gamma frontier at date t_i, i.e., the optimal function $\gamma_{t_i}(\cdot)$, does not fall within the parameterized set $\{\gamma_{t_i}(\cdot; \theta_i)|\theta_i \in \mathbb{R}^2\}$. Hence, the parametric approach is less accurate than it proved to be in the cases of the call spread and the digital options. Nevertheless, it gives lower bounds around 55, that is, it captures the correct magnitude of the exact move, from 40.7 in the Black-Scholes model to 58.4 in the UVM. If we choose $\theta_i \in \mathbb{R}$ and $\gamma_{t_i}(X, A; \theta_i) = \theta_i - \ln(X/X_0)$ with $\Delta = 1/4$, we get $(\theta_0^*, \theta_1^*, \theta_2^*, \theta_3^*) = (0.075, 0.075, 0.025, -0.0125)$ and a lower bound estimate of 53.85, a value close to 55.55. This shows that the realized variance A^1 hardly affects the gamma frontier. Since the conditional expectations in the BSDE approach were computed as if they only depended on X, the BSDE lower bound prices can be compared to this 53.85.

REMARK 9.9 Continuously computed variance The BSDE approach can be easily adapted to the case where the realized variance V_t can change continuously. In Remark 9.3, we have shown that in this case the price of the option in the UVM can be written $u(t, X_t, V_t)$ where u is a solution to

$$\partial_t u(t, x, v) + H(x, \partial_x^2 u(t, x, v), \partial_v u(t, x, v)) = 0, \quad (t, x, v) \in [0, T) \times \mathbb{R}_+^* \times \mathbb{R}_+^*$$

$$H(x, \partial_x^2 u, \partial_v u) = \max_{\underline{\sigma} \leq \sigma \leq \overline{\sigma}} \sigma^2 \left(\frac{1}{2} x^2 \partial_x^2 u + \partial_v u \right)$$

We can associate a two-dimensional 2-BSDE on the (X, V) plane to this fully nonlinear PDE:

$$dX_t = \hat{\sigma} X_t \, dW_t^0$$
$$dV_t = \hat{\sigma}^2 \, dt + \eta \, dW_t^1$$
$$dY_t = \left(-H(X_t, \Gamma_t^{XX}, Z_t^V) + \mathcal{L}^{X,V} u(t, X_t, V_t) \right) dt + Z_t^S \hat{\sigma} X_t \, dW_t^0 + Z_t^V \eta \, dW_t^1$$

with

$$\mathcal{L}^{X,V} u(t, x, v) = \frac{1}{2} \hat{\sigma}^2 \left(x^2 \partial_x^2 u + 2 \partial_v u \right) + \frac{1}{2} \eta^2 \partial_v^2 u$$

We used $\hat{\sigma}^2$ as the (forward) drift for the variance V, but this is arbitrary. We have introduced a diffusion term for V_t. Here, η is a constant and W^1 a Brownian motion orthogonal to W^0. Adding this (purely numerical) volatility term allows us to compute $Z_t^V \equiv \partial_v u$. Like the solution u to the PDE, the 2-BSDE is independent of η, but the numerical scheme depends on it. Too small or too big a choice for η would lead to a bad regression-based estimation of Z_t^V. ▯

9.5.2 Options with two underlyings and no uncertainty on correlation

We consider (projective) payoffs that depend on two correlated assets with a constant certain correlation ρ, which can be written as

$$g(X_T^1, X_T^2) = X_T^1 \mathcal{G} \left(\frac{X_T^2}{X_T^1} \right)$$

This simple payoff form together with the certainty on the correlation parameter allow to reduce the two-dimensional BSB to a simple one-dimensional BSB. Indeed, by changing from the risk-neutral measure $\mathbb{Q}$ to the measure $\mathbb{Q}^1$ associated to the numéraire X^1, we obtain

$$\mathbb{E}^{\mathbb{Q}}[g(X_T^1, X_T^2)|\mathcal{F}_t] = X_t^1 \mathbb{E}^{\mathbb{Q}^1}[\mathcal{G}(X_T)|\mathcal{F}_t]$$

Here, $X_t \equiv \frac{X_t^2}{X_t^1}$ is a local martingale under $\mathbb{Q}^1$, $(\tilde{W}_t^1, \tilde{W}_t^2) = (W_t^1 - \int_0^t \sigma_s^1 \, ds, W_t^2 - \rho \int_0^t \sigma_s^1 \, ds)$ is a Brownian motion with correlation ρ under $\mathbb{Q}^1$, and

$$dX_t = X_t \left(\sigma_t^2 d\tilde{W}_t^2 - \sigma_t^1 d\tilde{W}_t^1 \right)$$
$$\overset{\text{Law}}{=} X_t \sigma_t d\tilde{W}_t$$

with the control $(\sigma_t)^2$ taking values in the interval I,

$$I = \begin{cases} \left[\sum_{\alpha=1}^2 (\underline{\sigma}^\alpha)^2 - 2\rho\underline{\sigma}^1\underline{\sigma}^2, \sum_{\alpha=1}^2 (\overline{\sigma}^\alpha)^2 - 2\rho\overline{\sigma}^1\overline{\sigma}^2 \right] & \text{if } \rho \leq 0 \\ \left[\sum_{\alpha=1}^2 (\underline{\sigma}^\alpha)^2 - 2\rho\overline{\sigma}^1\overline{\sigma}^2, \sum_{\alpha=1}^2 (\overline{\sigma}^\alpha)^2 - 2\rho\underline{\sigma}^1\underline{\sigma}^2 \right] & \text{otherwise} \end{cases} \quad (9.11)$$

Finally, we have

$$\max_{\sigma_s^1, \sigma_s^2, \, s \in [t,T]} \mathbb{E}^{\mathbb{Q}}[g(X_T^1, X_T^2)|\mathcal{F}_t] = X_t^1 \max_{(\sigma_s)^2 \in I, \, s \in [t,T]} \mathbb{E}^{\mathbb{Q}^1}[\mathcal{G}(X_T)|\mathcal{F}_t]$$

Outperformer option. As an example of projective payoff, we consider an outperformer option $(X_T^1 - X_T^2)^+$. Since $\mathcal{G}$ is the put payoff, when $\rho \leq 0$, the true value is given by the Black-Scholes price with high volatilities. **Param:** We choose $\theta_i = (\theta_i^1, \theta_i^2) \in \mathbb{R}^2$, $\sigma_{t_i}^1(X; \theta_i) = \overline{\sigma}^1$ if $\theta_i^1 - \ln(X^1/X_0^1) \geq 0$, $= \underline{\sigma}^1$ otherwise, and $\sigma_{t_i}^2(X; \theta_i) = \overline{\sigma}^2$ if $\theta_i^2 - \ln(X^2/X_0^2) \geq 0$, $= \underline{\sigma}^2$ otherwise. **BSDE:** We try two choices of basis functions. First, we use X^1 and X^2. Second, we also add the symmetric second order polynomials $X^\alpha X^\beta, \alpha, \beta = 1, 2$. In all cases, a constant is included in the regression. First, we test the method assuming zero correlation (see Tables 9.7 and 9.8). Then, we assume a correlation $\rho = -0.5$ (see Table 9.9).

Table 9.7: Outperformer option with 2 uncorrelated assets valued using the parametric approach (column 2) and the BSDE approach (columns 4 to 9). The true price is $\mathcal{C} = 11.25$. Basis functions $= \{1, X^1, X^2\}$.

Δ	Param	M_1	12	13	14	15	16	17
1/2	11.26	Price	11.24	11.24	11.24	11.24	11.24	11.24
1/4	11.26	Price	9.65	10.11	10.04	10.16	10.28	9.69
1/8	11.26	Price	9.17	9.45	9.14	9.26	9.40	9.47
1/12	11.26	Price	9.17	9.67	9.47	9.38	9.32	9.84

The parametric method allows us to select the high volatilities everywhere and hence gives excellent results. In the BSDE approach, the choice of basis functions clearly affects the price estimate. In particular for the trivial basis

Table 9.8: Outperformer option with 2 uncorrelated assets valued using the parametric approach (column 2) and the BSDE approach (columns 4 to 9). The true price is $\mathcal{C} = 11.25$. Basis functions $= \{1, X^1, X^2, (X^1)^2, (X^2)^2, X^1 X^2\}$.

Δ	Param	M_1	12	13	14	15	16	17
1/2	**11.26**	Price	11.09	11.15	11.15	11.15	11.15	11.15
1/4	**11.26**	Price	10.95	11.07	11.10	11.13	11.16	**11.19**
1/8	**11.26**	Price	10.35	10.71	10.88	10.94	11.07	11.14
1/12	**11.26**	Price	10.39	10.68	10.74	10.91	11.02	11.13

Table 9.9: Outperformer option with 2 correlated assets ($\rho = -0.5$) valued using the parametric approach (column 2) and the BSDE approach (columns 4 to 9). The true price is $\mathcal{C} = 13.75$. Basis functions $= \{1, X^1, X^2, (X^1)^2, (X^2)^2, X^1 X^2\}$.

Δ	Param	M_1	12	13	14	15	16	17
1/2	**13.77**	Price	13.58	13.66	**13.68**	13.66	13.66	13.65
1/4	13.74	Price	13.13	13.48	13.50	13.58	13.64	13.68
1/8	13.74	Price	12.45	12.87	13.12	13.20	13.48	13.63
1/12	13.75	Price	12.52	12.82	12.98	13.27	13.44	13.59

$\{1, X_1, X_2\}$, when $\Delta \geq 1/4$, the numerical gammas resulting from regressions often take negative values. This does not happen with $\Delta = 1/2$ nor with the second basis. Furthermore, as observed previously, as Δ gets smaller, we need more and more simulations to obtain an accurate price. However, as we obtained a low-biased estimator, for a fixed Δ, we should increase M_1 as long as the price increases.

Outperformer spread option. Finally, we move to a more complex projective payoff $(X_T^2 - K_1 X_T^1)^+ - (X_T^2 - K_2 X_T^1)^+$ with $K_1 = 0.9$ and $K_2 = 1.1$. As explained above, this option can be valued using our Monte Carlo algorithms or a numerical solution to the one-dimensional BSB PDE (see the beginning of this section) with $\underline{\sigma} = 17.32\%$ and $\overline{\sigma} = 34.64\%$. We found $\mathcal{C}_{\text{PDE}} = 11.41$ with $\rho = -0.5$. **Param:** We choose $\theta_i = (\theta_i^1, \theta_i^2) \in \mathbb{R}^2$, $\sigma_{t_i}^1(X; \theta_i) = \overline{\sigma}^1$ if $\theta_i^1 - \ln(X^2/X^1) \geq 0$, $= \underline{\sigma}^1$ otherwise, and $\sigma_{t_i}^2(X; \theta_i) = \overline{\sigma}^2$ if $\theta_i^2 - \ln(X^2/X^1) \geq 0$, $= \underline{\sigma}^2$ otherwise. **BSDE:** We use two sets of basis functions (see Tables 9.10 and 9.11). The second set respects the fact that the exact price can be written as $X^1 u(t, X^2/X^1)$ and clearly improves the price estimator.

The parametric method performs very well, because the optimal volatilities

Table 9.10: Outperformer spread option with 2 correlated assets ($\rho = -0.5$) valued using the parametric approach (column 2) and the BSDE approach (columns 4 to 9). The true price is $C = 11.41$. The Black-Scholes price with mid-volatilities is $C_{\text{BS}} = 9.04$. Basis functions $= \{1, X^1, X^2, (X^1)^2, (X^2)^2, X^1X^2\}$.

Δ	Param	M_1	12	13	14	15	16	17
1/2	**11.37**	Price	11.07	11.07	11.10	**11.11**	11.09	11.10

Table 9.11: Outperformer spread option with 2 correlated assets ($\rho = -0.5$) valued using the parametric approach (column 2) and the BSDE approach (columns 4 to 9). The true price is $C = 11.41$. The Black-Scholes price with mid-volatilities is $C_{\text{BS}} = 9.04$. Basis functions $= \{X^1, X^2, \frac{(X^2)^2}{X^1}, \frac{(X^2)^3}{(X^1)^2}\}$.

Δ	Param	M_1	12	13	14	15	16	17
1/2	**11.37**	Price	11.24	11.25	11.26	11.26	11.27	11.27
1/4	**11.37**	Price	10.85	11.12	11.20	11.27	**11.29**	11.28

belong to the parameterized set we input. Indeed, we know that the true price reads $P(t, x^1, x^2) = x^1 u(t, x^2/x^1)$ with u the value of a call spread on X^2/X^1 in the one-dimensional UVM with

$$I = \left[\sum_{\alpha=1}^{2}(\underline{\sigma}^\alpha)^2 - 2\rho\underline{\sigma}^1\underline{\sigma}^2, \sum_{\alpha=1}^{2}(\overline{\sigma}^\alpha)^2 - 2\rho\overline{\sigma}^1\overline{\sigma}^2\right],$$

because we have chosen $\rho \leq 0$; see (9.11). Hence the optimal volatilities are $\underline{\sigma}^1, \underline{\sigma}^2$ when $\partial_x^2 u(t, x^2/x^1) < 0$, and $\overline{\sigma}^1, \overline{\sigma}^2$ when $\partial_x^2 u(t, x^2/x^1) \geq 0$. Now, $\text{sign}(\partial_x^2 u(t, x^2/x^1)) = \text{sign}(c_t - x^2/x^1)$ for some constant c_t. This shows that we could even have enforced $\theta_i^1 = \theta_i^2$ in our parameterization of the optimal volatilities, without deteriorating the price.

9.5.3 Options with two underlyings and uncertainty on correlation

When the correlation ρ is uncertain, one cannot reduce the two-dimensional BSB PDE to a simple one-dimensional BSB PDE as in Section 9.5.2, even for projective payoffs. To test the accuracy of our two algorithms when correlation is uncertain, we price the outperformer spread option with the same parameters as above, except that the time-t correlation ρ_t is not constant anymore but is assumed to stay within $[\underline{\rho}, \overline{\rho}]$. We pick $\underline{\rho} = -0.5$, $\overline{\rho} = 0.5$ and $\hat{\rho} = 0$. The (two-dimensional) PDE price is 12.83. When ρ_t was assumed to be constantly equal to -0.5, the price was 11.41. **Param:** We choose $\theta_i = (\theta_i^1, \theta_i^2, \theta_i^3) \in \mathbb{R}^3$, $\sigma_{t_i}^1(X; \theta_i) = \overline{\sigma}^1$ if $\theta_i^1 - \ln(X^2/X^1) \geq 0$, $= \underline{\sigma}^1$ otherwise; $\sigma_{t_i}^2(X; \theta_i) = \overline{\sigma}^2$ if $\theta_i^2 - \ln(X^2/X^1) \geq 0$, $= \underline{\sigma}^2$ otherwise; and $\rho_{t_i}(X; \theta_i) = \overline{\rho}$ if $-\theta_i^3 + \ln(X^2/X^1) \geq 0$, $= \underline{\rho}$ otherwise. **BSDE:** We use the basis functions $\{1, X^1, X^2, \frac{(X^2)^2}{X^1}, \frac{(X^2)^3}{(X^1)^2}\}$ (see Table 9.12).

Table 9.12: Outperformer spread option valued using the parametric approach (column 2) and the BSDE approach (columns 4 to 9). The correlation is uncertain, with $\rho = -0.5$ and $\rho = 0.5$. The true price is $\mathcal{C}_{\text{PDE}} = 12.83$. Basis functions $\{1, X^1, X^2, \frac{(X^2)^2}{X^1}, \frac{(X^2)^3}{(X^1)^2}\}$.

Δ	Param	M_1	12	13	14	15	16	17
1/2	12.50	Price	12.37	12.40	12.41	12.41	12.39	12.38
1/4	**12.67**	Price	11.58	11.79	12.38	**12.44**	12.44	12.41

It is noteworthy that even with our simple parameterization for the optimal volatilities and correlation, the parametric method performs very well.

Conclusion

In this chapter, we have provided two efficient, accurate, easily implementable Monte Carlo approaches to the pricing of derivatives in the UVM. In the first method, which consists of a series of backward low-dimensional maximizations, the main ingredient is a parameterization of the optimal covariance matrix. The second method uses a recent connection between fully nonlinear second order PDEs and 2-BSDEs. In addition, these two algorithms can be combined efficiently. We may use the BSDE algorithm to choose an efficient parameterization for the parametric approach.

As a word of caution, we have illustrated in our numerical experiments that results depend greatly on the parameterization/choice of regressors, which may require numerical experimentations and good understanding of the financial derivatives under consideration. This is a common feature in the pricing of American options. An upper bound à la Rogers (see Section 6.8.4) should help in assessing the accuracy of the lower bound. Finally, as indicated in Remark 9.2, a similar methodology could be used to include transaction costs in general diffusion models.

9.6   Exercise: UVM with penalty term

We generalize the uncertain volatility model by adding a penalty term on the volatility

$$u_t = \sup_{[t,T]} \mathbb{E}^{\mathbb{Q}} \left[g(X_T) - \int_t^T \mathcal{L}(s, \sigma_s^2) \, ds \Big| \mathcal{F}_t \right]$$

with $\mathcal{L}(s, \cdot)$ a convex function. As usual, $\sup_{[t,T]}$ means that the supremum is taken over all $(\mathcal{F}_t)$-adapted processes $(\sigma_s)_{t \leq s \leq T}$ such that for all $s \in [t, T]$, σ_s belongs to the domain $[\underline{\sigma}, \overline{\sigma}]$. This generalization was first considered in [37, 38].

1. We note $H(s, p)$ the Hamiltonian $H(s, p) \equiv \sup_{\sigma^2 \in [\underline{\sigma}^2, \overline{\sigma}^2]} \{ p\sigma^2 - \mathcal{L}(s, \sigma_s^2) \}$, corresponding to the Legendre-Fenchel transform of $\mathcal{L}$. Prove that the HJB PDE writes

$$\partial_t u(t, x) + H \left(t, \frac{1}{2} x^2 \partial_x^2 u \right) = 0, \quad u(T, x) = g(x)$$

2. We want to consider a UVM where X_t is constrained to match the density $\mathbb{Q}_t^{\mathrm{mkt}}$ for all $t \in [0, T]$, which is implied from t-vanilla options:

$$u_0 = \sup_{[0,T], X_s \sim \mathbb{Q}_s^{\mathrm{mkt}} \, \forall s \in [0,T]} \mathbb{E}^{\mathbb{Q}}[g(X_T)]$$

Using linear duality, prove that u_0 can be written as

$$u_0 = \inf_{\lambda(\cdot, \cdot)} \sup_{[0,T]} \mathbb{E}^{\mathbb{Q}} \left[g(X_T) + \int_0^T \lambda(s, X_s) \left(\sigma_s^2 - \sigma_{\mathrm{loc}}^2(s, X_s) \right) ds \right]$$

with σ_{loc} the Dupire local volatility.

3. Give a financial interpretation of the Lagrange multiplier λ in terms of robust super-replication.

4. Write the HJB PDE.

Chapter 10

McKean Nonlinear Stochastic Differential Equations

The advantage of being smart is that you can always make a fool, while the opposite is quite impossible.

— Woody Allen

As will be explained in Chapters 11 and 12, the calibration of local stochastic volatility models and local correlation models to market smiles leads to the so-called McKean nonlinear stochastic differential equations (SDEs). In such nonlinear SDEs, in contrast with classical Itô SDEs, the drift and volatility coefficients depend on the (unknown) marginal law of the process. The Fokker-Planck equation associated to this SDE is therefore nonlinear, hence the denomination. In this chapter, we review some basic properties of McKean SDEs and introduce the particle method, an elegant stochastic simulation of such processes, which will prove to be extremely efficient for calibration purposes (see Chapters 11 and 12).

10.1 Definition

McKean nonlinear stochastic differential equations were introduced by Henry McKean in 1966 [161]. A McKean equation for an n-dimensional process X is an SDE in which the drift and volatility depend not only on the current value X_t of the process, but also on the probability distribution $\mathbb{P}_t$ of X_t:

$$dX_t = b(t, X_t, \mathbb{P}_t)\, dt + \sigma(t, X_t, \mathbb{P}_t)\, dW_t, \quad \mathbb{P}_t = \text{Law}(X_t), \quad X_0 \in \mathbb{R}^n \quad (10.1)$$

where W_t is a d-dimensional Brownian motion.

Example 10.1 McKean-Vlasov SDEs
The basic prototype for a McKean equation is given by the McKean-Vlasov

SDE, where for $1 \leq i \leq n$ and $1 \leq j \leq d$,

$$b^i\left(t, x, \mathbb{P}_t\right) = \int b^i(t, x, y)\mathbb{P}_t(dy) = \mathbb{E}\left[b^i(t, x, X_t)\right] \qquad (10.2)$$

$$\sigma^i_j\left(t, x, \mathbb{P}_t\right) = \int \sigma^i_j(t, x, y)\mathbb{P}_t(dy) = \mathbb{E}\left[\sigma^i_j(t, x, X_t)\right]$$

Here the drift is just the mean value $\int b^i(t, X_t, y)\mathbb{P}_t(dy)$ of some function $b^i(t, X_t, \cdot)$ with respect to the distribution $\mathbb{P}_t$ of X_t, and likewise for the volatility. We will also assume that $b(t, x, y)$ and $\sigma(t, x, y)$ are Lipschitz-continuous in x and y. □

In [23, 163], uniqueness and existence are proved for Equation (10.1) if the drift and volatility coefficients are Lipschitz-continuous functions of x (and a linear growth condition) *and* $\mathbb{P}_t$, with respect to the so-called Wasserstein distance (this is verified for the McKean-Vlasov SDEs):

THEOREM 10.1 [23, 163]
Let $b : \mathbb{R}_+ \times \mathbb{R}^n \times \mathcal{P}_2(\mathbb{R}^n) \to \mathbb{R}^n$ and $\sigma : \mathbb{R}_+ \times \mathbb{R}^n \times \mathcal{P}_2(\mathbb{R}^n) \to \mathbb{R}^{n \times d}$ be Lipschitz-continuous functions and satisfy a linear growth condition

$$|b(t, X, \mathbb{P}) - b(t, Y, \mathbb{Q})| + |\sigma(t, X, \mathbb{P}) - \sigma(t, Y, \mathbb{Q})| \leq C\left(|X - Y| + d(\mathbb{P}, \mathbb{Q})\right)$$
$$|b(t, X, \mathbb{P})| + |\sigma(t, X, \mathbb{P})| \leq C\left(1 + |X|\right)$$

for the sum of the canonical metric on $\mathbb{R}^n$ and the Wasserstein distance d on the set $\mathcal{P}_2(\mathbb{R}^n)$ of probability measures with finite second order moment:

$$d\left(\mu, \nu\right) \equiv \inf_{\tau \in \mathcal{P}(\mathbb{R}^n \times \mathbb{R}^n) \text{ with marginals } \mu \text{ and } \nu} \left(\int_{\mathbb{R}^n \times \mathbb{R}^n} |x - y|^2 \tau(dx, dy)\right)^{\frac{1}{2}}$$

C is a positive constant. Then the nonlinear SDE

$$dX_t = b(t, X_t, \mathbb{P}_t)\, dt + \sigma(t, X_t, \mathbb{P}_t)\, dW_t, \qquad X_0 \in \mathbb{R}^n \qquad (10.3)$$

where $\mathbb{P}_t$ denotes the probability distribution of X_t admits a unique solution such that $\mathbb{E}(\sup_{0 \leq t \leq T} |X_t|^2) < \infty$.

The Wasserstein metric, also called Monge-Kantorovich metric, is related to model-independent bounds on multi-asset European option prices. Section 10.A is a short reminder on basic properties of this distance and its link with model-independent robust super-replication. In particular, this metric (on the probability space $\mathcal{P}_2(\mathbb{R}^n)$) induces the topology of weak convergence (see Chapter 7 in [26]).
The proof of this theorem, based on a pathwise fixed point technique and detailed in [23, 163], is quite instructive and will guide us in suggesting our Markovian projection algorithm in Section 11.5.

PROOF (idea)   Our proof is based on [23, 163]. Let $\mathcal{C} = C([0,T], \mathbb{R}^n)$ denote the space of continuous functions from $[0,T]$ to $\mathbb{R}^n$ and $\mathcal{P}_2(\mathcal{C})$ be the space of probability measures $\mathbb{P}$ on $\mathcal{C}$ such that $\mathbb{E}^{\mathbb{P}}[\|Y\|_\infty^2] \equiv \mathbb{E}^{\mathbb{P}}[\sup_{0 \le t \le T} |Y_t|^2] < \infty$ for all $Y \in \mathcal{C}$. The space $\mathcal{P}_2(\mathcal{C})$ is complete when endowed with the Wasserstein metric

$$D_t(\mathbb{P}, \mathbb{Q}) \equiv \inf \int_{\mathcal{C} \times \mathcal{C}} \sup_{0 \le s \le t} |Y_s - X_s|^2 \, R(dX, dY)$$

where the infimum is taken over all $R \in \mathcal{P}_2(\mathcal{C} \times \mathcal{C})$ with marginals $\mathbb{P}$ and $\mathbb{Q}$. Consider the map Φ, which associates to $\mathbb{P} \in \mathcal{P}_2(\mathcal{C})$ the law of $X_t^{\mathbb{P}}$ defined by

$$dX_t^{\mathbb{P}} = b(t, X_t^{\mathbb{P}}, \mathbb{P}_t) \, dt + \sigma(t, X_t^{\mathbb{P}}, \mathbb{P}_t) \, dW_t$$

with $\mathbb{P}_t$ the t-marginal of $\mathbb{P}$. Since b and σ are Lipschitz-continuous in x and satisfy a linear growth condition, the above (classical) SDE admits a unique strong solution and Φ takes its values in $\mathcal{P}_2(\mathcal{C})$ (see e.g., [13]). We observe that the process $X_t^{\mathbb{P}}$ solves Equation (10.3) if and only if its law is a fixed point of Φ. In order to complete the proof, it suffices to check that an iterate Φ^N of Φ is a strictly contractive map. This guarantees that Φ admits a unique fixed point.
For $\mathbb{P}, \mathbb{Q} \in \mathcal{P}_2(\mathcal{C})$, we denote $\delta X_s \equiv X_s^{\mathbb{P}} - X_s^{\mathbb{Q}}$, $\delta b_s = b(s, X_s^{\mathbb{P}}, \mathbb{P}_s) - b(s, X_s^{\mathbb{Q}}, \mathbb{Q}_s)$, and $\delta \sigma_s = \sigma(s, X_s^{\mathbb{P}}, \mathbb{P}_s) - \sigma(s, X_s^{\mathbb{Q}}, \mathbb{Q}_s)$. We first decompose

$$|\delta X_s|^2 \le 2 \left(\left| \int_0^s \delta b_u \, du \right|^2 + \left| \int_0^s \delta \sigma_u \, dW_u \right|^2 \right)$$

$$\le 2 \left(s \int_0^s |\delta b_u|^2 \, du + \left| \int_0^s \delta \sigma_u \, dW_u \right|^2 \right)$$

Then it follows from Doob's maximal inequality and the Lipschitz property of the coefficients b and σ that

$$h(t) \equiv \mathbb{E} \left[\sup_{s \le t} |\delta X_s|^2 \right] \le 2 \left(t \int_0^t \mathbb{E}[|\delta b_u|^2] \, du + 4 \int_0^t \mathbb{E}[|\delta \sigma_u|^2] \, du \right)$$

$$\le C \left(\int_0^t \mathbb{E}[|\delta X_u|^2] \, du + \int_0^t d^2(\mathbb{P}_u, \mathbb{Q}_u) \, du \right)$$

$$\le C \left(\int_0^t h(u) \, du + \int_0^t d^2(\mathbb{P}_u, \mathbb{Q}_u) \, du \right)$$

where C denotes a positive constant depending only on T. By Gronwall's lemma, one deduces that there exists a positive constant K depending only on T such that for $t \le T$,

$$h(t) \le K \int_0^t d^2(\mathbb{P}_u, \mathbb{Q}_u) \, du$$

Since $D_t^2(\Phi(\mathbb{P}), \Phi(\mathbb{Q})) \leq \mathbb{E}[\sup_{s \leq t} |\delta X_s|^2]$ and $d(\mathbb{P}_s, \mathbb{Q}_s) \leq D_s(\mathbb{P}, \mathbb{Q})$, the last inequality implies that for $t \leq T$

$$D_t^2(\Phi(\mathbb{P}), \Phi(\mathbb{Q})) \leq K \int_0^t D_s^2(\mathbb{P}, \mathbb{Q}) \, ds$$

By iterating this inequality, we obtain for $N \in \mathbb{N}^*$,

$$D_T^2(\Phi^N(\mathbb{P}), \Phi^N(\mathbb{Q})) \leq K^N \int_0^T \frac{(T-s)^{N-1}}{(N-1)!} D_s^2(\mathbb{P}, \mathbb{Q}) \, ds \leq \frac{K^N T^N}{N!} D_T^2(\mathbb{P}, \mathbb{Q})$$

For large N, Φ^N is a contraction, so Φ admits a unique fixed point. □

By a formal computation, the probability density function $p(t, y) \, dy \equiv \mathbb{P}_t(dy)$ of X_t is a solution to the Fokker-Planck PDE:

$$- \partial_t p(t, x) - \sum_{i=1}^n \partial_i \left(b^i(t, x, \mathbb{P}_t) p(t, x) \right)$$

$$+ \frac{1}{2} \sum_{i,j=1}^n \partial_{ij} \left(\sum_{k=1}^d \sigma_k^i(t, x, \mathbb{P}_t) \sigma_k^j(t, x, \mathbb{P}_t) p(t, x) \right) = 0 \quad (10.4)$$

with the initial condition

$$\lim_{t \to 0} p(t, x) = \delta(x - X_0)$$

It is nonlinear because $b^i(t, x, \mathbb{P}_t)$ and $\sigma_k^i(t, x, \mathbb{P}_t)$ depend on the unknown p. More precisely, one proves that the nonlinear process X_t has time marginals which satisfy in the weak sense the nonlinear PDE (10.4): For $f \in C_b^2(\mathbb{R}^n)$, by applying Itô's lemma,

$$f(X_t) = f(X_0) + \int_0^t \sum_{i=1}^n \sigma^i(s, X_s, \mathbb{P}_s) \partial_i f(X_s) \, dW_s$$

$$+ \int_0^t \left(\frac{1}{2} \sum_{i,j=1}^n \sum_{k=1}^d \sigma_k^i(s, X_s, \mathbb{P}_s) \sigma_k^j(s, X_s, \mathbb{P}_s) \partial_{ij} f(X_s) \right.$$

$$\left. + \sum_{i=1}^n b^i(s, X_s, \mathbb{P}_s) \partial_i f(X_s) \right) ds \quad (10.5)$$

Taking the expectation and integrating by parts yields a weak version of (10.4).

Example 10.2 Burgers PDE

We consider the McKean-Vlasov SDE with $b(x, y) = \mathbf{1}_{x>y}$, $\sigma(x, y) = 1$, and $n = d = 1$. Note that b is not Lipschitz-continuous in x and y here, and proving an existence and uniqueness result is difficult. For such a process, the nonlinear Fokker-Planck PDE is

$$\partial_t p(t, x) = \frac{1}{2}\partial_x^2 p(t, x) - \partial_x \left(p(t, x) \int_{-\infty}^{x} p(t, y)\, dy \right) = 0$$

By setting $u(t, x) = \int_{-\infty}^{x} p(t, y)\, dy$, we obtain the Burgers PDE

$$\partial_t u(t, x) = \frac{1}{2}\partial_x^2 u(t, x) - u(t, x)\partial_x u(t, x)$$

$\Box$

10.2 The particle method in a nutshell

The stochastic simulation of the McKean SDE (10.1) is very natural. It consists of replacing the law $\mathbb{P}_t$, which appears explicitly in the drift and diffusion coefficients, by its approximation given by the empirical distribution

$$\mathbb{P}_t^N = \frac{1}{N}\sum_{i=1}^{N}\delta_{X_t^{i,N}}$$

where the $(X_t^{i,N})_{1 \le i \le N}$ are solutions to the $(\mathbb{R}^n)^N$-dimensional classical (linear) SDE

$$dX_t^{i,N} = b\left(t, X_t^{i,N}, \mathbb{P}_t^N\right)\, dt + \sigma\left(t, X_t^{i,N}, \mathbb{P}_t^N\right)\, dW_t^i, \qquad \text{Law}\left(X_0^{i,N}\right) = \mathbb{P}_0$$

$\{W_t^i\}_{1 \le i \le N}$ are N independent d-dimensional Brownian motions; $\mathbb{P}_t^N$ is a random measure on $\mathbb{R}^n$. In the case of the prototype SDE (10.2), $\{X_t^{i,N}\}_{1 \le i \le N}$ are n-dimensional Itô processes given by

$$dX_t^{i,N} = \left(\int b(t, X_t^{i,N}, y)\, d\mathbb{P}_t^N(y) \right) dt + \left(\int \sigma(t, X_t^{i,N}, y)\, d\mathbb{P}_t^N(y) \right) dW_t^i$$

which is equivalent to

$$dX_t^{i,N} = \frac{1}{N}\sum_{j=1}^{N} b\left(t, X_t^{i,N}, X_t^{j,N}\right) dt + \frac{1}{N}\sum_{j=1}^{N}\sigma\left(t, X_t^{i,N}, X_t^{j,N}\right) dW_t^i \quad (10.6)$$

As a result, the paths $X^{i,N}$ *interact* with each other: the knowledge of t and $X_t^{i,N}$ is not enough to determine the drift and volatility of path i at time t and simulate $X^{i,N}$ at a later date $t + \delta t$; the knowledge of the position of the other paths $X^{j,N}$, $j \neq i$ is also required. Using an analogy with statistical physics, we use the word *particles* instead of paths and we speak of the *particle method*. The particle method differs from classical Monte Carlo methods as it involves a system of N interacting (bosonic) particles.

One can then try to show the *chaos propagation property* (see [23, 25] and Section 10.3). Briefly, if at $t = 0$, the $X_0^{i,N}$ are independent particles, then as $N \to \infty$, for any fixed $t > 0$, the $X_t^{i,N}$ are asymptotically independent and their empirical measure $\mathbb{P}_t^N$ converges in distribution toward the true measure $\mathbb{P}_t$. This means that, in the space of probabilities on the space of probabilities, the distribution of the random measure $\mathbb{P}_t^N$ converges toward a Dirac mass at the deterministic measure $\mathbb{P}_t$. Practically, it means (see [23, 25] and Theorem 10.2 (iii)) that for all functions $\varphi \in C_b(\mathbb{R}^n)$

$$\frac{1}{N} \sum_{i=1}^{N} \varphi(X_t^{i,N}) \xrightarrow[N\to\infty]{L^1} \int_{\mathbb{R}^n} \varphi(x) p(t, x) \, dx$$

where $p(t, \cdot)$ is the fundamental solution to the nonlinear Fokker-Planck PDE (10.4). Hence, if propagation of chaos holds, the particle method is convergent.

This is typical of mean-field approximations in statistical physics: In the large N limit, the $(\mathbb{R}^n)^N$-dimensional (linear) Fokker-Planck PDE approximates the nonlinear low-dimensional (n-dimensional) Fokker-Planck PDE (10.4). Then, the resulting drift and diffusion coefficients of $X_t^{i,N}$ depend not only on the position of the particle $X_t^{i,N}$, but also on the interaction with the other $N-1$ particles.

10.3 Propagation of chaos and convergence of the particle method

10.3.1 Building intuition: The BBGKY hierarchy

In order to justify the definition of propagation of chaos (see Section 10.3.2), we analyze the asymptotic behavior of the particle method for McKean-Vlasov SDEs from a PDE point of view. McKean-Vlasov SDEs have been defined in Example 10.1. For the sake of simplicity of this presentation, we consider the scalar case, i.e., $n = d = 1$, and we assume that b and σ do not depend on t. We denote by $\mu_t^{(N)}$ the density of $(X_t^{1,N}, \ldots, X_t^{N,N})$ with $X_t^{i,N}$ given by (10.6). We introduce the marginal laws:

$$\mu_t^{(k)}(x_1, \ldots, x_k) = \int \mu_t^{(N)}(x_1, \ldots, x_N) \, dx_{k+1} \ldots dx_N$$

By definition, we have the following consistency properties

$$\int \mu_t^{(1)}(x_1) \, dx_1 = 1$$

$$\int \mu_t^{(k)}(x_1, \dots, x_k) \, dx_k = \mu_t^{(k-1)}(x_1, \dots, x_{k-1}), \qquad k \in \{2, \dots, N\}$$

and $\mu_t^{(1)} = \text{Law}(X_t^{i,N})$. The PDE approach consists of deriving a hierarchy, called the Bogoliubov-Born-Green-Kirkwood-Yvon (BBGKY) hierarchy, of N-dimensional linear PDEs satisfied by the marginals $\mu_t^{(k)}$. The Fokker-Planck PDE associated to (10.6) is

$$\partial_t \mu_t^{(N)}(x_1, \dots, x_N) = -\frac{1}{N} \sum_{i,j=1}^{N} \partial_{x_i} \{b(x_i, x_j) \mu_t^{(N)}\}$$

$$+ \frac{1}{2N^2} \sum_{i,p,q=1}^{N} \partial_{x_i}^2 \{\sigma(x_i, x_p) \sigma(x_i, x_q) \mu_t^{(N)}\}$$

By integrating both sides of this equation with respect to the last $N - 1$ variables $(x_2, \dots, x_N)$, we get

$$\partial_t \int \mu_t^{(N)}(x_1, \dots, x_N) \prod_{l=2}^{N} dx_l = -\frac{1}{N} \sum_{i,j=1}^{N} \int \partial_{x_i} \{b(x_i, x_j) \mu_t^{(N)}\} \prod_{l=2}^{N} dx_l$$

$$+ \frac{1}{2N^2} \sum_{i,p,q=1}^{N} \int \partial_{x_i}^2 \{\sigma(x_i, x_p) \sigma(x_i, x_q) \mu_t^{(N)}\} \prod_{l=2}^{N} dx_l$$

We treat below in detail the drift term

$$D \equiv -\frac{1}{N} \sum_{i,j=1}^{N} \int \partial_{x_i} \{b(x_i, x_j) \mu_t^{(N)}\} \prod_{l=2}^{N} dx_l$$

the diffusion term being treated similarly. The drift term in ∂_{x_i} with $i \geq 2$ can be integrated out and cancels as the density vanishes at infinity. It remains

$$D = -\frac{1}{N} \int \prod_{l=2}^{N} dx_l \sum_{j=1}^{N} \partial_{x_1} \{b(x_1, x_j) \mu_t^{(N)}\}$$

Since the particles are indistinguishable (i.e., bosons), the density $\mu_t^{(N)}$ is totally symmetric in its arguments. As a consequence,

$$D = -\frac{1}{N} \partial_{x_1} \{b(x_1, x_1) \mu_t^{(1)}(x_1)\} - \frac{N-1}{N} \partial_{x_1} \left\{ \int b(x_1, y) \mu_t^{(2)}(x_1, y) \, dy \right\}$$

By treating the diffusion term similarly, we obtain the first hierarchy:

$$\partial_t \mu_t^{(1)}(x) = -\frac{1}{N}\partial_x\{b(x,x)\mu_t^{(1)}(x)\} - \frac{N-1}{N}\partial_x\left\{\int b(x,y)\mu_t^{(2)}(x,y)\,dy\right\}$$

$$+ \frac{1}{2N^2}\partial_x^2\{\sigma(x,x)^2\mu_t^{(1)}(x)\}$$

$$+ \frac{N-1}{2N^2}\partial_x^2\left\{\int \sigma(x,y)^2\mu_t^{(2)}(x,y)\,dy\right\}$$

$$+ \frac{N-1}{N^2}\partial_x^2\left\{\int \sigma(x,x)\sigma(x,y)\mu_t^{(2)}(x,y)\,dy\right\}$$

$$+ \frac{(N-1)(N-2)}{2N^2}\partial_x^2\left\{\int \sigma(x,y)\sigma(x,z)\mu_t^{(3)}(x,y,z)\,dy\,dz\right\}$$

The equation satisfied by $\mu_t^{(1)}(x)$ depends on the marginals $\mu_t^{(2)}$ and $\mu_t^{(3)}$. Proceeding similarly, we derive the complete hierarchy system where the equation satisfied by $\mu_t^{(k)}$ involves the marginals $\mu_t^{(k+1)}$ and $\mu_t^{(k+2)}$:

$$\partial_t \mu_t^{(k)}(x_1,\dots,x_k) \tag{10.7}$$

$$= -\frac{1}{N}\sum_{i,j=1}^{k}\partial_{x_i}\{b(x_i,x_j)\mu_t^{(k)}\}$$

$$-\frac{N-k}{N}\sum_{i=1}^{k}\partial_{x_i}\left\{\int b(x_i,y)\mu_t^{(k+1)}(x_1,\dots,x_k,y)\,dy\right\}$$

$$+\frac{1}{2N^2}\sum_{i,p,q=1}^{k}\partial_{x_i}^2\{\sigma(x_i,x_p)\sigma(x_i,x_q)\mu_t^{(k)}\}$$

$$+\frac{N-k}{2N^2}\sum_{i=1}^{k}\partial_{x_i}^2\left\{\int \sigma(x_i,y)^2\mu_t^{(k+1)}(x_1,\dots,x_k,y)\,dy\right\}$$

$$+\frac{N-k}{N^2}\sum_{i,p=1}^{k}\partial_{x_i}^2\left\{\sigma(x_i,x_p)\int \sigma(x_i,y)\mu_t^{(k+1)}(x_1,\dots,x_k,y)\,dy\right\}$$

$$+\frac{(N-k)(N-k-1)}{2N^2}$$

$$\sum_{i=1}^{k}\partial_{x_i}^2\left\{\int \sigma(x_i,y)\sigma(x_i,z)\mu_t^{(k+2)}(x_1,\dots,x_k,y,z)\,dy\,dz\right\}$$

We will now check that in the large N limit, i.e., $N \to \infty$, the solution to the hierarchy is

$$\mu_t^{(k)}(x_1,\dots,x_k) = \prod_{i=1}^{k}\mu_t^{(1)}(x_i) + \frac{1}{N}\delta\mu_t^{(k)}(x_1,\dots,x_k) + O(1/N^2) \tag{10.8}$$

By plugging this expression into Equation (10.7) and by identifying the leading term in $1/N^0$, we obtain

$$\sum_{i=1}^{k} \partial_t \mu_t^{(1)}(x_i) \prod_{\substack{r=1 \\ r \neq i}}^{k} \mu_t^{(1)}(x_r)$$

$$= -\sum_{i=1}^{k} \partial_{x_i} \left\{ \mu_t^{(1)}(x_i) \int b(x_i, y) \mu_t^{(1)}(y) \, dy \right\} \prod_{\substack{r=1 \\ r \neq i}}^{k} \mu_t^{(1)}(x_r)$$

$$+ \frac{1}{2} \sum_{i=1}^{k} \partial_{x_i}^2 \left\{ \left(\int \sigma(x_i, y) \mu_t^{(1)}(y) \, dy \right)^2 \mu_t^{(1)}(x_i) \right\} \prod_{\substack{r=1 \\ r \neq i}}^{k} \mu_t^{(1)}(x_r) \quad (10.9)$$

In particular, for $k = 1$, this reads

$$\partial_t \mu_t^{(1)}(x) = -\partial_x \left\{ \mu_t^{(1)}(x) \int b(x, y) \mu_t^{(1)}(y) \, dy \right\}$$

$$+ \frac{1}{2} \partial_x^2 \left\{ \mu_t^{(1)}(x) \left(\int \sigma(x, y) \mu_t^{(1)}(y) \, dy \right)^2 \right\} \quad (10.10)$$

This last equation is nothing other than the nonlinear Fokker-Planck equation (10.4) satisfied by the McKean-Vlasov process, i.e., $\mu_t^{(1)}(x) = p(t, x)$. By using Equation (10.10), Equation (10.9) is automatically satisfied, and this justifies the ansatz (10.8). Equation (10.8) shows that

$$\mu_t^{(k)}(x_1, \ldots, x_k) \xrightarrow[N \to \infty]{} \prod_{i=1}^{k} \mu_t^{(1)}(x_i)$$

This means that, when k is fixed and $N \to \infty$, the k particles become independent, and their asymptotic distribution is $\mu_t^{(1)}$. This property is called the *propagation of chaos*. As we will see in the next section (Theorem 10.2 (iii)), it is equivalent to the convergence of the particle method.

For completeness, we give the PDE satisfied by the subleading term $\delta \mu_t^{(k)}$ in expansion (10.8):

$$\partial_t \delta \mu_t^{(k)} = -\sum_{i,j=1}^{k} \partial_{x_i} \left\{ b(x_i, x_j) \prod_{r=1}^{k} \mu_t^{(1)}(x_r) \right\} \quad (10.11)$$

$$- \sum_{i=1}^{k} \partial_{x_i} \left\{ \int b(x_i, y) \delta \mu_t^{k+1}(x_1, \ldots, x_k, y) \, dy \right\}$$

$$+ k \sum_{i=1}^{k} \partial_{x_i} \left\{ \int b(x_i, y) \mu_t^{(1)}(y) \, dy \prod_{r=1}^{k} \mu_t^{(1)}(x_r) \right\}$$

$$+ \frac{1}{2} \sum_{i=1}^{k} \partial_{x_i}^2 \left\{ \int \sigma(x_i, y) \mu_t^{(1)}(y) \, dy \prod_{r=1}^{k} \mu_t^{(1)}(x_r) \right\}$$

$$+ \sum_{i,p=1}^{k} \partial_{x_i}^2 \left\{ \sigma(x_i, x_p) \int \sigma(x_i, y) \mu_t^{(1)}(y) \, dy \prod_{r=1}^{k} \mu_t^{(1)}(x_r) \right\}$$

$$+ \frac{1}{2} \sum_{i=1}^{k} \partial_{x_i}^2 \left\{ \int \sigma(x_i, y) \sigma(x_i, z) \delta \mu_t^{(k+2)}(x_1, \ldots, x_k, y, z) \, dy \, dz \right\}$$

$$- \left(k + \frac{1}{2} \right) \sum_{i=1}^{k} \partial_{x_i}^2 \left\{ \left(\int \sigma(x_i, y) \mu_t^{(1)}(y) \, dy \right)^2 \prod_{r=1}^{k} \mu_t^{(1)}(x_r) \right\}$$

for which the solution is

$$\delta \mu_t^{(k)}(x_1, \ldots, x_k) = \sum_{i=1}^{k} \delta \mu_t^{(1)}(x_i) \prod_{\substack{r=1 \\ r \neq i}}^{k} \mu_t^{(1)}(x_r)$$

with

$$\partial_t \delta \mu_t^{(1)}(x) = - \partial_x \left\{ b(x, x) \mu_t^{(1)}(x) \right\}$$

$$- \partial_x \left\{ \int b(x, y) \left(\mu_t^{(1)}(x) \delta \mu_t^{(1)}(y) + \mu_t^{(1)}(y) \delta \mu_t^{(1)}(x) \right) dy \right\}$$

$$+ \partial_x \left\{ \mu_t^{(1)}(x) \int b(x, y) \mu_t^{(1)}(y) \, dy \right\}$$

$$+ \frac{1}{2} \partial_x^2 \left\{ \mu_t^{(1)}(x) \int \sigma(x, y)^2 \mu_t^{(1)}(y) \, dy \right\}$$

$$+ \partial_x^2 \left\{ \sigma(x, x) \mu_t^{(1)}(x) \int \sigma(x, y) \mu_t^{(1)}(y) \, dy \right\}$$

$$+ \frac{1}{2} \partial_x^2 \left\{ \int \sigma(x, y) \sigma(x, z) \left(\mu_t^{(1)}(x) \mu_t^{(1)}(y) \delta \mu_t^{(1)}(z) + \mu_t^{(1)}(y) \mu_t^{(1)}(z) \delta \mu_t^{(1)}(x) \right. \right.$$

$$\left. \left. + \mu_t^{(1)}(z) \mu_t^{(1)}(x) \delta \mu_t^{(1)}(y) \right) dy \, dz \right\}$$

$$- \frac{3}{2} \partial_x^2 \left\{ \mu_t^{(1)}(x) \left(\int \sigma(x, y) \mu_t^{(1)}(y) \, dy \right)^2 \right\}$$

In the next section we give a formal definition of the propagation of chaos and relate it to the convergence of the particle method. In Section 10.3.3 we will rigorously prove that McKean-Vlasov SDEs propagate the chaos and hence that the particle method is convergent for this particular class of nonlinear SDEs.

10.3.2   Propagation of chaos and convergence of the particle method

DEFINITION 10.1 Empirical measure   *Let $X_1, \ldots, X_N$ be i.i.d. random variables with law μ. The empirical measure associated to the configuration $(X_1, \ldots, X_N)$ is*

$$\hat{\mu}^N = \frac{1}{N} \sum_{i=1}^{N} \delta_{X_i} \tag{10.12}$$

Note that it is a random probability measure with expectation $\mathbb{E}_\mu[\hat{\mu}^N] = \mu$, meaning that for all events A, $\mathbb{E}_\mu[\hat{\mu}^N(A)] = \mu(A)$.

DEFINITION 10.2 μ-chaotic distribution   *Let $\{\mu^N\}_{N \in \mathbb{N}}$ be a sequence of symmetric probabilities on $(\mathbb{R}^n)^N$. Let μ be a probability measure on $\mathbb{R}^n$. We say that μ^N is μ-chaotic if for each integer $k \geq 1$ and for all test functions $\varphi_1, \ldots, \varphi_k \in C_b(\mathbb{R}^n)$ (i.e., for all bounded continuous functions), we have*

$$\int \varphi_1(x_1) \cdots \varphi_k(x_k) \, d\mu^N(x_1, \ldots, x_N) \xrightarrow[N \to \infty]{} \int \varphi_1 \, d\mu \cdots \int \varphi_k \, d\mu$$

Stated otherwise, k particles (within N) are asymptotically independent and identically distributed as $N \to \infty$ (k being fixed). We say that μ^N is chaotic if there exists μ such that μ^N is μ-chaotic.

DEFINITION 10.3 Propagation of chaos   *Let us consider an N-dimensional SDE flow that associates to an initial probability measure μ_0^N a probability μ_t^N at time t. We say that this flow propagates the chaos if, for any initial chaotic measure μ_0^N and any $t > 0$, μ_t^N is chaotic.*

THEOREM 10.2
Let $\{\mu^N\}_{N \in \mathbb{N}}$ be a sequence of symmetric probabilities on $(\mathbb{R}^n)^N$, and μ be a probability measure on $\mathbb{R}^n$. The following four properties are equivalent:

(i) $\{\mu^N\}_{N \in \mathbb{N}}$ is μ-chaotic.

(ii) For all test functions $\varphi_1, \varphi_2 \in C_b(\mathbb{R}^n)$:

$$\int \varphi_1(x_1)\varphi_2(x_2) \, d\mu^N(x_1, \ldots x_N) \xrightarrow[N \to \infty]{} \int \varphi_1 \, d\mu \int \varphi_2 \, d\mu$$

(iii) Let $X_1, \ldots, X_N$ be random variables such that $\text{Law}(X_1, \ldots, X_N) = \mu^N$. Then, for all $\varphi \in C_b(\mathbb{R}^n)$,

$$\frac{1}{N} \sum_{i=1}^{N} \varphi(X_i) \xrightarrow[N \to \infty]{L^1} \int \varphi \, d\mu$$

(iv) Let $\hat{\mu}^N$ be the empirical measure associated to μ^N. Then

$$\mathbb{E}_{\mu^N} \left[\left| \int \varphi \, d\hat{\mu}^N - \int \varphi \, d\mu \right| \right] \xrightarrow[N \to \infty]{} 0$$

for all $\varphi \in C_b(\mathbb{R}^n)$.

This theorem proves that the convergence of the particle method, i.e., the fact that for all t and all $\varphi \in C_b(\mathbb{R}^n)$

$$\frac{1}{N} \sum_{i=1}^{N} \varphi(X_t^{i,N}) \xrightarrow[N \to \infty]{L^1} \int \varphi(x) p(t, x) \, dx$$

is equivalent to the propagation of chaos for the SDE $(X_t^{1,N}, \ldots, X_t^{N,N})$.

PROOF

(i) $\Rightarrow$ (ii): Follows obviously from the definition, with $k = 2$.

(ii) $\Rightarrow$ (iii): We have

$$\mathbb{E}_{\mu^N} \left[\left(\frac{1}{N} \sum_{i=1}^{N} \varphi(X_i) - \int \varphi \, d\mu \right)^2 \right]$$

$$= \mathbb{E}_{\mu^N} \left[\frac{1}{N^2} \sum_{i,j=1}^{N} \varphi(X_i) \varphi(X_j) - 2 \int \varphi \, d\mu \frac{1}{N} \sum_{i=1}^{N} \varphi(X_i) + \left(\int \varphi \, d\mu \right)^2 \right]$$

$$= \frac{1}{N} \mathbb{E}_{\mu^N}[\varphi(X_1)^2] + \frac{N-1}{N} \mathbb{E}_{\mu^N}[\varphi(X_1)\varphi(X_2)]$$

$$- 2 \int \varphi \, d\mu \, \mathbb{E}_{\mu^N}[\varphi(X_1)] + \left(\int \varphi \, d\mu \right)^2$$

$$\xrightarrow[N \to \infty]{} 0 + \left(\int \varphi \, d\mu \right)^2 - 2 \left(\int \varphi \, d\mu \right)^2 + \left(\int \varphi \, d\mu \right)^2 = 0$$

as from (ii) $\mathbb{E}_{\mu^N}[\varphi(X_1)\varphi(X_2)] \xrightarrow[N \to \infty]{} \left(\int \varphi \, d\mu \right)^2$ and $\mathbb{E}_{\mu^N}[\varphi(X_1)] \xrightarrow[N \to \infty]{} \int \varphi \, d\mu$.

(iii) $\Rightarrow$ (iv): Follows obviously from the definition of convergence in mean.

(iv) $\Rightarrow$ (i): Let $k \geq 1$ and $\varphi_1, \ldots, \varphi_k \in C_b(\mathbb{R}^n)$. We have

$$\left| \mathbb{E}_{\mu^N}[\varphi_1(X_1) \cdots \varphi_k(X_k)] - \int \varphi_1 \, d\mu \cdots \int \varphi_k \, d\mu \right|$$

$$\leq \left| \mathbb{E}_{\mu^N}[\varphi_1(X_1) \cdots \varphi_k(X_k)] - \mathbb{E}_{\mu^N}\left[\int \varphi_1 \, d\hat{\mu}^N \cdots \int \varphi_k \, d\hat{\mu}^N \right] \right|$$

$$+ \left| \mathbb{E}_{\mu^N}\left[\int \varphi_1 \, d\hat{\mu}^N \cdots \int \varphi_k \, d\hat{\mu}^N \right] - \int \varphi_1 \, d\mu \cdots \int \varphi_k \, d\mu \right|$$

The second term on the r.h.s. converges to zero thanks to hypothesis (iv). The first term on the r.h.s. reads

$$\left| \mathbb{E}_{\mu^N}[\varphi_1(X_1) \cdots \varphi_k(X_k)] - \frac{1}{N^k} \sum_{i_1, \ldots, i_k = 1}^{N} \mathbb{E}_{\mu^N}[\varphi_1(X_{i_1}) \cdots \varphi_k(X_{i_k})] \right|$$

In the above sum, $\frac{N!}{(N-k)!}$ terms are such that the indices $i_1, \ldots, i_k$ are all different. By symmetry, they are equal to $\mathbb{E}_{\mu^N}[\varphi_1(X_1) \cdots \varphi_k(X_k)]$ and can be added to the first term. The other terms can be bounded by M^k with $M = \sum_j \|\varphi_j\|_\infty$, so the first term is bounded by

$$\mathbb{E}_{\mu^N}[\varphi_1(X_1) \cdots \varphi_k(X_k)] \left(1 - \frac{1}{N^k} \frac{N!}{(N-k)!} \right) + \left(1 - \frac{1}{N^k} \frac{N!}{(N-k)!} \right) M^k$$

$$\leq 2M^k \left(1 - \frac{1}{N^k} \frac{N!}{(N-k)!} \right) \xrightarrow[N \to \infty]{} 0$$

□

10.3.3   The McKean-Vlasov SDE propagates the chaos

We now prove that the McKean-Vlasov SDE, introduced in Example 10.1, propagates the chaos. For the sake of clarity, we assume here that b and σ do not depend on t.

THEOREM 10.3
The propagation of chaos holds for the McKean-Vlasov SDE (10.2).

Theorem 10.1 shows that the process (10.2) exists and that its law is a (weak) solution to (10.4) if b and σ are Lipschitz-continuous in (x, y). The proof of the above theorem relies on the coupling method introduced by [23]. The coupling method consists of introducing N processes $\{Y_t^i\}_{1 \leq i \leq N}$ defined as

$$dY_t^i = b(Y_t^i, \mathbb{P}_t) \, dt + \sigma(Y_t^i, \mathbb{P}_t) \, dW_t^i, \qquad Y_0^i = X_0$$

with $b(y, \mathbb{P}_t) \equiv \int b(y, z) \mathbb{P}_t(dz)$, $\sigma(y, \mathbb{P}_t) \equiv \int \sigma(y, z) \mathbb{P}_t(dz)$, and $\mathbb{P}_t = \text{Law}(X_t)$. These are standard SDEs which admit a strong solution as b and σ are

Lipschitz-continuous functions. The density $q_i(t,x)$ of Y_t^i satisfies the *linear* Fokker-Planck equation

$$- \partial_t q_i(t,x) - \sum_{i=1}^{n} \partial_i \left(b^i(t,x,\mathbb{P}_t) q_i(t,x) \right)$$

$$+ \frac{1}{2} \sum_{i,j=1}^{n} \partial_{ij} \left(\sum_{k=1}^{d} \sigma_k^i(t,x,\mathbb{P}_t) \sigma_k^j(t,x,\mathbb{P}_t) q_i(t,x) \right) = 0$$

with Dirac initial condition. From (10.4), the density $p(t,x)$ of X_t is also a solution. By uniqueness, $\mathbb{P}_t = \text{Law}(Y_t^i)$.

PROPOSITION 10.1
Let $(X^{i,N})_{1 \leq i \leq N}$ be defined by (10.6). Then

$$\mathbb{E}[|X_t^{1,N} - Y_t^1|] \leq \frac{C(t)}{\sqrt{N}}$$

where $C(t)$ is a smooth function of time independent of N.

PROOF   For clarity, we write X^i instead of $X^{i,N}$. From Itô's lemma, we get

$$\mathbb{E}\left[(X_t^1 - Y_t^1)^2\right] = 2\mathbb{E}\left[\int_0^t (X_s^1 - Y_s^1) \left(\frac{1}{N} \sum_{j=1}^{N} \sigma(X_s^1, X_s^j) - \sigma(Y_s^1, \mathbb{P}_s) \right) dW_s^1 \right]$$

$$+ \int_0^t \mathbb{E}\left[\left(\frac{1}{N} \sum_{j=1}^{N} \sigma(X_s^1, X_s^j) - \sigma(Y_s^1, \mathbb{P}_s) \right)^2 \right] ds$$

$$+ 2 \int_0^t \mathbb{E}\left[(X_s^1 - Y_s^1) \left(\frac{1}{N} \sum_{j=1}^{N} b(X_s^1, X_s^j) - b(Y_s^1, \mathbb{P}_s) \right) \right] ds$$

The first term vanishes because the stochastic integral is a martingale. Let us look at the second term:

$$\left(\frac{1}{N} \sum_{j=1}^{N} \sigma(X_s^1, X_s^j) - \sigma(Y_s^1, \mathbb{P}_s) \right)^2 \leq 3 \left(\frac{1}{N} \sum_{j=1}^{N} \left(\sigma(X_s^1, X_s^j) - \sigma(X_s^1, Y_s^j) \right) \right)^2$$

$$+ 3 \left(\sum_{j=1}^{N} \frac{1}{N} \left(\sigma(X_s^1, Y_s^j) - \sigma(Y_s^1, Y_s^j) \right) \right)^2$$

$$+ 3 A_s(\sigma)$$

with $A_s(\sigma) \equiv \left(\frac{1}{N}\sum_{j=1}^{N}\sigma(Y_s^1, Y_s^j) - \sigma(Y_s^1, \mathbb{P}_s)\right)^2$. We have used the identity $(a + b + c)^2 \leq 3\left(a^2 + b^2 + c^2\right)$. Then using Cauchy-Schwartz's inequality and the Lipschitz property of σ w.r.t. x and y, we deduce the upper bound

$$\mathbb{E}\left[\left(\sum_{j=1}^{N}\frac{1}{N}\sigma(X_s^1, X_s^j) - \sigma(Y_s^1, \mathbb{P}_s)\right)^2\right] \leq 6||\sigma||_{\mathrm{Lip}}^2\mathbb{E}[|X_s^1 - Y_s^1|^2] + 3\mathbb{E}[A_s(\sigma)]$$

As for the third term, we have

$$2(X_s^1 - Y_s^1)\left(\frac{1}{N}\sum_{j=1}^{N}b(X_s^1, X_s^j) - b(Y_s^1, \mathbb{P}_s)\right)$$

$$\leq (X_s^1 - Y_s^1)^2 + \left(\frac{1}{N}\sum_{j=1}^{N}b(X_s^1, X_s^j) - b(Y_s^1, \mathbb{P}_s)\right)^2$$

whence

$$\mathbb{E}\left[2(X_s^1 - Y_s^1)\left(\frac{1}{N}\sum_{j=1}^{N}b(X_s^1, X_s^j) - b(Y_s^1, \mathbb{P}_s)\right)\right]$$

$$\leq \mathbb{E}\left[(X_s^1 - Y_s^1)^2\right] + 6||b||_{\mathrm{Lip}}^2\mathbb{E}[|X_s^1 - Y_s^1|^2] + 3\mathbb{E}[A_s(b)]$$

Eventually we get the inequality

$$\mathbb{E}[(X_t^1 - Y_t^1)^2] \leq \left(6||\sigma||_{\mathrm{Lip}}^2 + 6||b||_{\mathrm{Lip}}^2 + 1\right)\int_0^t \mathbb{E}[(X_s^1 - Y_s^1)^2]\,ds$$

$$+ 3\int_0^t \mathbb{E}[A_s(\sigma) + A_s(b)]\,ds$$

From Gronwall's inequality, we deduce that

$$\mathbb{E}[(X_t^1 - Y_t^1)^2] \leq 3e^{(6||\sigma||_{\mathrm{Lip}}^2 + 6||b||_{\mathrm{Lip}}^2 + 1)t}\int_0^t \mathbb{E}[A_s(\sigma) + A_s(b)]\,ds$$

To conclude, it is enough to show that $\mathbb{E}[A_s(h)] = O\left(\frac{1}{N}\right)$ for $h \in \{b, \sigma\}$. Now,

$$\mathbb{E}[A_s(h)] = \frac{1}{N^2}\sum_{i,j=1}^{N}\mathbb{E}[h(Y_s^1, Y_s^i)h(Y_s^1, Y_s^j)] + \mathbb{E}[h(Y_s^1, \mathbb{P}_s)^2]$$

$$- \frac{2}{N}\sum_{i=1}^{N}\mathbb{E}[h(Y_s^1, Y_s^i)h(Y_s^1, \mathbb{P}_s)]$$

From the independence of the processes $\{Y_s^i\}_{1\leq i\leq N}$ and the fact that $\text{Law}(Y_s^i) = \mathbb{P}_s$, we get

$$\mathbb{E}[A_s(h)] = \frac{2-N}{N^2}\mathbb{E}[h(Y_s^1,\mathbb{P}_s)^2] + \frac{1}{N^2}\mathbb{E}[h(Y_s^1,Y_s^1)^2]$$
$$+ \frac{N-1}{N^2}\mathbb{E}[h(Y_s^1,Y_s^2)^2] - \frac{2}{N^2}\mathbb{E}[h(Y_s^1,\mathbb{P}_s)h(Y_s^1,Y_s^1)] \leq \frac{K(t)}{N}$$

The four expectations above are finite due to the Lipschitz conditions on b and σ and the fact that $\mathbb{E}[\sup_{0\leq s\leq t}|Y_s|^2] < \infty$. ☐

PROOF of Theorem 10.3   Let us denote by μ_t the law of the solution X_t of the McKean SDE (10.2). From Theorem 10.2 (iii), it is enough to show that for all $\varphi \in C_b(\mathbb{R}^n)$,

$$\frac{1}{N}\sum_{i=1}^{N}\varphi(X_t^{i,N}) \xrightarrow[N\to\infty]{L^1} \int \varphi \, d\mu_t \qquad (10.13)$$

where the $X_t^{i,N}$ are defined by (10.6). Now,

$$\mathbb{E}\left[\left|\frac{1}{N}\sum_{i=1}^{N}\varphi(X_t^{i,N}) - \int \varphi \, d\mu_t\right|\right]$$
$$\leq \mathbb{E}_{\mu^N}\left[\left|\frac{1}{N}\sum_{i=1}^{N}\left(\varphi(X_t^{i,N}) - \varphi(Y_t^i)\right)\right|\right] + \mathbb{E}_{\mu^N}\left[\left|\frac{1}{N}\sum_{i=1}^{N}\varphi(Y_t^i) - \int \varphi \, d\mu_t\right|\right]$$

As by construction the processes $\{Y_t^i\}_{1\leq i\leq N}$ are i.i.d. with law μ_t, the second term above goes to zero as $N\to\infty$ from the law of large numbers.

For all $\epsilon > 0$, there exists a Lipschitz-continuous function φ_ϵ such that $|\varphi - \varphi_\epsilon| \leq \epsilon$. The first term is then bounded by

$$2\epsilon + \mathbb{E}_{\mu^N}\left[\left|\frac{1}{N}\sum_{i=1}^{N}\left(\varphi_\epsilon(X_t^{i,N}) - \varphi_\epsilon(Y_t^i)\right)\right|\right]$$
$$\leq 2\epsilon + ||\varphi_\epsilon||_{\text{Lip}}\mathbb{E}[|X_t^{1,N} - Y_t^1|] \leq 2\epsilon + ||\varphi_\epsilon||_{\text{Lip}}\frac{C(t)}{\sqrt{N}}$$

from Proposition 10.1. ☐

In particular, Equation (10.13) means that the particle method is convergent for McKean-Vlasov SDEs.

Conclusion

The particle method is a nice, powerful, fast, robust, and easily implementable algorithm for solving (multi-dimensional) nonlinear McKean SDEs. For McKean-Vlasov SDEs, we have shown the propagation of chaos property, which implies the convergence of the method. In the next two chapters, we will apply this technique to various calibration problems arising in quantitative finance. Even though we do not prove convergence, numerical tests will show that the particle method is surprisingly fast and accurate.

10.4   Exercise: Random matrices, Dyson Brownian motion, and McKean SDEs

Random matrices

Let $A = [A_{ij}]_{1 \leq i,j \leq n}$ be a Hermitian matrix with $A_{ij} = \bar{A}_{ji} = \frac{1}{\sqrt{n}}(x_{ij} + \imath y_{ij})$ and $A_{ii} = \sqrt{\frac{2}{n}}x_{ii}$ where x_{ij}, y_{ij} $(i < j)$ and x_{ii} are independent centered Gaussian r.v. with variance $1/2$. The set of such matrices is called the *Gaussian unitary ensemble* (GUE). Below, we denote with $\lambda_1 \leq \cdots \leq \lambda_n$ the (real) eigenvalues. In this exercise, we will study the statistical properties of the eigenvalues and show that in the limit $n \to \infty$ universal patterns emerge. A nice McKean SDE will then appear. This exercise is built from [77].

We recall that by the spectral theorem for finite-dimensional Hermitian matrices, A can be decomposed as

$$A = \sum_{i=1}^{n} \lambda_i P_{u_i}$$

where P_{u_i} denotes the projector on the space generated by the eigenvector u_i associated to the (real) eigenvalue λ_i. The eigenvectors can be chosen orthogonal: $(u_i, u_j) = \delta_{ij}$.

Prove that the spectrum of $A + \epsilon \delta A$, where δA is an Hermitian matrix, is given at the second order in ϵ by

$$\lambda_i(A + \epsilon \delta A) = \lambda_i(A) + \epsilon(u_i, \delta A u_i) + 2\epsilon^2 \sum_{j \neq i} \frac{|(u_i, \delta A u_j)|^2}{\lambda_i - \lambda_j} \qquad (10.14)$$

(Hint: look at your favorite textbook in quantum mechanics.)

The Dyson Brownian motion

Let A_t be an Hermitian matrix process. We assume that $(A_{t_1} - A_0, A_{t_2} - A_{t_1}, \ldots, A_{t_n} - A_{t_{n-1}})$ with $t_0 \equiv 0 < t_1 < \cdots < t_n$ are independent r.v. and the distribution of $A_{t_i} - A_{t_{i-1}}$ is the same as the distribution of $A\sqrt{t_i - t_{i-1}}$ with $A \sim$ GUE. By applying Itô's lemma, we have

$$d\lambda_i(A_t) = \partial_A \lambda_i(A_t) dA_t + \frac{1}{2}\partial_A^2 \lambda_i(A_t) d\langle A \rangle_t$$

1. From Equation (10.14), compute $\partial_A \lambda_i(A_t)$ and $\partial_A^2 \lambda_i(A_t)$:

$$d\lambda_t^i = (u_i, dA_t u_i) + \sum_{j \neq i} \frac{\mathbb{E}[|(u_i, Au_j)|^2]}{\lambda_t^i - \lambda_t^j} dt$$

where $\lambda_t^i \equiv \lambda_i(A_t)$ and $(u_i, dA_t u_i)$ is distributed as $(u_i, Au_i)\sqrt{dt}$ with $A \sim$ GUE. Note that as the distribution of A is invariant under unitary conjugation $A \mapsto U^{-1}AU$ with $U^\dagger = U^{-1}$ (hence the name GUE), we can choose the basis $\{u_i\}$ as the orthonormal basis of $\mathbb{C}^n$. Therefore, $(u_i, Au_i) = A_{ii}$ and $\mathbb{E}[(u_i, Au_j)^2] = \mathbb{E}[|A_{ij}|^2]$. But as the $\sqrt{n}A_{ii}/\sqrt{2}$ are i.i.d. copies of $\mathcal{N}(0, 1/2)_{\mathbb{R}}$, and $\sqrt{n}A_{ij}$ are i.i.d. copies of $\mathcal{N}(0, 1/2)_{\mathbb{C}}$, we get

$$d\lambda_t^i = \frac{1}{\sqrt{n}}dB_t^i + \frac{1}{n}\sum_{j \neq i} \frac{1}{\lambda_t^i - \lambda_t^j} dt \qquad (10.15)$$

with $(B_t^i)_{1 \leq i \leq n}$ n independent Brownian motions. This solution of this SDE is called the *Dyson Brownian motion*.

2. Write the Fokker-Planck PDE on $\mathbb{R}^n$ associated to (10.15).

3. Prove that the solution to this Fokker-Planck PDE is (Johansson's formula (2001)):

$$p(t, \lambda | \nu) = \frac{1}{(2\pi t)^{\frac{n}{2}}} \frac{\Delta_n(\lambda)}{\Delta_n(\nu)} \det\left[e^{-\frac{(\lambda_i - \nu_j)^2}{2t}} \right]_{1 \leq i,j \leq n}$$

with $\Delta_n(\lambda) \equiv \prod_{i<j} |\lambda_i - \lambda_j|$.

McKean SDE

1. Justify that SDE (10.15) can be seen as the particle approximation of the following McKean SDE (with no volatility term)

$$d\lambda_t = \int \frac{p(t, x)\, dx}{\lambda_t - x} dt$$

with $p(t, \cdot)$ the density of λ_t, solution to the nonlinear Fokker-Planck PDE:

$$\partial_t p(t, \lambda) = -\partial_\lambda \left(p(t, \lambda) \int \frac{p(t, x)}{\lambda - x} \, dx \right)$$

2. We denote by Hp the Hilbert transform of p:

$$Hp(t, \lambda) = \int \frac{p(t, x)}{\lambda - x} \, dx$$

for $\lambda \in \mathbb{H} = \{z \in \mathbb{C} \,|\, \Im(z) > 0\}$. Show that $Hp(t, \lambda)$ satisfies the complex Burgers equation:

$$\partial_t G(t, \lambda) = -G(t, \lambda)\partial_\lambda G(t, \lambda)$$

3. Compute the Hilbert transform of the Wigner semi-circle law $\rho(x) = \frac{1}{2\pi} \left(4 - x^2\right)_+^{\frac{1}{2}}$.

4. Conclude that $A \sim$ GUE converges in law when $n \to \infty$ to the Wigner semi-circle law.

10.A   The Monge-Kantorovich distance and its financial interpretation

The Monge-Kantorovich (or Wasserstein) distance $d_{\mathrm{MK}}(\mathbb{P}_1, \mathbb{P}_2)$ between two probability measures $\mathbb{P}_1$ and $\mathbb{P}_2$ on $\mathbb{R}^n$ equipped with the distance $d(\cdot, \cdot)$ (or more generally on a polish space X) is defined by

$$d_{\mathrm{MK}}(\mathbb{P}_1, \mathbb{P}_2)^p = \inf_{\tau \in \mathcal{P}(\mathbb{R}^n, \mathbb{R}^n) \text{ with marginals } \mathbb{P}_1 \text{ and } \mathbb{P}_2} \left(\int_{\mathbb{R}^n \times \mathbb{R}^n} d(x, y)^p \, \tau(dx, dy) \right)$$

$$(10.16)$$

where $p \geq 1$. It consists of finding a probability measure τ (called copula) on $\mathbb{R}^n \times \mathbb{R}^n$ with marginals $\mathbb{P}_1$ on the first coordinates and $\mathbb{P}_2$ on the second coordinates that attains the infimum of the expectation of $d(\cdot, \cdot)^p$. The Monge-Kantorovich distance defines a distance on the set of probability measures $\mathcal{P}_p(\mathbb{R}^n)$ with finite moments of order p, i.e., those measures $\mathbb{P}$ such that for some $x_0 \in \mathbb{R}^n$,

$$\int d(x_0, x)^p \, \mathbb{P}(dx) < \infty$$

This distance induces the weak convergence: A sequence $(\mathbb{P}_k)_{k\in\mathbb{N}}$ of probability measures in $\mathcal{P}_p(\mathbb{R}^n)$ satisfies for $\mathbb{P}\in\mathcal{P}(\mathbb{R}^n)$

$$d_{\mathrm{MK}}(\mathbb{P}_k,\mathbb{P}) \xrightarrow[k\to\infty]{} 0$$

if and only if $\mathbb{P}_k \longrightarrow \mathbb{P}$ in the weak sense:

$$\forall\varphi\in C_b(\mathbb{R}^n), \qquad \int\varphi(x)\,\mathbb{P}_k(dx) \xrightarrow[k\to\infty]{} \int\varphi(x)\,\mathbb{P}(dx)$$

This distance has a clear economic interpretation in terms of model-independent bounds on European option prices depending on two assets, say $X = S_T^1$ and $Y = S_T^2$, at a maturity date T. For simplicity, we take zero interest rates. The payoff reads $d(X,Y)^p$. Then, we impose that our model is calibrated to the vanilla smiles at maturity T for both assets, which implies the marginal risk-neutral distributions $\mathbb{P}_1$ and $\mathbb{P}_2$ of X and Y at T. Indeed, from the price $\mathcal{C}(T,K)$ of T-vanilla options with strike K, the risk-neutral density at T is given by $\mathbb{P}(K) = \partial_K^2 \mathcal{C}(T,K)$. The pricing model producing the lower bound consists, therefore, of choosing a joint distribution τ of (X,Y) with marginals $\mathbb{P}_1$ and $\mathbb{P}_2$ that minimizes the option fair value

$$\mathbb{E}^\tau[d(X,Y)^p]$$

This is precisely the Monge-Kantorovich distance.

Dual problem

The minimization problem in (10.16) can be seen as an infinite-dimensional linear programming problem, and this linear program can be dualized by introducing Lagrange's multipliers associated to the marginal constraints. We rewrite (10.16) as

$$d_{\mathrm{MK}}(\mathbb{P}_1,\mathbb{P}_2)^p = \inf_{\mathbb{P}\in M_+}\sup_{u_1,u_2} \left\{ \mathbb{E}^\mathbb{P}[d^p(X,Y) - u_1(X) - u_2(Y)] \right.$$

$$\left. + \mathbb{E}^{\mathbb{P}_1}[u_1(X)] + \mathbb{E}^{\mathbb{P}_2}[u_2(X)] \right\} \quad (10.17)$$

where the infimum is now taken over the set M_+ of positive measures on $\mathbb{R}^n\times\mathbb{R}^n$. For a given $\mathbb{P}$, if the first marginal differs from $\mathbb{P}^1$, then the supremum over u_1 is $+\infty$. If the first marginal is $\mathbb{P}^1$, then the supremum over u_1 of the terms in (10.17) that depend on u_1 is 0. By using an identical argument for u_2, we conclude that our relaxed form (10.17) is equivalent to (10.16). Then, by using a minimax argument, we can switch the infimum and the supremum (this can be rigorously justified using a Hahn-Banach theorem) and get

$$d_{\mathrm{MK}}(\mathbb{P}_1,\mathbb{P}_2)^p = \sup_{u_1,u_2} \left\{ \inf_{\mathbb{P}\in M_+} \left\{ \mathbb{E}^\mathbb{P}[d(X,Y)^p - u_1(X) - u_2(Y)] \right\} \right.$$

$$\left. + \mathbb{E}^{\mathbb{P}_1}[u_1(X)] + \mathbb{E}^{\mathbb{P}_2}[u_2(X)] \right\}$$

Eventually, we compute the infimum over M_+. If at some point, say $(X_0, Y_0) \in \mathbb{R}^n \times \mathbb{R}^n$, $u_1(X_0) + u_2(Y_0) > d(X_0, Y_0)^p$, then the infimum is $-\infty$ and is "attained" by $\mathbb{P} = \lambda \delta_{X_0} \delta_{Y_0}$ with $\lambda \longrightarrow +\infty$. This case can be disregarded as we compute a supremum over u_1, u_2 in a second step. If for all (X, Y), $u_1(X) + u_2(Y) \leq d(X, Y)^p$, then the infimum is zero. Therefore, we conclude that

$$d_{\mathrm{MK}}(\mathbb{P}_1, \mathbb{P}_2)^p = \sup_{u_1, u_2} \left\{ \mathbb{E}^{\mathbb{P}_1}[u_1(X)] + \mathbb{E}^{\mathbb{P}_2}[u_2(Y)] \right\}$$

subject to the constraints

$$\forall (x, y) \in \mathbb{R}^n \times \mathbb{R}^n, \qquad u_1(x) + u_2(y) \leq d(x, y)^p$$

This dual problem has a clear economic interpretation: the objective function $u_1(X) + u_2(Y)$ represents the value of the portfolio composed of two options written respectively on X and Y with market prices $\mathbb{E}^{\mathbb{P}_1}[u_1(X)]$ and $\mathbb{E}^{\mathbb{P}_2}[u_2(Y)]$. The constraints mean that the intrinsic value of the portfolio should be lower than the payoff $c(X, Y) = d(X, Y)^p$. The dual problem consists of maximizing the value of such portfolios $u_1(X) + u_2(Y)$.

Chapter 11

Calibration of Local Stochastic Volatility Models to Market Smiles

You're never fully dressed without a smile.

— Martin Charnin

The calibration of local stochastic volatility models to market smiles leads to McKean nonlinear stochastic differential equations (SDEs), introduced in Chapter 10. As shown in Section 10.1, the Fokker-Planck equations associated to McKean SDEs are nonlinear. In this chapter, we review various methods for solving such nonlinear PDEs. For one-factor stochastic volatility models, we can rely on finite difference schemes. But, as PDEs suffer from the curse of dimensionality, we must turn to probabilistic methods to handle multi-factor stochastic volatility models. In this chapter, we present several such Monte Carlo methods, such as the particle algorithm introduced in Section 10.2. We also devote ample space to the case of stochastic interest rates, which can be easily handled by the particle method. Parts of this research have been published in [128, 122].

11.1   Introduction

The trading of exotic options has called for the development of stochastic volatility models (SVMs). Exotic options bear exotic risks such as volatility-of-volatility risk, forward smile risk, or spot/volatility correlation risk. To price and hedge these risks, one should not use the Black-Scholes model, or the local volatility model, because such models give no control on them. One would rather use an SVM

$$df_t = a_t f_t \, dW_t \tag{11.1}$$

with f_t the forward (associated to some given maturity T) and a_t the stochastic volatility, which allows us to handle exotic risks through parameters such as volatility-of-volatility, spot/volatility correlation, or mean-reversion.

It is well known that SVMs produce a smile of implied volatilities (see the seminal papers [136, 176]; see also [59] for an analysis of the smile in general multi-factor second-generation stochastic volatility models). Of course, because they have a finite number of parameters (typically, 3 to 10), SVMs cannot be perfectly calibrated to full (inter- and extrapolated) market smiles, indexed by all strikes and maturities (until some final maturity T). Now, vanilla options often provide a good hedge of exotic options. Dynamic or static trading of vanilla options often results in reducing the variance of the final P&L. In those cases, it is important that the SVM incorporates the correct initial prices of the hedging instruments—the vanilla options.

How do we build an SVM that calibrates exactly to a full surface of implied volatilities? Due to the double infinity of constraints, indexed by strikes and maturities, one needs to introduce a double infinity of parameters. A natural way to do so, and a common practice in the foreign exchange market, is to embed a local volatility $\sigma(t, f)$ into the SVM:

$$df_t = a_t f_t \sigma(t, f_t) \, dW_t \tag{11.2}$$

We speak of local stochastic volatility models (in short LSVMs). The main issue we address in this chapter is how to build the local volatility $\sigma(t, f)$ to ensure that the market smile is exactly calibrated. Note that this local volatility function differs from Dupire's local volatility. For instance, if the smile produced by the naked SVM (11.1) is close to the market smile, one expects the calibrated local volatility $\sigma(t, f)$ to be uniformly close to 1.

The chapter is organized as follows. We first consider the case of deterministic interest rates. In Section 11.2, we show that the calibrated model follows a nonlinear SDE in the sense of McKean. Such SDEs were introduced in Chapter 10. The existence of the calibrated LSVM is discussed in Section 11.3. Then, in Section 11.4, we explain why for old-fashioned one-factor SVMs such as the Heston model, the calibration of the local volatility function can be achieved using a two-dimensional PDE solver of the nonlinear Fokker-Planck equation associated to this McKean SDE. For multi-factor SVMs, the PDE is at least three-dimensional, and because of the curse of dimensionality, one must turn to probabilistic methods. For more than a decade, all attempts at designing Monte Carlo algorithms resulted in approximate methods: even if one used an infinite number of Monte Carlo paths, and an infinitesimal discretization time step, one would not exactly reprice the market smile. As an example of such approximate techniques, we present the Markovian projection method in Section 11.5. The particle algorithm is an elegant simulation technique for McKean SDEs that we introduced in Section 10.2. We use this powerful tool to build an exact calibration algorithm: if one uses this algorithm with an infinite number of Monte Carlo paths, and an infinitesimal discretization time step, one exactly reprices the market smile. We describe the particular form of the particle method for the calibration of LSVMs to market smiles in Section 11.6. One of the great features of the particle method is that it easily extends to models where interest rates are also stochastic, as we show in Section 11.7. The numerical tests in Section 11.8 indicate the great speed and accuracy of the particle method.

11.2 The calibration condition

In the case of deterministic interest rates, repos, and dividend yield, a local stochastic volatility model (LSVM) is defined by the following SDE for the forward f_t of maturity T

$$df_t = a_t f_t \sigma(t, f_t) \, dW_t \tag{11.3}$$

where a_t is a (possibly multi-factor) stochastic process. It can be seen as an extension of the Dupire local volatility model (LVM), or as an extension of the stochastic volatility model (SVM). In an SVM, one handles only a finite number of parameters (volatility-of-volatility, spot/volatility correlations, etc.). As a consequence, one is not able to perfectly calibrate to the whole implied volatility surface. In order to be able to exactly calibrate market smiles, one "decorates" the volatility of the forward with a local volatility function $\sigma(t, f)$ as in (11.2).

From [96] (see also the proof of Proposition 11.1 in Section 11.7.1 which also handles the case of stochastic interest rates), this model is exactly calibrated to market smiles if and only if

$$\sigma_{\text{Dup}}(t, f)^2 = \sigma(t, f)^2 \mathbb{E}^{\mathbb{Q}}[a_t^2 | f_t = f] \tag{11.4}$$

where

$$\sigma_{\text{Dup}}(t, f)^2 = \frac{\partial_t \mathcal{C}(t, f)}{\frac{1}{2} f^2 \partial_f^2 \mathcal{C}(t, f)} \tag{11.5}$$

is the Dupire local volatility that is inferred from the vanilla market smile. $\mathcal{C}(t, f)$ is the market price (at time 0) of a call option with strike f and maturity t written on the forward of maturity T. For the sake of simplicity, we assume at this stage that there are no dividends. We explain how to deal with (discrete) dividends in our calibration procedure in Section 11.C.

Once the requirement that market marginals have to be calibrated exactly has been taken into account, SDE (11.3) can be rewritten as

$$df_t = f_t \frac{\sigma_{\text{Dup}}(t, f_t)}{\sqrt{\mathbb{E}^{\mathbb{Q}}[a_t^2 | f_t]}} a_t \, dW_t \tag{11.6}$$

The local volatility function depends on the joint pdf $p(t, f, a)$ of (f_t, a_t):

$$\sigma(t, f, p) = \sigma_{\text{Dup}}(t, f) \sqrt{\frac{\int p(t, f, a')\, da'}{\int a'^2 p(t, f, a')\, da'}} \tag{11.7}$$

This is an example of McKean SDEs. We introduced such nonlinear SDEs in Chapter 10; see Equation (10.1). Here, $X_t = (f_t, a_t)$. The volatility of f_t depends on the joint distribution of (f_t, a_t) through the conditional expectation $\mathbb{E}^{\mathbb{Q}}[a_t^2 | f_t]$.

11.3   Existence of the calibrated local stochastic volatility model

In (11.7), the Lipschitz condition of Theorem 10.1 is not satisfied. Hence a uniqueness and existence result for Equation (11.6) is not at all obvious. In particular, given a set of stochastic volatility parameters, it is not clear at all whether an LSVM exists for a given arbitrary arbitrage-free implied volatility surface: some smiles may not be attainable by the model.

However, a partial result exists: in [27] it is shown that the calibration problem for a LSVM is well posed but only (a) until some maturity T^*, (b) if the volatility-of-volatility is small enough, and (c) in the case of suitably regularized initial conditions—hence the result does not apply to Equation (10.4) because of the initial Dirac mass.

Our numerical experiments show that the calibration does not work for large enough volatility-of-volatility, whatever the algorithm used: PDE, Markovian projection method, or particle method. This may come from numerical issues, or from non-existence of a solution. The problem of deriving the set of stochastic volatility parameters for which the LSVM does exist for a given market smile, is very challenging and open. We illustrate this in Section 11.8.3.

11.4   The PDE method

For a one-factor LSVM, the nonlinear Fokker-Planck equation (10.4) can be solved using a PDE finite difference scheme. Such a model is defined by the process

$$df_t = a_t f_t \sigma(t, f_t)(\rho\, dZ_t + \sqrt{1 - \rho^2}\, dB_t) \tag{11.8}$$

$$da_t = b(a_t)\, dt + \sigma(a_t)\, dZ_t \tag{11.9}$$

Table 11.1: Examples of SVMs ($\sigma(t, f) = 1$).

Name	SDE
Stein-Stein	$\frac{df_t}{f_t} = a_t\,dW_t$ $da_t = \lambda(a_t - \bar{a})\,dt + \zeta\,dZ_t, \quad d\langle W, Z\rangle_t = \rho\,dt$
Geometric	$\frac{df_t}{f_t} = a_t\,dW_t$ $da_t = \lambda a_t\,dt + \zeta a_t\,dZ_t, \quad d\langle W, Z\rangle_t = \rho\,dt$
3/2-model	$\frac{df_t}{f_t} = a_t\,dW_t$ $da_t^2 = \lambda(a_t^2 - \bar{v}a_t^4)\,dt + \zeta a_t^3\,dZ_t, \quad d\langle W, Z\rangle_t = \rho\,dt$
SABR	$\frac{df_t}{f_t} = a_t f_t^{\beta-1}\,dW_t$ $da_t = \nu a_t\,dZ_t, \quad d\langle W, Z\rangle_t = \rho\,dt$
Scott-Chesney	$\frac{df_t}{f_t} = e^{y_t}\,dW_t$ $dy_t = \lambda\left(\bar{y} - y_t\right)dt + \zeta\,dZ_t, \quad d\langle W, Z\rangle_t = 0$
Heston	$\frac{df_t}{f_t} = a_t\,dW_t$ $da_t^2 = \lambda(\bar{v} - a_t^2)\,dt + \zeta a_t\,dZ_t, \quad d\langle W, Z\rangle_t = \rho\,dt$

with the initial values f_0 and $a_0 = \alpha$. B_t and Z_t are two uncorrelated $\mathbb{Q}$-Brownian motions. Practitioners call $\sigma(a)$ the *volatility-of-volatility* (or in short, vol-of-vol). Note that as a_t is not a traded asset, a_t is not a local martingale and we have, a priori, a drift term $b(a)$. We assume that $b(\cdot)$ and $\sigma(\cdot)$ are only functions of the volatility process a_t. We could assume that $b(\cdot)$ and $\sigma(\cdot)$ depend on the forward f_t as well, but we do not pursue this route as examples we look at do not exhibit this dependence. We have listed commonly used SVMs in Table 11.1. As explained in [156], the nonlinear Fokker-Planck equation can be solved using the following algorithm.

The PDE algorithm

We assume that we want to calibrate the one-factor LSVM on market smiles up to the maturity T. We divide the interval $(0, T)$ into M subintervals of length $\Delta t = \frac{T}{M}$ and denote $t_k = k\Delta t$. The idea is to solve a series of *linear* Fokker-Planck equations for the density p over the intervals $[t_k, t_{k+1}]$, with the local volatility $\sigma(t, f)$ being fixed at date t_k and then updated at date t_{k+1} using the new approximate density $p(t_{k+1}, f, a)$.

1. Initialize $k = 0$ and set $\alpha^{-1}f^{-1}\sigma(t, f)^{-1} \equiv \partial_f\left(\frac{\ln\frac{f}{f_0}}{\sigma_{\mathrm{BS}}(0, f)}\right)$ for all $t \in$

   $[t_k, t_{k+1}]$. $\sigma_{\mathrm{BS}}(0, f)$ is the short-time implied volatility for strike f. This specification of the local volatility $\sigma(t, f)$ is an approximation close to the exact short-time asymptotics of the implied volatility where one can show that (see e.g., [11], Chapter 6)

$$\lim_{t \to 0} \sigma_{BS}(t, f) = \lim_{t \to 0} \frac{\ln \frac{f}{f_0}}{\int_{f_0}^{f} \frac{dx}{x\sqrt{\sigma(t,x)^2 \mathbb{E}^{\mathbb{Q}}[a_t^2 | f_t = x]}}}$$

Our specification of $\sigma(0, f)$ has been obtained by using the approximation $\mathbb{E}^{\mathbb{Q}}[a_t^2 | f_t = x] \approx a^2$. A better approximation can be found in [11] (see Remark 6.8).

2. Solve the *linear* two-dimensional Fokker-Planck equation on this interval $[t_k, t_{k+1}]$:

$$\partial_t p(t, f, a) = \frac{1}{2}\partial_f^2(\sigma(t, f)^2 f^2 a^2 p(t, f, a)) + \frac{1}{2}\partial_a^2(\sigma(a)^2 p(t, f, a))$$
$$+ \rho \partial_{fa}(\sigma(a)\sigma(t, f) f p(t, f, a)) - \partial_a(b(a)p(t, f, a))$$

and store the vector $p(t_{k+1}, \cdot, \cdot)$ evaluated on the two-dimensional space grid.

3. Compute the local volatility at time t_{k+1}:

$$\sigma(t_{k+1}, f) = \frac{\sigma_{\text{Dup}}(t_{k+1}, f)}{\sqrt{\mathbb{E}^{\mathbb{Q}}[a_{t_{k+1}}^2 | f_{t_{k+1}} = f]}}$$
$$= \frac{\sigma_{\text{Dup}}(t_{k+1}, f)}{\sqrt{\frac{\int_0^\infty y^2 p(t_{k+1}, f, y)\, dy}{\int_0^\infty p(t_{k+1}, f, y)\, dy}}}$$

by doing a numerical integration of the joint density $p(t_{k+1}, f, y)$ (times y^2). Set $\sigma(t, f) \equiv \sigma(t_{k+1}, f)$ for all $t \in [t_{k+1}, t_{k+2}]$.

4. Set $k := k + 1$. Reiterate Steps 2 and 3 up to $k = M$.

11.5   The Markovian projection method

For multi-factor SVMs, the PDE solver described in Section 11.4 is in three or more dimensions, and because of the curse of dimensionality, one must turn to probabilistic methods. First attempts at calibrating local volatility extensions of arbitrary multi-factor stochastic volatility models to market smiles resulted in approximate solutions; see for instance [11, 128, 172, 171].

In this section, we present one of these techniques: the Markovian projection method [128]. Though approximate, it is interesting in and of itself.

Let us define a local martingale X_t that is driven by the stochastic volatility a_t only, and let us write again the SDE of the full dynamics for f_t:

$$dX_t = X_t a_t \, dW_t \qquad (11.10)$$

$$df_t = f_t \sigma(t, f_t) a_t \, dW_t \qquad (11.11)$$

We denote $\sigma_{\text{loc}}^{\text{SV}}$ as the effective local volatility associated to the SVM defined by (11.10):

$$\sigma_{\text{loc}}^{\text{SV}}(t, X)^2 \equiv \mathbb{E}^{\mathbb{Q}}[a_t^2 | X_t = X] \qquad (11.12)$$

In Chapter 9 of [11], the computation of $\sigma_{\text{loc}}^{\text{SV}}(t, X)$ is explained:

$$\sigma_{\text{loc}}^{\text{SV}}(T, X)^2 = \frac{\mathbb{E}^{\mathbb{Q}}[a_T^2 \Pi_T]}{\mathbb{E}^{\mathbb{Q}}[\Pi_T]}, \qquad \Pi_T = \frac{e^{-\frac{K^2}{2(1-\rho^2) \int_0^T a_s^2 ds}}}{\sqrt{\int_0^T a_s^2 \, ds}} \qquad (11.13)$$

with $K \equiv \ln \frac{X}{X_0} + \frac{1}{2} \int_0^T a_s^2 \, ds - \rho \int_0^T a_s \, dZ_s$. Note that this expression only requires the simulation of the stochastic volatility process a_t.

REMARK 11.1 Quick derivation   Expression (11.13) can be easily derived by observing that conditional on the filtration $\mathcal{F}^a$ generated by the Brownian motions that drive a_t, f_T has a lognormal density $\frac{1}{f\sqrt{2\pi V}} e^{-\frac{\left(\ln \frac{f}{m} + \frac{V}{2}\right)^2}{2V}}$ with $\ln m = \ln f_0 + \int_0^T \left(a_t \rho \, dZ_t - \frac{\rho^2}{2} a_t^2 dt\right) dt$ and $V = (1 - \rho^2) \int_0^T a_s^2 \, ds$.   □

Recall that $\sigma_{\text{Dup}}(t, f)$ denotes the Dupire local volatility. We assume that our LSVM is calibrated to the vanilla smile, and therefore we require that Equation (11.4) be satisfied:

$$\sigma_{\text{Dup}}(t, f)^2 = \sigma(t, f)^2 \mathbb{E}^{\mathbb{Q}}[a_t^2 | f_t = f] \qquad (11.14)$$

Proceeding as in the Markovian projection technique [172], we take the ratio of Equations (11.14) and (11.12) and we get

$$\frac{\sigma_{\text{Dup}}(t, f)^2}{\sigma_{\text{loc}}^{\text{SV}}(t, X)^2} = \sigma(t, f)^2 \frac{\mathbb{E}^{\mathbb{Q}}[a_t^2 | f_t = f]}{\mathbb{E}^{\mathbb{Q}}[a_t^2 | X_t = X]}$$

Following [172], if we assume that we are able to find a smooth monotonic mapping between f_t and X_t, say $X_t = H(t, f_t)$, we then get:

$$\sigma_{\text{Dup}}(t, f) = \sigma(t, f) \sigma_{\text{loc}}^{\text{SV}}(t, H(t, f)) \qquad (11.15)$$

as $f_t = f$ is equivalent to $X_t = H(t, f)$. If such a mapping exists, from Itô's formula, the dynamics of X_t is given by

$$dX_t = f_t \partial_f H(t, f_t) \sigma(t, f_t) a_t \, dW_t$$

$$+ \left(\partial_t H(t, f_t) + \frac{1}{2} f_t^2 \partial_f^2 H(t, f_t) \sigma(t, f_t)^2 a_t^2\right) dt \qquad (11.16)$$

This also satisfies (11.10):

$$dX_t = H(t, f_t)a_t\, dW_t \tag{11.17}$$

The volatility and drift in both equations must then coincide.

Volatility term

Identifying the volatility terms, we deduce that

$$f\partial_f H(t, f)\sigma(t, f) = H(t, f) \tag{11.18}$$

From (11.18), Equation (11.15) is equivalent to

$$\frac{\partial_f \ln H(t, f)}{\sigma_{\text{loc}}^{\text{SV}}(t, H(t, f))} = \frac{1}{f\sigma_{\text{Dup}}(t, f)}$$

Integrating the equation above, we get

$$\int_{f_0}^{f} \frac{df}{f\sigma_{\text{Dup}}(t, f)} = \Phi_t(H(t, f))$$

with

$$\Phi_t(x) = \int_{\Lambda(t)}^{x} \frac{dy}{y\sigma_{\text{loc}}^{\text{SV}}(t, y)}$$

and $\Lambda(t) = H(t, f_0)$, an integration constant. Using (11.15), we deduce that

$$\sigma(t, f) = \frac{\sigma_{\text{Dup}}(t, f)}{\sigma_{\text{loc}}^{\text{SV}}\left(t, \Phi_t^{-1}\left(\int_{f_0}^{f} \frac{dx}{x\sigma_{\text{Dup}}(t, x)}\right)\right)} \tag{11.19}$$

This formula involves the geodesic distance, i.e., the harmonic average, of the Dupire and effective local volatilities σ_{Dup} and $\sigma_{\text{loc}}^{\text{SV}}$. As a sanity check, when the local volatility produced by the SVM matches the Dupire local volatility, the solution to Equation (11.5) is $\sigma(t, f) = 1$ as expected (here $\Lambda(t) = f_0$, see below).

Drift term: optimal choice for $\Lambda(\cdot)$

Integrating the first order ODE (11.18), we obtain

$$\ln \frac{H(t, f)}{\Lambda(t)} = \int_{f_0}^{f} \frac{dy}{y\sigma(t, y)} \tag{11.20}$$

As a consequence, the drift term in (11.16) is

$$H(t, f_t)\left(\partial_t \left(\ln \Lambda(t) + \int_{f_0}^{f_t} \frac{dy}{y\sigma(t, y)}\right) + \frac{1}{2}a_t^2(1 - (f_t\sigma(t, f_t))')\right) \tag{11.21}$$

where the prime means a derivative with respect to the forward f. The calibration method would be exact if this drift term (11.21) vanished. It does not in general, so the method is only approximate: even if one used an infinite number of Monte Carlo paths, and an infinitesimal discretization time step, one would not exactly reprice the market smile, using the algorithm described below. However, we can choose the integration constant $\Lambda(t)$ in order to minimize the drift. When we replace f_t by f_0 and a_t^2 by its conditional average $\mathbb{E}^{\mathbb{Q}}[a_t^2|f = f_0]$, the drift becomes $H(t, f_0)$ times

$$\partial_t \ln \Lambda(t) + \frac{1}{2}\sigma_{\text{loc}}^{\text{SV}}(t, f_0)^2(1 - \sigma(t, f_0) - f_0\sigma(t, f_0)')$$

We set $\Lambda(t)$ such that the above expression vanishes:

$$\ln \frac{\Lambda(t)}{f_0} = -\frac{1}{2}\int_0^t \sigma_{\text{loc}}^{\text{SV}}(s, f_0)^2 \left(1 - \sigma(s, f_0) - f_0\sigma(s, f_0)'\right) ds \qquad (11.22)$$

Note that when $\sigma(t, f) \equiv 1$, the drift vanishes and $\Lambda(t) = f_0$. This is consistent with the fact that in this case, $H(t, f) = f$.

Markovian projection algorithm

The final meta-algorithm can be summarized by the following steps:

1. We calibrate the volatility function $\sigma_{\text{Dup}}(t, f)$ to the market vanilla smile using the Dupire formula (11.5).

2. We simulate the stochastic volatility process $\{a_t(\omega)\}$, whose dynamics do not involve the local volatility function $\sigma(t, f)$, and we compute the local volatility $\sigma_{\text{loc}}^{\text{SV}}$ on a space–time grid using Formula (11.13). This formula is very smooth (in particular, it involves no Heaviside function) and does not require many Monte Carlo paths to achieve convergence. In practice, we use $N = 2^{11}$ paths.

3. Using (11.19), we get the local volatility function $\sigma(t, f)$ by first setting $\Lambda(t) = f_0$. The numerical integrations are performed using a quadrature method on the interval $[0, 1]$ and the numerical inversion involved in Φ_t^{-1} uses a Brent algorithm. We repeat this step by using (11.22) for the computation of $\Lambda(\cdot)$, where $\sigma(t, f_0)$ and $\sigma'(t, f_0)$ have been computed previously. In our experience, repeating this step is not necessary for achieving convergence.

The reader should remark that this calibration method does not involve any optimization routine. However, this algorithm is only approximate because the drift in (11.16) should vanish and does not. We now introduce an exact calibration algorithm for LSVMs.

11.6   The particle method

The particle method is the first exact calibration method, in the sense that it is the first method that converges to the true local volatility function when computational effort increases (see the convergence results in Sections 10.2 and 10.3).

11.6.1   The particle method for the calibration of LSVMs to market smiles

The particle method, a technique of simulation of McKean SDEs, was introduced in Section 10.2. It consists of replacing the unknown marginal distribution $\mathbb{Q}_t$ of a McKean SDE at time t by its empirical distribution $\mathbb{Q}_t^N$; see Definition 10.1.

We have seen in Section 11.2 that the dynamical evolution of the calibrated LSVM is described by the following McKean SDE:

$$df_t = f_t \sigma(t, f_t, \mathbb{Q}_t) a_t \, dW_t \tag{11.23}$$

where $\mathbb{Q}_t$ denotes the distribution of (f_t, a_t) under $\mathbb{Q}$ and

$$\sigma(t, f_t, \mathbb{Q}_t) = \frac{\sigma_{\mathrm{Dup}}(t, f_t)}{\sqrt{\mathbb{E}^{\mathbb{Q}}[a_t^2 | f_t]}} \tag{11.24}$$

The N interacting particles cannot follow the dynamics

$$df_t^{i,N} = f_t^{i,N} \sigma(t, f_t^{i,N}, \mathbb{Q}_t^N) a_t^{i,N} \, dW_t^i$$

Indeed, $\mathbb{Q}_t^N$ is atomic; it admits no density $p_N(t, f, a)$, so the conditional expectation

$$\mathbb{E}^{\mathbb{Q}_t^N}[a_t^2 | f_t = f] = \frac{\int a'^2 p_N(t, f, a') da'}{\int p_N(t, f, a') da'} = \frac{\sum_{i=1}^N (a_t^{i,N})^2 \delta(f_t^{i,N} - f)}{\sum_{i=1}^N \delta(f_t^{i,N} - f)}$$

is not properly defined. As in the Nadaraya-Watson regression, instead of the Dirac function $\delta(\cdot)$, we use a regularizing kernel $\delta_{t,N}(\cdot)$, define

$$\sigma_N(t, f) = \sigma_{\mathrm{Dup}}(t, f) \sqrt{\frac{\sum_{i=1}^N \delta_{t,N}\left(f_t^{i,N} - f\right)}{\sum_{i=1}^N \left(a_t^{i,N}\right)^2 \delta_{t,N}\left(f_t^{i,N} - f\right)}} \tag{11.25}$$

and simulate

$$df_t^{i,N} = f_t^{i,N} \sigma_N(t, f_t^{i,N}) a_t^{i,N} \, dW_t^i \tag{11.26}$$

A similar algorithm has been detailed in [147] in the case of the calibration of a correlation smile.

11.6.2 Regularizing kernel

It is natural to take $\delta_{t,N}(x) = \frac{1}{h_{t,N}} K\left(\frac{x}{h_{t,N}}\right)$ where K is a fixed, symmetric kernel with a bandwidth $h_{t,N}$ that tends to zero as N grows to infinity. The exponential kernel $K(x) = \frac{1}{\sqrt{2\pi}} \exp\left(-x^2/2\right)$ and the quartic kernel $K(x) = \frac{15}{16}\left(1 - x^2\right)^2 \mathbf{1}_{\{|x| \leq 1\}}$ are typical examples. We use the latter because it saves computational time. We take

$$h_{t,N} = \kappa f_0 \sigma_{\mathrm{VS},t} \sqrt{\max(t, t_{\min})} N^{-\frac{1}{5}}$$

with $\sigma_{\mathrm{VS},t}$ the variance swap volatility at maturity t. The factor $N^{-\frac{1}{5}}$ comes from the minimization of the asymptotic mean integrated squared error of the Nadaraya-Watson estimator, which is the sum of two terms: bias and variance. The smaller the bandwidth, the smaller the bias, but the larger the variance. The critical bandwidth that minimizes the sum of bias and variance decreases as $N^{-\frac{1}{5}}$ for large N. Following Silverman's rule of thumb [22], the prefactor $\kappa f_0 \sigma_{\mathrm{VS},t} \sqrt{\max(t, t_{\min})}$ is on the order of the standard deviation of the regressor f_t. The fine-tuning of κ is crucial. In practice, we take $\kappa \simeq 1.5$, $t_{\min} = 1/4$.

11.6.3 Acceleration techniques

It is not necessary to compute $\sigma_N(t, f_t^{i,N})$ for all i using (11.25). One can save much time by computing $\sigma_N(t, f)$ for a grid $G_{f,t}$ of values of f, of size much smaller than N, say $N_{f,t}$, and then inter- and extrapolating. We use cubic splines, with a flat extrapolation, and $N_{f,t} = \max(N_f \sqrt{t}, N_f')$; typical values are $N_f = 30$ and $N_f' = 15$. The range of the grid can be inferred from the prices of digital options: $\mathbb{E}[\mathbf{1}_{f_t > \max G_{f,t}}] = \mathbb{E}[\mathbf{1}_{f_t < \min G_{f,t}}] = \alpha$. In practice, we take $\alpha = 10^{-3}$. Alternatively, one can use the $\lfloor \alpha N \rfloor$th smallest $f_t^{i,N}$ for $\min G_{f,t}$, and the $\lfloor \alpha N \rfloor$th largest $f_t^{i,N}$ for $\max G_{f,t}$.

Moreover, in the sums in (11.25), a large number of terms are negligible: we can disregard $f_t^{i,N}$ when it is far from f, say, when $\delta_{t,N}(f_t^{i,N} - f)$ is smaller than some threshold η. In practice, this requires sorting the particles according to the spot value. The cost of sorting, $O(N \ln N)$, is more than compensated by the acceleration in the $N_{f,t}$ evaluations of (11.25).

11.6.4 The algorithm

Let $\{t_k\}$ denote a time discretization of $[0, T]$. The particle algorithm for the calibration of LSVMs to market smiles can now be described by the following steps:

1. Initialize $k = 1$ and set $\sigma_N(t, f) = \frac{\sigma_{\mathrm{Dup}}(0, f)}{\alpha}$ for all $t \in [t_0 = 0, t_1]$.

2. Simulate the N processes $\{f_t^{i,N}, a_t^{i,N}\}_{1 \leq i \leq N}$ from t_{k-1} to t_k using a discretization scheme for (11.26), say, an Euler scheme.

3. Sort the particles according to the forward value. For $f \in G_{f,t_k}$, find the smallest index $\underline{i}(f)$ and the largest index $\bar{i}(f)$ for which $\delta_{t_k,N} \left(f_{t_k}^{i,N} - f \right) > \eta$, and compute the local volatility

$$
\sigma_N(t_k, f) = \sigma_{\text{Dup}}(t_k, f) \sqrt{\frac{\sum_{i=\underline{i}(f)}^{\bar{i}(f)} \delta_{t_k,N} \left(f_{t_k}^{i,N} - f \right)}{\sum_{i=\underline{i}(f)}^{\bar{i}(f)} \left(a_{t_k}^{i,N} \right)^2 \delta_{t_k,N} \left(f_{t_k}^{i,N} - f \right)}}
$$

Interpolate the local volatility using cubic splines, and extrapolate flat. Set $\sigma_N(t, f) \equiv \sigma_N(t_k, f)$ for all $t \in [t_k, t_{k+1}]$.

4. Set $k := k + 1$. Iterate steps 2 and 3 up to the maturity date T.

Step 2 is easily parallelizable. Note that we need no extra simulation to price an option in the calibrated model. We only need to ensure that all the dates needed in the calculation of the payout are included in the time discretization t_k. Then we estimate the price of the option by $\frac{1}{N} \sum_{i=1}^{N} H^{i,N}$ where $H^{i,N}$ is the discounted payout evaluated on the path of particle i.

REMARK 11.2 Local volatility extrapolation   In the above algorithm, we mentioned a flat extrapolation of the local volatility function for very low and very high strikes. This is very crude. One can obviously build clever extrapolations, in particular for matching market quotes of variance swaps. For instance, one may extrapolate the ratio of the local volatility to the Dupire local volatility, or equivalently, $\mathbb{E}^{\mathbb{Q}}[a_t^2 | f_t = f]$, either in a flat way or in a non-flat, more complex way. Importance sampling methods may be very helpful in building the local volatility for strikes that are far from the money. ▯

11.7   Adding stochastic interest rates

One of the great features of the particle method is that it easily extends to the case where interest rates are also stochastic. An LSVM with stochastic interest rates (SIR-LSVM) is defined in a risk-neutral measure $\mathbb{Q}$ by

$$
\frac{dS_t}{S_t} = r_t \, dt + \sigma(t, S_t) a_t \, dW_t \tag{11.27}
$$

where the short term rate r_t and the stochastic volatility a_t are Itô processes. Here as well, for the sake of simplicity, we assume for now that there are no dividends. We explain how to include (discrete) dividends in our calibration procedure in Section 11.C.

11.7.1 The calibration condition

In the proposition below, we give a necessary and sufficient condition for an SIR-LSVM to be calibrated to a given market smile. $\mathcal{C}(t, K)$ denotes the fair market value of a European call option with strike K and maturity t, and P_{st} the time-s value of the bond of maturity t.

PROPOSITION 11.1 Calibration of SIR-LSVMs to market smiles
Model (11.27) is exactly calibrated to the market smile if and only if

$$\sigma(t, K)^2 \frac{\mathbb{E}^{\mathbb{Q}}[D_{0t} a_t^2 | S_t = K]}{\mathbb{E}^{\mathbb{Q}}[D_{0t} | S_t = K]} = \sigma_{\text{Dup}}(t, K)^2 - \frac{\mathbb{E}^{\mathbb{Q}}[D_{0t} (r_t - r_t^0) \mathbf{1}_{S_t > K}]}{\frac{1}{2} K \partial_K^2 \mathcal{C}(t, K)} \quad (11.28)$$

for all (t, K), with $r_t^0 = -\partial_t \ln P_{0t}$ and

$$\sigma_{\text{Dup}}(t, K)^2 = \frac{\partial_t \mathcal{C}(t, K) + r_t^0 K \partial_K \mathcal{C}(t, K)}{\frac{1}{2} K^2 \partial_K^2 \mathcal{C}(t, K)}$$

REMARK 11.3   Equation (11.28) can be restated using the t-forward measure $\mathbb{Q}^t$ defined by $\frac{d\mathbb{Q}^t}{d\mathbb{Q}} = \frac{D_{0t}}{P_{0t}}$:

$$\sigma(t, K)^2 \mathbb{E}^{\mathbb{Q}^t}[a_t^2 | S_t = K] = \sigma_{\text{Dup}}(t, K)^2 - P_{0t} \frac{\mathbb{E}^{\mathbb{Q}^t}[(r_t - r_t^0) \mathbf{1}_{S_t > K}]}{\frac{1}{2} K \partial_K^2 \mathcal{C}(t, K)} \quad (11.29)$$

$\Box$

REMARK 11.4   Note that Equation (11.28) is a self-consistency equation because all the expectations depend on the unknown local volatility σ. Note also that a solution $\sigma(t, K)^2$ of (11.28) can be negative. In this case, the prices cannot be explained by the SIR-SLVM. $\Box$

PROOF   By applying Itô-Tanaka's formula on a discounted vanilla call payoff with maturity t and strike K, $\mathcal{P}_t \equiv D_{0t}(S_t - K)^+$, we have:

$$d\mathcal{P}_t = -D_{0t}(S_t - K)^+ r_t \, dt + D_{0t} \mathbf{1}_{S_t > K} S_t (r_t \, dt + a_t \sigma(t, S_t) \, dW_t)$$
$$+ \frac{1}{2} S_t^2 a_t^2 \sigma(t, S_t)^2 D_{0t} \delta(S_t - K) \, dt$$
$$= D_{0t} \mathbf{1}_{S_t > K} r_t K \, dt + D_{0t} \mathbf{1}_{S_t > K} a_t \sigma(t, S_t) S_t \, dW_t$$
$$+ \frac{1}{2} K^2 a_t^2 \sigma(t, K)^2 D_{0t} \delta(S_t - K) \, dt$$

By taking the expectation $\mathbb{E}^{\mathbb{Q}}[\cdot]$ on both sides of the above equation and by assuming that $M_t = \int_0^t D_{0s} \mathbf{1}_{S_s > K} a_s \sigma(s, S_s) S_s \, dW_s$ is a true martingale, we get

$$\partial_t \mathcal{C}_{\mathrm{m}}(t, K) = K \mathbb{E}^{\mathbb{Q}}[D_{0t} r_t \mathbf{1}_{S_t > K}] + \frac{1}{2} K^2 \sigma(t, K)^2 \mathbb{E}^{\mathbb{Q}}[D_{0t} a_t^2 \delta(S_t - K)]$$

where $\mathcal{C}_{\mathrm{m}}(t, K) = \mathbb{E}^{\mathbb{Q}}[\mathcal{P}_t]$ denotes the price of the call option in the model. Then, by using $\partial_K \mathcal{C}_{\mathrm{m}}(t, K) = -\mathbb{E}^{\mathbb{Q}}[D_{0t} \mathbf{1}_{S_t > K}]$ and $\partial_K^2 \mathcal{C}_{\mathrm{m}}(t, K) = \mathbb{E}^{\mathbb{Q}}[D_{0t} \delta(S_t - K)]$, we deduce that

$$\partial_t \mathcal{C}_{\mathrm{m}}(t, K) = K \mathbb{E}^{\mathbb{Q}}[D_{0t}(r_t - r_t^0) \mathbf{1}_{S_t > K}] - r_t^0 K \partial_K \mathcal{C}_{\mathrm{m}}(t, K)$$
$$+ \frac{1}{2} K^2 \partial_K^2 \mathcal{C}_{\mathrm{m}}(t, K) \sigma(t, K)^2 \frac{\mathbb{E}^{\mathbb{Q}}[D_{0t} a_t^2 | S_t = K]}{\mathbb{E}^{\mathbb{Q}}[D_{0t} | S_t = K]}$$

so the model is calibrated to the market smile if and only if

$$\partial_t \mathcal{C}(t, K) = K \mathbb{E}^{\mathbb{Q}}[D_{0t}(r_t - r_t^0) \mathbf{1}_{S_t > K}] - r_t^0 K \partial_K \mathcal{C}(t, K)$$
$$+ \frac{1}{2} K^2 \partial_K^2 \mathcal{C}(t, K) \sigma(t, K)^2 \frac{\mathbb{E}^{\mathbb{Q}}[D_{0t} a_t^2 | S_t = K]}{\mathbb{E}^{\mathbb{Q}}[D_{0t} | S_t = K]}$$

From the definition of $\sigma_{\mathrm{Dup}}(t, K)$, this is equivalent to

$$\sigma(t, K)^2 \frac{\mathbb{E}^{\mathbb{Q}}[D_{0t} a_t^2 | S_t = K]}{\mathbb{E}^{\mathbb{Q}}[D_{0t} | S_t = K]} = \sigma_{\mathrm{Dup}}(t, K)^2 - \frac{\mathbb{E}^{\mathbb{Q}}[D_{0t}(r_t - r_t^0) \mathbf{1}_{S_t > K}]}{\frac{1}{2} K \partial_K^2 \mathcal{C}(t, K)}$$

which completes the proof. □

From Proposition 11.1, the dynamics of the calibrated SIR-LSVM read as the following nonlinear McKean diffusion:

$$\frac{dS_t}{S_t} = r_t \, dt + \sigma(t, S_t, \mathbb{Q}_t) \, a_t \, dW_t$$

where

$$\sigma(t, K, \mathbb{Q}_t)^2 = \left(\sigma_{\mathrm{Dup}}(t, K)^2 - \frac{\mathbb{E}^{\mathbb{Q}}[D_{0t}(r_t - r_t^0) \mathbf{1}_{S_t > K}]}{\frac{1}{2} K \partial_K^2 \mathcal{C}(t, K)} \right)$$
$$\times \frac{\mathbb{E}^{\mathbb{Q}}[D_{0t} | S_t = K]}{\mathbb{E}^{\mathbb{Q}}[D_{0t} a_t^2 | S_t = K]} \quad (11.30)$$

Alternatively, it reads as the following nonlinear McKean diffusion for the forward $f_t = S_t/P_{tT}$ in the forward measure $\mathbb{Q}^T$, where T denotes the last maturity date for which we want to calibrate the market smile:

$$\frac{df_t}{f_t} = \sigma\left(t, P_{tT} f_t, \mathbb{Q}_t^T\right) a_t \, dW_t^T - \sigma_{\mathrm{P}}^T(t) . dB_t^T$$

where[1]

$$\sigma\left(t, K, Q_t^T\right)^2 = \left(\sigma_{\mathrm{Dup}}(t, K)^2 - P_{0T}\frac{\mathbb{E}^{Q^T}\left[P_{tT}^{-1}\left(r_t - r_t^0\right)\mathbf{1}_{S_t > K}\right]}{\frac{1}{2}K\partial_K^2 \mathcal{C}(t, K)}\right)$$

$$\times \frac{\mathbb{E}^{Q^T}\left[P_{tT}^{-1}|S_t = K\right]}{\mathbb{E}^{Q^T}\left[P_{tT}^{-1}a_t^2|S_t = K\right]} \quad (11.31)$$

$\sigma_{\mathrm{P}}^T(t)$ is the volatility of the bond P_{tT}, and B_t^T the (possibly multi-dimensional) $\mathbb{Q}^T$-Brownian motion that drives the interest rate curve.

11.7.2   The PDE method

In the case of a local volatility model with stochastic interest rates, i.e., with no stochastic volatility component ($a_t \equiv 1$), with a one-factor short rate model, the effective volatility of the forward can be exactly computed using a 2-dimensional PDE solver of the Fokker-Planck equation. The algorithm goes along the lines presented in Section 11.4. One needs to compute at each time t_k the following two-dimensional integral, once the probability density function $p(t_k, r, f)$ has been computed:

$$\mathbb{E}^{Q^T}\left[P_{tT}^{-1}\left(r_t - r_t^0\right)\mathbf{1}_{P_{tT}f_t > K}\right] = \int_{\mathbb{R}} dr \int_{K/P_{tT}(r)}^{\infty} df\, P_{tT}^{-1}(r)\left(r - r_t^0\right)p(t_k, r, f)$$

11.7.3   The particle method

In the case of SIR-LSVMs, a particle is described by three processes (S_t, a_t, r_t). If we use the representation (11.30) of the local volatility, we define

$$\sigma_N(t, S)^2 = \left(\sigma_{\mathrm{Dup}}(t, S)^2 - \frac{\frac{1}{N}\sum_{i=1}^{N} D_{0t}^{i,N}\left(r_t^{i,N} - r_t^0\right)\mathbf{1}_{S_t^{i,N} > S}}{\frac{1}{2}S\partial_K^2 \mathcal{C}(t, S)}\right)$$

$$\times \frac{\sum_{i=1}^{N} D_{0t}^{i,N}\delta_{t,N}\left(S_t^{i,N} - S\right)}{\sum_{i=1}^{N} D_{0t}^{i,N}\left(a_t^{i,N}\right)^2\delta_{t,N}\left(S_t^{i,N} - S\right)} \quad (11.32)$$

and simulate

$$\frac{dS_t^{i,N}}{S_t^{i,N}} = r_t^{i,N}\, dt + \sigma_N\left(t, S_t^{i,N}\right)a_t^{i,N}\, dW_t^i$$

where the W^i are $\mathbb{Q}$-Brownian motions.

In many commonly used short rate models, the price of the zero coupon P_{tT} bond has a closed formula, i.e., is a deterministic function of (t, r_t), and

---

[1]We have used the identity: $\mathbb{E}^{\mathbb{Q}}[D_{0t}X_t] = P_{0T}\mathbb{E}^{Q^T}\left[P_{tT}^{-1}X_t\right]$.

a particle can also be described by three processes (f_t, a_t, r_t). Using the representation (11.31), we then define

$$
\sigma_N(t,S)^2 = \left(\sigma_{\mathrm{Dup}}(t,S)^2 - P_{0T} \frac{\frac{1}{N}\sum_{i=1}^N \left(P_{tT}^{i,N}\right)^{-1}\left(r_t^{i,N} - r_t^0\right)\mathbf{1}_{S_t^{i,N}>S}}{\frac{1}{2}S\partial_K^2 \mathcal{C}(t,S)} \right)
$$
$$
\times \frac{\sum_{i=1}^N \left(P_{tT}^{i,N}\right)^{-1}\delta_{t,N}\left(S_t^{i,N}-S\right)}{\sum_{i=1}^N \left(P_{tT}^{i,N}\right)^{-1}\left(a_t^{i,N}\right)^2 \delta_{t,N}\left(S_t^{i,N}-S\right)} \tag{11.33}
$$

and simulate

$$
df_t^{i,N} = f_t^{i,N}\sigma_N\left(t, f_t^{i,N}P_{tT}^{i,N}\right)a_t^{i,N}\,dW_t^i - f_t^{i,N}\sigma_{\mathrm{P}}^{T,i,N}(t).\,dB_t^i
$$

where W^i and B^i are $\mathbb{Q}^T$-Brownian motions.

11.7.4   First example: The particle method for the hybrid Ho-Lee/Dupire model

For simplicity, here we assume that $a_t \equiv 1$. Let us consider the case where (r_t) follows a Ho-Lee model:

$$
dr_t = \theta(t)\,dt + \sigma_{\mathrm{r}}\,dB_t \tag{11.34}
$$

B is a $\mathbb{Q}$-Brownian motion, correlated with W, the correlation being constant and equal to ρ. We can choose the drift $\theta(t)$ such that the market zero-coupon curve is calibrated:

$$
P_{tT} = \frac{P_{0T}^{\mathrm{mkt}}}{P_{0t}^{\mathrm{mkt}}}\exp\left(\frac{1}{2}\sigma_{\mathrm{r}}^2(T-t)^2 t - \sigma_{\mathrm{r}}(T-t)B_t^T\right) \tag{11.35}
$$

where

$$
B_t^T = B_t + \sigma_{\mathrm{r}}Tt - \sigma_{\mathrm{r}}^2\frac{t^2}{2}
$$

is a $\mathbb{Q}^T$-Brownian motion. Since in the Ho-Lee model the drift is deterministic, we have

$$
r_t - r_t^0 = \sigma_{\mathrm{r}}B_t^t = \sigma_{\mathrm{r}}\left(B_t^T - \sigma_{\mathrm{r}}t(T-t)\right) \tag{11.36}
$$

In the particle algorithm for the hybrid Ho-Lee/Dupire model, a particle can then be described by (S_t, B_t^T) and one can simulate it under the final forward measure $\mathbb{Q}^T$ by plugging (11.35) and (11.36) into (11.33), with $a_t \equiv 1$. To be precise, we simulate the T-forward price $f_t = S_t/P_{tT}$ under the final forward measure as follows:

$$
df_t^{i,N} = f_t^{i,N}\sigma_N\left(t, f_t^{i,N}P_{tT}^{i,N}\right)a_t^{i,N}\,dW_t^i + f_t^{i,N}\sigma_{\mathrm{r}}(T-t)\,dB_t^i
$$

where W^i and B^i are $\mathbb{Q}^T$-Brownian motions, and

$$\sigma_N(t,S)^2 = \sigma_{\mathrm{Dup}}(t,S)^2 - \sigma_{\mathrm{r}} P_{0t}^{\mathrm{mkt}}$$
$$\times \frac{\frac{1}{N}\sum_{i=1}^N \exp\left(\sigma_{\mathrm{r}}(T-t)B_t^i - \frac{1}{2}\sigma_{\mathrm{r}}^2(T-t)^2 t\right)\left(B_t^i - \sigma_{\mathrm{r}}t(T-t)\right)\mathbf{1}_{S_t^{i,N}>S}}{\frac{1}{2}S\partial_K^2 \mathcal{C}(t,S)}$$

To simulate $f_{t_{k+1}}$ from f_{t_k}, one can use the following log-Euler discretization scheme

$$f_{t_{k+1}} = f_{t_k}\exp\left(\sigma(t_k, P_{t_k T}f_{t_k})(W_{t_{k+1}}^T - W_{t_k}^T) + \sigma_{\mathrm{r}}\int_{t_k}^{t_{k+1}}(T-t)\,dB_t^T\right.$$
$$\left. -\frac{1}{2}(t_{k+1}-t_k)\left(\sigma(t_k, P_{t_k T}f_{t_k})^2 - 2\rho\sigma_{\mathrm{r}}\sigma(t_k, P_{t_k T}f_{t_k})I_k^1 - \sigma_{\mathrm{r}}^2 I_k^2\right)\right)$$

with

$$I_k^1 = \frac{1}{t_{k+1}-t_k}\int_{t_k}^{t_{k+1}}(T-t)\,dt = T - \frac{t_k + t_{k+1}}{2}$$
$$I_k^2 = \frac{1}{t_{k+1}-t_k}\int_{t_k}^{t_{k+1}}(T-t)^2\,dt = T^2 - T(t_k + t_{k+1}) + \frac{t_k^2 + t_k t_{k+1} + t_{k+1}^2}{3}$$

and simulate exactly the Gaussian vector

$$\left(W_{t_{k+1}}^T - W_{t_k}^T, B_{t_{k+1}}^T - B_{t_k}^T, \int_{t_k}^{t_{k+1}}(T-t)\,dB_t^T\right)$$

whose covariance matrix simply reads $(t_{k+1}-t_k)\Sigma_k$, where

$$\Sigma_k = \begin{pmatrix} 1 & \rho & \rho I_k^1 \\ \rho & 1 & I_k^1 \\ \rho I_k^1 & I_k^1 & I_k^2 \end{pmatrix}$$

11.7.5 Second example: The particle method for the hybrid Hull-White/Dupire model

Here again we assume for simplicity that $a_t \equiv 1$. We look at the particular case where (r_t) follows the Hull-White model:

$$dr_t = \kappa(\theta(t) - r_t)\,dt + \sigma_{\mathrm{r}}\,dB_t \tag{11.37}$$

B is a $\mathbb{Q}$-Brownian motion, correlated with W, the correlation being ρ. The explicit solution is:

$$r_t = e^{-\kappa t}\left(r_0 + \kappa\int_0^t e^{\kappa s}\theta(s)\,ds + \sigma_{\mathrm{r}}\int_0^t e^{\kappa s}\,dB_s\right) \tag{11.38}$$

The function $\theta(t)$ is chosen such that the zero coupon curve observed at $t = 0$ is matched:

$$\theta(t) = \frac{1}{\kappa} \frac{\partial f^{\text{mkt}}}{\partial t}(0, t) + f^{\text{mkt}}(0, t) + \frac{\sigma_{\text{r}}^2}{2\kappa^2}\left(1 - e^{-2\kappa t}\right)$$

where

$$f^{\text{mkt}}(0, t) = -\frac{\partial \ln P_{0t}^{\text{mkt}}}{\partial t} = r_t^0$$

The bond prices are given by

$$P_{tT} = \frac{P_{0T}^{\text{mkt}}}{P_{0t}^{\text{mkt}}} \exp\left(-A(t, T)(r_t - r_t^0) - \frac{\sigma^2}{4\kappa^3}(e^{-\kappa T} - e^{-\kappa t})^2(e^{2\kappa t} - 1)\right) \quad (11.39)$$

where

$$A(t, T) = \frac{1 - e^{-\kappa(T-t)}}{\kappa}$$

and the volatility of the zero-coupon bond is

$$\sigma_{\text{P}}^T(t) = -\sigma_{\text{r}} A(t, T)$$

When $\kappa \to 0$, we are back to the Ho-Lee model for which $\sigma_{\text{P}}^T(t) = -\sigma_{\text{r}}(T - t)$. In view of simulating all processes under the terminal forward measure $\mathbb{Q}^T$, we aim at writing $r_t - r_t^0$ in terms of the $\mathbb{Q}^T$-Brownian motion

$$B_t^T = B_t - \int_0^t \sigma_{\text{P}}^T(s)\,ds = B_t + \sigma_{\text{r}} \int_0^t A(s, T)\,ds$$

From (11.38), we have for all T

$$r_t = e^{-\kappa t}\left(r_0 + \int_0^t e^{\kappa s}(\kappa\theta(s) - \sigma_{\text{r}}^2 A(s, T))\,ds + \sigma_{\text{r}} \int_0^t e^{\kappa s}\,dB_s^T\right)$$

so that, taking $T = t$,

$$r_t^0 = \mathbb{E}^{\mathbb{Q}^t}[r_t] = e^{-\kappa t}\left(r_0 + \int_0^t e^{\kappa s}(\kappa\theta(s) - \sigma_{\text{r}}^2 A(s, t))\,ds\right)$$

and

$$r_t - r_t^0 = \sigma_{\text{r}} e^{-\kappa t} \int_0^t e^{\kappa s}\,dB_s^t$$

Now,

$$B_s^t = B_s^T - \sigma_{\text{r}} \int_0^s (A(u, T) - A(u, t))\,du$$

so

$$r_t - r_t^0 = \sigma_r e^{-\kappa t} \int_0^t e^{\kappa s} \left(dB_s^T - \sigma_r (A(s,T) - A(s,t)) \, ds \right)$$

$$= e^{-\kappa t} \left(\sigma_r \int_0^t e^{\kappa s} \, dB_s^T - \frac{\sigma_r^2}{\kappa^2} \sinh(\kappa t) \left(1 - e^{-\kappa(T-t)} \right) \right) \quad (11.40)$$

In the particle algorithm for the hybrid Hull-White/Dupire model, a particle can then be described by $(S_t, r_t - r_t^0)$ and one can simulate it under the final forward measure $\mathbb{Q}^T$ by plugging (11.39) and (11.40) into (11.33), with $a_t \equiv 1$. To be precise, we simulate the T-forward price $f_t = S_t/P_{tT}$ under the final forward measure as follows:

$$df_t^{i,N} = f_t^{i,N} \sigma_N \left(t, f_t^{i,N} P_{tT}^{i,N} \right) a_t^{i,N} \, dW_t^i + f_t^{i,N} \sigma_r A(t,T) \, dB_t^i$$

where W^i and B^i are $\mathbb{Q}^T$-Brownian motions, and

$$\sigma_N(t,S)^2 = \sigma_{\mathrm{Dup}}(t,S)^2 - \sigma_r P_{0t}^{\mathrm{mkt}} \frac{\frac{1}{N} \sum_{i=1}^N \Pi_t^{i,N} \left(r_t^{i,N} - r_t^0 \right) \mathbf{1}_{S_t^{i,N} > S}}{\frac{1}{2} S \partial_K^2 \mathcal{C}(t,S)}$$

$$\Pi_t^{i,N} = \exp \left(A(t,T)(r_t^{i,N} - r_t^0) + \frac{\sigma^2}{4\kappa^3} (e^{-\kappa T} - e^{-\kappa t})^2 (e^{2\kappa t} - 1) \right)$$

In order to simulate $(f_{t_{k+1}}, r_{t_{k+1}} - r_{t_{k+1}}^0)$ from $(f_{t_k}, r_{t_k} - r_{t_k}^0)$, one can observe that

$$r_{t_{k+1}} - r_{t_{k+1}}^0 = e^{-\kappa(t_{k+1}-t_k)} (r_{t_k} - r_{t_k}^0) + \sigma_r \int_{t_k}^{t_{k+1}} e^{-\kappa(t_{k+1}-t)} \, dB_t^T$$

$$- \frac{\sigma_r^2}{\kappa^2} e^{-\kappa t_{k+1}} \left(\sinh(\kappa t_{k+1})(1 - e^{-\kappa(T-t_{k+1})}) - \sinh(\kappa t_k)(1 - e^{-\kappa(T-t_k)}) \right)$$

and use the following approximation scheme for f_t:

$$f_{t_{k+1}} = f_{t_k} \exp \left(\sigma(t_k, P_{t_k T} f_{t_k})(W_{t_{k+1}}^T - W_{t_k}^T) + \sigma_r \int_{t_k}^{t_{k+1}} \frac{1 - e^{-\kappa(T-t)}}{\kappa} \, dB_t^T \right.$$

$$\left. - \frac{1}{2} (t_{k+1} - t_k) \left(\sigma(t_k, P_{t_k T} f_{t_k})^2 + 2\rho \sigma_r \sigma(t_k, P_{t_k T} f_{t_k}) I_k^1 + \sigma_r^2 I_k^2 \right) \right)$$

with

$$I_k^1 = \frac{1}{t_{k+1} - t_k} \int_{t_k}^{t_{k+1}} \frac{1 - e^{-\kappa(T-t)}}{\kappa} \, dt = \frac{\kappa(t_{k+1} - t_k) - e^{-\kappa T}(e^{\kappa t_{k+1}} - e^{\kappa t_k})}{\kappa^2(t_{k+1} - t_k)}$$

$$I_k^2 = \frac{1}{t_{k+1} - t_k} \int_{t_k}^{t_{k+1}} \left(\frac{1 - e^{-\kappa(T-t)}}{\kappa} \right)^2 \, dt$$

$$= \frac{\kappa(t_{k+1} - t_k) - 2e^{-\kappa T}(e^{\kappa t_{k+1}} - e^{\kappa t_k}) + \frac{1}{2} e^{-2\kappa T}(e^{2\kappa t_{k+1}} - e^{2\kappa t_k})}{\kappa^3(t_{k+1} - t_k)}$$

We recover the Ho-Lee expressions when $\kappa \to 0$. Then the simulation of $(f_{t_{k+1}}, r_{t_{k+1}} - r^0_{t_{k+1}})$ boils down to simulating exactly the Gaussian vector

$$\left(W^T_{t_{k+1}} - W^T_{t_k}, \int_{t_k}^{t_{k+1}} e^{-\kappa(t_{k+1}-t)} \, dB^T_t, \int_{t_k}^{t_{k+1}} \frac{1 - e^{-\kappa(T-t)}}{\kappa} \, dB^T_t \right)$$

whose covariance matrix is $(t_{k+1} - t_k)\Sigma_k$ where

$$\Sigma_k = \begin{pmatrix} 1 & \rho J^1_k & \rho I^1_k \\ \rho J^1_k & J^2_k & L_k \\ \rho I^1_k & L_k & I^2_k \end{pmatrix}$$

with

$$J^1_k = \frac{1}{t_{k+1} - t_k} \int_{t_k}^{t_{k+1}} e^{-\kappa(t_{k+1}-t)} \, dt = \frac{1 - e^{-\kappa(t_{k+1}-t_k)}}{\kappa(t_{k+1} - t_k)}$$

$$J^2_k = \frac{1}{t_{k+1} - t_k} \int_{t_k}^{t_{k+1}} e^{-2\kappa(t_{k+1}-t)} \, dt = \frac{1 - e^{-2\kappa(t_{k+1}-t_k)}}{2\kappa(t_{k+1} - t_k)}$$

$$L_k = \frac{1}{t_{k+1} - t_k} \int_{t_k}^{t_{k+1}} e^{-\kappa(t_{k+1}-t)} \frac{1 - e^{-\kappa(T-t)}}{\kappa} \, dt$$

$$= \frac{J^1_k}{\kappa} - e^{-\kappa(T+t_{k+1})} \frac{e^{2\kappa t_{k+1}} - e^{2\kappa t_k}}{2\kappa^2(t_{k+1} - t_k)}$$

11.7.6   Malliavin representation of the local volatility

We now give another expression of the contribution of stochastic interest rates to local volatility:

$$\frac{\mathbb{E}^{\mathbb{Q}}[D_{0t}\left(r_t - r^0_t\right)\mathbf{1}_{S_t > K}]}{\frac{1}{2}K\partial^2_K \mathcal{C}(t, K)} \equiv P_{0t} \frac{\mathbb{E}^{\mathbb{Q}^t}[\left(r_t - r^0_t\right)\mathbf{1}_{S_t > K}]}{\frac{1}{2}K\partial^2_K \mathcal{C}(t, K)}$$

Numerical implementation of the particle algorithm using the alternative formula proves to produce a much more accurate and smooth estimation of the local volatility for strikes that are far from the money. As a consequence, it is very useful for extrapolation purposes. To derive this new formula, we will make use of the Malliavin calculus.

11.7.6.1   The general formula

Recall that $r^0_t = \mathbb{E}^{\mathbb{Q}^t}[r_t]$. From the martingale representation theorem [13]

$$r_t - r^0_t = \int_0^t \sigma^t_r(s) \, dB^t_s$$

with $\sigma^t_r(s)$ an adapted process. Note that, from Clark-Ocone's formula [19], $\sigma^t_r(s) = \mathbb{E}^{\mathbb{Q}^t}_s[D^{B^t}_s r_t]$ with $D^{B^t}_s$, the Malliavin derivative with respect to the

Brownian motion B^t, and $\mathbb{E}_s$ the conditional expectation given $\mathcal{F}_s$, the natural filtration of all the Brownian motions used. The application of Clark-Ocone's formula to the process $\mathbf{1}_{S_t > K}$ gives

$$P_{0t}\mathbb{E}^{\mathbb{Q}^t}\left[(r_t - r_t^0)\,\mathbf{1}_{S_t>K}\right]$$

$$= P_{0t}\mathbb{E}^{\mathbb{Q}^t}\left[\left(\int_0^t \sigma_r^t(s).\,dB_s^t\right)\left(\mathbb{E}^{\mathbb{Q}^t}\left[\mathbf{1}_{S_t>K}\right] + \int_0^t \mathbb{E}_s^{\mathbb{Q}^t}[D_s^{B^t}\mathbf{1}_{S_t>K}].\,dB_s^t\right.\right.$$

$$\left.\left.+ \int_0^t \mathbb{E}_s^{\mathbb{Q}^t}[D_s^{Z^t}\mathbf{1}_{S_t>K}]\,dZ_s^t\right)\right]$$

$$= P_{0t}\mathbb{E}^{\mathbb{Q}^t}\left[\left(\int_0^t \sigma_r^t(s).\,dB_s^t\right)\left(\int_0^t \mathbb{E}_s^{\mathbb{Q}^t}[D_s^{B^t}\mathbf{1}_{S_t>K}].\,dB_s^t\right)\right]$$

where we have expressed the $\mathbb{Q}^t$-Brownian motion W^t that drives the spot dynamics in terms of the (possibly multi-dimensional) $\mathbb{Q}^t$-Brownian motion B^t that drives the short rate dynamics and another $\mathbb{Q}^t$-Brownian motion Z^t orthogonal to B^t. Then, from Itô's isometry,

$$P_{0t}\mathbb{E}^{\mathbb{Q}^t}\left[(r_t - r_t^0)\,\mathbf{1}_{S_t>K}\right] = P_{0t}\mathbb{E}^{\mathbb{Q}^t}\left[\int_0^t \sigma_r^t(s).\mathbb{E}_s^{\mathbb{Q}^t}[D_s^{B^t}\mathbf{1}_{S_t>K}]\,ds\right]$$

$$= P_{0t}\int_0^t \mathbb{E}^{\mathbb{Q}^t}\left[\sigma_r^t(s).D_s^{B^t}\mathbf{1}_{S_t>K}\right]ds$$

$$= P_{0t}\int_0^t \mathbb{E}^{\mathbb{Q}^t}\left[\sigma_r^t(s).D_s^{B^t}S_t\,\delta_K(S_t)\right]ds$$

$$= P_{0t}\int_0^t \mathbb{E}^{\mathbb{Q}^t}[\sigma_r^t(s).D_s^{B^t}S_t|S_t = K]\mathbb{E}^{\mathbb{Q}^t}\left[\delta_K(S_t)\right]ds$$

$$= \partial_K^2 \mathcal{C}(t,K)\int_0^t \mathbb{E}^{\mathbb{Q}^t}[\sigma_r^t(s).D_s^{B^t}S_t|S_t = K]\,ds$$

As a consequence, the contribution of stochastic interest rates to local volatility reads

$$P_{0t}\frac{\mathbb{E}^{\mathbb{Q}^t}\left[(r_t - r_t^0)\,\mathbf{1}_{S_t>K}\right]}{\frac{1}{2}K\partial_K^2\mathcal{C}(t,K)} = \frac{2}{K}\int_0^t \mathbb{E}^{\mathbb{Q}^t}[\sigma_r^t(s).D_s^{B^t}S_t|S_t = K]\,ds \qquad (11.41)$$

We call this trick a Malliavin "disintegration by parts," because it transforms an unconditional expectation involving the Heaviside function $\mathbf{1}_{S_t>K}$ into a conditional expectation given $S_t = K$. The Malliavin integration by parts formula (6.38) goes the other way round. Note that the second derivative $\partial_K^2\mathcal{C}(t,K)$ of the call option with respect to strike cancels out in the right-hand side of Equation (11.41). This is fortunate as the computation of this term is sensitive to the strike interpolation/extrapolation method. Also, both $\mathbb{E}^{\mathbb{Q}^t}\left[(r_t - r_t^0)\,\mathbf{1}_{S_t>K}\right]$ and $K\partial_K^2\mathcal{C}(t,K)$ are very small for strikes K that are far from the money. Numerically, this $0/0$ ratio can be problematic. There is no such problem in the right-hand side of Equation (11.41), because of the Malliavin disintegration by parts. This makes the Malliavin representation of the contribution of stochastic interest rates to local volatility very useful in practice, in particular when one wants to design an accurate extrapolation of the contribution of stochastic interest rates to local volatility for strikes that are far from the money.

11.7.6.2   The case of one-factor short rate models

For the sake of simplicity, let us assume here that the short rate r_t follows a one-factor Itô diffusion

$$dr_t = \mu_\mathrm{r}(t, r_t)\, dt + \sigma_\mathrm{r}(t, r_t)\, dB_t$$

where $\mu_\mathrm{r}(t, r_t)$ and $\sigma_\mathrm{r}(t, r_t)$ are deterministic functions of the time t and the short rate r_t, and B_t is a one-dimensional $\mathbb{Q}$-Brownian motion with $d\langle B, W\rangle_t = \rho\, dt$. Then $\sigma_\mathrm{P}^T(t)$, the volatility of the bond P_{tT}, is also a deterministic function $\sigma_\mathrm{P}^T(t, r_t)$ of the time t and the short rate r_t. Moreover, we assume that the stochastic volatility is not correlated with the stochastic rate r_t. Both assumptions can be easily relaxed but at the cost of additional computations. By explicitly computing $D_s^{B^t} S_t$, we get the following expression:[2]

PROPOSITION 11.2
Under the assumptions above, the contribution of stochastic interest rates to local volatility reads

$$P_{0t}\frac{\mathbb{E}^{\mathbb{Q}^t}\left[(r_t - r_t^0)\, \mathbf{1}_{S_t > K}\right]}{\frac{1}{2}K\partial_K^2 \mathcal{C}(t, K)} = 2\mathbb{E}^{\mathbb{Q}^t}\left[V_t\left(\rho U_t^t + \Theta_t^t \Xi_t^t - \Lambda_t^t\right) | S_t = K\right] \quad (11.42)$$

with

$$\frac{dV_t}{V_t} = S_t \partial_S \sigma(t, S_t) a_t\left(dW_t - a_t \sigma(t, S_t)\, dt\right), \qquad V_0 = 1 \quad (11.43)$$

$$dU_t^T = \sigma_\mathrm{r}^T(t) a_t \frac{\sigma(t, S_t)}{V_t}\, dt, \qquad U_0^T = 0 \quad (11.44)$$

$$\frac{dR_t^T}{R_t^T} = \left(\partial_r \mu_\mathrm{r}(t, r_t) + \sigma_\mathrm{r}(t, r_t)\partial_r \sigma_\mathrm{P}^T(t, r_t)\right) dt + \partial_r \sigma_\mathrm{r}(t, r_t)\, dB_t, \; R_0^T = 1 \; (11.45)$$

$$d\Theta_t^T = \frac{R_t^T}{V_t}\left(1 + \rho\partial_r \sigma_\mathrm{P}^T(t, r_t)\sigma(t, S_t) a_t\right) dt, \qquad \Theta_0^T = 0 \quad (11.46)$$

$$d\Xi_t^T = \frac{\sigma_\mathrm{r}^T(t)\sigma_\mathrm{r}(t, r_t)}{R_t^T}\, dt, \qquad \Xi_0^T = 0 \quad (11.47)$$

---

[2]Equations (11) and (14) in [122] for U and Ξ are actually erroneous in general - our mistake - and are replaced respectively by Equations (11.44) and (11.47). However, they are correct for the Ho-Lee model - the model we used in our numerical experiments in [122] - because in this case $\sigma_\mathrm{r}^T(t) = \sigma_\mathrm{r}(t, r_t)$.

$$dΛ_t^T = Θ_t^T \, dΞ_t^T, \qquad Λ_0^T = 0 \tag{11.48}$$

and

$$σ_r^T(t) = \mathbb{E}_t^{\mathbb{Q}^T}[R_T^T](R_t^T)^{-1}σ_r(t, r_t) \tag{11.49}$$

PROOF   See Section 11.A. ▯

11.7.6.3   The particular cases of the Ho-Lee and Hull-White models

Proposition 11.2 is not completely satisfactory in two ways. First, the extra processes U_t^T, R_t^T, $Θ_t^T$, $Ξ_t^T$, and $Λ_t^T$ depend on T, which means that in (11.42), one has to simulate 5 processes for each value of t! Second, $σ_r^T(t)$ has still to be evaluated in closed form. Considering constant short rate volatility and affine short rate drift:

$$dr_t = (λ(t) - κr_t) \, dt + σ_r \, dB_t \tag{11.50}$$

solves those two issues at one time. This extra hypothesis is restrictive, but actually encompasses the cases of commonly used short rate models, such as the Ho-Lee (11.34) and Hull-White (11.37) models, so the results below are very useful in practice. In the Ho-Lee model, $κ = 0$ and $λ(t) = θ(t)$; in the Hull-White model, $κ > 0$ and $λ(t) = κθ(t)$.

Under (11.50), the volatility of the bond is deterministic so that $∂_r σ_P^T(t, r_t) = 0$. Then the process $R_t = e^{-κt}$ is independent of T, it coincides with the tangent process of r_t,

$$σ_r^T(t) = σ_r e^{-κ(T-t)}$$

and (11.42) reads

$$P_{0t} \frac{\mathbb{E}^{\mathbb{Q}^t}[(r_t - r_t^0) \, \mathbf{1}_{S_t > K}]}{\frac{1}{2}K∂_K^2 \mathcal{C}(t, K)} = 2σ_r e^{-κt} \mathbb{E}^{\mathbb{Q}^t}[V_t(ρU_t + Θ_tΞ_t - Λ_t) | S_t = K] \tag{11.51}$$

with

$$\frac{dV_t}{V_t} = S_t ∂_S σ(t, S_t) a_t (dW_t - a_t σ(t, S_t) \, dt), \qquad V_0 = 1$$

$$dU_t = e^{κt} a_t \frac{σ(t, S_t)}{V_t} \, dt, \qquad U_0 = 0$$

$$dΘ_t = \frac{e^{-κt}}{V_t} \, dt, \qquad Θ_0 = 0$$

$$Ξ_t = \begin{cases} σ_r \frac{e^{2κt} - 1}{2κ} & \text{if } κ \neq 0 \\ σ_r t & \text{otherwise} \end{cases}$$

$$dΛ_t = σ_r e^{2κt} Θ_t \, dt, \qquad Λ_0 = 0$$

The computation of (11.51), for all t, requires the simulation of only 3 processes f_t, r_t, V_t, and 3 integrals U_t, Θ_t, Λ_t.
In this case, we eventually obtain the following representation of the local volatility (11.31):

$$
\sigma\left(t, K, \mathbb{Q}_t^T\right)^2 = \frac{\mathbb{E}^{\mathbb{Q}^T}\left[P_{tT}^{-1}|S_t = K\right]}{\mathbb{E}^{\mathbb{Q}^T}\left[P_{tT}^{-1} a_t^2 |S_t = K\right]}
$$
$$
\times \left(\sigma_{\text{Dup}}(t, K)^2 - 2\sigma_r e^{-\kappa t} \frac{\mathbb{E}^{\mathbb{Q}^T}\left[P_{tT}^{-1} V_t \left(\rho U_t + \Theta_t \Xi_t - \Lambda_t\right)|S_t = K\right]}{\mathbb{E}^{\mathbb{Q}^T}\left[P_{tT}^{-1}|S_t = K\right]}\right)
$$
$$(11.52)$$

where the dynamics for V_t is[3]

$$
\frac{dV_t}{V_t} = S_t \partial_S \sigma(t, S_t) a_t \left(dW_t^T + \left(\rho \sigma_{\text{P}}^T(t, r_t) - a_t \sigma(t, S_t)\right) dt\right), \qquad V_0 = 1
$$

11.7.7 Local stochastic volatility combined with Libor Market Models

Let us now look at the case where the stochastic interest rates are described by Libor Market Models (LMMs). The risk-neutral measure $\mathbb{Q}$ does not exist in LMMs as it is not possible to invest in an ultra-short Libor. The above discussion must then be refined. In the measure $\mathbb{Q}^{T_N}$ associated to the bond P_{tT_N} of maturity $T_N = T$ (the last maturity considered), the forward $f_t = S_t/P_{tT}$ is a local martingale

$$
\frac{df_t}{f_t} = \sigma(t, S_t, \mathbb{Q}_t^T) a_t\, dW_t^T - \sigma_{\text{P}}^T(t).dB_t^T
$$

with $\sigma_{\text{P}}^T(t)$ the volatility of the bond P_{tT}. Note that because of the bond $P_{tT_{\beta(t)-1}}$, with $T_{\beta(t)-1}$ the nearest future Libor fixing date, the dynamics of the bond

$$
P_{tT} = P_{tT_{\beta(t)-1}} \prod_{i=\beta(t)}^{N} \frac{1}{1 + \tau_i L_i(t)}
$$

(hence its volatility) is not defined in a LMM. Here $L_i(t) \equiv L_i(t, T_{i-1}, T_i)$ is the value at t of the Libor fixing at T_{i-1} and operating between T_{i-1} and T_i. Only the volatility of a ratio of bonds maturing at the Libor fixing dates arising, for example, in a change of numéraire, is well defined. In order to get the evolution of $P_{tT_{\beta(t)-1}}$, which cannot be deduced from our discrete forward rates, we need to model the instantaneous forward rate volatility. This is an additional arbitrary input to the discrete-tenor setting. As a convenient

---

[3]U_t, Θ_t, Ξ_t, and Λ_t have finite variation and are not affected by the change of measure from $\mathbb{Q}$ to $\mathbb{Q}^T$.

choice, we assume that the volatility of the short term instantaneous forward rate vanishes (zero ultra-short volatility). More specifically, we use a linear interpolation for the value of $P_{tT_{\beta(t)-1}}$

$$P_{tT_{\beta(t)-1}} = \frac{1}{1 + (T_{\beta(t)-1} - t)L_{\beta(t)-1}(T_{\beta(t)-2})} \tag{11.53}$$

Another possible choice, in which the volatility of the instantaneous forward rate does not vanish, is

$$P_{tT_{\beta(t)-1}} = \frac{1}{1 + (T_{\beta(t)-1} - t)\left(\theta L_{\beta(t)-1}(T_{\beta(t)-2}) + (1 - \theta)L_{\beta(t)}(t)\right)} \tag{11.54}$$

Then, from our choice (11.53), we get that the volatility $\sigma_{\mathrm{P}}^T(t)$ (assuming here a one-dimensional LMM) is

$$\sigma_{\mathrm{P}}^T(t) = -\sum_{i=\beta(t)}^N \frac{\tau_i \sigma_i(t)}{1 + \tau_i L_i(t)} \tag{11.55}$$

with $\sigma_i(t)$ the (normal) volatility of the Libor L_i, i.e., $dL_i(t) = \sigma_i(t)\, dB_t^{T_i}$. As the drift of $\frac{dP_{tT}}{P_{tT}}$ under any risk-neutral measure is $r_t\, dt$, from (11.53), the instantaneous rate r_t is

$$r_t = \frac{L_{\beta(t)-1}(T_{\beta(t)-2})}{1 + (T_{\beta(t)-1} - t)L_{\beta(t)-1}(T_{\beta(t)-2})} \tag{11.56}$$

This expression requires only the computation of the Libors L_i at each tenor date $\{T_{i-1}\}$.[4] Expression (11.41) can then be applied with $\sigma_{\mathrm{P}}^T(t)$ given by (11.55) and

$$\sigma_r^t(s) = \mathbb{E}_s^{\mathbb{Q}^t}[D_s^{B^t} r_t] = \mathbb{E}_s^{\mathbb{Q}^t}\left[\frac{D_s^{B^t} L_{\beta(t)-1}(T_{\beta(t)-2})}{\left(1 + (T_{\beta(t)-1} - t)L_{\beta(t)-1}(T_{\beta(t)-2})\right)^2}\right]$$

One can explicitly compute $D_s^{B^t} S_t$, and assuming here a one-dimensional LMM with a local volatility $\sigma_i(t) = \sigma_i(t, L_i(t))$, (11.41) can then be approximated by (see Section 11.B)

$$P_{0t} \frac{\mathbb{E}^{\mathbb{Q}^t}\left[(r_t - r_t^0)\, \mathbf{1}_{S_t > K}\right]}{\frac{1}{2} K \partial_K^2 \mathcal{C}(t, K)}$$
$$\approx 2\mathbb{E}^{\mathbb{Q}^t}\left[V_t\left(\rho U_t^{\beta(t)-1} + \Theta_t^{\beta(t)-1} \Xi_t^{\beta(t)-1} - \Lambda_t^{\beta(t)-1}\right)\Big| S_t = K\right] \tag{11.57}$$

---

[4]Note that in a Markov functional model (see for example [138]), Libors $L_i(t) \equiv L_i(t, B_t^T)$ are functions of a one-dimensional $\mathbb{Q}^T$-Brownian motion B_t^T and are (usually) stored as splines at each tenor date.

with

$$\frac{dV_t}{V_t} = S_t \partial_S \sigma(t, S_t) a_t \left(dW_t - a_t \sigma(t, S_t) \, dt \right), \qquad V_0 = 1$$

$$dU_t^{\beta(T)-1} = \sigma_{\beta(T)-1}(t) a_t \frac{\sigma(t, S_t)}{V_t} \, dt, \qquad U_0^{\beta(T)-1} = 0$$

$$\frac{dR_t^{\beta(T)-1}}{R_t^{\beta(T)-1}} = \partial_L \beta(T)-1 \sigma_{\beta(T)-1}(t) \, dB_t^{T_{\beta(T)-1}}, \qquad R_0^{\beta(T)-1} = 1$$

$$d\Theta_t^{\beta(T)-1} = \frac{R_t^{\beta(T)-1}}{V_t} \, dt, \qquad \Theta_0^{\beta(T)-1} = 0$$

$$d\Xi_t^{\beta(T)-1} = \frac{(\sigma_t^{\beta(T)-1})^2}{R_t^{\beta(T)-1}} \, dt, \qquad \Xi_0^{\beta(T)-1} = 0$$

$$d\Lambda_t^{\beta(T)-1} = \Theta_t^{\beta(T)-1} \, d\Xi_t^{\beta(T)-1}, \qquad \Lambda_0^{\beta(T)-1} = 0$$

This expression requires the simulation of the processes f_t, L_t^i, V_t, R_t^i and the integrals U_t^i, Θ_t^i, Ξ_t^i, Λ_t^i.

11.8   The particle method: Numerical tests

11.8.1   Hybrid Ho-Lee/Dupire model

We consider a local volatility model ($a_t \equiv 1$) with stochastic interest rates where the short rate follows a Ho-Lee model, for which the volatility $\sigma_r(s) = \sigma_r$ is a constant. A bond of maturity T is given in this model by

$$P_{tT} = \frac{P_{0T}^{\text{mkt}}}{P_{0t}^{\text{mkt}}} \exp\left(\frac{\sigma_r^2 (T-t)^2 t}{2} - \sigma_r (T-t) B_t^T \right) \qquad (11.58)$$

with a volatility $\sigma_P^T(t) = -\sigma_r(T-t)$. From (11.52), the local volatility is

$$\sigma\left(t, K, \mathbb{Q}_t^T\right)^2 = \sigma_{\text{Dup}}(t, K)^2 - 2\rho\sigma_r \frac{\mathbb{E}^{\mathbb{Q}^T}[P_{tT}^{-1} V_t U_t | S_t = K]}{\mathbb{E}^{\mathbb{Q}}[P_{tT}^{-1} | S_t = K]}$$

$$- 2\sigma_r^2 \frac{\mathbb{E}^{\mathbb{Q}^T}[P_{tT}^{-1} V_t \left(t\Theta_t - \Lambda_t \right) | S_t = K]}{\mathbb{E}^{\mathbb{Q}^T}[P_{tT}^{-1} | S_t = K]} \qquad (11.59)$$

with

$$\frac{dV_t}{V_t} = S_t \partial_S \sigma(t, S_t) \left(dW_t^T + \left(\rho \sigma_P^T(t) - \sigma(t, S_t) \right) dt \right), \qquad V_0 = 1$$

$$U_t = \int_0^t \frac{\sigma(s, S_s)}{V_s} \, ds, \qquad \Theta_t = \int_0^t \frac{ds}{V_s}, \qquad \Lambda_t = \int_0^t \Theta_s \, ds$$

As a sanity check, when σ_{Dup} depends only on the time t, we obtain the exact expression for $\sigma(\cdot)$ as expected:

$$\sigma(t)^2 = \sigma_{\text{Dup}}(t)^2 - 2\rho\sigma_r \int_0^t \sigma(s)\,ds - \sigma_r^2 t^2 \tag{11.60}$$

Note that in [53], the local volatility (11.58) is approximated by

$$\sigma(t,K)^2 \approx \sigma_{\text{Dup}}(t,K)^2 - 2\rho\sigma_r \int_0^t \sigma(s,K)\,ds$$

i.e.,

$$\frac{d}{dt}\sigma(t,K)^2 = \frac{d}{dt}\sigma_{\text{Dup}}(t,K)^2 - 2\rho\sigma_r\sigma(t,K), \quad \sigma(0,K) = \sigma_{\text{Dup}}(0,K) \tag{11.61}$$

The approximation consists of first replacing U_t by $\frac{\sigma(t,S_t)t}{V_t}$ and then neglecting the last term of order $\sigma_r^2 t^2$. One computes the approximated $\sigma(t,K)$ using an ODE solver. Practitioners typically use such approximations for $\sigma(t,K)$ whose quality deteriorates significantly far out of the money or for long maturities. We emphasize that even in the simple case where σ_{Dup} depends only on the time t, the above approximation is not exact because of the missing term $\sigma_r^2 t^2$. Our algorithm achieves exact calibration in this case with a single particle as this expression does not depend on conditional expectations anymore (see Equation (11.60)).

We have checked the accuracy of our calibration procedure on the DAX market smile (30-May-11). We have chosen $\sigma_r(s) = 6.3$ bps per day (1% per year) and set the correlation between the stock and the rate to $\rho = 40\%$. The time discretization $\Delta t = t_{k+1} - t_k$ has been set to $\Delta t = 1/100$ and we have used $N = 2^{10}$ or $N = 2^{12}$ particles. After calibrating the model using the particle algorithm, we have computed vanilla smiles using a (quasi) Monte Carlo pricer with $N = 2^{15}$ paths and a time step of $1/250$. Figure 11.1 shows the implied volatility for the market smile (DAX, 30-May-11) and the hybrid Ho-Lee/Dupire model for maturities of 4 years and 10 years. When we use the Malliavin representation, the computational time is around 4 seconds for maturities up to 10 years with $N = 2^{10}$ particles (12 seconds with $N = 2^{12}$). Our algorithm definitively outperforms a (two-dimensional) PDE implementation and has already converged with $N = 2^{10}$ particles. Note that the calibration is also exact using Equation (11.33), i.e., with no use of the Malliavin representation, but with a larger computational time of 8 seconds with $N = 2^{10}$ particles (26 seconds with $N = 2^{12}$). As shown in Table 11.2, the absolute error in implied volatility is of a few basis points. For completeness, we have plotted the smile obtained from the hybrid Ho-Lee/Dupire model without any calibration, i.e., $\sigma(t,K) = \sigma_{\text{Dup}}(t,K)$, to materialize the impact of the stochastic rates.

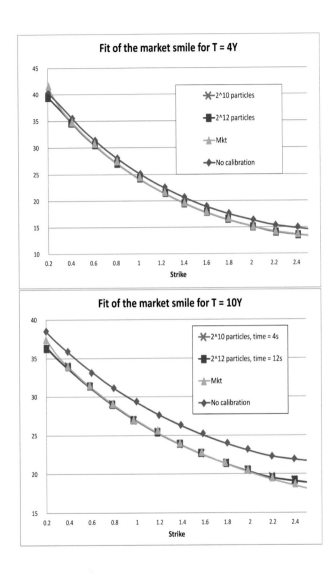

Figure 11.1: Implied volatilities of the DAX as of 30-May-11 for $T = 4Y$ (up) and $T = 10Y$ (bottom). Ho-Lee parameters: $\sigma_r(s) = 6.3$ bps per day, $\rho = 40\%$. Computation time with $\Delta t = 1/100$, $N = 2^{10}$ on a full 10Y implied volatility surface with a Intel Core Duo©, 3 Ghz, 3 GB of Ram: 4s.

Strike	0.5	0.7	0.8	0.9	1	1.1	1.2	1.3	1.5	1.8
w/ Malliavin, 4s	14	10	10	9	8	7	6	5	3	1
w/o Malliavin, 8s	16	8	7	4	1	1	1	3	3	5

Table 11.2: Implied volatilities of the DAX as of 30-May-11 for $T = 10Y$. Errors in bps using the particle method with $N = 2^{10}$ particles.

11.8.2 Bergomi's local stochastic volatility model

As a next example, we consider Bergomi's LSV model [58, 128]:

$$df_t = f_t \sigma(t, f_t) \sqrt{\xi_t^t}\, dW_t$$
$$\xi_t^T = \xi_0^T f^T(t, x_t^T)$$
$$f^T(t, x) = \exp(2\sigma x - 2\sigma^2 h(t, T))$$
$$x_t^T = \alpha_\theta \left((1 - \theta) e^{-k_X(T-t)} X_t + \theta e^{-k_Y(T-t)} Y_t \right)$$
$$\alpha_\theta = \left((1 - \theta)^2 + \theta^2 + 2\rho_{XY}\theta(1 - \theta) \right)^{-1/2}$$
$$dX_t = -k_X X_t\, dt + dW_t^X$$
$$dY_t = -k_Y Y_t\, dt + dW_t^Y$$

where

$$h(t, T) = (1 - \theta)^2 e^{-2k_X(T-t)} \mathbb{E}\left[X_t^2 \right] + \theta^2 e^{-2k_Y(T-t)} \mathbb{E}\left[Y_t^2 \right]$$
$$+ 2\theta(1 - \theta) e^{-(k_X + k_Y)(T-t)} \mathbb{E}[X_t Y_t]$$

$$\mathbb{E}\left[X_t^2 \right] = \frac{1 - e^{-2k_X t}}{2k_X}$$

$$\mathbb{E}\left[Y_t^2 \right] = \frac{1 - e^{-2k_Y t}}{2k_Y}$$

$$\mathbb{E}[X_t Y_t] = \rho_{XY} \frac{1 - e^{-(k_X + k_Y)t}}{k_X + k_Y}$$

This model, commonly used by practitioners, is a variance swap curve model which admits a two-dimensional Markovian representation. We have performed similar tests as in the previous section (see Figure 11.2). The Bergomi model parameters are $\sigma = 200\%$ (the volatility of an ultra-short volatility), $\theta = 22.65\%$, $k_X = 4$, $k_Y = 12.5\%$, $\rho_{XY} = 30\%$, $\rho_{SX} = -50\%$, $\rho_{SY} = -50\%$. The time discretization has been fixed to $\Delta t = 1/100$ and we have used $N = 2^{10}$, $N = 2^{12}$, or $N = 2^{13}$ particles. Figure 11.2 shows the implied volatility for the market smile (DAX, 30-May-11) and the LSV model for maturities of 4 years and 10 years. The computational time is 4 seconds for maturities up to 10 years with $N = 2^{10}$ particles (11 seconds with $N = 2^{12}$). This should be compared with the approximate calibration [128] (detailed in

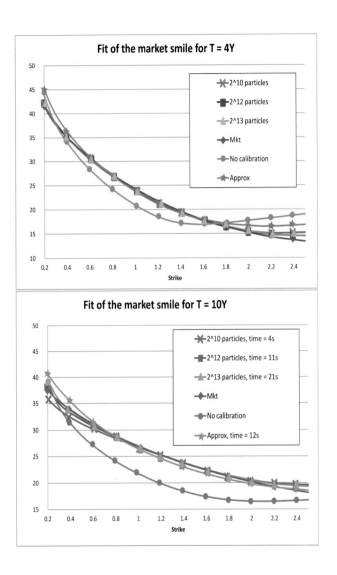

Figure 11.2: Implied volatilities of the DAX as of 30-May-11 for $T = 4Y$ (up) and $T = 10Y$ (bottom). Bergomi model parameters: $\sigma = 200\%$, $\theta = 22.65\%$, $k_X = 4$, $k_Y = 12.5\%$, $\rho_{XY} = 30\%$, $\rho_{SX} = -50\%$, $\rho_{SY} = -50\%$.

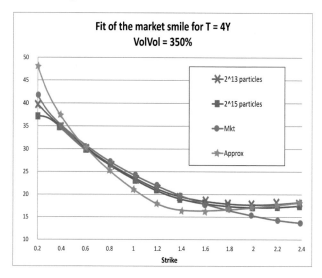

Figure 11.3: Implied volatilities of the DAX as of 30-May-11 for $T = 4Y$. Bergomi model parameters: $\sigma = \mathbf{350\%}$, $\theta = 22.65\%$, $k_X = 4$, $k_Y = 12.5\%$, $\rho_{XY} = 30\%$, $\rho_{SX} = -50\%$, $\rho_{SY} = -50\%$.

Section 11.5) with a computational time around 12 seconds. In order to illustrate that we have used stressed parameters to check the efficiency of our algorithm, we have plotted the smile produced by the naked SVM, which significantly differs from the market smile.

11.8.3 Existence under question

As highlighted in Section 11.3, the existence of LSV models for a given market smile is not at all obvious, although this seems to be a common belief in the quant community. In order to illustrate this mathematical question, we have decided to check our algorithm with a volatility-of-volatility $\sigma = 350\%$ (see Figure 11.3). This large value of volatility-of-volatility is sometimes needed in order to generate typical levels of forward skew for indices, in a model where skew is generated by volatility-of-volatility only.

Our algorithm seems to converge with $N = 2^{13}$ particles around the money, but the market smile is not calibrated. For the maturity $T = 4$ years, we have an error of around 61 bps at the money, which increases to 245 bps for $K = 2$. This may come from numerical issues, or from non-existence of a solution. For comparison, we graph the result of the Markovian projection approximate calibration (see Section 11.5), which definitively breaks down for high levels of volatility-of-volatility.

11.8.4   Hybrid Ho-Lee/Local Bergomi model

We now go one step further in complexity and consider a Bergomi LSV model with Ho-Lee stochastic interest rates. We should emphasize that since this model is driven by four Brownian motions, a calibration relying on a PDE solver is out of the question. The Bergomi model parameters are those used in the previous section. Additionally, we have chosen $\sigma_r(s) = 6.3$ bps per day and set the correlation between the stock and the rate to $\rho = 40\%$. The stochastic volatility and the interest rate are assumed to be uncorrelated. In this case, a PDE solver would be four-dimensional, hence not an option, and no approximate formulas were known at the time of writing this book, to the best of our knowledge. Figure 11.4 shows that the particle method achieves excellent calibration in just a few seconds, or a few tens of seconds. The computational time is 4 seconds for maturities up to 10 years with $N = 2^{10}$ particles (20 seconds with $N = 2^{12}$). The particle method is an important breakthrough, because it is the first method allowing calibration of SIR-LSVMs to market smiles.

Conclusion

At this point, we hope that we have convinced the reader that the particle method is a very powerful, robust, fast (as fast as a standard Monte Carlo), and easy-to-implement method for calibrating a local stochastic volatility model to the market smile of a single asset. One of the great features of the particle method is that it easily handles stochastic interest rates, as well as the addition of extra stochastic factors. In the next chapter, we will show how the particle method can be used in a multi-asset framework to calibrate local correlation models to the market smile of a basket.

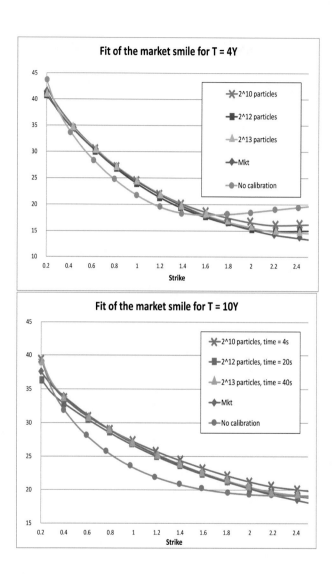

Figure 11.4: Implied volatilities of the DAX as of 30-May-11 for $T = 4\mathrm{Y}$ (up) and $T = 10\mathrm{Y}$ (bottom). Bergomi model parameters: $\sigma = 200\%$, $\theta = 22.65\%$, $k_X = 4$, $k_Y = 12.5\%$, $\rho_{XY} = 30\%$, $\rho_{SX} = -50\%$, $\rho_{SY} = -50\%$. Ho-Lee parameters: $\sigma_{\mathrm{r}}(s) = 6.3$ bps per day, $\rho = 40\%$.

11.9   Exercise: Dynamics of forward variance swaps for the double lognormal SVM

We consider the double lognormal SVM defined by Equations (1.15)–(1.17). We want to compute the dynamics of forward variance swaps defined by $\text{VS}_t^{t_1,\Delta} \equiv \frac{1}{\Delta} \int_{t_1}^{t_1+\Delta} \mathbb{E}^{\mathbb{Q}}[V_s|\mathcal{F}_t]\, ds$.

1. Using Clark-Ocone's formula, prove that

$$
d\mathbb{E}^{\mathbb{Q}}[V_s|\mathcal{F}_t] = \sum_{i=1}^{3} \mathbb{E}^{\mathbb{Q}}[D_t^i V_s|\mathcal{F}_t]\, dW_t^i
$$

   where D^i denotes the Malliavin derivative w.r.t. the Brownian W^i.

2. Compute explicitly $\mathbb{E}^{\mathbb{Q}}[D_t^i V_s|\mathcal{F}_t]$ and deduce the dynamics of forward variance swaps.

3. Perform a similar computation for a local double lognormal SVM.

11.A   Proof of Proposition 11.2

Under the assumptions of Proposition 11.2, let us prove that for all T, K,

$$
P_{0T} \frac{\mathbb{E}^{\mathbb{Q}^T}\left[\left(r_T - r_T^0\right) \mathbf{1}_{S_T > K}\right]}{\frac{1}{2} K \partial_K^2 \mathcal{C}(T,K)} = 2\mathbb{E}^{\mathbb{Q}^T}\left[V_T \left(\rho U_T^T + \Theta_T^T \Xi_T^T - \Lambda_T^T\right) | S_T = K\right]
$$

where the processes V, U^T, Θ^T, Ξ^T, and Λ^T are defined in Proposition 11.2. The change of measure from $\mathbb{Q}$ to $\mathbb{Q}^T$ is given by

$$
\frac{d\mathbb{Q}^T}{d\mathbb{Q}}|\mathcal{F}_t = \frac{\tilde{P}_{tT}}{P_{0T}} = \exp\left(\int_0^t \sigma_{\mathrm{P}}^T(s,r_s)\, dB_s - \frac{1}{2}\int_0^t \sigma_{\mathrm{P}}^T(s,r_s)^2\, ds\right)
$$

so

$$
B_t^T = B_t - \int_0^t \sigma_{\mathrm{P}}^T(s,r_s)\, ds, \qquad Z_t^T = Z_t
$$

are two independent $\mathbb{Q}^T$-Brownian motions. We set $W^T = \rho B_t^T + \sqrt{1-\rho^2} Z_t^T$. From the "disintegration by parts" formula (11.41), in order to prove (11.42), we need to compute $D_s^{B^T} S_T$. This is done in the next two lemmas.

LEMMA 11.1

For all $s \le t \le T$,

$$D_s^{B^T} r_t = R_t^T (R_s^T)^{-1} \sigma_r(s, r_s) \tag{11.62}$$

where

$$R_t^T = \exp\left(\int_0^t \partial_r \left(\mu_r(u, r_u) + \sigma_P^T(u, r_u)\sigma_r(u, r_u)\right) du + \int_0^t \partial_r \sigma_r(u, r_u) \, dB_u^T\right)$$

PROOF From

$$dr_t = \mu_r(t, r_t) \, dt + \sigma_r(t, r_t) \, dB_t$$
$$= \left(\mu_r(t, r_t) + \sigma_P^T(t, r_t)\sigma_r(t, r_t)\right) dt + \sigma_r(t, r_t) \, dB_t^T$$

we deduce that

$$\frac{dD_s^{B^T} r_t}{D_s^{B^T} r_t} = \mathbf{1}_{s \le t} \left(\partial_r \left(\mu_r(t, r_t) + \sigma_P^T(t, r_t)\sigma_r(t, r_t)\right) dt + \partial_r \sigma_r(t, r_t) \, dB_t^T\right)$$

Since $D_s^{B^T} r_s = \sigma_r(s, r_s)$, we eventually get (11.62). □

REMARK 11.5 Note that R_t^T depends on T and is *not* the tangent process of r_t, unless $\partial_r \sigma_P^T(u, r_u) = 0$. Indeed, the tangent process is

$$T_t = \exp\left(\int_0^t \partial_r \mu_r(u, r_u) \, du + \int_0^t \partial_r \sigma_r(u, r_u) \, dB_u\right)$$
$$= \exp\left(\int_0^t \left(\partial_r \mu_r(u, r_u) + \partial_r \sigma_r(u, r_u)\sigma_P^T(u, r_u)\right) du + \int_0^t \partial_r \sigma_r(u, r_u) \, dB_u^T\right)$$
$$= R_t^T \exp\left(-\int_0^t \sigma_r(u, r_u)\partial_r \sigma_P^T(u, r_u) \, du\right)$$

□

LEMMA 11.2

For all $s \le t \le T$,

$$D_s^{B^T} S_t = \rho \sigma(s, S_s) \, a_s \frac{V_t}{V_s} S_t + (R_s^T)^{-1} \sigma_r(s, r_s) V_t S_t \left(\Theta_t^T - \Theta_s^T\right) \tag{11.63}$$

where

$$d\Theta_t^T = \frac{R_t^T}{V_t} \left(1 + \rho \partial_r \sigma_P^T(t, r_t)\sigma(t, S_t) a_t\right) dt$$
$$\frac{dV_t}{V_t} = \sigma'(t, S_t) S_t a_t \left(dW_t - \sigma(t, S_t) a_t \, dt\right)$$

PROOF   For all $s \leq t \leq T$, we have

$$S_t = S_s + \int_s^t S_u \left(r_u + \rho \sigma_P^T(u, r_u) \sigma(u, S_u) a_u \right) du$$

$$+ \int_s^t \sigma(u, S_u) a_u S_u \left(\rho \, dB_u^T + \sqrt{1 - \rho^2} \, dZ_u^T \right)$$

Using the chain rule, we get

$$D_s^{B^T} S_t = \rho \sigma(s, S_s) a_s S_s + \int_s^t D_s^{B^T} S_u \left(r_u + \rho \sigma_P^T(u, r_u) \sigma(u, S_u) a_u \right) du$$

$$+ \int_s^t S_u \left(D_s^{B^T} r_u + \rho \partial_r \sigma_P^T(u, r_u) D_s^{B^T} r_u \sigma(u, S_u) a_u \right.$$

$$\left. + \rho \sigma_P^T(u, r_u) \sigma'(u, S_u) D_s^{B^T} S_u a_u + \rho \sigma_P^T(u, r_u) \sigma(u, S_u) D_s^{B^T} a_u \right) du$$

$$+ \int_s^t \sigma'(u, S_u) D_s^{B^T} S_u a_u S_u \, dW_u^T + \int_s^t \sigma(u, S_u) D_s^{B^T} a_u S_u \, dW_u^T$$

$$+ \int_s^t \sigma(u, S_u) a_u D_s^{B^T} S_u \, dW_u^T$$

For simplicity, we have assumed in Proposition 11.2 that rates and stochastic volatility are uncorrelated, so $D_s^{B^T} a_u \equiv 0$. Plugging this equality, as well as (11.62), into the above formula, we get

$$D_s^{B^T} S_t = \rho \sigma(s, S_s) a_s S_s$$

$$+ \int_s^t D_s^{B^T} S_u \Big\{ \left(r_u + \rho \sigma_P^T(u, r_u) \right) \left(\sigma(u, S_u) + \sigma'(u, S_u) S_u \right) a_u \, du$$

$$+ \left(\sigma(u, S_u) + \sigma'(u, S_u) S_u \right) a_u \, dW_u^T \Big\}$$

$$+ \int_s^t D_s^{B^T} r_u S_u \left(1 + \rho \partial_r \sigma_P^T(u, r_u) \sigma(u, S_u) a_u \right) du$$

$$= \rho \sigma(s, S_s) a_s S_s$$

$$+ \int_s^t D_s^{B^T} S_u \Big\{ r_u \, du + \left(\sigma(u, S_u) + \sigma'(u, S_u) S_u \right) a_u \, dW_u \Big\}$$

$$+ \int_s^t D_s^{B^T} r_u S_u \left(1 + \rho \partial_r \sigma_P^T(u, r_u) \sigma(u, S_u) a_u \right) du$$

As a consequence, $(D_s^{B^T} S_t, t \in [s, T])$ is the unique solution $L = (L_t, s \leq t \leq T)$ to

$$dL_t = L_t \left\{ r_t \, dt + \left(\sigma(t, S_t) + \sigma'(t, S_t) S_t \right) a_t \, dW_t \right\} \tag{11.64}$$

$$+ R_t^T (R_s^T)^{-1} \sigma_r(s, r_s) S_t \left(1 + \rho \partial_r \sigma_P^T(t, r_t) \sigma(t, S_t) a_t \right) dt$$

$$L_s = \rho \sigma(s, S_s) a_s S_s$$

Let us assume that $D_s^{B^T} S_t = A_s Y_t + C_s \theta_t$ for some processes A, Y, C, and θ, which may depend on T. Then

$$dD_s^{B^T} S_t = A_s dY_t + C_s d\theta_t$$
$$= (A_s Y_t + C_s \theta_t) \{ r_t dt + (\sigma(t, S_t) + \sigma'(t, S_t) S_t) a_t \, dW_t \}$$
$$+ R_t^T (R_s^T)^{-1} \sigma_r(s, r_s) S_t \left(1 + \rho \partial_r \sigma_P^T(t, r_t) \sigma(t, S_t) a_t \right) dt$$

Let us further assume that Y follows the dynamics

$$dY_t = Y_t \{ r_t \, dt + (\sigma(t, S_t) + \sigma'(t, S_t) S_t) a_t \, dW_t \}$$

with some nonzero initial condition Y_s, which is left unspecified. Then

$$C_s d\theta_t = C_s \theta_t \frac{dY_t}{Y_t} + R_t^T (R_s^T)^{-1} \sigma_r(s, r_s) S_t \left(1 + \rho \partial_r \sigma_P^T(t, r_t) \sigma(t, S_t) a_t \right) dt$$

In order for $d\theta_t$ to be independent of s, let

$$C_s = (R_s^T)^{-1} \sigma_r(s, r_s)$$

Then

$$d\theta_t = \theta_t \frac{dY_t}{Y_t} + R_t^T S_t \left(1 + \rho \partial_r \sigma_P^T(t, r_t) \sigma(t, S_t) a_t \right) dt$$

Eventually, $D_s^{B^T} S_s = \rho \sigma(s, S_s) a_s S_s$, so the values of A_s, $Y_s \neq 0$, and θ_s must be chosen such that

$$A_s Y_s + C_s \theta_s = \rho \sigma(s, S_s) a_s S_s$$

i.e.,

$$A_s = \frac{\rho \sigma(s, S_s) a_s S_s - C_s \theta_s}{Y_s} = \frac{\rho \sigma(s, S_s) a_s S_s - (R_s^T)^{-1} \sigma_r(s, r_s) \theta_s}{Y_s}$$

Note that A, C, and θ depend on T, but Y does not. With this choice of A, Y, C, θ, for any $s \leq T$, both processes $(D_s^{B^T} S_t, t \in [s, T])$ and $(A_s Y_t + C_s \theta_t, t \in [s, T])$ are solutions to (11.64), so they must coincide:

$$D_s^{B^T} S_t = \rho \sigma(s, S_s) a_s S_s \frac{Y_t}{Y_s} + (R_s^T)^{-1} \sigma_r(s, r_s) \left(\theta_t - \theta_s \frac{Y_t}{Y_s} \right) \quad (11.65)$$

Letting $\Theta_t^T = \theta_t / Y_t$ and $V_t = Y_t / S_t$ completes the proof. ∎

REMARK 11.6 One can easily check on (11.65) that the result does not depend on the particular choice of $Y_s \neq 0$ and θ_s. Indeed, Y_t / Y_s is independent of the specification of Y_s, and

$$M_t = \theta_t - \theta_s \frac{Y_t}{Y_s}$$

is independent of θ_s, because

$$dM_t = M_t \frac{dY_t}{Y_t} + R_t^T S_t \left(1 + \rho \partial_r \sigma_P^T(t, r_t) \sigma(t, S_t) a_t\right) dt$$

with initial condition $M_s = 0$.   □

Eventually, combining (11.41) and (11.63), we get

$$P_{0T} \frac{\mathbb{E}^{\mathbb{Q}^T} \left[\left(r_T - r_T^0\right) \mathbf{1}_{S_T > K}\right]}{\frac{1}{2} K \partial_K^2 \mathcal{C}(T, K)}$$

$$= \frac{2}{K} \int_0^T \mathbb{E}^{\mathbb{Q}^T} \left[\sigma_r^T(s) D_s^{B^T} S_T | S_T = K\right] ds$$

$$= \frac{2}{K} \int_0^T \mathbb{E}^{\mathbb{Q}^T} \left[\sigma_r^T(s) V_T S_T \left(\frac{\rho \sigma(s, S_s) a_s}{V_s} + \sigma_r(s, r_s) \frac{\Theta_T^T - \Theta_s^T}{R_s^T}\right) \Big| S_T = K\right] ds$$

$$= 2 \int_0^T \mathbb{E}^{\mathbb{Q}^T} \left[\sigma_r^T(s) V_T \left(\frac{\rho \sigma(s, S_s) a_s}{V_s} + \sigma_r(s, r_s) \frac{\Theta_T^T - \Theta_s^T}{R_s^T}\right) \Big| S_T = K\right] ds$$

$$= 2 \mathbb{E}^{\mathbb{Q}^T} \left[V_T \left(\rho U_T^T + \Theta_T^T \Xi_T^T - \Lambda_T^T\right) | S_T = K\right]$$

with

$$U_t^T = \int_0^t \frac{\sigma_r^T(s) \sigma(s, S_s) a_s}{V_s} ds$$

$$\Xi_t^T = \int_0^t \frac{\sigma_r^T(s) \sigma_r(s, r_s)}{R_s^T} ds$$

$$\Lambda_t^T = \int_0^t \Theta_s^T d\Xi_s^T$$

and, from (11.62), we have

$$\sigma_r^T(t) = \mathbb{E}_t^{\mathbb{Q}^T} [D_t^{B^T} r_T] = \mathbb{E}_t^{\mathbb{Q}^T} [R_T^T](R_t^T)^{-1} \sigma_r(t, r_t)$$

11.B   Proof of Formula (11.57)

PROOF   Proceeding as in Section 11.A, we get

$$dD_s^{B^T} S_t = \left(S_t D_s^{B^T} r_t + \rho(D_s^{B^T} \sigma_P^T(t)) \sigma(t, S_t) S_t a_t\right) dt + \frac{dY_t}{Y_t} D_s^{B^T} S_t$$

where r_t is given by (11.56) and

$$\sigma_{\mathrm{P}}^T(t) = - \sum_{i=\beta(t)}^{\beta(T)-1} \frac{\hat{\tau}_i \sigma_i(t)}{1 + \hat{\tau}_i L_i(t)}$$

with $\hat{\tau}_i = \tau_i$, $i = \beta(t), \ldots, \beta(T) - 2$, and $\hat{\tau}_{\beta(T)-1} = T - T_{\beta(T)-2}$. Recall that $\sigma_i(t)$ is just a short notation for $\sigma_i(t, L_i(t))$. Here, as an approximation at first order in the volatility, we have $D_s^{B^T} L_i(t) \simeq D_s^{B^{T_i}} L_i(t) = \sigma_i(s) \frac{R_t^i}{R_s^i} \mathbf{1}_{t \geq s}$ with R_t^i the tangent process to L_t^i. Then, by neglecting the factor $1/\left(1 + (T_{\beta(t)-1} - t) L_{\beta(t)-1}(T_{\beta(t)-2})\right)$ in r_t (see (11.56)), we obtain

$$D_s^{B^T} r_t \simeq \sigma_{\beta(t)-1}(s) \frac{R_{T_{\beta(t)-2}}^{\beta(t)-1}}{R_s^{\beta(t)-1}} \mathbf{1}_{T_{\beta(t)-2} \geq s}$$

$$\sigma_{\mathrm{r}}^T(s) \simeq \sigma_{\beta(T)-1}(s) \mathbf{1}_{T_{\beta(T)-2} \geq s}$$

$$D_s^{B^T} \sigma_{\mathrm{P}}^T(t) \simeq - \sum_{i=\beta(t)}^{\beta(T)-1} \partial_{L_i} \left(\frac{\hat{\tau}_i \sigma_i(t)}{1 + \hat{\tau}_i L_i(t)} \right) \sigma_i(s) \frac{R_t^i}{R_s^i} \mathbf{1}_{t \geq s}$$

The reader can easily check that the solution $D_s^{B^T} S_T$ can then be written as

$$D_s^{B^T} S_T = \left(\rho \sigma(s, S_s) S_s a_s - \frac{\sigma_{\beta(T)-1}(s)}{R_s^{\beta(T)-1}} \theta_s^{\beta(T)-1} \right) \frac{Y_T}{Y_s} + \frac{\sigma_{\beta(T)-1}(s)}{R_s^{\beta(T)-1}} \theta_T^{\beta(T)-1}$$

(11.66)

for $s \leq T$, with

$$d\theta_t^{\beta(t)-1} = \frac{dY_t}{Y_t} \theta_t^{\beta(t)-1} + S_t R_{T_{\beta(t)-2}}^{\beta(t)-1} \, dt$$

By plugging (11.66) into (11.41), we get our final expression (11.57). $\quad\square$

11.C   Including (discrete) dividends

We assume that the spot process S_t jumps down by the dividend amount $D(t_i, S_{t_i^-})$ paid at time t_i, and that between dividend dates $\{t_i\}$ it follows a SIR-LSVM under a risk-neutral measure $\mathbb{Q}$

$$S_t = S_0 + \int_0^t r_s S_s \, ds + \int_0^t S_s \sigma(s, S_s) a_s \, dW_s - \sum_{t_i \leq t} D(t_i, S_{t_i^-})$$

Market prices are in agreement with dividends if and only if

$$C(S_0, t_i, K) = \mathbb{E}^{\mathbb{Q}^{\text{mkt}}} \left[D_{0t_i} \left(S_{t_i^-} - D(t_i, S_{t_i^-}) - K \right)^+ \right] \tag{11.67}$$

with D_{st} the discount factor from t to s and $C(S_0, t_i, K)$ the market price of a call option with maturity t_i and strike K. Furthermore, we assume that dividends are part cash, part yield:

$$D(t, S) = \alpha(t)S_0 + \beta(t)S$$

11.C.1   Calibration of the Dupire local volatility

Reference [128] explains how to build an implied volatility surface satisfying (11.67) with deterministic rates by introducing a *continuous* local martingale X_t defined by $S_t = A(t)S_0 + B(t)X_t$ with

$$B(t) = B_0 P_{0t}^{-1} \prod_{i:\, t>t_i} \left(1 - \beta(t_i^-) \right) \tag{11.68}$$

$$A(t) = P_{0t}^{-1} \prod_{i:\, t>t_i} \left(1 - \beta(t_i^-) \right) \left(A_0 - \sum_{k:\, t>t_k} \alpha(t_k) P_{0t_k} \prod_{j:\, t_k>t_j} \left(1 - \beta(t_j^-) \right)^{-1} \right) \tag{11.69}$$

$A(t)$ and $B(t)$ have been chosen in order to make X_t driftless. Using this mapping between the processes S_t and X_t, the fair values of call options with strike K and maturity T written respectively for S_t and X_t are related by

$$\mathcal{C}^X(X_0, K, T) = B(T)^{-1} C(S_0, A(T)S_0 + B(T)K, T) \tag{11.70}$$

We impose the initial condition $S_0 = X_0$. This gives $B_0 = 1 - A_0$. We take $A_0 = 0$.

From the market prices $C(S_0, \cdot, \cdot)$ (see Section 11.C.3 for how to compute this value from the market implied volatility), we can get the prices for vanilla options $\mathcal{C}^X(S_0, K, T)$ written on X from Equation (11.70).

Since X is driftless, the Dupire local volatility for X consistent with these prices is

$$\bar{\sigma}(T, K)^2 = \frac{\partial_T \mathcal{C}^X(X_0, K, T)}{\frac{1}{2} K^2 \partial_K^2 \mathcal{C}^X(X_0, K, T)}$$

Note that the local volatility for S_t is then

$$\sigma_{\text{Dup+Div}}(t, S) = \bar{\sigma} \left(t, \frac{S - A(t)S_0}{B(t)} \right) \frac{S - A(t)S_0}{S} \tag{11.71}$$

11.C.2 Calibration of the SIR-LSVM

By closely following the proof of Proposition 11.1, the SIR-LSVM is calibrated exactly to the market smile if and only if

$$
\frac{1}{2}K^2\partial_K^2\mathcal{C}(t,K)\sigma(t,K)^2\mathbb{E}^{\mathbb{Q}^t}[a_t^2|S_t = K]
$$
$$
= \partial_t\mathcal{C}(t,K) + r_t^0 K\partial_K\mathcal{C}(t,K) - P_{0t}\mathbb{E}^{\mathbb{Q}^t}[(r_t - r_t^0)\,\mathbf{1}_{S_t>K}] \quad (11.72)
$$

between two dividend dates and satisfies the following matching condition at a dividend date:

$$
\mathcal{C}(t_i,K) = \mathbb{E}^{\mathbb{Q}}\left[D_{0t_i}\left(S_{t_i^-} - D(t_i,S_{t_i^-}) - K\right)^+\right] \quad (11.73)
$$

For a local volatility model with dividends and deterministic rate ($a_t \equiv 1$, $r_t \equiv r_t^0$), the necessary and sufficient condition is

$$
\frac{1}{2}K^2\partial_K^2\mathcal{C}(t,K)\sigma_{\text{Dup+Div}}(t,K)^2 = \partial_t\mathcal{C}(t,K) + r_t^0 K\partial_K\mathcal{C}(t,K) \quad (11.74)
$$

with the same matching condition (11.73) and $\sigma_{\text{Dup+Div}}$ given by (11.71). Therefore, by subtracting (11.74) from (11.72), we get that the SIR-LSVM with discrete dividends is calibrated exactly to the market smile if and only if

$$
\sigma(t,K)^2\mathbb{E}^{\mathbb{Q}^t}[a_t^2|S_t = K] = \sigma_{\text{Dup+Div}}(t,K)^2 - P_{0t}\frac{\mathbb{E}^{\mathbb{Q}^t}[(r_t - r_t^0)\,\mathbf{1}_{S_t>K}]}{\frac{1}{2}K\partial_K^2\mathcal{C}(t,K)}
$$

between two dividend dates and satisfies (11.73) at dividend dates t_i, where we have used that the quantity $\mathbb{E}^{\mathbb{Q}}[D_{0t_i}(S_{t_i^-} - D(t_i,S_{t_i^-}) - K)^+]$ is the same when computed in the SIR-LSVM or when computed in the local volatility model with deterministic interest rates, because both models are calibrated to the market smile. Finally, by closely following the proof of Proposition 11.2, we obtain formula (11.42) with σ_{Dup} replaced by $\sigma_{\text{Dup+Div}}$ and the extra processes (V_t, Θ_t^T) replaced by

$$
V_t^{\text{div}} = V_t \prod_i \left(1 + \alpha(t_i)\frac{S_0}{S_{t_i^-}}\right)\mathbf{1}_{t_i \leq t}
$$
$$
\Theta_t^{T,\text{div}} = \Theta_t^T \prod_i \frac{1}{1 - \beta(t_i)}\mathbf{1}_{t_i \leq t}
$$

11.C.3 Approximate formula for the call option in the Black-Scholes model with discrete dividends

In this section, we derive an efficient approximation for a European call option in the case of a lognormal Black-Scholes model with discrete dividends:

$$
S_t = S_0 + \int_0^t r(s)S_s\,ds + \int_0^t S_s\sigma_{\text{BS}}\,dW_s - \sum_{t_i \leq t} D(t_i, S_{t_i^-})
$$

Other approximations can be found in [64]. The exact price can be obtained by solving the forward PDE

$$\partial_T \mathcal{C}(T, K) = \frac{1}{2}\sigma_{\mathrm{BS}}^2 K^2 \partial_K^2 \mathcal{C}(T, K), \qquad \mathcal{C}(0, K) = (S_0 - K)^+$$

with the matching condition at dividend dates:

$$\partial_K^2 \mathcal{C}(t_i^+, K) = \frac{\partial_K^2 \mathcal{C}(t_i^-, \bar{K})}{|1 - \partial_S D(t_i, \bar{K})|}$$

with $\bar{K} - D(t_i, \bar{K}) = K$.

PROPOSITION 11.3
$\mathcal{C}(S_0, K, T)$ *can be approximated by*

$$\mathcal{C}(S_0, K, T) \approx B(T)\bar{\gamma}^{-1}$$
$$\mathrm{BS}\left(\bar{\gamma}^2\sigma_{\mathrm{BS}}^2 \int_0^T \gamma(t)^{-2}\, dt, \frac{K - A(T)S_0}{B(T)}\bar{\gamma} + (1 - \bar{\gamma})S_0 \Big| S_0\right) \quad (11.75)$$

with

$$\bar{\gamma} = \frac{2}{\left(\int_0^T \gamma^{-2}(s)\, ds\right)^2} \int_0^T \gamma(t)^{-1}\, dt \int_0^t \gamma(s)^{-2}\, ds \quad (11.76)$$

and

$$\gamma(t)^{-1} = \left(1 - \sum_{k:t>t_k} \alpha(t_k)P_{0t_k} \prod_{i:\, t_k>t_i} \left(1 - \beta(t_i^-)\right)^{-1}\right) \quad (11.77)$$

PROOF   With $\sigma(t, S) = \sigma_{\mathrm{BS}}$, we obtain from (11.71) that X_t follows the SDE

$$dX_t = \sigma_{\mathrm{BS}}\left(X + \lambda(t)S_0\right) dW_t$$

with $\lambda(t) = A(t)B(t)^{-1}$. This can be written as

$$dX_t = \sigma(t)\left(\gamma(t)X + (1 - \gamma(t))X_0\right) dW_t$$

with $\sigma(t) = \sigma_{\mathrm{BS}}\gamma(t)^{-1}$ and $\gamma(t) = (1 + \lambda(t)\frac{S_0}{X_0})^{-1}$. With the change of time, $t' = \int_0^t \sigma(s)^2\, ds$, we get

$$dX_{t'} = \left(\gamma(t')X + (1 - \gamma(t'))X_0\right) dW_{t'}$$

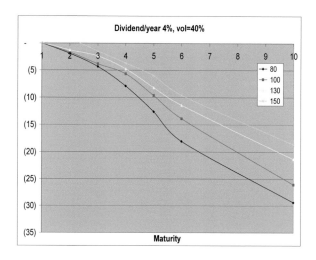

Figure 11.5: Differences in implied volatility $(\times 10^4)$ between the approximate formula (11.75) and a PDE pricer as a function of the maturity (in years); $\sigma_{\mathrm{BS}} = 0.4$; dividends in cash each year $= 4\%$; zero interest rate.

Using skew averaging (see [172]), this diffusion can be approximated by a displaced diffusion model

$$dX = (\bar{\gamma} X + (1 - \bar{\gamma}) X_0) \, dW_t$$

with $\bar{\gamma}$ defined by (11.76). For such a model, the fair value of a call option with strike K and maturity T is given by

$$\mathcal{C}^X(X_0, K, T) = \bar{\gamma}^{-1} \mathrm{BS}\left(\bar{\gamma}^2 T, K\bar{\gamma} + (1 - \bar{\gamma}) X_0 | X_0\right)$$

Using (11.70), we get our final result. $\quad\Box$

The accuracy of our approximation is shown in Figure 11.5.

Chapter 12

Calibration of Local Correlation Models to Market Smiles

The answer is yes, but what was the question?

— Woody Allen

Allowing correlation to be local, i.e., state-dependent, in multi-asset models allows better hedging by incorporating correlation moves in the delta. When options on a basket, be it a stock index, a cross FX rate, or an interest rate spread, are liquidly traded, one may want to calibrate a local correlation to these option prices. In this chapter we discuss calibration methods for local correlation models. In particular, we introduce a new general technique that produces a whole family of local correlation models. With this new family at hand, one can pick a model that not only calibrates to the smile of basket options but also has extra desirable properties, like fitting a view on correlation skew, mimicking historical correlation, or matching prices of exotic options. The models are described by nonlinear SDEs and built using the particle method, which was introduced in Section 10.2. We also show how this technique generalizes at no extra cost to (i) models that combine stochastic interest rates, stochastic dividend yield, local stochastic volatility, and local correlation; and (ii) single-asset path-dependent volatility models. Our numerical tests in the FX context show the wide variety of admissible correlations and give insight on lower bounds/upper bounds on general multi-asset option prices given the smile of a basket and the smiles of its constituents. Parts of this research have been published in [123].

12.1 Introduction

Many practitioners use a multi-asset version of the Dupire local volatility model to price multi-asset derivatives. Most of the time, the correlation matrix is assumed to be constant, e.g., some constant historical correlation ρ^{hist}. In the equity market, since banks usually "sell correlation," i.e., sell products that have a positive sensitivity to correlation, they tend to overprice correlation and often use a convex combination of ρ^{hist} and the matrix $\mathbf{1}$ representing full correlation of the assets, whose all entries are equal to one:

$$\rho = (1 - \lambda)\rho^{\text{hist}} + \lambda\mathbf{1}, \qquad \lambda \in [0, 1]$$

However, such a model is not able to reproduce the market smile of implied volatilities of stock index options: typically, when a constant correlation allows us to match the price of the at-the-money implied volatility of the index, it generates a skew which is much smaller than the market skew (roughly twice as small). Stated otherwise, the smile of index options contains information on how much more correlated its constituents are in a bearish market, and how less correlated they are in a bullish market.

Local correlation models, where the correlation matrix is allowed to be state-dependent:

$$\rho(t, S_t^1, \dots, S_t^N)$$

are able to capture this information. They are of high practical importance, not only because they include correlation variability in option prices, but mainly because they allow better hedging by incorporating correlation moves in the delta. This is crucial for short cross-gamma positions, where an underestimation of correlation in periods of crises yields a daily P&L bleeding that can only be stopped by incurring a large remarking-to-market loss. Many investment banks were affected by this effect in 2008 after the bankruptcy of Lehman Brothers. Like the local volatility model, local correlation models do not aim at describing the real-world dynamics of the assets, but at helping traders risk-manage their correlation positions, especially during crises. Local correlation models are also very useful in the context of foreign exchange (FX) options. They allow us to build models that are consistent with the market smiles of two FX rates, and the market smile of the cross rate. They are also used in interest rates to calibrate to spread option prices.

Let us say that a local correlation model is *admissible* if it calibrates to the market smile of the index. To the best of our knowledge, only two admissible models have been suggested so far in the literature. In both models the correlation matrix $\rho = (1 - \lambda)\rho^0 + \lambda\rho^1$ is assumed to lie on the line defined by two fixed correlation matrices ρ^0 and ρ^1. The first model, proposed by Langnau [152], assumes that the instantaneous variance of a stock index in the multi-asset local volatility–local correlation model is *local in index*, i.e., it depends on the stocks only through the index value. The second model, presented in Reghai [175] and Guyon and Henry-Labordère [122], assumes that the instantaneous correlation itself (or equivalently λ) is local in index. It may seem enough to have those two models at hand. However, they both have drawbacks. First, both models may actually fail to be admissible. In [152] and [122], a unique correlation candidate is explicitly built and may fail to be positive semi-definite (PSD). Then one projects the candidate onto the set of correlation matrices, and the resulting model does not perfectly calibrate. (The correlation candidate has more chances to be PSD in the second model; see Section 12.6). In [175], the correlation matrix is built by solving a fixed point problem that may have no solution. (Precisely, it has no solution when the correlation candidate exhibited in [122] is not PSD.) Second, even if both models are admissible, there is no reason why one would undergo either correlation structure. For instance, the resulting correlation may have a weird skew (dependence on the asset values), or its skew may be far from the one historically observed, or it may generate prices of other options that are far from market quotes.

In this chapter, we build a new family of local correlation models using the particle method. This family is parameterized by two functions a and b that depend on time and on the values of all the underlying assets, and may depend as well on any set of path-dependent variables. Instead of assuming that the basket variance or the correlation (or equivalently λ) is local in index, we assume that $a + b\lambda$ is. The two above models are just two particular points in this family: they correspond to two particular choices of (a, b). Table 12.1 helps compare the two existing models with the new family of models.

Using the new family, one can now design one's favorite local correlation model in order to match a view on a correlation skew, and/or reproduce some features of historical data, and/or calibrate to other option prices, *on top of* reproducing the market smile of the basket, be it a stock index, a cross-FX rate, or an interest rate spread. Not all models in the family are admissible: for a given (a, b), the particle method generates an explicit local correlation candidate, and admissible models correspond to those pairs (a, b) for which the candidate is PSD at all times and for all asset values. If for some time and asset values a correlation candidate fails to be PSD, we project it onto the set of correlation matrices and carry on using the particle method. The resulting model is not perfectly admissible, but the imperfect calibration may be accurate enough for trading purposes. This typically happens when the correlation candidate fails to be PSD only for unlikely asset values (see examples in Section 12.10).

Another way of calibrating jointly to the smile of a basket and to the smiles of its components is described in [147] where Jourdain and Sbai build an incomplete stochastic volatility–stochastic correlation model by following a top-down approach in which the level of a stock index induces some feedback on the dynamics of its constituents. Attempts to approximately calibrate to a triangle of FX market smiles in a symmetric way include [90], where a multi-Heston model with constant correlations is used. Reghai [175] considers the pricing of options on worst-of in a model where the local correlation depends on the stocks only through the worst performance of the basket constituents and suggests a historical calibration procedure. Delanoe [91] addresses the question of calibrating such a model to option prices and discusses stochastic volatility extensions of local correlation models. In the context of constant

Table 12.1: Summary of models and methods for calibrating to basket smile. The correlation matrix $\rho = (1-\lambda)\rho^0 + \lambda\rho^1$ lies on the line defined by two fixed correlation matrices ρ^0 and ρ^1.

	Langnau [152]	Reghai [175]	Guyon and Henry-Labordère [122]	New model [123]
Function of basket value	Basket variance	Correlation, λ	Correlation, λ	$a + b\lambda$
Function of all underlying assets	Correlation, λ	Basket variance	Basket variance	a, b, basket variance and correlation
Possibly function of any path-dependent variable	/	/	/	a, b, basket variance and correlation
Calibration method	Closed form	Fixed point	Particle method	Particle method
Correlation candidate built explicitly	Yes	No	Yes, time step by time step	Yes, time step by time step
Avoids computing implied volatilities	Yes	No	Yes	Yes
Number of degrees of freedom	0	0	0	An infinity: all functions a and b

correlation, Avellaneda *et al.* [40] are first to give the formula for the equivalent local volatility of a basket of stocks (see (12.25)), and estimate it using short term asymptotics at order zero, namely Varadhan's formula and the method of steepest descent. The expansion at order one, as well as an extension to local in index correlation models, are proved in [11]. Durrleman and El Karoui [98] price options written on a domestic asset based on implied volatilities of options on the same asset expressed in a foreign currency and the exchange rate and, given a local correlation, derive explicit formulas to compute the at-the-money implied volatility, skew, convexity, and term structure for short maturities. In [84], Cont and Deguest use a random mixture of reference models to build a multi-asset model consistent with a set of observed single- and multi-asset derivative prices. Austing [35] provides an analytic formula for a joint probability density such that all three market smiles in a FX triangle are repriced. A few stochastic correlation models have also been suggested and analyzed in the literature, including [117, 88, 28].

The chapter is structured as follows. In Section 12.3, we introduce our new family of local correlation models in the simple context of the FX triangle smile calibration problem, briefly recalled in Section 12.2. In Section 12.4 we show how easy it is to build this new family of local correlations step by step from inception to maturity using the particle method. We highlight some key examples in Section 12.5. Important links between the various admissible local correlations are investigated in Section 12.6. In Section 12.7 we give an intuition of the reason why an inadequate joint extrapolation of local volatilities may lead to the non-existence of (strictly) admissible local correlation models. Section 12.8 deals with the impact of correlation on the prices of multi-asset options, with a reminder on implied correlation *à la* Dupire [97] and a new formula *à la* Gatheral [6]. In Section 12.9 we show how to build our new family of local correlation models in the context of the N-dimensional stock index smile calibration problem. In Section 12.10, our numerical examples in the FX context show the wide variety of admissible correlations and give insight on lower bounds and upper bounds on prices of multi-asset options when the smile of a basket and the smiles of its constituents are given. In Section 12.11 we generalize to models that combine stochastic interest rates, stochastic dividend yield, local stochastic volatility, and local correlation. Finally, in Section 12.12, we show how to easily adapt the main idea in this chapter to build single-asset, path-dependent volatility models that calibrate to the smile. The proofs are gathered in Section 12.A.

12.2 The FX triangle smile calibration problem

Let us introduce our new family of local correlation models in the simple context of the FX triangle smile calibration problem. Section 12.9 deals with the general N-dimensional basket case. Let S^1, S^2 be two FX rates, and $S^{12} = S^1/S^2$ be the cross rate. One can think of $S^1 = $ EUR/USD, $S^2 = $ GBP/USD, and $S^{12} = $ EUR/GBP. Assume we know from the market the surfaces of implied volatility for S^1, S^2, and S^{12} until some maturity T, and that those surfaces are jointly arbitrage-free. They correspond to three local volatility surfaces that we denote by $\sigma_1(t, S^1)$, $\sigma_2(t, S^2)$, and $\sigma_{12}(t, S^{12})$. Assume the following model $\mathcal{M}_\rho$ for the dynamics of S^1 and S^2:

$$dS_t^1 = \left(r_t^d - r_t^1\right) S_t^1 \, dt + \sigma_1(t, S_t^1) S_t^1 \, dW_t^1$$
$$dS_t^2 = \left(r_t^d - r_t^2\right) S_t^2 \, dt + \sigma_2(t, S_t^2) S_t^2 \, dW_t^2 \qquad (12.1)$$
$$d\langle W^1, W^2 \rangle_t = \rho(t, S_t^1, S_t^2) \, dt$$

All interest rates are deterministic; both rates S^1 and S^2 follow local volatility dynamics; the two driving processes W^1 and W^2 are Brownian motions under the risk-neutral measure $\mathbb{Q}$ associated to the anchor (domestic) currency (USD in our example); they have a local instantaneous correlation $\rho(t, S_t^1, S_t^2) \in [-1, 1]$.

Let $\mathbb{E}^{\mathbb{Q}^f}$ denote the expectation under the risk-neutral measure $\mathbb{Q}^f$ associated to the foreign currency in S^2 (GBP in our example):

$$\frac{d\mathbb{Q}^f}{d\mathbb{Q}} = \frac{S_T^2}{S_0^2} \exp\left(\int_0^T \left(r_t^2 - r_t^d\right) dt\right) \equiv \frac{S_T^2}{S_0^2} \frac{D_{0T}^d}{D_{0T}^2} \qquad (12.2)$$

where $D_{0T}^i = \exp\left(-\int_0^T r_t^i \, dt\right)$. Then we have the following:

PROPOSITION 12.1

Model $\mathcal{M}_\rho$ is calibrated to the market smile of the cross rate S^{12} if and only if for all $t \in [0, T]$,

$$\mathbb{E}_\rho^{\mathbb{Q}^f}\left[\sigma_1^2(t, S_t^1) + \sigma_2^2(t, S_t^2) - 2\rho(t, S_t^1, S_t^2)\sigma_1(t, S_t^1)\sigma_2(t, S_t^2) \,\Big|\, \frac{S_t^1}{S_t^2}\right] = \sigma_{12}^2\left(t, \frac{S_t^1}{S_t^2}\right) \qquad (12.3)$$

Equation (12.3) is equivalent to

$$\frac{\mathbb{E}^{\mathbb{Q}}_{\rho}\left[S_t^2\left(\sigma_1^2(t,S_t^1)+\sigma_2^2(t,S_t^2)-2\rho(t,S_t^1,S_t^2)\sigma_1(t,S_t^1)\sigma_2(t,S_t^2)\right)\left|\frac{S_t^1}{S_t^2}\right.\right]}{\mathbb{E}^{\mathbb{Q}}_{\rho}\left[S_t^2\left|\frac{S_t^1}{S_t^2}\right.\right]}$$

$$= \sigma_{12}^2\left(t,\frac{S_t^1}{S_t^2}\right) \quad (12.4)$$

Note that the left hand side of Equation (12.3) depends on the correlation in two ways: (i) explicitly through the random variable $\sigma_1^2 + \sigma_2^2 - 2\rho\sigma_1\sigma_2$, and (ii) implicitly through the conditional distribution of (S_t^1, S_t^2) given $\frac{S_t^1}{S_t^2}$ under $\mathbb{Q}^f$. To emphasize point (ii), we have written $\mathbb{E}^{\mathbb{Q}^f}_{\rho}$ instead of $\mathbb{E}^{\mathbb{Q}^f}$.

PROOF   Under Model $\mathcal{M}_{\rho}$, the dynamics of the cross rate $S^{12} = S^1/S^2$ reads

$$\frac{dS_t^{12}}{S_t^{12}} = (r_t^2 - r_t^1)\,dt + \sigma_1(t,S_t^1)\,dW_t^{1,f} - \sigma_2(t,S_t^2)\,dW_t^{2,f} = (r_t^2 - r_t^1)\,dt + a_t\,dW_t^f$$

where $a_t^2 = \sigma_1^2(t,S_t^1) + \sigma_2^2(t,S_t^2) - 2\rho(t,S_t^1,S_t^2)\sigma_1(t,S_t^1)\sigma_2(t,S_t^2)$ and

$$W_t^{1,f} = W_t^1 - \int_0^t \rho(s,S_s^1,S_s^2)\sigma_2(s,S_s^2)\,ds$$

$$W_t^{2,f} = W_t^2 - \int_0^t \sigma_2(s,S_s^2)\,ds$$

$$W_t^f = \int_0^t \frac{\sigma_1(s,S_s^1)\,dW_s^{1,f} - \sigma_2(s,S_s^2)\,dW_s^{2,f}}{a_s}$$

are three $\mathbb{Q}^f$-Brownian motions. To conclude, it is enough to apply Proposition 12.8 in Section 12.A in the particular case where the interest rate and dividend yield are deterministic ($r_t = r_t^0 = r_t^2$ and $q_t = q_t^0 = r_t^1$). □

Let us denote by $\mathcal{C}$ the set of functions $\rho : [0,T] \times \mathbb{R}_+^* \times \mathbb{R}_+^* \to [-1,1]$. Any $\rho \in \mathcal{C}$ satisfying (12.3) will be called an *admissible correlation*. Two important and difficult theoretical questions are the following ones:

1. How do we verify that the three *surfaces* of implied volatility for S^1, S^2, and S^{12} are jointly arbitrage-free? How do we detect joint arbitrages? (Exercise 12.13.1 deals with the simpler case where only one maturity T is considered, and Exercise 12.13.2 with the case where finitely many maturities T are considered.)

2. Assuming no arbitrage, under which condition on $\sigma_1(t,S^1)$, $\sigma_2(t,S^2)$, and $\sigma_{12}(t,S^{12})$ does there exist an admissible correlation?

The non-existence of an admissible correlation may be due to the extrapolations of the three local volatilities (see Section 12.7). In practice, this means that "good" correlation candidates $\rho(t, S^1, S^2)$, such as the ones exhibited in [152, 122] may fail to be true correlations, i.e., to belong to $[-1, 1]$, but only for very small or very large values of S^1 or S^2 or S^{12}. In practice, this may not be a problem: the "good" correlation candidates, when capped at $+1$ and floored at -1, become "almost" admissible correlations, meaning that the smile of the cross rate is correctly reproduced almost everywhere, except maybe very far from the money. One might also want to modify the extrapolations of the local volatilities that appear to be problematic. For instance, in the situation of Figure 12.6, one may want to modify the low-strike extrapolations of σ_1 and σ_2.

In this chapter, we are interested in the following important practical question: How do we build an almost (if not strictly) admissible correlation $\rho(t, S^1, S^2)$ having desirable properties, such as matching a view on some correlation skew, reproducing historical features, calibrating to other option prices, etc.?

REMARK 12.1 We may have started with a general stochastic process (ρ_t) for the correlation, which possibly depends on extra sources of randomness. In this situation, the calibration condition (12.3) still holds with $\rho(t, S_t^1, S_t^2) \equiv \mathbb{E}_{\rho_t}^{\mathbb{Q}^f} \left[\rho_t \, | \, S_t^1, S_t^2 \right] = \mathbb{E}_{\rho_t}^{\mathbb{Q}} \left[\rho_t \, | \, S_t^1, S_t^2 \right]$. This result is not completely trivial, as one needs to show that $\mathbb{E}_{\rho_t}^{\mathbb{Q}^f}$ can be replaced by $\mathbb{E}_{\rho}^{\mathbb{Q}^f}$. This follows from Gyöngy's theorem; see Section 12.8.2 for a simple derivation. As a consequence, if there exists an admissible correlation process ρ_t, then there exists an admissible local correlation $\rho(t, S^1, S^2)$: as far as calibration to the market smile of the cross rate is concerned, assuming a local correlation $\rho(t, S^1, S^2)$ is not restrictive. □

12.3 A new representation of admissible correlations

To answer the above question, we now introduce a new representation of admissible correlations. We say that a function is *local in X* if it is a function of (t, X) only, say $f(t, X)$. When $X = S^1/S^2$, we also say *local in cross*. The idea is to consider all functions $a(t, S^1, S^2)$ and $b(t, S^1, S^2)$ such that $a(t, S^1, S^2) + b(t, S^1, S^2)\rho(t, S^1, S^2)$ is local in cross.

PROPOSITION 12.2 Local in cross $a + b\rho$ representation of admissible correlations

Let $\rho \in \mathcal{C}$. It is an admissible correlation if and only if there exists two functions $a(t, S^1, S^2)$ and $b(t, S^1, S^2)$ such that b does not vanish and

$$\rho(t, S_t^1, S_t^2) = \frac{1}{b(t, S_t^1, S_t^2)} \times$$

$$\left\{ \frac{\mathbb{E}_\rho^{\mathbb{Q}^f} \left[\sigma_1^2(t, S_t^1) + \sigma_2^2(t, S_t^2) + 2\frac{a(t, S_t^1, S_t^2)}{b(t, S_t^1, S_t^2)} \sigma_1(t, S_t^1)\sigma_2(t, S_t^2) \left| \frac{S_t^1}{S_t^2} \right. \right] - \sigma_{12}^2\left(t, \frac{S_t^1}{S_t^2}\right)}{2\mathbb{E}_\rho^{\mathbb{Q}^f} \left[\frac{\sigma_1(t, S_t^1)\sigma_2(t, S_t^2)}{b(t, S_t^1, S_t^2)} \left| \frac{S_t^1}{S_t^2} \right. \right]} \right.$$

$$\left. - a(t, S_t^1, S_t^2) \right\} \quad (12.5)$$

We will denote by $\rho_{(a,b)}$ a solution to (12.5).

PROOF Let $\rho \in \mathcal{C}$ be an admissible correlation. Let us pick two functions $a(t, S^1, S^2)$ and $b(t, S^1, S^2)$ such that b does not vanish and

$$a(t, S^1, S^2) + b(t, S^1, S^2)\rho(t, S^1, S^2) \equiv f\left(t, \frac{S^1}{S^2}\right)$$

is local in cross. We can always do so, by choosing for instance $b \equiv 1$ and $a(t, S^1, S^2) = f\left(t, \frac{S^1}{S^2}\right) - \rho(t, S^1, S^2)$ for some function f. Then

$$\sigma_{12}^2\left(t, \frac{S_t^1}{S_t^2}\right) = \mathbb{E}_\rho^{\mathbb{Q}^f}\left[\sigma_1^2(t, S_t^1) + \sigma_2^2(t, S_t^2) - 2\rho(t, S_t^1, S_t^2)\sigma_1(t, S_t^1)\sigma_2(t, S_t^2) \left| \frac{S_t^1}{S_t^2} \right. \right]$$

$$= \mathbb{E}_\rho^{\mathbb{Q}^f}\left[\sigma_1^2(t, S_t^1) + \sigma_2^2(t, S_t^2) + 2\frac{a(t, S_t^1, S_t^2)}{b(t, S_t^1, S_t^2)}\sigma_1(t, S_t^1)\sigma_2(t, S_t^2) \left| \frac{S_t^1}{S_t^2} \right. \right]$$

$$-2\left(a + b\rho\right)\left(t, \frac{S_t^1}{S_t^2}\right)\mathbb{E}_\rho^{\mathbb{Q}^f}\left[\frac{\sigma_1(t, S_t^1)\sigma_2(t, S_t^2)}{b(t, S_t^1, S_t^2)} \left| \frac{S_t^1}{S_t^2} \right. \right]$$

As a consequence ρ satisfies (12.5). Conversely, if $\rho \in \mathcal{C}$ satisfies (12.5), then it satisfies (12.3) and is thus an admissible correlation. □

We call (12.5) the *local in cross $a + b\rho$ representation* of admissible correlations. Two common approaches for trying to build admissible correlations correspond to two special cases of this assumption: when $a \equiv 0$ and $b \equiv 1$, one assumes that the correlation itself is local in cross (see [122, 175]); when $a = \sigma_1^2 + \sigma_2^2$ and $b = -2\sigma_1\sigma_2$, one assumes that the instantaneous variance of the cross rate is local in cross (see [150, 152]). We will come back to both examples and introduce new ones in Sections 12.5 and 12.10.

Our main question can now be restated as follows: Can we find a pair of functions (a, b) such that $\rho_{(a,b)}$ is (at least almost) admissible, and on top of that, a view on some correlation skew is matched, or some historical features of correlation are reproduced, or extra option prices are calibrated, etc.?

Note that (12.5) is a circular equation: the right-hand side of (12.5) depends on $\rho_{(a,b)}$ through the two conditional expectations. To the best of our knowledge, the existence of the nonlinear stochastic differential equations (SDEs), or McKean SDEs, describing the calibrated models

$$dS_t^1 = \left(r_t^d - r_t^1\right) S_t^1 \, dt + \sigma_1(t, S_t^1) S_t^1 \, dW_t^1$$

$$dS_t^2 = \left(r_t^d - r_t^2\right) S_t^2 \, dt + \sigma_2(t, S_t^2) S_t^2 \, dW_t^2$$

$$d\langle W^1, W^2 \rangle_t = \left(\frac{\mathbb{E}^{\mathbb{Q}}\left[S_t^2 \left(\sigma_1^2 + \sigma_2^2 + 2\frac{a}{b}\sigma_1\sigma_2\right) \Big| \frac{S_t^1}{S_t^2} \right] - \sigma_{12}^2\left(t, \frac{S_t^1}{S_t^2}\right) \mathbb{E}^{\mathbb{Q}}\left[S_t^2 \Big| \frac{S_t^1}{S_t^2} \right]}{2\mathbb{E}^{\mathbb{Q}}\left[S_t^2 \frac{\sigma_1\sigma_2}{b} \Big| \frac{S_t^1}{S_t^2} \right]} \right.$$

$$\left. -a(t, S_t^1, S_t^2) \right) \frac{dt}{b(t, S_t^1, S_t^2)}$$

(where we have dropped the arguments of the functions in the conditional expectations for clarity) is still an open mathematical question (see Section 11.3). In practice, one may try to build a solution $\rho_{(a,b)} \in \mathcal{C}$ using the particle method [122], as we explain in the next section.

REMARK 12.2   As stated above, one can always require that $b \equiv 1$. Consequently, *any* correlation candidate is also of the subtype $\rho_{(a,1)}$:

$$\rho_{(a,1)}(t, S_t^1, S_t^2) = \frac{\mathbb{E}^{\mathbb{Q}^f}_{\rho_{(a,1)}}\left[\sigma_1^2 + \sigma_2^2 + 2a\sigma_1\sigma_2 \Big| \frac{S_t^1}{S_t^2} \right] - \sigma_{12}^2\left(t, \frac{S_t^1}{S_t^2}\right)}{2\mathbb{E}^{\mathbb{Q}^f}_{\rho_{(a,1)}}\left[\sigma_1\sigma_2 \Big| \frac{S_t^1}{S_t^2} \right]} - a(t, S_t^1, S_t^2)$$

$$(12.6)$$

The advantage of dealing with $a + b\rho$ instead of $a + \rho$ is that it includes the common approach where $\sigma_1^2 + \sigma_2^2 - 2\rho\sigma_1\sigma_2$ is assumed to be local in cross, with both $a = \sigma_1^2 + \sigma_2^2$ and $b = -2\sigma_1\sigma_2$ being *independent of* ρ. In general one cannot require that $a \equiv 0$, because it would require that $\rho(t, S^1, S^2) = 0 \Rightarrow \rho(t, \lambda S^1, \lambda S^2) = 0$ for all $\lambda > 0$. Any admissible correlation satisfying the above condition is also of the subtype $\rho_{(0,b)}$:

$$\rho_{(0,b)}(t, S_t^1, S_t^2) = \frac{\mathbb{E}^{\mathbb{Q}^f}_{\rho_{(0,b)}}\left[\sigma_1^2(t, S_t^1) + \sigma_2^2(t, S_t^2) \Big| \frac{S_t^1}{S_t^2} \right] - \sigma_{12}^2\left(t, \frac{S_t^1}{S_t^2}\right)}{2b(t, S_t^1, S_t^2)\mathbb{E}^{\mathbb{Q}^f}_{\rho_{(0,b)}}\left[\frac{\sigma_1(t,S_t^1)\sigma_2(t,S_t^2)}{b(t,S_t^1,S_t^2)} \Big| \frac{S_t^1}{S_t^2} \right]} \quad (12.7)$$

In particular, any non-vanishing admissible correlation is also of the subtype $\rho_{(0,b)}$.   □

REMARK 12.3   Our method allows us to consider more general models where the instantaneous correlation depends on path-dependent variables as well. For instance, we can handle situations where ρ depends not only on (t, S_t^1, S_t^2) but also on the running averages $\frac{1}{t}\int_0^t S_u^1 du$ and $\frac{1}{t}\int_0^t S_u^2 du$, or on the running minimums and maximums of S^1 and S^2, or on the realized correlation over the past few days, or on the realized volatilities over the past few days, etc. All one has to do is to also include the path-dependent variables in the arguments of the functions a and b. □

12.4 The particle method for local correlation

The particle method for solving various smile calibration problems, including calibration of local stochastic volatility models, with or without stochastic interest rates, and of local correlation models, has been presented in [122]. It was also used in [147] to calibrate a model coupling an index and its constituents. In the context presented in Section 12.3, the particle algorithm can be described as follows. Let $\{t_k\}$ denote a time discretization of $[0, T]$. We simulate N processes $(S_t^{1,i}, S_t^{2,i})_{1 \le i \le N}$ starting from (S_0^1, S_0^2) at time 0 using N independent Brownian motions *under the domestic measure* $\mathbb{Q}$ as follows:

1. Initialize $k = 1$ and set $\rho_{(a,b)}(t, S^1, S^2) = \dfrac{\sigma_1^2(0, S^1) + \sigma_2^2(0, S^2) - \sigma_{12}^2\left(0, \frac{S^1}{S^2}\right)}{2\sigma_1(0, S^1)\sigma_2(0, S^2)}$ for all $t \in [t_0 = 0; t_1]$. (At $t = 0$, no conditional expectation is computed so $\rho_{(a,b)}$ does not depend on (a, b).)

2. Simulate $(S_t^{1,i}, S_t^{2,i})_{1 \le i \le N}$ from t_{k-1} to t_k using a discretization scheme, say, a log-Euler scheme.

3. For all S^{12} in a grid G_{t_k} of cross-rate values, compute

$$
\begin{aligned}
Y_{t_k}^i = {}& \sigma_1^2(t_k, S_{t_k}^{1,i}) + \sigma_2^2(t_k, S_{t_k}^{2,i}) \\
&+ 2\frac{a(t_k, S_{t_k}^{1,i}, S_{t_k}^{2,i})}{b(t_k, S_{t_k}^{1,i}, S_{t_k}^{2,i})}\sigma_1(t_k, S_{t_k}^{1,i})\sigma_2(t_k, S_{t_k}^{2,i})
\end{aligned}
$$

$$
E_{t_k}^{\mathrm{num}}(S^{12}) = \frac{\sum_{i=1}^N S_{t_k}^{2,i} Y_{t_k}^i \delta_{t_k,N}\left(\frac{S_{t_k}^{1,i}}{S_{t_k}^{2,i}} - S^{12}\right)}{\sum_{i=1}^N S_{t_k}^{2,i} \delta_{t_k,N}\left(\frac{S_{t_k}^{1,i}}{S_{t_k}^{2,i}} - S^{12}\right)}
$$

$$
E_{t_k}^{\mathrm{den}}(S^{12}) = \frac{\sum_{i=1}^N S_{t_k}^{2,i} \frac{\sigma_1(t_k, S_{t_k}^{1,i})\sigma_2(t_k, S_{t_k}^{2,i})}{b(t_k, S_{t_k}^{1,i}, S_{t_k}^{2,i})} \delta_{t_k,N}\left(\frac{S_{t_k}^{1,i}}{S_{t_k}^{2,i}} - S^{12}\right)}{\sum_{i=1}^N S_{t_k}^{2,i} \delta_{t_k,N}\left(\frac{S_{t_k}^{1,i}}{S_{t_k}^{2,i}} - S^{12}\right)}
$$

$$
f(t_k, S^{12}) = \frac{E_{t_k}^{\mathrm{num}}(S^{12}) - \sigma_{12}^2\left(t_k, S^{12}\right)}{2E_{t_k}^{\mathrm{den}}(S^{12})}
$$

interpolate and extrapolate $f(t_k, \cdot)$, for instance using cubic splines, and, for all $t \in [t_k, t_{k+1}]$, set

$$\rho_{(a,b)}(t, S^1, S^2) = \frac{1}{b(t, S^1, S^2)} \left(f\left(t_k, \frac{S^1}{S^2}\right) - a(t, S^1, S^2) \right)$$

4. Set $k := k + 1$. Iterate steps 2 and 3 up to the maturity date T.

Here, $\delta_{t,N}(x) = \frac{1}{h_{t,N}} K\left(\frac{x}{h_{t,N}}\right)$ is an approximation of the delta Dirac function; K is a fixed, symmetric, nonnegative kernel; $h_{t,N}$ is a bandwidth that tends to zero as N grows to infinity. $E_t^{\mathrm{num}}(S^{12})$ and $E_t^{\mathrm{den}}(S^{12})$ approximate the conditional expectations

$$\mathbb{E}^{\mathbb{Q}^f}_{\rho_{(a,b)}} \left[\sigma_1^2(t, S_t^1) + \sigma_2^2(t, S_t^2) + 2\frac{a(t, S_t^1, S_t^2)}{b(t, S_t^1, S_t^2)} \sigma_1(t, S_t^1)\sigma_2(t, S_t^2) \,\bigg|\, \frac{S_t^1}{S_t^2} = S^{12} \right]$$

and

$$\mathbb{E}^{\mathbb{Q}^f}_{\rho_{(a,b)}} \left[\frac{\sigma_1(t, S_t^1)\sigma_2(t, S_t^2)}{b(t, S_t^1, S_t^2)} \,\bigg|\, \frac{S_t^1}{S_t^2} = S^{12} \right]$$

respectively. Alternative methods for estimating such conditional expectations include B-spline techniques as explained in a recent work by Corlay [85]. Implementation details can be found in Section 12.10.

12.5 Some examples of pairs of functions (a, b)

As already mentioned in Section 12.3, the two existing approaches for trying to build admissible correlations are special cases of the local in cross $a + b\rho$ representation:

- $a \equiv 0$ and $b \equiv 1$: In this case [122, 175], one assumes that the correlation itself is local in cross:

$$\rho_{(0,1)}(t, S_t^1, S_t^2) = \frac{\mathbb{E}^{\mathbb{Q}^f}_{\rho_{(0,1)}} \left[\sigma_1^2(t, S_t^1) + \sigma_2^2(t, S_t^2) \,\big|\, \frac{S_t^1}{S_t^2} \right] - \sigma_{12}^2\left(t, \frac{S_t^1}{S_t^2}\right)}{2\mathbb{E}^{\mathbb{Q}^f}_{\rho_{(0,1)}} \left[\sigma_1(t, S_t^1)\sigma_2(t, S_t^2) \,\big|\, \frac{S_t^1}{S_t^2} \right]} \tag{12.8}$$

We then speak of the *local in cross ρ model* or *local in cross correlation model*. If at some date $t < T$, $\rho_{(0,1)}(t, S^1, S^2) \notin [-1, 1]$ for some FX rate values S^1, S^2, then the trial is a failure: $\rho_{(0,1)}$ is not admissible.

- $a = \sigma_1^2 + \sigma_2^2$ and $b = -2\sigma_1\sigma_2$: In this case, one assumes that the instantaneous variance of the cross rate is local in cross. This is in the

Table 12.2: Examples of simple but symmetry-breaking choices of (a, b) and the corresponding correlation candidates.

a	b	$\rho_{(a,b)}(t, S_t^1, S_t^2)$
0	σ_1	$\dfrac{\mathbb{E}_{\rho_{(a,b)}}^{\mathbb{Q}^f}\left[\sigma_1^2(t,S_t^1)+\sigma_2^2(t,S_t^2)\Big\|\frac{S_t^1}{S_t^2}\right]-\sigma_{12}^2\left(t,\frac{S_t^1}{S_t^2}\right)}{2\sigma_1(t,S_t^1)\mathbb{E}_{\rho_{(a,b)}}^{\mathbb{Q}^f}\left[\sigma_2(t,S_t^2)\Big\|\frac{S_t^1}{S_t^2}\right]}$
0	σ_2	$\dfrac{\mathbb{E}_{\rho_{(a,b)}}^{\mathbb{Q}^f}\left[\sigma_1^2(t,S_t^1)+\sigma_2^2(t,S_t^2)\Big\|\frac{S_t^1}{S_t^2}\right]-\sigma_{12}^2\left(t,\frac{S_t^1}{S_t^2}\right)}{2\mathbb{E}_{\rho_{(a,b)}}^{\mathbb{Q}^f}\left[\sigma_1(t,S_t^1)\Big\|\frac{S_t^1}{S_t^2}\right]\sigma_2(t,S_t^2)}$
σ_1^2	$-2\sigma_1\sigma_2$	$\dfrac{\sigma_1^2(t,S_t^1)+\mathbb{E}_{\rho_{(a,b)}}^{\mathbb{Q}^f}\left[\sigma_2^2(t,S_t^2)\Big\|\frac{S_t^1}{S_t^2}\right]-\sigma_{12}^2\left(t,\frac{S_t^1}{S_t^2}\right)}{2\sigma_1(t,S_t^1)\sigma_2(t,S_t^2)}$
σ_2^2	$-2\sigma_1\sigma_2$	$\dfrac{\mathbb{E}_{\rho_{(a,b)}}^{\mathbb{Q}^f}\left[\sigma_1^2(t,S_t^1)\Big\|\frac{S_t^1}{S_t^2}\right]+\sigma_2^2(t,S_t^2)-\sigma_{12}^2\left(t,\frac{S_t^1}{S_t^2}\right)}{2\sigma_1(t,S_t^1)\sigma_2(t,S_t^2)}$

spirit of [152] and has been studied in this FX context in [150]. In this case we speak of the *local in cross volatility model*, and denote by ρ^* the correlation candidate:

$$\rho^*(t, S_t^1, S_t^2) = \frac{\sigma_1^2(t, S_t^1) + \sigma_2^2(t, S_t^2) - \sigma_{12}^2\left(t, \frac{S_t^1}{S_t^2}\right)}{2\sigma_1(t, S_t^1)\sigma_2(t, S_t^2)} \tag{12.9}$$

Note that this is the only situation where no estimation of conditional expectation (given the value of S_t^1/S_t^2) is needed. As a consequence, ρ^* is well defined even if it exits the interval $[-1, 1]$. If at some date $t < T$, $\rho^*(t, S^1, S^2) \notin [-1, 1]$ for some S^1, S^2, then the trial is a failure: ρ^* is not admissible.

Another natural choice of (a, b) is the following:

- $a \equiv 0$ and $b = \sigma_1\sigma_2$: In this case, one assumes that the local covariance $\rho(t, S^1, S^2)\sigma_1(t, S^1)\sigma_2(t, S^2)$ of increments of S^1 and S^2 is local in cross. We then speak of the *local in cross covariance model*. This choice defines a model calibrated to the three FX smiles if and only if

$$\frac{\mathbb{E}_{\rho_{(0,\sigma_1\sigma_2)}}^{\mathbb{Q}^f}\left[\sigma_1^2(t, S_t^1) + \sigma_2^2(t, S_t^2)\Big\|\frac{S_t^1}{S_t^2}\right] - \sigma_{12}^2\left(t, \frac{S_t^1}{S_t^2}\right)}{2\sigma_1(t, S_t^1)\sigma_2(t, S_t^2)} \in [-1, 1]$$

Other simple but symmetry-breaking choices of (a, b) are given in Table 12.2, together with the corresponding correlation candidates.

REMARK 12.4   Had we only considered the types $\rho_{(a,1)}$ (which span the space of all admissible correlations), by blindly applying (12.6), we would have got for the local in cross volatility assumption $a = \sigma_1^2 + \sigma_2^2 - 2\rho\sigma_1\sigma_2 - \rho$ and the following correlation candidate

$$\rho_{(a,1)}(t, S_t^1, S_t^2) = \frac{\mathbb{E}_{\rho_{(a,1)}}^{\mathbb{Q}^f}\left[\sigma_1^2 + \sigma_2^2 + 2a\sigma_1\sigma_2 \left|\frac{S_t^1}{S_t^2}\right.\right] - \sigma_{12}^2\left(t, \frac{S_t^1}{S_t^2}\right)}{2\mathbb{E}_{\rho_{(a,1)}}^{\mathbb{Q}^f}\left[\sigma_1\sigma_2 \left|\frac{S_t^1}{S_t^2}\right.\right]} - a(t, S_t^1, S_t^2)$$

Though we can recover (12.9) from the above equation, it is much more convenient to consider the $a + b\rho$ formulation, because in this case both a and b are independent of ρ.   ▯

REMARK 12.5   In general, the instantaneous volatility of the cross $\sqrt{\sigma_1^2 + \sigma_2^2 - 2\rho\sigma_1\sigma_2}$ in the model depends jointly on S^1 and S^2, and not only on S^{12}, whereas the instantaneous volatility of S^1 (resp. S^2) depends only on S^1 (resp. S^2). This means that in general the model is not symmetric in the three FX rates. If ρ^* takes values in $[-1, 1]$, then the local in cross volatility model is the unique local volatility–local correlation model that is symmetric in the three FX rates. Otherwise, no such model exists. The asymmetry may not necessarily be seen as a drawback: in cases where a currency is "stronger" than the other two, it is natural to choose it as the anchor rate and it might make sense to price and hedge in a model where the volatility of the cross is actually a function of the two rates against the strong currency separately, not only of the cross rate. Note, however, that if one takes EUR or GBP as anchors, i.e., domestic currency, instead of USD, one gets another family of admissible local correlation models, and different prices for exotic options. See for instance [90] for a symmetric way to *approximately* calibrate to a triangle of FX market smiles.   ▯

12.6   Some links between local correlations

Assume that $\rho_{(a,b)}$ is an *admissible* correlation. We can express the affine transform $a + b\rho_{(a,b)}$ of $\rho_{(a,b)}$ as an average of the same affine transform of the correlation candidate ρ^* (even if ρ^* takes values outside $[-1, 1]$):

$$\left(a + b\rho_{(a,b)}\right)\left(t, \frac{S_t^1}{S_t^2}\right) = \frac{\mathbb{E}_{\rho_{(a,b)}}^{\mathbb{Q}^f}\left[(a + b\rho^*)\left(t, S_t^1, S_t^2\right)\frac{\sigma_1(t,S_t^1)\sigma_2(t,S_t^2)}{b(t,S_t^1,S_t^2)} \left|\frac{S_t^1}{S_t^2}\right.\right]}{\mathbb{E}_{\rho_{(a,b)}}^{\mathbb{Q}^f}\left[\frac{\sigma_1(t,S_t^1)\sigma_2(t,S_t^2)}{b(t,S_t^1,S_t^2)} \left|\frac{S_t^1}{S_t^2}\right.\right]}$$

$$\equiv \mathbb{E}_{\rho_{(a,b)}}^{\mathbb{Q}^{\frac{\sigma_1\sigma_2}{b}}}\left[(a + b\rho^*)\left(t, S_t^1, S_t^2\right) \left|\frac{S_t^1}{S_t^2}\right.\right]$$

where

$$\frac{dQ_{b}^{\sigma_1\sigma_2}}{dQ^f} \equiv \frac{\frac{\sigma_1(t,S_t^1)\sigma_2(t,S_t^2)}{b(t,S_t^1,S_t^2)}}{\mathbb{E}_{\rho_{(a,b)}}^{Q^f}\left[\frac{\sigma_1(t,S_t^1)\sigma_2(t,S_t^2)}{b(t,S_t^1,S_t^2)}\right]}$$

In particular, if $\rho_{(0,1)}$ is an admissible correlation, $\rho_{(0,1)}$ is a weighted average of ρ^* on each line where $\frac{S_t^1}{S_t^2}$ is constant:

$$\rho_{(0,1)}\left(t,\frac{S_t^1}{S_t^2}\right) = \frac{\mathbb{E}_{\rho_{(0,1)}}^{Q^f}\left[\rho^*\left(t,S_t^1,S_t^2\right)\sigma_1(t,S_t^1)\sigma_2(t,S_t^2)\left|\frac{S_t^1}{S_t^2}\right.\right]}{\mathbb{E}_{\rho_{(0,1)}}^{Q^f}\left[\sigma_1(t,S_t^1)\sigma_2(t,S_t^2)\left|\frac{S_t^1}{S_t^2}\right.\right]}$$

$$\equiv \mathbb{E}_{\rho_{(a,b)}}^{Q^{\sigma_1\sigma_2}}\left[\rho^*\left(t,S_t^1,S_t^2\right)\left|\frac{S_t^1}{S_t^2}\right.\right] \tag{12.10}$$

This has two consequences:

- If $\rho_{(0,1)}$ is an admissible correlation, then its image, i.e., the range of values it takes, is included in the image of ρ^*.

- The smallest time at which $\rho_{(0,1)}$ fails to be a correlation function

$$\tau_{\rho_{(0,1)}} = \inf\left\{t \in [0,T] \mid \exists S^1, S^2 > 0, \rho_{(0,1)}(t,S^1,S^2) \notin [-1,1]\right\}$$

is larger than or equal to the smallest time τ_{ρ^*} at which ρ^* fails to be a correlation function.

As for the volatility of S^1/S^2, if we denote

$$\sigma_{(a,b)}^2(t,S_t^1,S_t^2) = \sigma_1^2(t,S_t^1) + \sigma_2^2(t,S_t^2) - 2\rho_{(a,b)}(t,S_t^1,S_t^2)\sigma_1(t,S_t^1)\sigma_2(t,S_t^2)$$

we have by construction

$$\sigma_{12}^2\left(t,\frac{S_t^1}{S_t^2}\right) = \mathbb{E}_{\rho_{(a,b)}}^{Q^f}\left[\sigma_{(a,b)}^2(t,S_t^1,S_t^2)\left|\frac{S_t^1}{S_t^2}\right.\right] \tag{12.11}$$

In the particular case where σ_1 and σ_2 depend only on t (no skew on S^1 nor S^2), then (12.11) simply reads

$$\sigma_{12}^2\left(t,\frac{S_t^1}{S_t^2}\right) = \sigma_1(t)^2 + \sigma_2(t)^2 - 2\mathbb{E}_{\rho_{(a,b)}}^{Q^f}\left[\rho_{(a,b)}(t,S_t^1,S_t^2)\left|\frac{S_t^1}{S_t^2}\right.\right]\sigma_1(t)\sigma_2(t)$$

i.e., using (12.8) and (12.9),

$$\rho^*(t,S_t^1,S_t^2) = \rho_{(0,1)}\left(t,\frac{S_t^1}{S_t^2}\right) = \mathbb{E}_{\rho_{(a,b)}}^{Q^f}\left[\rho_{(a,b)}(t,S_t^1,S_t^2)\left|\frac{S_t^1}{S_t^2}\right.\right] \tag{12.12}$$

Note that in this case, all the seven examples of Section 12.5 boil down to the same correlation model, where the correlation is local in cross. In this situation, $\rho_{(0,1)} = \rho^*$ is, among all the admissible correlations $\rho(t, S^1, S^2)$, the one with the smallest image. This means that if $\rho_{(0,1)} = \rho^*$ is not admissible, then no correlation $\rho(t, S^1, S^2)$ is admissible; $\rho_{(0,1)}\left(t, \frac{S_t^1}{S_t^2}\right) = \rho^*\left(t, \frac{S_t^1}{S_t^2}\right) > 1$ corresponds to the situation where

$$|\sigma_1(t) - \sigma_2(t)| > \sigma_{12}\left(t, \frac{S_t^1}{S_t^2}\right)$$

and $\rho_{(0,1)}\left(t, \frac{S_t^1}{S_t^2}\right) = \rho^*\left(t, \frac{S_t^1}{S_t^2}\right) < -1$ corresponds to the situation where

$$\sigma_1(t) + \sigma_2(t) < \sigma_{12}\left(t, \frac{S_t^1}{S_t^2}\right)$$

Equation (12.12) tells us that in the case where S^1 and S^2 have no skew, all admissible correlations have same average value under $\mathbb{Q}^f$ over each line where S^1/S^2 is constant, and this common average value is given by $\rho_{(0,1)}$.

12.7   Joint extrapolation of local volatilities

As stated in Section 12.2, failure to be a correlation function, i.e., the fact that $\tau_\rho < T$, where

$$\tau_\rho = \inf\left\{t \in [0, T] \,|\, \exists S^1, S^2 > 0, \, \rho(t, S^1, S^2) \notin [-1, 1]\right\}$$

may be the consequence of an inadequate joint extrapolation of the three local volatilities σ_1, σ_2, and σ_{12}. To give an intuition of this, let us look at the particular case of the correlation candidate ρ^*. It takes values in $[-1, 1]$ if and only if for all t, S^1, S^2,

$$\left|\sigma_1(t, S^1) - \sigma_2(t, S^2)\right| \leq \sigma_{12}\left(t, \frac{S^1}{S^2}\right) \leq \sigma_1(t, S^1) + \sigma_2(t, S^2)$$

This is equivalent to saying that

$$\underline{\sigma}_{12} \leq \sigma_{12} \leq \overline{\sigma}_{12} \tag{12.13}$$

where the functions $\underline{\sigma}_{12}$ and $\overline{\sigma}_{12}$ are defined by

$$\underline{\sigma}_{12}(t, S^{12}) = \sup_{S^2 > 0} \left|\sigma_1(t, S^2 S^{12}) - \sigma_2(t, S^2)\right|$$

$$\overline{\sigma}_{12}(t, S^{12}) = \inf_{S^2 > 0} \left\{\sigma_1(t, S^2 S^{12}) + \sigma_2(t, S^2)\right\}$$

A first problem is that there is no guarantee that $\underline{\sigma}_{12} \leq \overline{\sigma}_{12}$. If this does not hold, ρ^* is guaranteed to be inadmissible, whatever the local volatility

σ^{12} of the cross rate. For instance, the extrapolations of σ_1 and σ_2 may be such that $\underline{\sigma}_{12}(t, S^{12}) \equiv +\infty$, if at least one of both local volatilities is unbounded, because in the definition of $\underline{\sigma}_{12}$ we take the supremum over *all* values of S^2, even extremely unlikely values. For instance, it is common to build extrapolations where asymptotically the squared local volatility is an affine function of the log-spot. When the asymptotic slopes of σ_1^2 and σ_2^2 differ, then $\underline{\sigma}_{12}(t, S^{12}) \equiv +\infty$. In such a case, ρ^* must cross the $+1$ boundary for S^1 or S^2 far enough from the money. However, this may not be a problem in practice, if ρ^* lies in $[-1, 1]$ for a broad range of likely values of S^1 and S^2. A second problem is that, even if $\underline{\sigma}_{12} \leq \overline{\sigma}_{12}$, the market local volatility of the cross rate may fail to lie in between the two. This may indicate an inadequate joint extrapolation of the three local volatilities involved.

For general (a, b), there is no necessary and sufficient condition as simple as (12.13) for $\tau_{\rho_{(a,b)}}$ to be greater than T, i.e., for $\rho_{(a,b)}$ to be well defined over $[0, T] \times \mathbb{R}_+^* \times \mathbb{R}_+^*$. The above reasoning shows that even with "good" candidates (a, b), one may have $\tau_{\rho_{(a,b)}} = 0$ in theory, because $\rho_{(a,b)}$ exits $[-1, 1]$ for extremely small or large values of S^1, S^2, or S^{12}. However, again, this may not be a problem in practice if, in the Monte Carlo procedure, only a small proportion of paths (if any) reaches the region where $\rho_{(a,b)}$ is capped to 1 or floored to -1. See Section 12.10 for many numerical examples.

12.8 Price impact of correlation

Different choices of admissible $\rho_{(a,b)}$ will lead to different prices for exotic options on S^1 and S^2, while still producing the same prices for vanilla options on S^1, vanilla options on S^2, and vanilla options on the cross rate S^1/S^2. In this section we analyze the impact of $\rho_{(a,b)}$ on the price of options on S^1 and S^2. This helps in developing an intuition of the model, and choosing the right model for pricing and hedging a given option.

12.8.1 The price impact formula

Here we follow a reasoning inspired by El Karoui *et al.* [103] and Dupire [97] to quantify the impact of the correlation model on the price of the option that has payout $g(S_T^1, S_T^2)$ in domestic currency (USD in our example) at maturity T.

PROPOSITION 12.3
Let $\rho_0(t, S^1, S^2)$ be a correlation function, and $P_0(t, S^1, S^2)$ be the corresponding pricing function in domestic currency (Model $\mathcal{M}_{\rho_0}$). We consider a pro-

cess (S_t^1, S_t^2) *whose dynamics derive, not from ρ_0, but from a general correlation process ρ_t (Model $\mathcal{M}_{\rho_t}$). Then under integrability assumptions the price difference between Model $\mathcal{M}_{\rho_t}$ and Model $\mathcal{M}_{\rho_0}$ is the expected value of the integrated discounted tracking error:*

$$D_{0T}^d \mathbb{E}_{\rho_t}^{\mathbb{Q}}[g(S_T^1, S_T^2)] - P_0(0, S_0^1, S_0^2) =$$

$$\mathbb{E}_{\rho_t}^{\mathbb{Q}}\left[\int_0^T D_{0t}^d(\rho_t - \rho_0(t, S_t^1, S_t^2))\sigma_1(t, S_t^1)\sigma_2(t, S_t^2)S_t^1 S_t^2 \partial_{S^1 S^2}^2 P_0(t, S_t^1, S_t^2)\, dt\right]$$

$$(12.14)$$

We have used the notation $\mathbb{E}_{\rho_t}^{\mathbb{Q}}$ to emphasize that the process (S_t^1, S_t^2) is simulated under Model $\mathcal{M}_{\rho_t}$. The instantaneous tracking error at date t

$$\epsilon_t \equiv (\rho_t - \rho_0(t, S_t^1, S_t^2))\sigma_1(t, S_t^1)\sigma_2(t, S_t^2)S_t^1 S_t^2 \partial_{S^1 S^2}^2 P_0(t, S_t^1, S_t^2)$$

consists of the spread of the two correlations times the $\mathcal{M}_{\rho_0}$-cross-gamma $\partial_{S^1 S^2}^2 P_0$, times the product of (normal) volatilities $\sigma_1 \sigma_2 S^1 S^2$, where all terms are evaluated at the spots (S_t^1, S_t^2) defined by Model $\mathcal{M}_{\rho_t}$. The interpretation as an error comes from the fact that $\epsilon_t\, dt$ is the infinitesimal P&L between t and $t + dt$ of a delta-hedged long position in one option, when one uses the pricing function P_0 and the corresponding deltas $\partial_{S^1} P_0$, $\partial_{S^2} P_0$, derived from Model $\mathcal{M}_{\rho_0}$ while the actual dynamics of the assets is given by Model $\mathcal{M}_{\rho_t}$.

PROOF P_0 is solution to the backward PDE

$$(\partial_t + \mathcal{L})P_0 = 0 \tag{12.15}$$

$$P_0(T, S^1, S^2) = g(S^1, S^2) \tag{12.16}$$

where

$$\mathcal{L} = \frac{1}{2}\sigma_1^2(t, S^1)(S^1)^2 \partial_{S^1}^2 + \frac{1}{2}\sigma_2^2(t, S^2)(S^2)^2 \partial_{S^2}^2$$

$$+\rho_0(t, S^1, S^2)\sigma_1(t, S^1)\sigma_2(t, S^2)S^1 S^2 \partial_{S^1 S^2}^2 + (r_t^d - r_t^1)S^1 \partial_{S^1} + (r_t^d - r_t^2)S^2 \partial_{S^2} - r_t^d.$$

Applying Itô's formula to the function P_0 and to the process (S_t^1, S_t^2), and using (12.15) and (12.16), we get

$$D_{0T}^d g(S_T^1, S_T^2) - P_0(0, S_0^1, S_0^2)$$

$$= M_t + \int_0^T D_{0t}^d \left(\partial_t P_0 + \frac{1}{2}\sigma_1^2(t, S_t^1)(S_t^1)^2 \partial_{S^1}^2 P_0\right.$$

$$+\frac{1}{2}\sigma_2^2(t, S_t^2)(S_t^2)^2 \partial_{S^2}^2 P_0 + \rho_t \sigma_1(t, S_t^1)\sigma_2(t, S_t^2)S_t^1 S_t^2 \partial_{S^1 S^2}^2 P_0$$

$$+(r_t^d - r_t^1)S_t^1 \partial_{S^1} P_0 + (r_t^d - r_t^2)S_t^2 \partial_{S^2} P_0 - r_t^d P_0\Big)\, dt$$

$$= M_t + \int_0^T D_{0t}^d(\rho_t - \rho_0)\sigma_1(t, S_t^1)\sigma_2(t, S_t^2)S_t^1 S_t^2 \partial_{S^1 S^2}^2 P_0\, dt$$

where $D_{0T}^d = \exp\left(-\int_0^T r_t^d\, dt\right)$ is the (deterministic) discount factor, we have omitted the arguments (t, S_t^1, S_t^2) of P_0 and its derivatives for the sake of clarity, and

$$M_t = \int_0^T D_{0t}^d \partial_{S^1} P_0(t, S_t^1, S_t^2) \sigma_1(t, S_t^1) S_t^1\, dW_t^1$$

$$+ \int_0^T D_{0t}^d \partial_{S^2} P_0(t, S_t^1, S_t^2) \sigma_2(t, S_t^2) S_t^2\, dW_t^2$$

is a local martingale under the risk-neutral measure $\mathbb{Q}$. Under integrability conditions, it is a true martingale and, taking expectations, we get (12.14). □

Equation (12.14) has several interesting consequences that we address in Sections 12.8.2, 12.8.3, and 12.8.4.

12.8.2 Equivalent local correlation

From (12.14), by conditioning on (S_t^1, S_t^2), we get

$$D_{0T}^d \mathbb{E}_{\rho_t}^{\mathbb{Q}}[g(S_T^1, S_T^2)] - P_0(0, S_0^1, S_0^2) =$$

$$\mathbb{E}_{\rho_t}^{\mathbb{Q}}\left[\int_0^T D_{0t}^d(\rho_{\mathrm{loc}}(t, S_t^1, S_t^2) - \rho_0(t, S_t^1, S_t^2))\right.$$

$$\left. \sigma_1(t, S_t^1)\sigma_2(t, S_t^2) S_t^1 S_t^2 \partial_{S^1 S^2}^2 P_0(t, S_t^1, S_t^2)\, dt \right] \quad (12.17)$$

where

$$\rho_{\mathrm{loc}}(t, S_t^1, S_t^2) = \mathbb{E}_{\rho_t}^{\mathbb{Q}}[\rho_t | S_t^1, S_t^2]$$

is the equivalent local correlation. From Gyöngy's theorem [124], we know that the model, say $\mathcal{M}_{\rho_{\mathrm{loc}}}$, that uses local correlation function ρ_{loc} generates the same distributions for (S_t^1, S_t^2) as Model $\mathcal{M}_{\rho_t}$, for all t. This can be easily rederived by applying (12.17) with the model $\mathcal{M}_{\rho_{\mathrm{loc}}}$ playing the role of $\mathcal{M}_{\rho_0}$:

$$D_{0T}^d \mathbb{E}_{\rho_t}^{\mathbb{Q}}[g(S_T^1, S_T^2)] - P_{\rho_{\mathrm{loc}}}(0, S_0^1, S_0^2) =$$

$$\mathbb{E}_{\rho_t}^{\mathbb{Q}}\left[\int_0^T D_{0t}^d(\rho_{\mathrm{loc}}(t, S_t^1, S_t^2) - \rho_{\mathrm{loc}}(t, S_t^1, S_t^2))\sigma_1\sigma_2 S_t^1 S_t^2 \partial_{S^1 S^2}^2 P_{\rho_{\mathrm{loc}}}\, dt \right] = 0$$

This proves that all vanilla payoffs $g(S_T^1, S_T^2)$ have identical prices in models $\mathcal{M}_{\rho_t}$ and $\mathcal{M}_{\rho_{\mathrm{loc}}}$, i.e., (S_T^1, S_T^2) have identical distributions under $\mathbb{Q}$ in both models.

12.8.3 Implied correlation

Equation (12.14), or equivalently Equation (12.17), also allows us to define the implied correlation. Given a general model $\mathcal{M}_{\rho_t}$ and a payoff $g(S_T^1, S_T^2)$, we define the implied correlation $\rho(T, g)$ as the value of the constant correlation such that the option has same price in Model $\mathcal{M}_{\rho_t}$ and in the model with constant correlation function, i.e., such that

$$\mathbb{E}^{\mathbb{Q}}_{\rho_t}\left[\int_0^T D_{0t}^d(\rho_t - \rho(T, g))\sigma_1(t, S_t^1)\sigma_2(t, S_t^2)S_t^1 S_t^2 \partial^2_{S^1 S^2} P_{\rho(T,g)}(t, S_t^1, S_t^2)\, dt\right] = 0$$

or, equivalently,

$$\rho(T, g) = \frac{\mathbb{E}^{\mathbb{Q}}_{\rho_t}\left[\int_0^T \rho_t D_{0t}^d \sigma_1(t, S_t^1)\sigma_2(t, S_t^2)S_t^1 S_t^2 \partial^2_{S^1 S^2} P_{\rho(T,g)}(t, S_t^1, S_t^2)\, dt\right]}{\mathbb{E}^{\mathbb{Q}}_{\rho_t}\left[\int_0^T D_{0t}^d \sigma_1(t, S_t^1)\sigma_2(t, S_t^2)S_t^1 S_t^2 \partial^2_{S^1 S^2} P_{\rho(T,g)}(t, S_t^1, S_t^2)\, dt\right]}$$

$$(12.18)$$

This is similar to Dupire's expression of implied volatility as a weighted average of the spot volatility [97]. Note that (12.18) is a fixed point equation, because the right-hand side depends on $\rho(T, g)$ as well through the cross-gamma $\partial^2_{S^1 S^2} P_{\rho(T,g)}$. If (12.18) admits a unique solution, then the implied correlation exists and is uniquely defined.

Following Guyon and Henry-Labordère [121], one can estimate the implied correlation by estimating the fixed point of the mapping

$$\rho \mapsto \int_0^T \int_0^\infty \int_0^\infty \rho_{\text{loc}}(t, S^1, S^2)q_\rho(t, S^1, S^2)\, dS^1\, dS^2\, dt$$

where

$$q_\rho(t, S^1, S^2) = \frac{1}{Z}D_{0t}^d \sigma_1(t, S^1)\sigma_2(t, S^2)S^1 S^2 \partial^2_{S^1 S^2} P_\rho(t, S^1, S^2)p(t, S^1, S^2)$$

with $p(t, S^1, S^2)$ the probability density function of (S_t^1, S_t^2) when the correlation is ρ_t, or, equivalently, $\rho_{\text{loc}}(t, S_t^1, S_t^2)$, and where Z is the normalizing constant such that $\int_0^T \int_0^\infty \int_0^\infty q_\rho(t, S^1, S^2)\, dS^1\, dS^2\, dt = 1$. Of course the density $p(t, S^1, S^2)$ is unknown—otherwise we could compute exactly the price of the option. One way to estimate the implied correlation is to compute the fixed point of the approximate mapping where $p(t, S^1, S^2)$ is replaced by some explicit estimate $\hat{p}(t, S^1, S^2)$. In the particular case where the instantaneous volatilities and correlation are constant, the weight $q_\rho(t, S^1, S^2)$ is known explicitly for the payoff $g(S_T^1, S_T^2) = (S_T^1 - K S_T^2)^+$. Figure 12.1 shows the graphs of $(S^1, S^2) \mapsto q_\rho(t, S^1, S^2)$ for increasing values of t, from 0 to T.

Following Gatheral [6] (see also [121]), we can get an alternative expression for the implied correlation by considering the situation where the local correlation

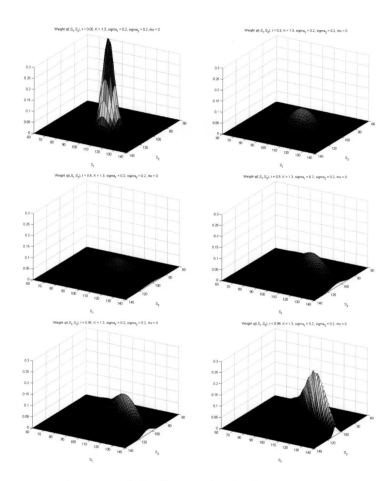

Figure 12.1: Graphs of $(S^1, S^2) \mapsto q_\rho(t, S^1, S^2)$ for increasing values of t. Black-Scholes model: $\sigma_1 = 20\%$, $\sigma_2 = 20\%$, $\rho = 0$, $S_0^1 = 100$, $S_0^2 = 100$. Payoff $g(S_T^1, S_T^2) = (S_T^1 - K S_T^2)^+$, $K = 1.3$, $T = 1$.

function $\rho(t)$ is a deterministic function of time only. The option has the same price in Model $\mathcal{M}_{\rho_t}$ and in this model if and only if

$$\int_0^T D_{0t}^d \mathbb{E}_{\rho_t}^{\mathbb{Q}} \left[(\rho_t - \rho(t)) \sigma_1(t, S_t^1) \sigma_2(t, S_t^2) S_t^1 S_t^2 \partial_{S^1 S^2}^2 P_{\rho(t)}(t, S_t^1, S_t^2) \right] dt = 0$$

There is a unique function $\rho(t) \equiv \rho(t; T, g)$ such that, not only is the time integral zero, but the integrand also vanishes for each time slice t:

$$\rho(t; T, g) = \frac{\mathbb{E}_{\rho_t}^{\mathbb{Q}} \left[\rho_t \sigma_1(t, S_t^1) \sigma_2(t, S_t^2) S_t^1 S_t^2 \partial_{S^1 S^2}^2 P_{\rho(t; T, g)}(t, S_t^1, S_t^2) \right]}{\mathbb{E}_{\rho_t}^{\mathbb{Q}} \left[\sigma_1(t, S_t^1) \sigma_2(t, S_t^2) S_t^1 S_t^2 \partial_{S^1 S^2}^2 P_{\rho(t; T, g)}(t, S_t^1, S_t^2) \right]}$$

Note that this is again a fixed point equation, because the right-hand side depends on $\rho(t; T, g)$ through the cross-gamma $\partial_{S^1 S^2}^2 P_{\rho(t; T, g)}$. In the particular case where $\sigma_1(t, S_t^1) = \sigma_1(t)$ and $\sigma_2(t, S_t^2) = \sigma_2(t)$ depend only on time (no volatility skew on S^1 and S^2), then

$$\rho(T, g) = \frac{\int_0^T \rho(t; T, g) \sigma_1(t) \sigma_2(t) \, dt}{\int_0^T \sigma_1(t) \sigma_2(t) \, dt}$$

$$= \frac{\int_0^T \frac{\mathbb{E}_{\rho_t}^{\mathbb{Q}} \left[\rho_t S_t^1 S_t^2 \partial_{S^1 S^2}^2 P_{\rho(t; T, g)}(t, S_t^1, S_t^2) \right]}{\mathbb{E}_{\rho_t}^{\mathbb{Q}} \left[S_t^1 S_t^2 \partial_{S^1 S^2}^2 P_{\rho(t; T, g)}(t, S_t^1, S_t^2) \right]} \sigma_1(t) \sigma_2(t) \, dt}{\int_0^T \sigma_1(t) \sigma_2(t) \, dt} \qquad (12.19)$$

In the even more particular case where σ_1 and σ_2 are constant, this reads

$$\rho(T, g) = \frac{1}{T} \int_0^T \frac{\mathbb{E}_{\rho_t}^{\mathbb{Q}} \left[\rho_t S_t^1 S_t^2 \partial_{S^1 S^2}^2 P_{\rho(t; T, g)}(t, S_t^1, S_t^2) \right]}{\mathbb{E}_{\rho_t}^{\mathbb{Q}} \left[S_t^1 S_t^2 \partial_{S^1 S^2}^2 P_{\rho(t; T, g)}(t, S_t^1, S_t^2) \right]} \, dt \qquad (12.20)$$

Equations (12.19) and (12.20) are similar to Gatheral's formula for the implied volatility (see [6, 121]). Note, however, that we had to assume no skew on S^1 and S^2 to derive them. When S^1 or S^2 have a skew, we can no longer easily link the implied volatility $\rho(T, g)$ to $\rho(t; T, g)$. Equation (12.20) looks like Equation (12.18), but actually differs in two ways: (i) Equation (12.20) involves a time average of a space average, whereas Equation (12.18) involves a joint time and space average, and (ii) the cross-gammas involved in both equations differ slightly: Equation (12.20) uses the cross-gamma computed with time-dependent correlation $\rho(t; T, g)$, whereas Equation (12.18) uses the cross-gamma computed with constant correlation $\rho(T, g)$.

12.8.4   Impact of correlation on price

For a given option, Equation (12.14) helps us to understand the impact on the option price of a particular choice of local correlation function. Basically, high option prices correspond to local correlation functions that are large in the region where the cross-gamma is positive, and small (i.e., close to -1) in the region where the cross-gamma is negative. Of course the cross-gamma depends on the particular model picked. Equation (12.14) states that to compute the price difference between Model $\mathcal{M}_{\rho_t}$ and reference Model $\mathcal{M}_{\rho_0}$ you may compute the right-hand side expectation where (S_t^1, S_t^2) is simulated under Model $\mathcal{M}_{\rho_t}$ and the cross-gamma is computed under Model $\mathcal{M}_{\rho_0}$. Actually, in the case when the process ρ_t is a local correlation process $\rho_1(t, S_t^1, S_t^2)$, the roles of $\mathcal{M}_{\rho_0}$ and $\mathcal{M}_{\rho_1}$ can be swapped:

$$D_{0T}^d \mathbb{E}_{\rho_0}^{\mathbb{Q}} [g(S_T^1, S_T^2)] - P_1(0, S_0^1, S_0^2) =$$
$$\mathbb{E}_{\rho_0}^{\mathbb{Q}} \left[\int_0^T D_{0t}^d (\rho_0(t, S_t^1, S_t^2) - \rho_1(t, S_t^1, S_t^2)) \right.$$
$$\left. \sigma_1(t, S_t^1)\sigma_2(t, S_t^2) S_t^1 S_t^2 \partial_{S^1 S^2}^2 P_1(t, S_t^1, S_t^2) \, dt \right] \quad (12.21)$$

or, equivalently,

$$P_1(0, S_0^1, S_0^2) - P_0(0, S_0^1, S_0^2) =$$
$$\mathbb{E}_{\rho_0}^{\mathbb{Q}} \left[\int_0^T D_{0t}^d (\rho_1(t, S_t^1, S_t^2) - \rho_0(t, S_t^1, S_t^2)) \right.$$
$$\left. \sigma_1(t, S_t^1)\sigma_2(t, S_t^2) S_t^1 S_t^2 \partial_{S^1 S^2}^2 P_1(t, S_t^1, S_t^2) \, dt \right] \quad (12.22)$$

Stated otherwise, to compute the price difference between two local correlation models $\mathcal{M}_{\rho_1}$ and $\mathcal{M}_{\rho_0}$, you may compute the expected value of the integrated tracking error where:

- either (S_t^1, S_t^2) is simulated under Model $\mathcal{M}_{\rho_1}$ and the cross-gamma is computed under Model $\mathcal{M}_{\rho_0}$,

- or (S_t^1, S_t^2) is simulated under Model $\mathcal{M}_{\rho_0}$ and the cross-gamma is computed under Model $\mathcal{M}_{\rho_1}$.

12.8.5 Uncertain correlation model

The highest possible price for payoff g, given local volatility dynamics for S^1 and S^2, is

$$D_{0T}^d \sup_{\rho_t \in \mathcal{R}} \mathbb{E}_{\rho_t}^{\mathbb{Q}} [g(S_T^1, S_T^2)]$$

where $\mathcal{R}$ denotes the set of all adapted stochastic processes taking values in $[-1, 1]$. It is given by the solution $P(0, S_0^1, S_0^2)$ to the (nonlinear) Hamilton-

Jacobi-Bellman (HJB) equation:

$$\partial_t P + \frac{1}{2}\sigma_1^2(t, S^1)(S^1)^2 \partial_{S^1}^2 P + \frac{1}{2}\sigma_2^2(t, S^2)(S^2)^2 \partial_{S^2}^2 P$$
$$+ \sup_{\rho \in [-1,1]} \left\{ \rho \sigma_1(t, S^1)\sigma_2(t, S^2) S^1 S^2 \partial_{S^1 S^2}^2 P \right\}$$
$$+ (r_t^d - r_t^1) S^1 \partial_{S^1} P + (r_t^d - r_t^2) S^2 \partial_{S^2} P - r_t^d P = 0$$
$$P(T, S^1, S^2) = g(S^1, S^2)$$

that is,

$$\partial_t P + \frac{1}{2}\sigma_1^2(t, S^1)(S^1)^2 \partial_{S^1}^2 P + \frac{1}{2}\sigma_2^2(t, S^2)(S^2)^2 \partial_{S^2}^2 P$$
$$+ \rho \left(\partial_{S^1 S^2}^2 P \right) \sigma_1(t, S^1)\sigma_2(t, S^2) S^1 S^2 \partial_{S^1 S^2}^2 P \qquad (12.23)$$
$$+ (r_t^d - r_t^1) S^1 \partial_{S^1} P + (r_t^d - r_t^2) S \partial_{S^2}^2 P - r_t^d P = 0$$
$$P(T, S^1, S^2) = g(S^1, S^2)$$

where

$$\rho(\Gamma) = \begin{cases} +1 & \text{if } \Gamma \geq 0 \\ -1 & \text{otherwise} \end{cases}$$

As expected from (12.14), the highest option price corresponds to the local correlation function that is worth $+1$ in the region where the cross-gamma is positive, and -1 in the region where the cross-gamma is negative. Here, for consistency, the cross-gamma must be computed within this extremal model $\mathcal{M}_{\mathrm{HJB}}$, i.e., by solving (12.23). Symmetrically, the lower bound

$$D_{0T}^d \inf_{\rho_t \in \mathcal{R}} \mathbb{E}_{\rho_t}[g(S_T^1, S_T^2)]$$

is given by $P(0, S_0^1, S_0^2)$, where P is solution to the (nonlinear) HJB equation

$$\partial_t P + \frac{1}{2}\sigma_1^2(t, S^1)(S^1)^2 \partial_{S^1}^2 P + \frac{1}{2}\sigma_2^2(t, S^2)(S^2)^2 \partial_{S^2}^2 P$$
$$- \rho \left(\partial_{S^1 S^2}^2 P \right) \sigma_1(t, S^1)\sigma_2(t, S^2) S^1 S^2 \partial_{S^1 S^2}^2 P$$
$$+ (r_t^d - r_t^1) S^1 \partial_{S^1} P + (r_t^d - r_t^2) S \partial_{S^2}^2 P - r_t^d P = 0$$
$$P(T, S^1, S^2) = g(S^1, S^2)$$

12.9 The equity index smile calibration problem

Let us now see to what extent the reasoning presented above in Sections 12.2 and 12.3 for the FX smile triangle calibration problem can be extended to the N-dimensional equity index smile calibration problem. Let us consider an index $I_t = \sum_{i=1}^{N} \alpha_i S_t^i$ made of N weighted stocks, each of which is modeled using its own local volatility:

$$dS_t^i = r_t S_t^i \, dt + \sigma_i(t, S_t^i) S_t^i \, dW_t^i, \qquad d\langle W^i, W^j \rangle_t = \rho_{ij}(t, S_t) \, dt \qquad (12.24)$$

The interest rate r_t is deterministic; $\{W^i\}$ denotes a multi-dimensional Brownian motion with an instantaneous correlation function of the time and the N stock values $S_t = (S_t^1, \ldots, S_t^N)$. This model is calibrated to the index smile if and only if (see proof in Section 12.A)

$$I_t^2 \sigma_{\text{Dup}}^I(t, I_t)^2 = \mathbb{E}_\rho \left[\sum_{i,j=1}^{N} \alpha_i \alpha_j \rho_{ij}(t, S_t) \sigma_i(t, S_t^i) \sigma_j(t, S_t^j) S_t^i S_t^j \,\bigg|\, I_t \right] \qquad (12.25)$$

where σ_{Dup}^I denotes the Dupire local volatility of the index. To ease notations, let us denote by

$$v_\rho(t, S_t) = \sum_{i,j=1}^{N} \alpha_i \alpha_j \rho_{ij}(t, S_t) \sigma_i(t, S_t^i) \sigma_j(t, S_t^j) S_t^i S_t^j$$

the instantaneous (normal) variance of the basket of stocks within Model (12.24). Then Equation (12.25) simply reads

$$I_t^2 \sigma_{\text{Dup}}^I(t, I_t)^2 = \mathbb{E}_\rho \left[v_\rho(t, S_t) \,|\, I_t \right] \qquad (12.26)$$

REMARK 12.6 It is a common mistake to believe that in order to exclude arbitrage opportunities we must have the stronger statement $I_t^2 \sigma_{\text{Dup}}^I(t, I_t)^2 = v_\rho(t, S_t)$ at each point in time. This is incorrect in two ways. First, only the conditional expected value of the basket variance given the index value matters. Second, if no ρ satisfies (12.26), it does not mean that arbitrage opportunities exist, but only that prices are inconsistent with local volatilities–local correlation modeling, and that one has to consider more general models, for instance models that include stochastic volatility. ⬚

Let $\mathcal{C}$ denote the set of all correlation matrices, i.e., the set of all real symmetric PSD matrices with all diagonal entries equal to one. Any function $\rho : [0, T] \times \mathbb{R}_+^* \times \mathbb{R}_+^* \to \mathcal{C}$ satisfying (12.26) will be called an *admissible correlation*. We aim at identifying some admissible correlations. Let ρ be an admissible correlation. It is made of $N(N-1)/2$ parameters (the off-diagonal entries) satisfying one scalar equation, so we reduce the dimension of the problem by assuming that $\rho(t, S)$ lies on the line defined by two given correlation matrices $\rho^0(t, S)$ and $\rho^1(t, S)$ that may depend on (t, S) but are usually taken to be constant:

$$\rho(t, S) = (1 - \lambda(t, S))\rho^0(t, S) + \lambda(t, S)\rho^1(t, S), \qquad \lambda(t, S) \in \mathbb{R} \qquad (12.27)$$

When $\lambda = 0$, $\rho = \rho^0$; when $\lambda = 1$, $\rho = \rho^1$. If $\lambda \in [0, 1]$, ρ is guaranteed to be a correlation matrix, because the set of correlation matrices is convex. When ρ^0 (resp. ρ^1) does not belong to the boundary of the set of correlation matrices, ρ may be a correlation matrix even if $\lambda < 0$ (resp. $\lambda > 1$). With this specification of $\rho(t, S)$, (12.26) reads

$$I_t^2 \sigma_{\text{Dup}}^I(t, I_t)^2 = \mathbb{E}_\rho \left[v_{\rho^0}(t, S_t) + \left(v_{\rho^1}(t, S_t) - v_{\rho^0}(t, S_t) \right) \lambda(t, S_t) \middle| I_t \right] \quad (12.28)$$

Similarly to Proposition 12.2, we have the following:

PROPOSITION 12.4
Assume that $\rho(t, S)$, as defined by (12.27), takes values in $\mathcal{C}$. Then it is an admissible correlation if and only if there exist two functions a and b such that b does not vanish and λ satisfies the self-consistency equation

$$\rho \equiv (1 - \lambda)\rho^0 + \lambda\rho^1 \in \mathcal{C}$$

$$\lambda(t, S_t) = \frac{1}{b(t, S_t)} \left(\frac{I_t^2 \sigma_{\text{Dup}}^I(t, I_t)^2 - \mathbb{E}_\rho \left[v_{\rho^0} - \frac{a}{b} \left(v_{\rho^1} - v_{\rho^0} \right) \middle| I_t \right]}{\mathbb{E}_\rho \left[\frac{1}{b} \left(v_{\rho^1} - v_{\rho^0} \right) \middle| I_t \right]} - a(t, S_t) \right)$$

$$(12.29)$$

We denote by $\lambda_{(a,b)}$ a solution to (12.29) and write $\rho_{(a,b)} \equiv (1 - \lambda_{(a,b)})\rho^0 + \lambda_{(a,b)}\rho^1$.

PROOF   Assume that $\rho(t, S) = (1 - \lambda(t, S))\rho^0(t, S) + \lambda(t, S)\rho^1(t, S)$ is an admissible correlation. Let us pick two functions a and b such that b does not vanish and

$$a(t, S_t) + b(t, S_t)\lambda(t, S_t) \equiv f(t, I_t)$$

is local in index, i.e., is a function of (t, I_t) only, say $f(t, I_t)$. We can always do so, by choosing for instance $b \equiv 1$ and $a(t, S_t) = f(t, I_t) - \lambda(t, S_t)$ for some function f. Then

$$I_t^2 \sigma_{\text{Dup}}^I(t, I_t)^2 = (a + b\lambda)(t, I_t)\mathbb{E}_\rho \left[\frac{1}{b(t, S_t)} \left(v_{\rho^1}(t, S_t) - v_{\rho^0}(t, S_t) \right) \middle| I_t \right]$$

$$+ \mathbb{E}_\rho \left[v_{\rho^0}(t, S_t) - \frac{a(t, S_t)}{b(t, S_t)} \left(v_{\rho^1}(t, S_t) - v_{\rho^0}(t, S_t) \right) \middle| I_t \right]$$

and λ satisfies (12.29). Conversely, if a function λ satisfies both conditions in (12.29), then ρ satisfies (12.28) and is an admissible correlation. ⬜

For a given (a, b), $\rho_{(a,b)}$ is guaranteed to be PSD if $\lambda_{(a,b)}$ takes values in $[0, 1]$. We call the resulting model the *local in index $a + b\lambda$ model*. Obviously, it also depends on the choice of ρ^0 and ρ^1.

The first two approaches that have been suggested in the literature for calibrating local correlation models to index option prices correspond to special cases of this formulation:

- $a \equiv 0$ and $b \equiv 1$: In this case one assumes that the linear combination parameter λ itself is local in index. Then [122]

$$\lambda_{(0,1)}(t, S_t) = \frac{I_t^2 \sigma_{\mathrm{Dup}}^I(t, I_t)^2 - \mathbb{E}_{\rho_{(0,1)}} \left[v_{\rho^0}(t, S_t) \big| I_t \right]}{\mathbb{E}_{\rho_{(0,1)}} \left[v_{\rho^1}(t, S_t) - v_{\rho^0}(t, S_t) \big| I_t \right]}$$

and we speak of the *local in index* λ *model*. If at some date $t < T$, $\rho_{(0,1)}(t, S)$ is not a correlation matrix for some S, then the trial is a failure: $\rho_{(0,1)}$ is not admissible. In [175], the same model is investigated but no explicit formula is given for $\lambda_{(0,1)}$; instead, $\lambda_{(0,1)}$ is computed as the fixed point of a mapping that requires computing basket implied volatilities at each iteration, which makes this method slower. Precisely, the mapping admits no fixed point when for some t, S, the candidate $\rho_{(0,1)}$ exhibited in [122] is not PSD.

- $a = v_{\rho^0}$ and $b = v_{\rho^1} - v_{\rho^0}$: In this case one assumes that the instantaneous variance of the index within Model (12.24) is local in index. In this case we denote $\lambda_{(a,b)} = \lambda^*$ [152]:

$$\lambda^*(t, S_t) = \frac{I_t^2 \sigma_{\mathrm{Dup}}^I(t, I_t)^2 - v_{\rho^0}(t, S_t)}{v_{\rho^1}(t, S_t) - v_{\rho^0}(t, S_t)}$$

and we speak of the *local in index volatility model*. This is the only situation where no estimation of conditional expectation (given the value of I_t) is needed. Note that λ^* is well defined even if the corresponding ρ^* is not PSD. If at some date $t < T$, $\rho^*(t, S)$ is not a correlation matrix for some S, then the trial is a failure: ρ^* is not admissible.

Another choice of (a, b) that respects the symmetry of the problem is the following:

- $a \equiv 0$ and $b = v_{\rho^1} - v_{\rho^0}$: In this case

$$\lambda_{(0, v_{\rho^1} - v_{\rho^0})}(t, S_t) = \frac{I_t^2 \sigma_{\mathrm{Dup}}^I(t, I_t)^2 - \mathbb{E}_{\rho_{(0, v_{\rho^1} - v_{\rho^0})}} \left[v_{\rho^0}(t, S_t) \big| I_t \right]}{v_{\rho^1}(t, S_t) - v_{\rho^0}(t, S_t)}$$

REMARK 12.7   As already mentioned in Remark 12.3, our method allows us to handle local correlations that depend on path-dependent variables, like some running averages, running maximums, running minimums, etc. It is enough to add those path-dependent variables to the arguments of the functions a, b, and λ. ⬚

12.10   Numerical experiments on the FX triangle problem

12.10.1   Calibration

We have tested several local $a + b\rho$ models on March 2012 market data involving the three currencies USD, EUR, and GBP, and using USD as domestic currency: $S^1 = $ EUR/USD, $S^2 = $ GBP/USD, $S^{12} = S^1/S^2 = $ EUR/GBP. For simplicity, we have assumed zero interest rates. The three surfaces of local volatilities are shown in Figure 12.2. Different pairs of functions (a, b) are tested. For each pair (a, b), the instantaneous correlation $(S^1, S^2) \mapsto \rho(T, S^1, S^2)$ at maturity (top left), and the instantaneous volatility

$$(S^1, S^2) \mapsto \sqrt{\sigma_1^2(T, S^1) + \sigma_2^2(T, S^2) - 2\rho_{(a,b)}(T, S^1, S^2)\sigma_1(T, S^1)\sigma_2(T, S^2)}$$

of the cross rate at maturity (top right), the repriced smile of the cross rate at maturity T (bottom left), and the scatter plot of Monte Carlo sampled paths (S_T^1, S_T^2) (bottom right) are shown in Figures 12.3 through 12.17. In the case when the instantaneous correlation $\rho_{(a,b)}(t, S^1, S^2)$ takes values above $+1$, we simply cap it to $+1$, we highlight the corresponding (S_T^1, S_T^2) on the scatter plot with circles, and we show the proportion of those paths as a function of time on a fifth graph.

We picked $T = 1$, and used the particle method described in Section 12.4 with $N = 10,000$ Monte Carlo paths and the time step $\Delta t = \frac{1}{80}$. For the nonparametric regressions, we used the quartic kernel $K(x) = (1 - x^2)^2 \mathbf{1}_{\{|x| \leq 1\}}$ and a bandwidth

$$h = \kappa \bar{\sigma}^{12} S_0^{12} \sqrt{\max(t, t_{\min})} N^{-\frac{1}{5}}$$

where $\bar{\sigma}^{12} = 10\%$ is a typical level for the volatility of S^{12}, $t_{\min} = 0.25$ and $\kappa = 3$. The conditional expectations are computed on a grid $G_{S,t}$ of $N_{S,t} = \max(N_S\sqrt{t}, N_S')$ values of the conditioning random variable $S^{12} = S^1/S^2$, with $N_S = 30$ and $N_S' = 15$. We use the 1% and 99% quantiles of the distribution of S_t^{12} as the minimum and maximum values of the grid $G_{S,t}$. Then the function

$$f\left(t, \frac{S^1}{S^2}\right) = a(t, S^1, S^2) + b(t, S^1, S^2)\rho(t, S^1, S^2)$$

is interpolated and extrapolated (in a flat way) using cubic splines.

The bottom left graphs also show the market implied volatilities of the cross rate at maturity, as well as the smiles produced by the constant correlation model for three values of constant correlation: 71%, 72%, and 73%. This allows us to translate the calibration error in terms of correlation points. 72%

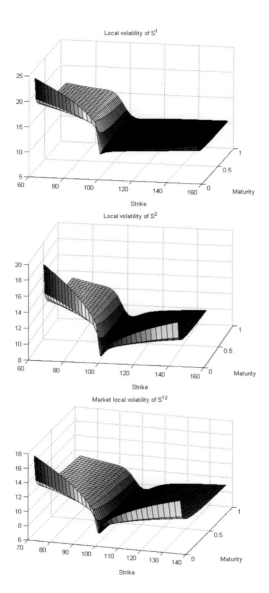

Figure 12.2: Surfaces of local volatilities σ_1, σ_2, and σ_{12}.

is the value of the constant correlation that fits the market value of the ATM implied volatility of the cross rate at maturity.

Figures 12.3 to 12.15 illustrate the variety of (at least almost) admissible correlations. Before we introduce the local in cross $a + b\rho$ representation, one would only choose between two local correlations: the local in cross correlation $\rho_{(0,1)}$ (Figure 12.3) or the correlation ρ^* corresponding to a local in cross volatility of the cross (Figure 12.4). Thanks to our local in cross $a + b\rho$ representation, among the variety of admissible correlations that it produces, one can now pick one's favorite depending one's criterion:

- Match a view on the correlation skew, for instance, match a value of

$$\Delta\rho \equiv \rho(T, 1.05S_0^1, 1.05S_0^2) - \rho(T, 0.95S_0^1, 0.95S_0^2)$$

- Reproduce some features of the historical shape of the correlation between the returns in S^1 and S^2, as a function of the values of S^1 and S^2. Figure 12.18 shows what this shape looks like for EUR/USD and GBP/USD over the periods January 2007–June 2013 (top) and January 2011–June 2013 (bottom).

- Fit the price of options on S^1 and S^2 (other than the payoffs $(S^1 - KS^2)^+$, which are automatically fitted to the market since the correlation is admissible).

For instance, one may fit a negative skew by using $a = 0$ and $b = (S^1S^2)^\alpha$ with $\alpha > 0$; see Figures 12.6 and 12.9. The larger α, the more negative the skew $\Delta\rho$. However, too large values of α produce correlation candidates that are not admissible. For instance, in Figure 12.6, we observe that we had to cap the correlation to 1 for some small values of S^1 and S^2. However, this is still acceptable in practice because only 0.3% of the simulated spots undergo this capped correlation. Conversely, one may fit a positive skew by using a negative value for α; see Figure 12.12.

If one observes that the historical correlation is large when the spots are low, which is typical in equity markets, then one may slightly transform function b and use for instance $b = \min(S^1, S^2)^{2\alpha}$; see Figures 12.7 and 12.10. This has almost no impact on the price of many products (see Table 12.3), but allows us to incorporate correlation impact in the delta hedge and avoids posting a large remarking-to-market loss in case of crisis. If, on the contrary, one wants to decrease correlation for low spot values, one may use, for instance, $b = \max(S^1, S^2)^{2\alpha}$; see Figures 12.8 and 12.11.

Choosing $\rho_{(0,1)}$ implies pricing vanishing correlation skew across lines where S^1/S^2 is constant, and may not be desirable. Choosing ρ^* may imply pricing and hedging with a correlation that varies strongly with the spot values and which is highly asymmetric (see Figure 12.4). As expected from (12.10), the image of $\rho_{(0,1)}$ is much narrower than the image of ρ^*: $\rho_{(0,1)}$ varies much less than ρ^*.

Note that, in our example, the third financially natural correlation, namely the local in cross covariance correlation, is characterized by highly varying correlation and volatility of S^1/S^2, and has to be capped at 1 for S^1 large and S^2 around the money, and for S^2 large and S^1 around the money. However, only around 0.5% of the simulated paths are affected by the cap (see Figure 12.5).

An extreme admissible correlation is shown in Figure 12.13. From (12.12), we know that all admissible correlations approximately share the same average value on each line where S^1/S^2 is constant. (This is exact only when the two rates have no skew, and when the average is taken under $\mathbb{Q}^f$.) The correlation in Figure 12.13 was built so that the local correlation is very high and roughly constant when $\frac{S^1}{S_0^1} + \frac{S^2}{S_0^2}$ is less than 2, and very low and roughly constant when $\frac{S^1}{S_0^1} + \frac{S^2}{S_0^2}$ is greater than 2, and has the correct average value on those lines where the cross is constant. A smoothed version using the tanh function is shown in Figure 12.14.

Eventually, Figure 12.15 shows that our new method allows to build very diverse admissible correlations; here, for instance, a local correlation which is peaked around (S_0^1, S_0^2). Figures 12.16 and 12.17 illustrate that a wrong choice of functions (a, b) can lead to inadmissible correlations, with a high proportion (resp. 22% and 16%) of correlations that have to be capped, resulting in a poor calibration of the smile of the cross rate.

12.10.2 Pricing

To illustrate the impact of the local correlation model on the price of options, we have considered the following three derivative products:

$$\text{Min of calls}: \quad g(S_T^1, S_T^2) = \min\left(\left(\frac{S_T^1}{K^1} - 1\right)^+, \left(\frac{S_T^2}{K^2} - 1\right)^+\right), \quad K^i = S_0^i$$

$$\text{Put on worst}: \quad g(S_T^1, S_T^2) = \left(K - \min\left(\frac{S_T^1}{S_0^1}, \frac{S_T^2}{S_0^2}\right)\right)^+, \quad K = 0.95$$

$$\text{Put on basket}: \quad g(S_T^1, S_T^2) = \left(K - \left(\frac{S_T^1}{S_0^1} + \frac{S_T^2}{S_0^2}\right)\right)^+, \quad K = 1.8$$

The prices are shown in Table 12.3 for different admissible correlations. For each of these products, we can build an intuition of the impact of the local correlation model on the price by looking at the instantaneous correlation surfaces in Figures 12.3–12.14 and using the price impact formula (12.14). To this end, note that the cross-gammas of these options at maturity are simply proportional to the following Dirac masses:

$$\text{Min of calls}: \quad \delta\left(\frac{S^2}{K^2} - \frac{S^1}{K^1}\right) \mathbf{1}_{\left\{\frac{S^1}{K^1} \geq 1\right\}}$$

$$\text{Put on worst}: \quad -\delta\left(\frac{S^2}{S_0^2} - \frac{S^1}{S_0^1}\right) \mathbf{1}_{\left\{\frac{S_T^1}{S_0^1} \leq K\right\}}$$

$$\text{Put on basket}: \quad \delta\left(\frac{S^1}{S_0^1} + \frac{S^2}{S_0^2} - K\right)$$

From (12.14), we expect the higher prices to correspond to local correlation surfaces that are:

- larger in the neighborhood of the half-line $\frac{S^1}{S^2} = \frac{K^1}{K^2}$, $S^1 \geq K^1$, for the min of calls;

- smaller in the neighborhood of the half-line $\frac{S^1}{S^2} = \frac{S_0^1}{S_0^2}$, $S^1 \leq KS_0^1$, for the put on worst;

- larger in the neighborhood of the segment $\frac{S^1}{S_0^1} + \frac{S^2}{S_0^2} = K$, $S^1, S^2 > 0$, for the put on basket.

This is indeed verified. For these products, the highest and lowest prices always correspond to the local in cross covariance correlation (Figure 12.5), and to the extreme correlation of Figure 12.13. This way, we have provided a numerical partial answer to the difficult problem of determining the lower and upper bounds of prices of options on (S^1, S^2) given the three surfaces of implied volatilities on S^1, S^2, and S^1/S^2, and the corresponding models. The answer is only partial because here we have only considered local volatility models (see Section 12.11 for a generalization to stochastic local volatility models) and a small number of admissible local correlations $\rho_{(a,b)}$. Note that the range of prices is already quite large, from 2.37 to 2.91, for instance, for the min of calls, despite the fact that the three surfaces of implied volatilities are calibrated. In Table 12.3 we have also reported prices of a digital call on the cross rate with strike $K = 1.1\frac{S_0^1}{S_0^2}$ and a double-no-touch on the cross rate with barriers $K_1 = 0.9\frac{S_0^1}{S_0^2}$ and $K_2 = 1.1\frac{S_0^1}{S_0^2}$. The derivation of lower and upper bounds of prices of calls on the cross rate S^1/S^2 of maturity T given the two smiles at maturity T of S^1 and S^2 and the at-the-money implied volatility and skew of S^1/S^2 can be found in [129].

Table 12.3: Price in percent of the min of calls, put on worst, put on basket, digital call (DC) on S^{12}, and double-no-touch (DNT) on S^{12} for different admissible (or almost admissible) correlations described by the pair of functions (a, b). We used the same $50{,}000 = 10{,}000 + 40{,}000$ Brownian paths for all choices of (a, b). The first $10{,}000$ paths are those used for calibration of the correlation.

a	b	Min of calls	Put on worst	Put on basket	DC on S^{12}	DNT on S^{12}
	Standard deviation	≈ 0.020	≈ 0.027	≈ 0.027	≈ 0.18	≈ 0.22
	Constant correlation 72%	2.59	3.47	1.88	21.18	57.97
0	1	2.65	3.49	1.91	**20.53**	**58.02**
$\sigma_1^2 + \sigma_2^2$	$-2\sigma_1\sigma_2$	2.53	3.37	1.99	20.46	58.17
0	$\sigma_1\sigma_2$	**2.91**	**3.70**	**1.78**	19.75	59.41
0	σ_1	2.81	3.62	1.83	20.22	58.67
0	σ_2	2.78	3.60	1.85	20.27	58.39
σ_1^2	$-2\sigma_1\sigma_2$	2.67	3.51	1.91	20.47	58.12
σ_2^2	$-2\sigma_1\sigma_2$	2.80	3.60	1.84	20.38	58.50
0	$\sqrt{S^1 S^2}$	2.56	3.41	1.95	20.48	58.11
0	$\max(S^1, S^2)$	2.56	3.40	1.95	20.50	58.10
0	$\min(S^1, S^2)$	2.56	3.41	1.95	20.46	58.15
0	$(S^1 S^2)^{1/4}$	2.61	3.45	1.93	20.45	58.09
0	$\sqrt{\max(S^1, S^2)}$	2.61	3.45	1.93	20.46	58.08
0	$\sqrt{\min(S^1, S^2)}$	2.61	3.45	1.93	20.44	58.10
0	$\dfrac{1}{\sqrt{S^1 S^2}}$	2.74	3.56	1.87	20.41	58.21
0	$1.5 + \mathbf{1}\left\{\frac{S^1}{S_0} + \frac{S^2}{S_0} > 2\right\}$	**2.37**	**3.25**	**2.06**	**19.71**	**59.26**
0	$2 + \frac{1}{2}\tanh\left(10\left(\frac{S^1}{S_0} + \frac{S^2}{S_0} - 2\right)\right)$	2.42	3.28	2.04	20.14	58.65

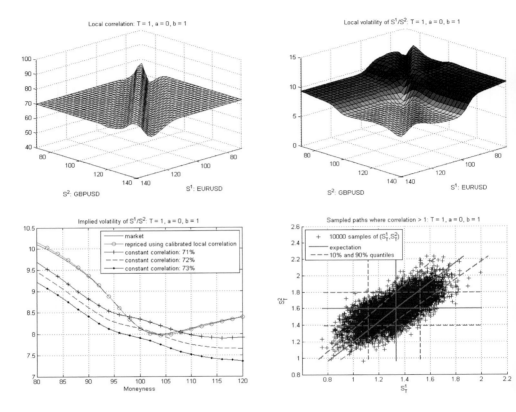

Figure 12.3: Local correlation (top left), local volatility of the cross rate (top right), implied volatility of the cross rate (bottom left), and scatter plot of (S^1, S^2) (bottom right) at maturity for $a = 0$, $b = 1$.

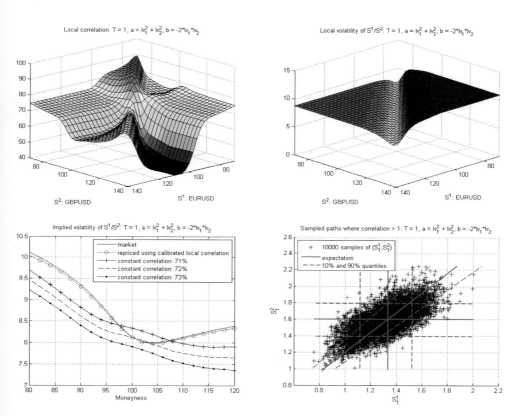

Figure 12.4: Local correlation (top left), local volatility of the cross rate (top right), implied volatility of the cross rate (bottom left), and scatter plot of (S^1, S^2) (bottom right) at maturity for $a(t, S^1, S^2) = \sigma_1^2(t, S^1) + \sigma_2^2(t, S^2)$, $b(t, S^1, S^2) = -2\sigma_1(t, S^1)\sigma_2(t, S^2)$.

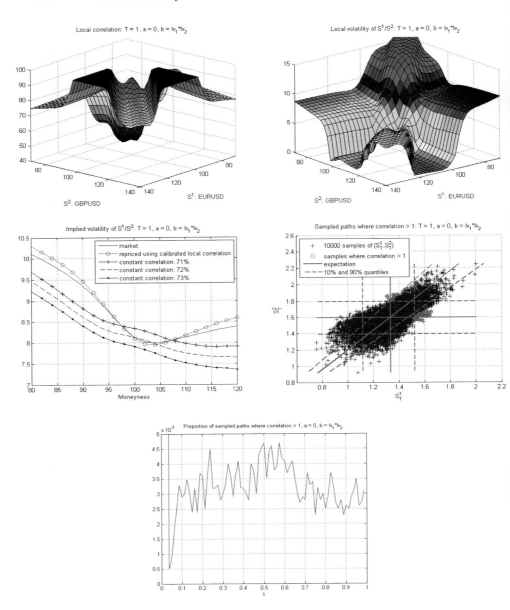

Figure 12.5: Local correlation (top left), local volatility of the cross rate (top right), implied volatility of the cross rate (bottom left), and scatter plot of (S^1, S^2) (bottom right) at maturity for $a = 0$, $b(t, S^1, S^2) = \sigma_1(t, S^1)\sigma_2(t, S^2)$. The fifth graph shows the proportion of capped correlations as a function of t.

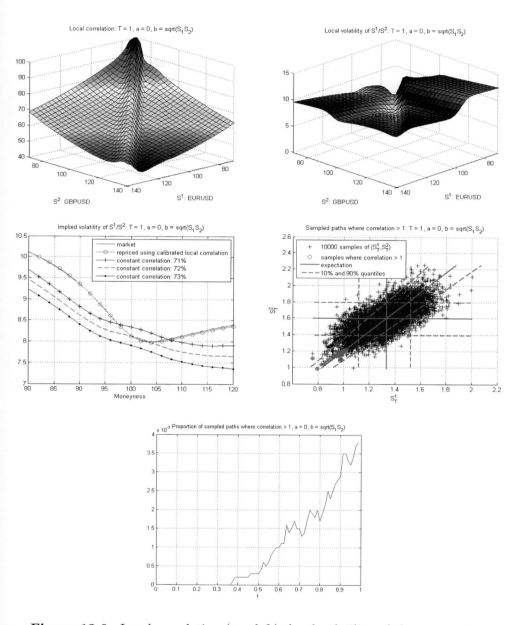

Figure 12.6: Local correlation (top left), local volatility of the cross rate (top right), implied volatility of the cross rate (bottom left), and scatter plot of (S^1, S^2) (bottom right) at maturity for $a = 0$, $b(t, S^1, S^2) = \sqrt{S^1 S^2}$. The fifth graph shows the proportion of capped correlations as a function of t.

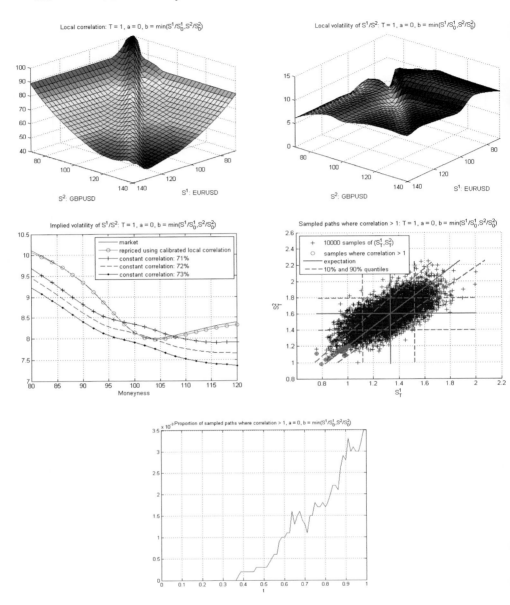

Figure 12.7: Local correlation (top left), local volatility of the cross rate (top right), implied volatility of the cross rate (bottom left), and scatter plot of (S^1, S^2) (bottom right) at maturity for $a = 0$, $b(t, S^1, S^2) = \min\left(\frac{S^1}{S_0^1}, \frac{S^2}{S_0^2}\right)$. The fifth graph shows the proportion of capped correlations as a function of t.

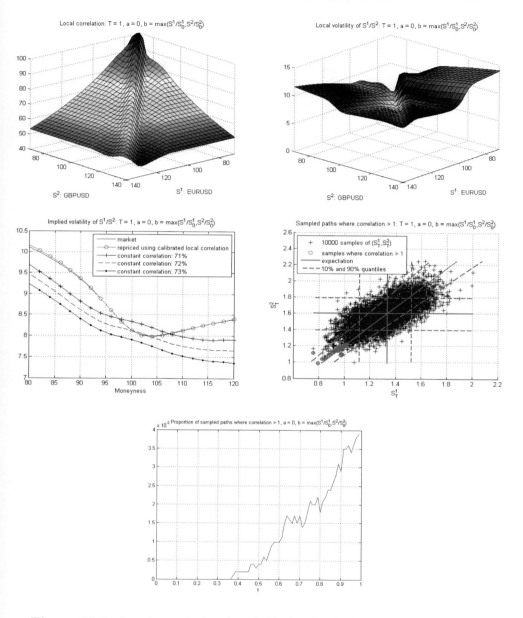

Figure 12.8: Local correlation (top left), local volatility of the cross rate (top right), implied volatility of the cross rate (bottom left), and scatter plot of (S^1, S^2) (bottom right) at maturity for $a = 0$, $b(t, S^1, S^2) = \max\left(\frac{S^1}{S_0^1}, \frac{S^2}{S_0^2}\right)$. The fifth graph shows the proportion of capped correlations as a function of t.

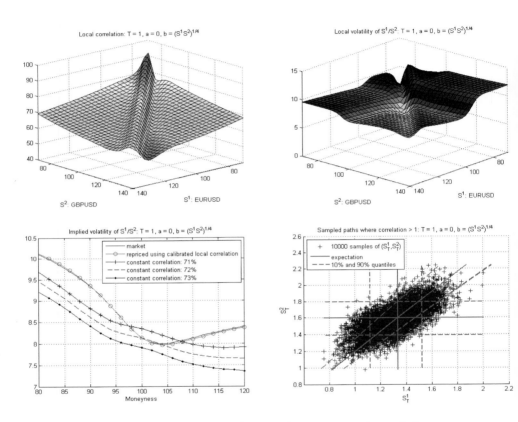

Figure 12.9: Local correlation (top left), local volatility of the cross rate (top right), implied volatility of the cross rate (bottom left), and scatter plot of (S^1, S^2) (bottom right) at maturity for $a = 0$, $b(t, S^1, S^2) = (S^1 S^2)^{1/4}$.

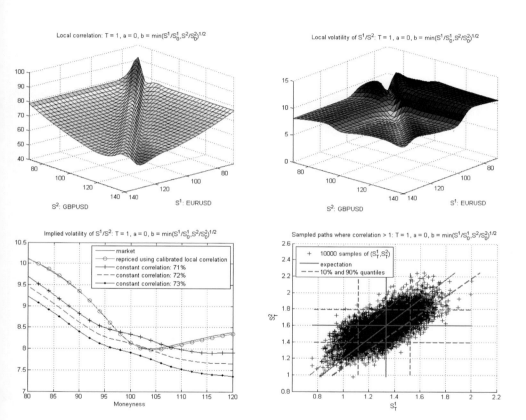

Figure 12.10: Local correlation (top left), local volatility of the cross rate (top right), implied volatility of the cross rate (bottom left), and scatter plot of (S^1, S^2) (bottom right) at maturity for $a = 0$, $b(t, S^1, S^2) = \sqrt{\min(S^1, S^2)}$.

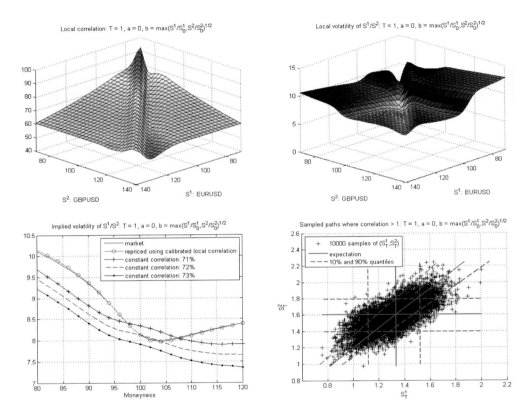

Figure 12.11: Local correlation (top left), local volatility of the cross rate (top right), implied volatility of the cross rate (bottom left), and scatter plot of (S^1, S^2) (bottom right) at maturity for $a = 0$, $b(t, S^1, S^2) = \sqrt{\max(S^1, S^2)}$.

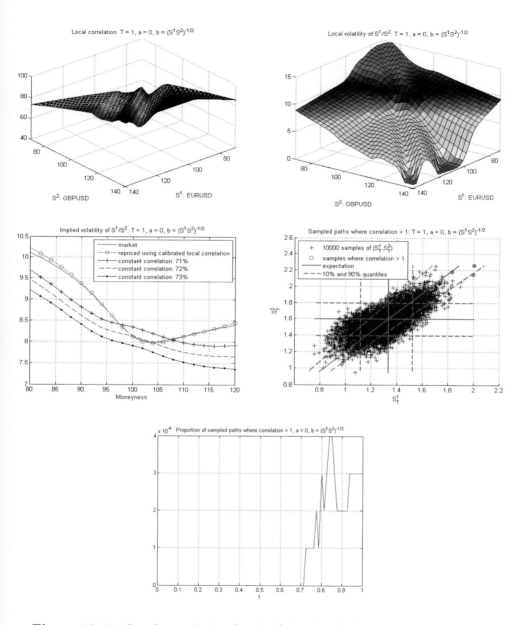

Figure 12.12: Local correlation (top left), local volatility of the cross rate (top right), implied volatility of the cross rate (bottom left), and scatter plot of (S^1, S^2) (bottom right) at maturity for $a = 0$, $b(t, S^1, S^2) = \frac{1}{\sqrt{S^1 S^2}}$. The fifth graph shows the proportion of capped correlations as a function of t.

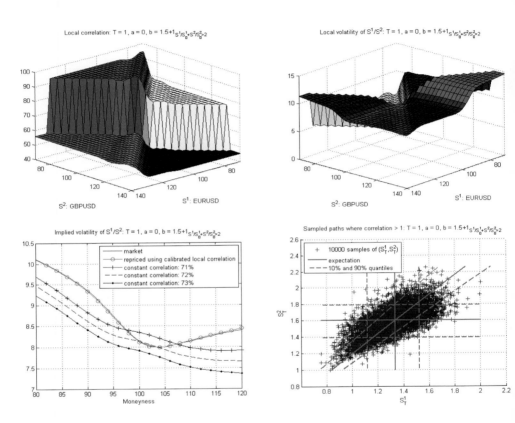

Figure 12.13: Local correlation (top left), local volatility of the cross rate (top right), implied volatility of the cross rate (bottom left), and scatter plot of (S^1, S^2) (bottom right) at maturity for $a = 0$, $b(t, S^1, S^2) = 1.5 + \mathbf{1}_{\left\{ \frac{S^1}{S_0^1} + \frac{S^2}{S_0^2} > 2 \right\}}$.

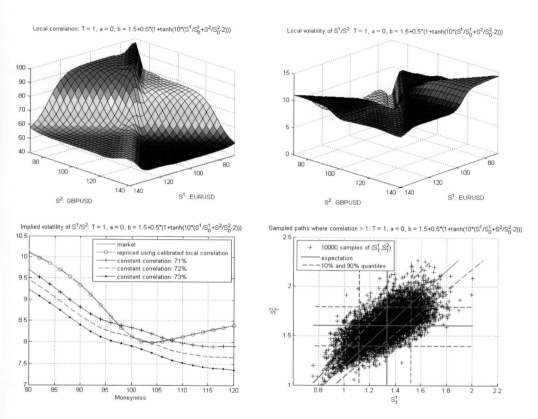

Figure 12.14: Local correlation (top left), local volatility of the cross rate (top right), implied volatility of the cross rate (bottom left), and scatter plot of (S^1, S^2) (bottom right) at maturity for $a = 0$, $b(t, S^1, S^2) = 1.5 + \frac{1}{2}\left(1 + \tanh\left(10\left(\frac{S^1}{S_0^1} + \frac{S^2}{S_0^2} - 2\right)\right)\right)$.

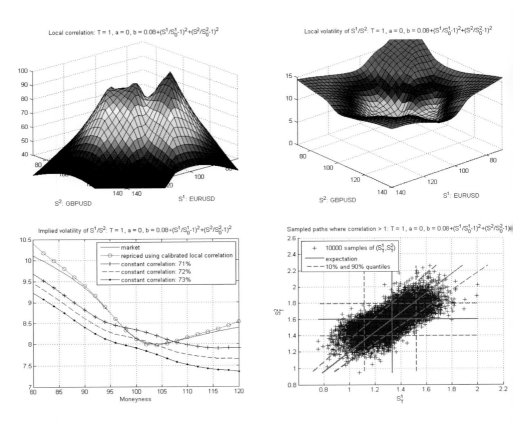

Figure 12.15: Local correlation (top left), local volatility of the cross rate (top right), implied volatility of the cross rate (bottom left), and scatter plot of (S^1, S^2) (bottom right) at maturity for $a = 0$, $b(t, S^1, S^2) = 0.08 + \left(\frac{S^1}{S_0^1} - 1\right)^2 + \left(\frac{S^2}{S_0^2} - 1\right)^2$.

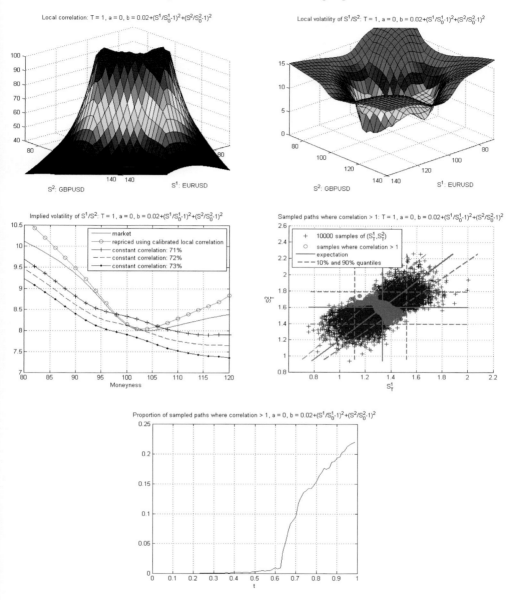

Figure 12.16: Local correlation (top left), local volatility of the cross rate (top right), implied volatility of the cross rate (bottom left), and scatter plot of (S^1, S^2) (bottom right) at maturity for $a = 0$, $b(t, S^1, S^2) = 0.02 + \left(\frac{S^1}{S_0^1} - 1\right)^2 + \left(\frac{S^2}{S_0^2} - 1\right)^2$. The fifth graph shows the proportion of capped correlations as a function of t.

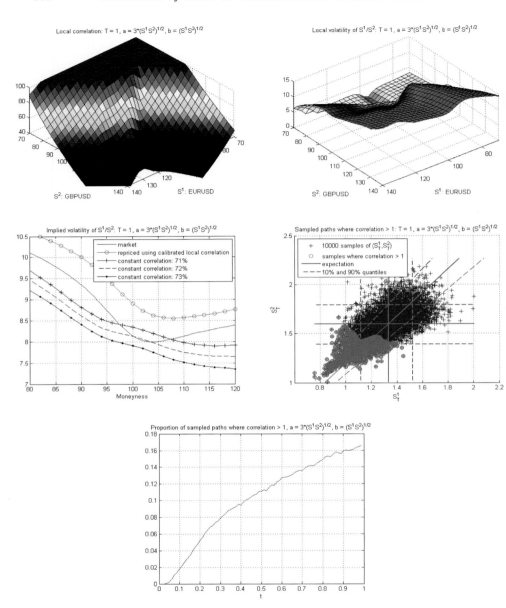

Figure 12.17: Local correlation (top left), local volatility of the cross rate (top right), implied volatility of the cross rate (bottom left), and scatter plot of (S^1, S^2) (bottom right) at maturity for $a(t, S^1, S^2) = 3\sqrt{S^1 S^2}$, $b(t, S^1, S^2) = \sqrt{S^1 S^2}$. The fifth graph shows the proportion of capped correlations as a function of t.

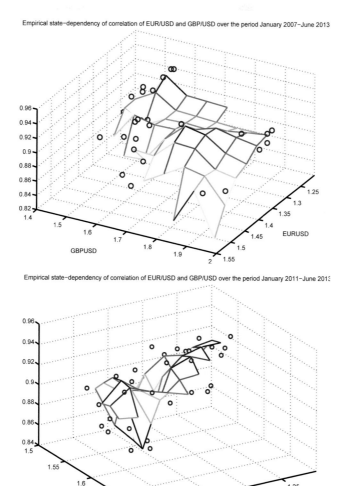

Figure 12.18: Empirical state-dependency of correlation of EUR/USD and GBP/USD over the period January 2007–June 2013 (top) and over the period January 2011–June 2013 (bottom).

12.11 Generalization to stochastic volatility, stochastic interest rates, and stochastic dividend yield

12.11.1 The FX triangle smile calibration problem

Let us show how to generalize the construction of families of local correlation models for the FX triangle smile calibration problem in the presence of local stochastic volatility, stochastic interest rates, and local correlation. For the sake of simplicity (see Remark 12.8 for a more general case), let us assume that the extra Brownian motions W^3, W^4, W^5... that drive the dynamics of the stochastic volatilities and the stochastic interest rates are *independent* of the two Brownian motions W^1 and W^2 that drive the dynamics of (S^1, S^2). The correlation matrix of W^3, W^4, W^5... is assumed to be known and constant. Only the correlation between W^1 and W^2 is unknown; it is assumed to be local:

$$dS_t^1 = \left(r_t^d - r_t^1\right) S_t^1 \, dt + \sigma_1(t, S_t^1) a_t^1 S_t^1 \, dW_t^1$$
$$dS_t^2 = \left(r_t^d - r_t^2\right) S_t^2 \, dt + \sigma_2(t, S_t^2) a_t^2 S_t^2 \, dW_t^2 \qquad (12.30)$$
$$d\langle W^1, W^2 \rangle_t = \rho(t, S_t^1, S_t^2, a_t^1, a_t^2, D_{0t}^d, D_{0t}^1, D_{0t}^2) \, dt$$

a_t^1, a_t^2, r_t^d, r_t^1 and r_t^2 are Itô processes driven by the extra Brownian motions W^3, W^4, W^5... The instantaneous correlation between W^1 and W^2 is assumed to depend not only on the FX rates S_t^1 and S_t^2, but also on the stochastic volatilities a_t^1 and a_t^2 and on the (stochastic) discount factors D_{0t}^d, D_{0t}^1, and D_{0t}^2:[1] $D_{0t}^i = \exp\left(-\int_0^t r_s^i ds\right)$. To keep notations short, we write $\rho(t, X_t)$ with $X_t = (S_t^1, S_t^2, a_t^1, a_t^2, D_{0t}^d, D_{0t}^1, D_{0t}^2)$.

First, the local volatility $\sigma_1(t, S^1)$ is calibrated to the market smile of S^1 using Proposition 12.8 in Section 12.A with $r_t = r_t^d$, $q_t = r_t^1$ and $a_t = \sigma_1(t, S_t^1) a_t^1$:

$$\sigma_1(t, K)^2 \frac{\mathbb{E}^{\mathbb{Q}}[D_{0t}^d (a_t^1)^2 | S_t^1 = K]}{\mathbb{E}^{\mathbb{Q}}[D_{0t}^d | S_t^1 = K]} = \sigma_{\text{Dup}}^1(t, K)^2$$

$$- \frac{\mathbb{E}^{\mathbb{Q}}\left[D_{0t}^d \left(r_t^d - r_t^1 - (r_t^{d,0} - r_t^{1,0})\right) \mathbf{1}_{S_t^1 > K}\right]}{\frac{1}{2} K \partial_K^2 \mathcal{C}(t, K)} + \frac{\mathbb{E}^{\mathbb{Q}}\left[D_{0t}^d \left(r_t^1 - r_t^{1,0}\right) (S_t^1 - K)^+\right]}{\frac{1}{2} K^2 \partial_K^2 \mathcal{C}(t, K)}$$

where $r_t^{d,0} = -\partial_t \ln P_{0t}^d$, $r_t^{1,0} = -\partial_t \ln P_{0t}^1$, and

$$\sigma_{\text{Dup}}^1(t, K)^2 = \frac{\partial_t \mathcal{C}_1(t, K) + (r_t^{d,0} - r_t^{1,0}) K \partial_K \mathcal{C}_1(t, K) + r_t^{1,0} \mathcal{C}_1(t, K)}{\frac{1}{2} K^2 \partial_K^2 \mathcal{C}_1(t, K)}$$

---

[1] The instantaneous correlation may also depend on r_t^d, r_t^1, r_t^2. It may actually depend on any $\mathcal{F}_t$-measurable random variable, including path-dependent variables (see Remark 12.3).

with $\mathcal{C}_1(t, K)$ the market price of the call option on S^1 with strike K and maturity t; and likewise for $\sigma_2(t, S^2)$. This is achieved using the particle algorithm (see [122]). The knowledge of the local correlation $\rho(t, X)$ between the two FX rates is not required at this step. Then $\rho(t, X)$ is calibrated to the market smile of the cross rate S^{12} by using the following proposition.

PROPOSITION 12.5
The local volatilities σ_1 and σ_2 being fixed, Model (12.30) is calibrated to the market smile of the cross FX rate if and only if

$$\frac{\mathbb{E}_\rho^{\mathbb{Q}^f}\left[D_{0t}^2\left(\left(\sigma_1(t, S_t^1)a_t^1\right)^2 + \left(\sigma_2(t, S_t^2)a_t^2\right)^2 - 2\rho(t, X_t)\sigma_1(t, S_t^1)a_t^1\sigma_2(t, S_t^2)a_t^2\right)\Big|S_t^{12} = K\right]}{\mathbb{E}_\rho^{\mathbb{Q}^f}\left[D_{0t}^2\big|S_t^{12} = K\right]}$$

$$= \sigma_{\mathrm{Dup}}^{12}(t, K)^2 - \frac{\mathbb{E}_\rho^{\mathbb{Q}^f}\left[D_{0t}^2\left(r_t^2 - r_t^1 - (r_t^{2,0} - r_t^{1,0})\right)\mathbf{1}_{S_t^{12} > K}\right]}{\frac{1}{2}K\partial_K^2\mathcal{C}(t, K)} + \frac{\mathbb{E}_\rho^{\mathbb{Q}^f}\left[D_{0t}^2(r_t^1 - r_t^{1,0})(S_t^{12} - K)^+\right]}{\frac{1}{2}K^2\partial_K^2\mathcal{C}(t, K)}$$

(12.31)

for all (t, K), where $D_{0t}^2 = \exp\left(-\int_0^t r_s^2 ds\right)$, $r_t^{1,0}$, and $r_t^{2,0}$ are deterministic interest rates, and

$$\sigma_{\mathrm{Dup}}^{12}(t, K)^2 = \frac{\partial_t\mathcal{C}(t, K) + (r_t^{2,0} - r_t^{1,0})K\partial_K\mathcal{C}(t, K) + r_t^{1,0}\mathcal{C}(t, K)}{\frac{1}{2}K^2\partial_K^2\mathcal{C}(t, K)}$$

is the market local volatility of the cross rate S^{12} computed using the deterministic interest rates $r_t^{1,0}$ and $r_t^{2,0}$. $\mathcal{C}(t, K)$ is the market price of the call option on S^{12} with strike K and maturity t.

PROOF Under Model (12.30), the dynamics of the cross rate $S^{12} = S^1/S^2$ reads

$$\frac{dS_t^{12}}{S_t^{12}} = (r_t^2 - r_t^1)\, dt + \sigma_1(t, S_t^1)a_t^1\, dW_t^{1,f} - \sigma_2(t, S_t^2)a_t^2\, dW_t^{2,f}$$

$$= (r_t^2 - r_t^1)\, dt + a_t\, dW_t^f$$

where $a_t^2 = \left(\sigma_1(t, S_t^1)a_t^1\right)^2 + \left(\sigma_2(t, S_t^2)a_t^2\right)^2 - 2\rho(t, X_t)\sigma_1(t, S_t^1)a_t^1\sigma_2(t, S_t^2)a_t^2$
and

$$W_t^{1,f} = W_t^1 - \int_0^t \rho(s, X_s)\sigma_2(s, S_s^2)a_s^2\, ds$$

$$W_t^{2,f} = W_t^2 - \int_0^t \sigma_2(s, S_s^2)a_s^2\, ds$$

$$W_t^f = \int_0^t \frac{\sigma_1(s, S_s^1)a_s^1\, dW_s^{1,f} - \sigma_2(s, S_s^2)a_s^2\, dW_s^{2,f}}{a_s}$$

are three $\mathbb{Q}^f$-Brownian motions, where $\mathbb{Q}^f$ is the risk-neutral measure associated to the foreign currency in S^2 (GBP in our example, see (12.2)). To conclude, it is enough to apply Proposition 12.8 in Section 12.A with $r_t = r_t^2$ and $q_t = r_t^1$. $\qquad\square$

Equation (12.31) is equivalent to

$$
\frac{\mathbb{E}_\rho^\mathbb{Q}\left[D_{0t}^d S_t^2\left(\left(\sigma_1(t,S_t^1)a_t^1\right)^2 + \left(\sigma_2(t,S_t^2)a_t^2\right)^2 - 2\rho(t,X_t)\sigma_1(t,S_t^1)a_t^1\sigma_2(t,S_t^2)a_t^2\right)\Big|S_t^{12} = K\right]}{\mathbb{E}_\rho^\mathbb{Q}\left[D_{0t}^d S_t^2\big|S_t^{12} = K\right]}
$$

$$
= \sigma_{\text{Dup}}^{12}(t,K)^2 - \frac{\mathbb{E}_\rho^\mathbb{Q}\left[D_{0t}^d \frac{S_t^2}{S_0^2}\left(r_t^2 - r_t^1 - (r_t^{2,0} - r_t^{1,0})\right)\mathbf{1}_{S_t^{12}>K}\right]}{\frac{1}{2}K\partial_K^2 \mathcal{C}(t,K)} + \frac{\mathbb{E}_\rho^\mathbb{Q}\left[D_{0t}^d \frac{S_t^2}{S_0^2}(r_t^1 - r_t^{1,0})(S_t^{12} - K)^+\right]}{\frac{1}{2}K^2\partial_K^2 \mathcal{C}(t,K)}
$$

$$(12.32)$$

and to

$$
\mathbb{E}_\rho^{\mathbb{Q}^{f,t}}\left[\left(\sigma_1(t,S_t^1)a_t^1\right)^2 + \left(\sigma_2(t,S_t^2)a_t^2\right)^2 - 2\rho(t,X_t)\sigma_1(t,S_t^1)a_t^1\sigma_2(t,S_t^2)a_t^2\Big|S_t^{12} = K\right]
$$

$$
= \sigma_{\text{Dup}}^{12}(t,K)^2 - P_{0t}^2\frac{\mathbb{E}_\rho^{\mathbb{Q}^{f,t}}\left[\left(r_t^2 - r_t^1 - (r_t^{2,0} - r_t^{1,0})\right)\mathbf{1}_{S_t^{12}>K}\right]}{\frac{1}{2}K\partial_K^2 \mathcal{C}(t,K)} + P_{0t}^2\frac{\mathbb{E}_\rho^{\mathbb{Q}^{f,t}}\left[(r_t^1 - r_t^{1,0})(S_t^{12} - K)^+\right]}{\frac{1}{2}K^2\partial_K^2 \mathcal{C}(t,K)}
$$

$$(12.33)$$

where $\mathbb{Q}^{f,t}$ denotes the foreign t-forward measure: $\frac{d\mathbb{Q}^{f,t}}{d\mathbb{Q}^f} = \frac{D_{0t}^2}{P_{0t}^2}$.

We say that a correlation ρ is admissible if Equation (12.31), or equivalently (12.32) or (12.33), holds. To build the set of all admissible correlations, we easily extend the local in cross $a + b\rho$ representation that was presented in Section 12.3 in the following way: for an admissible ρ, pick two functions $a(t,X)$ and $b(t,X)$ such that b does not vanish and

$$a(t,X) + b(t,X)\rho(t,X)$$

is local in cross, i.e., depends on X only through $S^{12} \equiv S^1/S^2$, where $X = (S^1, S^2, a^1, a^2, D_{0.}^d, D_{0.}^1, D_{0.}^2)$. One can always do so by choosing $b \equiv 1$ and $a(t,X) = f\left(t,S^{12}\right) - \rho(t,X)$ for some function f. Then (12.32) is equivalent to

$$
\frac{\mathbb{E}_\rho^\mathbb{Q}\left[D_{0t}^d S_t^2\left(\left(\sigma_1(t,S_t^1)a_t^1\right)^2 + \left(\sigma_2(t,S_t^2)a_t^2\right)^2 + 2\frac{a(t,X_t)}{b(t,X_t)}\sigma_1(t,S_t^1)a_t^1\sigma_2(t,S_t^2)a_t^2\right)\Big|S_t^{12} = K\right]}{\mathbb{E}_\rho^\mathbb{Q}\left[D_{0t}^d S_t^2\big|S_t^{12} = K\right]}
$$

$$
- 2\left(a + b\rho\right)(t,K)\frac{\mathbb{E}_\rho^\mathbb{Q}\left[D_{0t}^d S_t^2 \frac{\sigma_1(t,S_t^1)a_t^1\sigma_2(t,S_t^2)a_t^2}{b(t,X_t)}\Big|S_t^{12} = K\right]}{\mathbb{E}_\rho^\mathbb{Q}\left[D_{0t}^d S_t^2\big|S_t^{12} = K\right]}
$$

$$
= \sigma_{\text{Dup}}^{12}(t,K)^2 - \frac{\mathbb{E}_\rho^\mathbb{Q}\left[D_{0t}^d \frac{S_t^2}{S_0^2}\left(r_t^2 - r_t^1 - (r_t^{2,0} - r_t^{1,0})\right)\mathbf{1}_{S_t^{12}>K}\right]}{\frac{1}{2}K\partial_K^2 \mathcal{C}(t,K)} + \frac{\mathbb{E}_\rho^\mathbb{Q}\left[D_{0t}^d \frac{S_t^2}{S_0^2}(r_t^1 - r_t^{1,0})(S_t^{12} - K)^+\right]}{\frac{1}{2}K^2\partial_K^2 \mathcal{C}(t,K)}
$$

from which one gets $\rho(t,X) = \rho_{(a,b)}(t,X) \equiv \frac{f(t,S^{12})-a(t,X)}{b(t,X)}$ with $f\left(t,S^{12}\right) \equiv$

$\frac{N_f(t,S^{12})}{D_f(t,S^{12})}$ defined by

$$
N_f(t,K) = \frac{\mathbb{E}_\rho^Q\left[D_{0t}^d S_t^2\left((\sigma_1(t,S_t^1)a_t^1)^2 + (\sigma_2(t,S_t^2)a_t^2)^2 + 2\frac{a(t,X_t)}{b(t,X_t)}\sigma_1(t,S_t^1)a_t^1\sigma_2(t,S_t^2)a_t^2\right)\Big|S_t^{12} = K\right]}{\mathbb{E}_\rho^Q\left[D_{0t}^d S_t^2 \big| S_t^{12} = K\right]}
$$

$$
-\sigma_{\mathrm{Dup}}^{12}(t,K)^2 + \frac{\mathbb{E}_\rho^Q\left[D_{0t}^d \frac{S_t^2}{S_0^2}\left(r_t^2 - r_t^1 - (r_t^{2,0} - r_t^{1,0})\right)\mathbf{1}_{S_t^{12} > K}\right]}{\frac{1}{2}K\partial_K^2 \mathcal{C}(t,K)}
$$

$$
-\frac{\mathbb{E}_\rho^Q\left[D_{0t}^d \frac{S_t^2}{S_0^2}(r_t^1 - r_t^{1,0})(S_t^{12} - K)^+\right]}{\frac{1}{2}K^2\partial_K^2 \mathcal{C}(t,K)}
$$

$$
D_f(t,K) = 2\frac{\mathbb{E}_\rho^Q\left[D_{0t}^d S_t^2 \frac{\sigma_1(t,S_t^1)a_t^1\sigma_2(t,S_t^2)a_t^2}{b(t,X_t)}\Big|S_t^{12} = K\right]}{\mathbb{E}_\rho^Q\left[D_{0t}^d S_t^2 \big| S_t^{12} = K\right]}
$$

One can then compute $\rho_{(a,b)}$ using the particle method. Eventually one has to verify that $\rho_{(a,b)}(t,X) \in [-1,1]$. If this is not the case, one may cap and floor $\rho_{(a,b)}$ when needed and check how large the resulting smile calibration error is.

REMARK 12.8   One may wish to correlate the extra Brownian motions that drive the dynamics of the stochastic volatilities and the stochastic interest rates, and the two Brownian motions that drive the dynamics of the two FX rates. Here is one way to adapt the above method. Assume for simplicity that each of the extra processes a_t^1, a_t^2, r_t^d, r_t^1, and r_t^2 is driven by exactly one extra Brownian motion. Pick a set C^* of admissible constant values for the 12 correlations

$$
C(\rho) = \{\rho_{S^1a^1}, \rho_{S^1r^d}, \rho_{S^1r^1}, \rho_{a^1r^d}, \rho_{a^1r^1}, \rho_{r^dr^1},
$$
$$
\rho_{S^2a^2}, \rho_{S^2r^d}, \rho_{S^2r^2}, \rho_{a^2r^d}, \rho_{a^2r^2}, \rho_{r^dr^2}\}
$$

The first six correlations are used to calibrate σ_1; the last six to calibrate σ_2. Then one builds two full correlation matrices ρ^0 and ρ^1 as follows: first, one picks constant values for all the unspecified correlations in the matrix except $\rho_{S^1S^2}$. Those values can be arbitrary or inferred from historical data, and may make the matrix fail to be PSD. Then one chooses the extremal value $\rho_{S^1S^2} = -1$ (resp. 1) and projects the resulting matrix onto the space of correlation matrices to get ρ^0 (resp. ρ^1). The projection method *must leave* $C(\rho)$ *unchanged*. This can be done by using weighted norms on matrices (see for instance [133]). Then one assumes that the entire 7×7 correlation matrix $\rho(t,X)$ lies on the line defined by ρ^0 and ρ^1: $\rho(t,X) = (1 - \lambda(t,X))\rho^0 + \lambda(t,X)\rho^1$, picks two functions $a(t,X)$ and $b(t,X)$ (with b non-vanishing), and using the particle method, builds $\lambda_{(a,b)}(t,X)$ such that $a + b\lambda_{(a,b)}$ is local in cross and the calibration condition (12.31) is satisfied, with $\rho(t,X_t)$ replaced by $(1 - \lambda_{(a,b)}(t,X_t))\rho_{12}^0 + \lambda_{(a,b)}(t,X_t)\rho_{12}^1$. Then one has to verify that $\lambda_{(a,b)}$ takes values in $[0,1]$. Actually, any ρ^0 and ρ^1 for which $C(\rho^0) = C(\rho^1) = C^*$ will do the job: this guarantees that the knowledge of $\rho(t,X)$ is not needed during the first step of the calibration procedure, i.e., the calibration of σ_1 and σ_2, so that indeed the calibration procedure can be cut in two consecutive steps. It is indeed desirable to calibrate σ_1 and σ_2 independently of "cross-correlations" such as $\rho_{S^1 S^2}$, $\rho_{S^1 a^2}$, $\rho_{S^2 a^1} \ldots$ ☐

12.11.2   The equity index smile calibration problem

Let us consider a model that combines local stochastic volatility, stochastic interest rate, stochastic repo (inclusive of the dividend yield), and local correlation. Here again, let us assume for simplicity that the extra Brownian motions that drive the dynamics of the stochastic volatilities, the stochastic interest rate, and the stochastic repos are *independent* of the Brownian motions $(W^1, \ldots, W^N)$ that drive the dynamics of the N stocks $(S^1, \ldots, S^N)$. The correlation matrix of the extra Brownian motions is assumed to be known and constant. Only the correlation of $(W^1, \ldots, W^N)$ is unknown; it is assumed to be local:

$$\frac{dS_t^i}{S_t^i} = (r_t - q_t^i)\, dt + \sigma_i(t, S_t^i) a_t^i\, dW_t^i, \qquad d\langle W^i, W^j \rangle_t = \rho_{ij}(t, X_t)\, dt \quad (12.34)$$

where r_t, q_t^i, a_t^i are stochastic processes, $X_t = (S_t^1, \ldots, S_t^N, a_t^1, \ldots, a_t^N, D_{0t})$,[2] and $D_{0t} = \exp\left(-\int_0^t r_s ds\right)$.

First, the local volatilities $\sigma_i(t, S^i)$ are calibrated to the market smiles of the S^i's using Proposition 12.8 in Section 12.A:

$$\sigma_i(t, K)^2 \frac{\mathbb{E}[D_{0t}(a_t^i)^2 | S_t^i = K]}{\mathbb{E}[D_{0t} | S_t^i = K]} = \sigma_{\text{Dup}}^i(t, K)^2$$

$$- \frac{\mathbb{E}\left[D_{0t}\left(r_t - q_t^i - (r_t^0 - q_t^{i,0})\right) \mathbf{1}_{S_t^i > K}\right]}{\frac{1}{2} K \partial_K^2 \mathcal{C}_i(t, K)} + \frac{\mathbb{E}\left[D_{0t}\left(q_t^i - q_t^{i,0}\right)(S_t^i - K)^+\right]}{\frac{1}{2} K^2 \partial_K^2 \mathcal{C}_i(t, K)}$$

where $r_t^0 = -\partial_t \ln P_{0t}$, $q_t^{i,0} = r_t^0 - \partial_t \ln \frac{f_0^{i,t}}{S_0^i}$ (with $f_0^{i,t}$ the forward of maturity t), and

$$\sigma_{\text{Dup}}^i(t, K)^2 = \frac{\partial_t \mathcal{C}_i(t, K) + (r_t^0 - q_t^{i,0}) K \partial_K \mathcal{C}_i(t, K) + q_t^{i,0} \mathcal{C}_i(t, K)}{\frac{1}{2} K^2 \partial_K^2 \mathcal{C}_i(t, K)}$$

where $\mathcal{C}_i(t, K)$ is the market price of the call option on S^i with strike K and maturity t. This is achieved in practice thanks to the particle algorithm (see Chapter 11). At this step, the local correlation $\rho(t, X)$ does not need to be known. Then $\rho(t, X)$ is calibrated to the market smile of the index $I_t = \sum_{i=1}^N \alpha_i S_t^i$ by using the following:

---

[2] X_t may actually include any $\mathcal{F}_t$-measurable random variable (see Remark 12.3).

PROPOSITION 12.6

The local volatilities σ_i being fixed, Model (12.34) is calibrated to the market smile of the index if and only if

$$\frac{\mathbb{E}_\rho[D_{0t}v_\rho(t, X_t)|I_t = K]}{\mathbb{E}_\rho[D_{0t}|I_t = K]} = K^2 \sigma_{\text{Dup}}^I(t, K)^2$$

$$- K \frac{\mathbb{E}_\rho\left[D_{0t}\left(r_t - q_t - (r_t^0 - q_t^0)\right) \mathbf{1}_{I_t > K}\right]}{\frac{1}{2}\partial_K^2 \mathcal{C}(t, K)} + \frac{\mathbb{E}_\rho\left[D_{0t}\left(q_t - q_t^0\right)(I_t - K)^+\right]}{\frac{1}{2}\partial_K^2 \mathcal{C}(t, K)}$$

$$\tag{12.35}$$

for all (t, K), where

$$v_\rho(t, X_t) = \sum_{i,j=1}^N \alpha_i \alpha_j \rho_{ij}(t, X_t)\sigma_i(t, S_t^i)a_t^i\sigma_j(t, S_t^j)a_t^j S_t^i S_t^j \tag{12.36}$$

$$q_t = \frac{\sum_{i=1}^N \alpha_i S_t^i q_t^i}{\sum_{i=1}^N \alpha_i S_t^i}$$

r_t^0 *and* q_t^0 *are the deterministic interest rate and repo, and*

$$\sigma_{\text{Dup}}^I(t, K)^2 = \frac{\partial_t \mathcal{C}(t, K) + (r_t^0 - q_t^0)K\partial_K \mathcal{C}(t, K) + q_t^0 \mathcal{C}(t, K)}{\frac{1}{2}K^2\partial_K^2 \mathcal{C}(t, K)}$$

with $\mathcal{C}(t, K)$ the market price of the call option on I with strike K and maturity t.

PROOF   Under Model (12.34), the dynamics of the index $I_t = \sum_{i=1}^N \alpha_i S_t^i$ reads

$$dI_t = (r_t - q_t)I_t \, dt + \sqrt{v_\rho(t, X_t)} \, dW_t$$

W is a Brownian motion. To conclude, it is enough to apply Proposition 12.8 in Section 12.A. □

Following the lines of Section 12.9, if one assumes that the correlation matrix lies on the line defined by two correlation matrices ρ_0 and ρ_1, which may depend on (t, X_t),

$$\rho(t, X_t) = (1 - \lambda(t, X_t))\rho^0(t, X_t) + \lambda(t, X_t)\rho^1(t, X_t), \qquad \lambda(t, X_t) \in \mathbb{R}$$

then (12.35) reads

$$\frac{\mathbb{E}_\rho\left[D_{0t}\left(v_{\rho^0}(t, X_t) + (v_{\rho^1} - v_{\rho^0})(t, X_t)\lambda(t, X_t)\right)\Big|I_t = K\right]}{\mathbb{E}_\rho[D_{0t}|I_t = K]} = K^2 \sigma_{\text{Dup}}^I(t, K)^2$$

$$- K \frac{\mathbb{E}_\rho\left[D_{0t}\left(r_t - q_t - (r_t^0 - q_t^0)\right) \mathbf{1}_{I_t > K}\right]}{\frac{1}{2}\partial_K^2 \mathcal{C}(t, K)} + \frac{\mathbb{E}_\rho\left[D_{0t}\left(q_t - q_t^0\right)(I_t - K)^+\right]}{\frac{1}{2}\partial_K^2 \mathcal{C}(t, K)}$$

When one further assumes that there exist two functions $a(t, X)$ and $b(t, X)$ such that b does not vanish and $a + b\lambda$ is local in index, then one has

$$\frac{\mathbb{E}_\rho\left[D_{0t}\left(v_{\rho^0}(t, X_t) - \frac{a(t, X_t)}{b(t, X_t)}(v_{\rho^1}(t, X_t) - v_{\rho^0}(t, X_t))\right)\middle| I_t = K\right]}{\mathbb{E}_\rho[D_{0t}|I_t = K]}$$

$$+ (a + b\lambda)(t, K)\frac{\mathbb{E}_\rho\left[D_{0t}\frac{v_{\rho^1}(t, X_t) - v_{\rho^0}(t, X_t)}{b(t, X_t)}\middle| I_t = K\right]}{\mathbb{E}_\rho[D_{0t}|I_t = K]} = K^2 \sigma_{\text{Dup}}^I(t, K)^2$$

$$- K\frac{\mathbb{E}_\rho\left[D_{0t}\left(r_t - q_t - (r_t^0 - q_t^0)\right)\mathbf{1}_{I_t > K}\right]}{\frac{1}{2}\partial_K^2 \mathcal{C}(t, K)} + \frac{\mathbb{E}_\rho\left[D_{0t}\left(q_t - q_t^0\right)(I_t - K)^+\right]}{\frac{1}{2}\partial_K^2 \mathcal{C}(t, K)}$$

from which one gets $\lambda(t, X) = \lambda_{(a,b)}(t, X) \equiv \frac{f(t,I) - a(t,X)}{b(t,X)}$ with $f(t, I) \equiv \frac{N_f(t,I)}{D_f(t,I)}$ defined by

$$N_f(t, K) = K^2 \sigma_{\text{Dup}}^I(t, K)^2 - K\frac{\mathbb{E}_{\rho_{(a,b)}}\left[D_{0t}\left(r_t - q_t - (r_t^0 - q_t^0)\right)\mathbf{1}_{I_t > K}\right]}{\frac{1}{2}\partial_K^2 \mathcal{C}(t, K)}$$

$$+ \frac{\mathbb{E}_{\rho_{(a,b)}}\left[D_{0t}\left(q_t - q_t^0\right)(I_t - K)^+\right]}{\frac{1}{2}\partial_K^2 \mathcal{C}(t, K)}$$

$$- \frac{\mathbb{E}_{\rho_{(a,b)}}\left[D_{0t}\left(v_{\rho^0}(t, X_t) - \frac{a(t, X_t)}{b(t, X_t)}(v_{\rho^1}(t, X_t) - v_{\rho^0}(t, X_t))\right)\middle| I_t = K\right]}{\mathbb{E}_{\rho_{(a,b)}}[D_{0t}|I_t = K]}$$

$$D_f(t, K) = \frac{\mathbb{E}_{\rho_{(a,b)}}\left[D_{0t}\frac{v_{\rho^1}(t, X_t) - v_{\rho^0}(t, X_t)}{b(t, X_t)}\middle| I_t = K\right]}{\mathbb{E}_{\rho_{(a,b)}}[D_{0t}|I_t = K]}$$

where $\rho_{(a,b)} = (1 - \lambda_{(a,b)})\rho^0 + \lambda_{(a,b)}\rho^1$. One can then compute $\rho_{(a,b)}$ using the particle method. Eventually, one has to verify that $\rho_{(a,b)}(t, X)$ is a true correlation matrix. If this is not the case, one may "cap" and "floor" $\rho_{(a,b)}$ (to ρ^0 or ρ^1) when needed and check how large the resulting smile calibration error is. It is very easy to adapt Remark 12.8 to extend to cases where the extra Brownian motions are correlated with $(W^1, \dots, W^N)$.

12.12   Path-dependent volatility

In Remark 12.3 we noticed that the particle method easily accommodates path-dependent correlation. It is easy to adapt this remark to build single-asset path-dependent volatility models that calibrate to the smile. The dynamics of a single asset having path-dependent volatility reads

$$\frac{dS_t}{S_t} = (r_t - q_t)\, dt + \sigma(t, S_t, X_t)\, dW_t \tag{12.37}$$

where X_t stands for a set of path-dependent variables. The interest and repo rates r_t and q_t are assumed deterministic. The vector X_t may for instance include the running average, a moving average, the running minimum, the running maximum, the realized volatility on the past few days, etc. Hobson and Rogers [134] suggested a model where X_t is a collection of exponentially weighted moments of past returns.

Path-dependent volatility models of type (12.37) have the very nice property of being complete so that, unlike stochastic volatility models, prices are uniquely defined, independently of utility or preferences. The path-dependency allows for dynamics for the spot and the implied volatility that are richer than those produced by the local volatility model.

Assume that the market smile of S is arbitrage-free. How do we calibrate σ to it? If X_t is void, the answer is well known [95]: there is a unique solution $\sigma(t, S) = \sigma_{\mathrm{Dup}}(t, S)$, called the local volatility of S, given by (12.43) with $r_t^0 = r_t$ and $q_t^0 = q_t$. This is the famous local volatility model. What if X_t includes some information on the past of S?

PROPOSITION 12.7 Local in spot $a + b\sigma^2$ representation of admissible path-dependent volatilities
The path-dependent volatility $\sigma(t, S, X)$ calibrates to the market smile of S (we say it is admissible*) if and only if there exist two functions $a(t, S, X)$ and $b(t, S, X)$ such that b does not vanish and*

$$\sigma^2(t, S_t, X_t) = \frac{1}{b(t, S_t, X_t)} \times$$

$$\left(\frac{\sigma_{\mathrm{Dup}}^2(t, S_t) + \mathbb{E}_\sigma\left[\left. \frac{a(t,S_t,X_t)}{b(t,S_t,X_t)} \right| S_t \right]}{\mathbb{E}_\sigma\left[\left. \frac{1}{b(t,S_t,X_t)} \right| S_t \right]} - a(t, S_t, X_t) \right) \tag{12.38}$$

We call (12.38) the *local in spot $a + b\sigma^2$ representation* of admissible path-dependent volatilities.

PROOF Assume that σ calibrates to the market smile of S. From Proposition 12.8 in Section 12.A, this is equivalent to saying that

$$\mathbb{E}_\sigma[\sigma(t, S_t, X_t)^2 | S_t] = \sigma_{\mathrm{Dup}}^2(t, S_t) \tag{12.39}$$

Choose two functions $a(t, S, X)$ and $b(t, S, X)$ such that b does not vanish and

$$a(t, S, X) + b(t, S, X)\sigma^2(t, S, X) \equiv f(t, S)$$

is local in spot, i.e., depends on (S, X) only through S. One can always do so by picking $b \equiv 1$ and $a(t, S, X) = f(t, S) - \sigma^2(t, S, X)$ for some local in spot function f. If $\sigma(t, S, X)$ does not vanish, one can also pick $a \equiv 0$ and $b(t, S, X) = \frac{f(t,S)}{\sigma^2(t,S,X)}$. Then from (12.39)

$$
\left(a + b\sigma^2\right)(t, S_t)\mathbb{E}_\sigma \left[\left.\frac{1}{b(t, S_t, X_t)}\right| S_t\right] - \mathbb{E}_\sigma \left[\left.\frac{a(t, S_t, X_t)}{b(t, S_t, X_t)}\right| S_t\right] = \sigma^2_{\mathrm{Dup}}(t, S_t)
$$

from which we get that $\sigma = \sigma_{(a,b)}$ is a solution to

$$
\sigma^2_{(a,b)}(t, S_t, X_t) = \frac{1}{b(t, S_t, X_t)} \times
$$

$$
\left(\frac{\sigma^2_{\mathrm{Dup}}(t, S_t) + \mathbb{E}_{\sigma_{(a,b)}} \left[\left.\frac{a(t,S_t,X_t)}{b(t,S_t,X_t)}\right| S_t\right]}{\mathbb{E}_{\sigma_{(a,b)}} \left[\left.\frac{1}{b(t,S_t,X_t)}\right| S_t\right]} - a(t, S_t, X_t)\right) \quad (12.40)
$$

Conversely, if a function $\sigma_{(a,b)}$ satisfies (12.40), then it is satisfies (12.39) and thus calibrates to the market smile of S. $\quad\Box$

Note that, like (12.5), (12.40) is a circular equation: the two conditional expectations on the right-hand side depend on $\sigma_{(a,b)}$. To the best of our knowledge, the existence of the nonlinear SDEs, or Mc Kean SDEs, describing the calibrated models

$$
\frac{dS_t}{S_t} = (r_t - q_t)\, dt
$$

$$
+ \sqrt{\frac{1}{b(t, S_t, X_t)} \left(\frac{\sigma^2_{\mathrm{Dup}}(t, S_t) + \mathbb{E}\left[\left.\frac{a(t,S_t,X_t)}{b(t,S_t,X_t)}\right| S_t\right]}{\mathbb{E}\left[\left.\frac{1}{b(t,S_t,X_t)}\right| S_t\right]} - a(t, S_t, X_t)\right)}\, dW_t
$$

is still an open mathematical question.

In practice, one may try to build a solution $\sigma_{(a,b)}$ using the particle method:

1. Initialize $k = 1$ and set $\sigma_{(a,b)}(t, S, X) = \sigma_{\mathrm{Dup}}(0, S)$ for all $t \in [t_0 = 0; t_1]$.

2. Simulate $(S_t^i)_{1 \le i \le N}$ from t_{k-1} to t_k using a discretization scheme, say, a log-Euler scheme.

3. For all S in a grid G_{t_k} of spot values, compute non-parametric estimations $E_{t_k}^{\mathrm{num}}(S)$ and $E_{t_k}^{\mathrm{den}}(S)$ of

$$
\mathbb{E}\left[\left.\frac{a(t_k, S_{t_k}, X_{t_k})}{b(t_k, S_{t_k}, X_{t_k})}\right| S_{t_k} = S\right] \quad \text{and} \quad \mathbb{E}\left[\left.\frac{1}{b(t_k, S_{t_k}, X_{t_k})}\right| S_{t_k} = S\right]
$$

set $f(t_k, S) = \frac{\sigma^2_{\mathrm{Dup}}(t_k, S) + E_{t_k}^{\mathrm{num}}(S)}{E_{t_k}^{\mathrm{den}}(S)}$, interpolate and extrapolate $f(t_k, \cdot)$, for instance using cubic splines, and for all $t \in [t_k, t_{k+1}]$, set $\sigma_{(a,b)}(t, S, X)$ $= \sqrt{\frac{f(t_k, S) - a(t, S, X)}{b(t, S, X)}}$.

4. Set $k := k + 1$. Iterate Steps 2 and 3 up to the maturity date T.

For a given pair (a, b), if at some point in time and for some path the quantity $\frac{f(t_k, S) - a(t, S, X)}{b(t, S, X)}$ is negative, i.e., $\sigma^2_{(a,b)}$ is negative, this means that there is no admissible path-dependent volatility such that $a + b\sigma^2$ is local in spot. However, one can then floor $\sigma^2_{(a,b)}(t, S, X)$ to zero and carry on using the particle method until maturity. Then one must check how bad the smile calibration is. It may happen that the path-dependent volatility has to be floored on only a few paths, in which case the calibration error may be acceptable. We then say that the path-dependent volatility is *almost admissible*.

Given a set X of path-dependent variables, the method offers a huge number of degrees of freedom, namely the functions a and b, that can be used to build a path-dependent volatility that not only is (at least almost) admissible, but also is better than the local volatility model at reproducing some historical features of volatility, or calibrates to extra option prices, etc.

The generalization to stochastic volatility, stochastic interest rates, and stochastic dividend yield is straightforward. Assume that

$$\frac{dS_t}{S_t} = (r_t - q_t)\, dt + \sigma(t, S_t, X_t)\alpha_t\, dW_t$$

where now r_t and q_t are Itô processes, as well as the stochastic volatility α_t. From Proposition 12.8 in Section 12.A, this model is calibrated to the market smile of S if and only if for all t, K

$$\frac{\mathbb{E}[D_{0t}\sigma^2(t, S_t, X_t)\alpha_t^2 | S_t = K]}{\mathbb{E}[D_{0t} | S_t = K]} = \sigma^2_{\text{Dup}}(t, K)$$

$$-\frac{\mathbb{E}\left[D_{0t}\left(r_t - q_t - (r_t^0 - q_t^0)\right)\mathbf{1}_{S_t > K}\right]}{\frac{1}{2}K\partial_K^2 \mathcal{C}(t, K)} + \frac{\mathbb{E}\left[D_{0t}\left(q_t - q_t^0\right)(S_t - K)^+\right]}{\frac{1}{2}K^2\partial_K^2 \mathcal{C}(t, K)}$$

where σ_{Dup} is the Dupire local volatility computed using the deterministic rate r_t^0 and the deterministic dividend yield q_t^0. Following the same reasoning as above, we have that in this model a path-dependent volatility $\sigma(t, X)$ calibrates to the smile of S (we say it is *admissible*) if and only if there exist two functions $a(t, S, X)$ and $b(t, S, X)$ such that b does not vanish and σ satisfies the self-consistency equation $\sigma^2(t, S, X) = \frac{f(t, S) - a(t, S, X)}{b(t, S, X)}$ with

$f(t, K) \equiv \frac{N_f(t,K)}{D_f(t,K)}$ defined by

$$N_f(t, K) = \frac{\mathbb{E}_\sigma \left[\frac{a(t,S_t,X_t)}{b(t,S_t,X_t)} D_{0t} \alpha_t^2 \,\middle|\, S_t = K \right]}{\mathbb{E}_\sigma [D_{0t} | S_t = K]}$$

$$+ \sigma_{\mathrm{Dup}}^2(t, K) - \frac{\mathbb{E}_\sigma \left[D_{0t} \left(r_t - q_t - (r_t^0 - q_t^0) \right) \mathbf{1}_{S_t > K} \right]}{\frac{1}{2} K \partial_K^2 \mathcal{C}(t, K)}$$

$$+ \frac{\mathbb{E}_\sigma \left[D_{0t} \left(q_t - q_t^0 \right) (S_t - K)^+ \right]}{\frac{1}{2} K^2 \partial_K^2 \mathcal{C}(t, K)}$$

$$D_f(t, K) = \frac{\mathbb{E}_\sigma \left[\frac{D_{0t} \alpha_t^2}{b(t,S_t,X_t)} \,\middle|\, S_t = K \right]}{\mathbb{E}_\sigma [D_{0t} | S_t = K]}$$

Again, one can use the particle method to check if a pair (a, b) gives rise to an admissible path-dependent volatility. Some pairs (a, b) may be such that the self-consistency equation has no solution. Within the particle method, this is reflected in the quantity $\frac{f(t,S) - a(t,S,X)}{b(t,S,X)}$ being negative at some point in time and for some simulated path.

Conclusion

To the best of our knowledge, only two local correlation models have been proposed in the past in order to exactly calibrate to the smile of a basket, be it a stock index, a cross FX rate, an interest rate spread, etc. Both models may actually fail to calibrate to the basket smile, and even if they do not, they impose a particular shape of the correlation matrix which may be irrelevant. In this chapter we have suggested a general procedure that produces a whole family of local correlation models, many of which calibrate to the basket smile. The two existing models are just two particular points in the new family of models. We have also shown how to build admissible models that combine stochastic interest rates, stochastic dividend yield, local stochastic volatility, and local correlation. This generality is reached *at no cost*: in all cases the usual particle method does the job. Our procedure also easily adapts to build single-asset path-dependent volatility models that calibrate to the market smile.

The huge number of degrees of freedom, represented by the two functions a and b, allows one to pick one's favorite correlation with desirable properties among the new family of admissible correlations. This way we reconcile static calibration, i.e., calibration from a snapshot of prices of options on basket, and dynamic calibration, i.e., calibration from historical study of state-dependency of correlation. Our numerical tests show the wide variety of admissible correlations and give insight on lower bounds/upper bounds on general multi-asset option prices given the smile of a basket and the smiles of its constituents. The derivation of the exact bounds; the derivation of explicit conditions under which a triangle of *surfaces* of FX implied volatilities is jointly arbitrage-free; and, when so, the derivation of conditions under which an admissible local correlation does exist in theory, are three examples of important open questions. Exercises 12.13.1 and 12.13.2 below are related to the second question.

12.13   Exercises

12.13.1   Arbitrage-freeness condition on cross-currency smiles: The case of one maturity

In this exercise, we want to determine the necessary and sufficient conditions under which the smiles of the FX rates S^1, S^2, and the cross-currency rate $S^{12} \equiv S^1/S^2$ *at one maturity* T are *jointly* arbitrage-free. We assume that each of the three smiles is individually arbitrage-free. For simplicity, we also assume zero interest rates. This exercise is based on [129].

1. Explain how to infer the marginal distribution $\mathbb{Q}^1$ of S_T^1 from market prices of vanilla options on S_T^1.

2. Justify that there is no joint arbitrage if and only if $\mathrm{UB} = 0$ where

$$\mathrm{UB} \equiv \inf_{(u_1, u_2, u_{12}) \in \mathcal{M}} \left\{ \mathbb{E}^{\mathbb{Q}^1}[u_1(S_T^1)] + \mathbb{E}^{\mathbb{Q}^2}[u_2(S_T^2)] + \mathbb{E}^{\mathbb{Q}^{12}}[u_{12}(S_T^{12})] \right\}$$

   with

$$\mathcal{M} \equiv \{(u_1, u_2, u_{12}) \in \mathrm{L}^1(\mathbb{Q}^1) \times \mathrm{L}^1(\mathbb{Q}^2) \times \mathrm{L}^1(\mathbb{Q}^{12}) \,|$$
$$\forall (s^1, s^2) \in (\mathbb{R}_+^*)^2, \quad u_1(s^1) + u_2(s^2) + u_{12}(s^1/s^2) \geq 0\}$$

3. Note that this is a linear programming problem. How would you solve it in practice? (Hint: The payoffs u_1, u_2, u_{12} may be represented as linear combinations of call option payouts.)

4. Prove that

$$\mathrm{UB} = \inf_{u_1 \in \mathrm{L}^1(\mathbb{Q}^1), u_2 \in \mathrm{L}^1(\mathbb{Q}^2)} \left\{ \mathbb{E}^{\mathbb{Q}^1}[u_1(S_T^1)] + \mathbb{E}^{\mathbb{Q}^2}[u_2(S_T^2)] - \mathbb{E}^{\mathbb{Q}^{12}}[u_{12}^*(S_T^{12})] \right\}$$

   with u_{12}^* the inf-convolution of u_1 and u_2: $u_{12}^*(s^{12}) \equiv \inf_{s^1 \in \mathbb{R}_+} \{u_1(s^1) + u_2(s^1/s^{12})\}$.

5. Following closely the Monge-Kantorovich duality explained in Section 10.A, prove that UB can be written as

$$\text{UB} \equiv \sup_{\mathbb{Q} \in \mathcal{M}^*} \mathbb{E}^{\mathbb{Q}}[0]$$

(hence the interpretation of UB as an upper bound) where $\mathcal{M}^*$ denotes the space of joint probability measures on (S_T^1, S_T^2, S_T^{12}) with marginal distributions $\mathbb{Q}^1$, $\mathbb{Q}^2$, and $\mathbb{Q}^{12}$:

$$\mathcal{M}^* = \left\{ \mathbb{Q} \,\middle|\, S_T^1 \sim_{\mathbb{Q}} \mathbb{Q}^1, S_T^2 \sim_{\mathbb{Q}} \mathbb{Q}^2, S_T^{12} \equiv \frac{S_T^1}{S_T^2} \sim_{\mathbb{Q}} \mathbb{Q}^{12} \right\}$$

6. Prove that there is no joint arbitrage if and only if $\mathcal{M}^* \neq \emptyset$.

12.13.2   Arbitrage-freeness condition on cross-currency smiles: The case of finitely many maturities

We want to generalize the previous exercise in the case of finitely many maturities. We assume that vanilla options on the three assets S^1, S^2, S^{12} are traded on the market for all strikes and for maturities $(t_j)_{1 \leq j \leq n}$. For convenience, we write $S^3 \equiv S^{12}$. Below we denote by $(\mathbb{Q}_j^i)_{1 \leq i \leq 3, 1 \leq j \leq n}$ the marginal distribution of asset i at time t_j, and $S_j^i \equiv S_{t_j}^i$. In this exercise, again for simplicity, we assume zero interest rates. This exercise is based on [129, 52].

1. Justify that there is no joint arbitrage if and only if UB $= 0$ where

$$\text{UB} \equiv \inf_{(u_j^i)_{1 \leq i \leq 3, 1 \leq j \leq n}, (\Delta_j^i)_{1 \leq i \leq 3, 1 \leq j \leq n-1} \in \mathcal{M}} \sum_{i=1}^{3} \sum_{j=1}^{n-1} \mathbb{E}^{\mathbb{Q}_j^i}[u_j^i(S_j^i)]$$

with

$$\mathcal{M} \equiv \{ u_j^i \in L^1(\mathbb{Q}_j^i), \Delta_j^i \in C_b((\mathbb{R}_+)^j) \mid \forall (s_j^i)_{1 \leq i \leq 3, 1 \leq j \leq n} \in \mathbb{R}_+^*$$
$$\sum_{i=1}^{3} \sum_{j=1}^{n} u_j^i(s_j^i) + \sum_{i=1}^{3} \sum_{j=1}^{n-1} \Delta_j^i(s_1^i, \ldots, s_j^i)(s_{j+1}^i - s_j^i) \geq 0 \}$$

2. Explain why this linear programming problem is harder to solve in practice than the one in the previous exercise.

3. Give an interpretation in terms of (robust) super-replication.

4. Following closely the Monge-Kantorovich duality explained in Section 10.A, prove that UB can be written as

$$\text{UB} \equiv \sup_{\mathbb{Q} \in \mathcal{M}^*} \mathbb{E}^{\mathbb{Q}}[0]$$

(hence the interpretation of UB as an upper bound) with

$$\mathcal{M}^* \equiv \{\mathbb{Q} \mid S_j^i \sim_\mathbb{Q} \mathbb{Q}_j^i, \quad \mathbb{Q}_1^i \underset{\text{convex}}{\leq} \mathbb{Q}_2^i \underset{\text{convex}}{\leq} \cdots \underset{\text{convex}}{\leq} \mathbb{Q}_n^i,$$

$$\mathbb{E}^\mathbb{Q}[S_j^i \mid S_1^i, \ldots, S_{j-1}^i] = S_{j-1}^i, \quad \forall i \in \{1,2,3\}, \ \forall j \in \{1, \ldots, n\}\}$$

$\mathbb{Q}_j^i \underset{\text{convex}}{\leq} \mathbb{Q}_{j+1}^i$ means that the marginals $\mathbb{Q}_j^i$ and $\mathbb{Q}_{j+1}^i$ are in convex order:

$$\forall K \in \mathbb{R}_+, \qquad \mathbb{E}^{\mathbb{Q}_j^i}[(S-K)^+] \leq \mathbb{E}^{\mathbb{Q}_{j+1}^i}[(S-K)^+]$$

5. Prove that there is no joint arbitrage if and only if $\mathcal{M}^* \neq \emptyset$. What does it mean in terms of models?

12.A Calibration of an LSVM with stochastic interest rates and stochastic dividend yield

The following proposition gives a necessary and sufficient condition for a model to be calibrated to a given smile, in the presence of stochastic volatility (possibly including some local volatility component), stochastic interest rates, and stochastic dividend yield.

PROPOSITION 12.8

Let us consider the following dynamics for an asset S, where the volatility a_t, the interest rate r_t, and the repo q_t, inclusive of the dividend yield, are all stochastic processes:

$$\frac{dS_t}{S_t} = (r_t - q_t)\, dt + a_t\, dW_t \tag{12.41}$$

Model (12.41) is exactly calibrated to the market smile of S if and only if

$$\frac{\mathbb{E}[D_{0t} a_t^2 \mid S_t = K]}{\mathbb{E}[D_{0t} \mid S_t = K]} = \sigma_{\text{Dup}}(t,K)^2$$

$$- \frac{\mathbb{E}\left[D_{0t}\left(r_t - q_t - (r_t^0 - q_t^0)\right)\mathbf{1}_{S_t > K}\right]}{\frac{1}{2}K\partial_K^2\mathcal{C}(t,K)} + \frac{\mathbb{E}\left[D_{0t}\left(q_t - q_t^0\right)(S_t - K)^+\right]}{\frac{1}{2}K^2\partial_K^2\mathcal{C}(t,K)} \tag{12.42}$$

for all (t,K), where $D_{0t} = \exp\left(-\int_0^t r_s ds\right)$ is the discount factor, r_t^0 and q_t^0 are deterministic rates and repos, and

$$\sigma_{\text{Dup}}(t,K)^2 = \frac{\partial_t\mathcal{C}(t,K) + (r_t^0 - q_t^0)K\partial_K\mathcal{C}(t,K) + q_t^0\mathcal{C}(t,K)}{\frac{1}{2}K^2\partial_K^2\mathcal{C}(t,K)} \tag{12.43}$$

with $C(t, K)$ the market price of the call option on S with strike K and maturity t.

REMARK 12.9   The deterministic rate r_t^0 is typically taken to be equal to $-\partial_t \ln P_{0t}$, with P_{0t} the price at time 0 of a zero-coupon bond maturing at time t. Then one can infer a deterministic repo rate q_t^0 from the forward price f_0^t:

$$q_t^0 = r_t^0 - \partial_t \ln \frac{f_0^t}{S_0}$$

$\square$

PROOF   By applying Itô-Tanaka's formula on a discounted vanilla call payoff with maturity t and strike K, $\mathcal{P}_t \equiv D_{0t}(S_t - K)^+$, we have:

$$d\mathcal{P}_t = -D_{0t}(S_t - K)^+ r_t\,dt + D_{0t}\mathbf{1}_{S_t>K}S_t\left((r_t - q_t)\,dt + a_t dW_t\right)$$
$$+ \frac{1}{2}S_t^2 a_t^2 D_{0t}\delta(S_t - K)\,dt$$
$$= D_{0t}\mathbf{1}_{S_t>K}(r_t - q_t)K\,dt - D_{0t}q_t(S_t - K)^+\,dt + D_{0t}\mathbf{1}_{S_t>K}a_t S_t dW_t$$
$$+ \frac{1}{2}K^2 a_t^2 D_{0t}\delta(S_t - K)\,dt$$

By taking the expectation $\mathbb{E}[\cdot]$ on both sides of the above equation and by assuming that $M_t = \int_0^t D_{0s}\mathbf{1}_{S_s>K}a_s S_s dW_s$ is a true martingale, we get

$$\partial_t \mathcal{C}_{\mathrm{m}}(t, K) = K\mathbb{E}[D_{0t}(r_t - q_t)\mathbf{1}_{S_t>K}] - \mathbb{E}[D_{0t}q_t(S_t - K)^+]$$
$$+ \frac{1}{2}K^2 \sigma(t, K)^2 \mathbb{E}[D_{0t}a_t^2 \delta(S_t - K)]$$

where $\mathcal{C}_{\mathrm{m}}(t, K) = \mathbb{E}[\mathcal{P}_t]$ denotes the price of the call option in the model. Then, by using that $\partial_K \mathcal{C}_{\mathrm{m}}(t, K) = -\mathbb{E}[D_{0t}\mathbf{1}_{S_t>K}]$ and $\partial_K^2 \mathcal{C}_{\mathrm{m}}(t, K) = \mathbb{E}[D_{0t}\delta(S_t - K)]$, we deduce that

$$\partial_t \mathcal{C}_{\mathrm{m}}(t, K) = K\mathbb{E}[D_{0t}(r_t - q_t - (r_t^0 - q_t^0))\mathbf{1}_{S_t>K}] - (r_t^0 - q_t^0)K\partial_K \mathcal{C}_{\mathrm{m}}(t, K)$$
$$- \mathbb{E}[D_{0t}(q_t - q_t^0)(S_t - K)^+] - q_t^0 \mathcal{C}_{\mathrm{m}}(t, K) + \frac{1}{2}K^2 \partial_K^2 \mathcal{C}_{\mathrm{m}}(t, K)\frac{\mathbb{E}[D_{0t}a_t^2|S_t = K]}{\mathbb{E}[D_{0t}|S_t = K]}$$

with the initial condition $\mathcal{C}_{\mathrm{m}}(0, K) = (S_0 - K)^+$ so by uniqueness of the solution to this PDE the model is calibrated to the market smile of S if and only if

$$\partial_t \mathcal{C}(t, K) = K\mathbb{E}[D_{0t}(r_t - q_t - (r_t^0 - q_t^0))\mathbf{1}_{S_t>K}] - (r_t^0 - q_t^0)K\partial_K \mathcal{C}(t, K)$$
$$- \mathbb{E}[D_{0t}(q_t - q_t^0)(S_t - K)^+] - q_t^0 \mathcal{C}(t, K) + \frac{1}{2}K^2 \partial_K^2 \mathcal{C}(t, K)\frac{\mathbb{E}[D_{0t}a_t^2|S_t = K]}{\mathbb{E}[D_{0t}|S_t = K]}$$

From the definition of $\sigma_{\mathrm{Dup}}(t, K)$, this is equivalent to (12.42), which completes the proof.

$\square$

Chapter 13

Marked Branching Diffusions

Les arbres tardifs sont ceux qui portent les meilleurs fruits.[1]

— Molière, in *Le Malade Imaginaire*

In Chapter 7, we introduced first and second order (Markovian) backward stochastic differential equations (BSDEs). They provide a generalization of the Feynman-Kac theorem for nonlinear PDEs. Unfortunately, in practice, numerically solving BSDEs requires the computation of conditional expectations, typically using regression methods. Finding regressors of good quality is difficult, notably for multi-asset portfolios. This leads us to introduce a new promising method based on branching diffusions describing a marked Galton-Watson random tree. We first show how these branching diffusions can provide stochastic representations for solutions of a large class of semilinear parabolic PDEs in which the nonlinearity can be approximated by a polynomial function. We then briefly extend our method to fully nonlinear PDEs. Numerical examples, including the computation of the counterparty risk, illustrates the efficiency of our algorithm. Parts of this research have been published in [130] and [132], and presented in [131].

13.1 Nonlinear Monte Carlo algorithms for some semilinear PDEs

Before considering fully nonlinear PDEs, we first focus on a particular class of semilinear PDEs:

$$\partial_t u(t,x) + \mathcal{L}u(t,x) + f(u(t,x)) = 0, \qquad \forall(t,x) \in [0,T] \times \mathbb{R}^d \quad (13.1)$$
$$u(T,x) = g(x), \qquad \forall x \in \mathbb{R}^d$$

where $\mathcal{L}$ is the Itô generator of a (multi-dimensional) diffusion and f a (eventually smooth) function. For the sake of simplicity, we assume that f does

---

[1] "The trees that are slow to grow bear the best fruit."

not depend explicitly on t and x. Below, we review various Monte Carlo techniques for solving these semilinear PDEs and highlight their weaknesses. The BSDE method (Section 13.1.2) was presented in Chapter 7.

13.1.1 A brute force algorithm

Using Feynman-Kac's formula, the solution to PDE (13.1) can be represented stochastically as

$$u(t, x) = \mathbb{E}_{t,x}[g(X_T)] + \int_t^T \mathbb{E}_{t,x}[f(u(s, X_s))] \, ds \qquad (13.2)$$

with X an Itô diffusion with generator $\mathcal{L}$ and $\mathbb{E}_{t,x}[\cdot] = \mathbb{E}[\cdot | X_t = x]$. By assuming that $f(u) = \beta \tilde{f}(u)$ with β a small parameter, we get the first order approximation in β:

$$u(t, x) = \mathbb{E}_{t,x}[g(X_T)] + \int_t^T \mathbb{E}_{t,x}[f(\mathbb{E}_{s,X_s}[g(X_T)])] \, ds + o(\beta) \qquad (13.3)$$

This situation is common in practice where the nonlinearity appears as a small perturbation to the linear component. Then, at a next step, we discretize the Riemann integral into

$$u(t, x) \simeq \mathbb{E}_{t,x}[g(X_T)] + \sum_{i=1}^{n} \mathbb{E}_{t,x}[f(\mathbb{E}_{t_i,X_{t_i}}[g(X_T)])]\Delta t_i$$

This last expression can be numerically tackled by using a brute force nested Monte Carlo method. The embedded Monte Carlo (N_2 paths) is used to compute $\mathbb{E}_{t_i,X_{t_i}}[g(X_T)]$ on each of the N_1 paths generated by the Monte Carlo algorithm at the first level (see Figure 13.1). Although straightforward, this method suffers from two major drawbacks: (i) it is only approximate, and (ii) it requires generating $O(N_1 \times N_2 \times n)$ paths, where n is the number of discretization steps. Even with small numbers like $N_1 = N_2 = 10^3$ and $n = 10$, this means 10 million paths to simulate.

Can we design a simple (nonlinear) Monte Carlo algorithm which solves PDE (13.1), without relying on an approximation such as (13.3) and with a complexity of order $O(N_1)$?

13.1.2 Backward stochastic differential equations

A first approach, explained in Chapter 7, consists of simulating a 1-BSDE (see Exercise 7.4):

$$dX_t = b(t, X_t) \, dt + \sigma(t, X_t). \, dW_t, \qquad X_0 = x \qquad (13.4)$$
$$dY_t = -f(Y_t) \, dt + Z_t.(\sigma(t, X_t) \, dW_t) \qquad (13.5)$$
$$Y_T = g(X_T) \qquad (13.6)$$

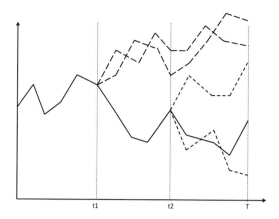

Figure 13.1: Brute force nested Monte Carlo method.

where (Y, Z) are required to be adapted processes and

$$\mathcal{L} = \sum_i b_i \partial_{x^i} + \frac{1}{2} \sum_{i,j} (\sigma \sigma^*)_{ij} \partial^2_{x^i x^j}$$

Let us divide $(0, T)$ into subintervals (t_{i-1}, t_i) and set $\Delta t_i = t_i - t_{i-1}$, $\Delta W_i = W_{t_i} - W_{t_{i-1}}$, and $\Delta = \max_i \Delta t_i$. Examples of numerical (explicit) schemes include

$$Y^{\Delta}_{t_n} = g(X^{\Delta}_{t_n})$$
$$Y^{\Delta}_{t_{i-1}} = \mathbb{E}_{i-1}[Y^{\Delta}_{t_i}] + f\left(\mathbb{E}_{i-1}[Y^{\Delta}_{t_i}]\right) \Delta t_i$$

and

$$Y^{\Delta}_{t_n} = g(X^{\Delta}_{t_n})$$
$$Y^{\Delta}_{t_{i-1}} = \mathbb{E}_{i-1}[Y^{\Delta}_{t_i}] + \mathbb{E}_{i-1}[f(Y^{\Delta}_{t_i})]\Delta t_i$$

with $\mathbb{E}_{i-1}[\cdot] = \mathbb{E}[\cdot | \mathcal{F}_{t_{i-1}}]$. They require the computation of the conditional expectation $\mathbb{E}_{t_{i-1}}[Y_{t_i}]$ or $\mathbb{E}_{i-1}[f(Y^{\Delta}_{t_i})]$ at each discretization date (in practice by regression methods), which could be quite difficult and/or time-consuming, especially for multi-asset portfolios.

13.1.3 Gradient representation

A more powerful approach, which does not rely on regression methods but is unfortunately only applicable in the scalar case $d = 1$, is synthesized by

Theorem 13.1 as proved in [56]. It relies on Kunita's theory of stochastic flows of diffeomorphisms that we summarize in the following proposition:

PROPOSITION 13.1 Stochastic flow [15]

Let $b, \sigma : [0, T] \times \mathbb{R} \to \mathbb{R}$ be bounded functions, together with their derivatives with respect to their second argument up to the order 3. Then for $t \leq s \leq T$, the one-dimensional SDEs[2]

$$dY_s^z = b(s, Y_s^z)\, ds + \sigma(s, Y_s^z)\, dW_s, \quad Y_t^z = z$$
$$d\zeta_s^a = (-b + \sigma\partial_2\sigma)(T + t - s, \zeta_s^a)\, ds + \sigma(T + t - s, \zeta_s^a)\, d\hat{W}_s, \quad \zeta_t^a = a \quad (13.7)$$

with W a Brownian motion and $\hat{W}_s ==W_{T+t-s} - W_T$ (a Brownian motion as well), admit a solution such that for each $(s, \omega) \in [t, T] \times \Omega$, the map $z \mapsto Y_s^z$ (resp. $a \mapsto \zeta_s^a$) is an increasing diffeomorphism of $\mathbb{R}$. The derivative $\partial_z Y_s^z$ solves the SDE

$$d\partial_z Y_s^z = \partial_2 b(s, Y_s^z)\partial_z Y_s^z\, ds + \partial_2 \sigma(s, Y_s^z)\partial_z Y_s^z\, dW_s$$

Moreover, ζ^a is the inverse flow of Y^z, i.e.,

$$\forall (s, z) \in [t, T] \times \mathbb{R}, \quad Y_s^{\zeta_{T+t}^a} = \zeta_{T+t-s}^a \quad -a.s.$$

PROOF (Sketch)   We are going to show that $Y_s^{\zeta_s^a} = a$. By introducing the process $dT_t = dt$, one can work with a time-homogeneous Itô diffusion. By taking $a \equiv \zeta_{T+t-s}^a$, this will imply that $Y_s^{\zeta_s^{\zeta_{T+t-s}^a}} = \zeta_{T+t-s}^a = Y_s^{\zeta_{T+t}^a}$ where we have used the Markov property and the time-homogeneity, $\zeta_s^{\zeta_{T+t-s}^a} = \zeta_{T+t}^a$. In the deterministic case (i.e., $\sigma = 0$) the result is obvious and corresponds to the time inversion $s \mapsto T + t - s$. In the stochastic case, the result can be obtained by switching to the Stratonovich form

$$dY_s^z = \left(b - \frac{1}{2}\sigma\partial_2\sigma\right)(s, Y_s^z)\, ds + \sigma(s, Y_s^z) \diamond dW_s$$

As Leibnitz's rule still applies with Stratonovich's calculus, we get as in the deterministic case

$$d\zeta_s^a = -\left(b - \frac{1}{2}\sigma\partial_2\sigma\right)(T + t - s, \zeta_s^a)\, ds - \sigma(T + t - s, \zeta_s^a) \diamond dW_s$$
$$= (-b + \sigma\partial_2\sigma)(T + t - s, \zeta_s^a)\, ds - \sigma(T + t - s, \zeta_s^a)\, dW_s$$
$$= (-b + \sigma\partial_2\sigma)(T + t - s, \zeta_s^a)\, ds + \sigma(T + t - s, \zeta_s^a)\, d\hat{W}_s$$

---

[2]∂_2 means the first order derivative w.r.t. Y.

which completes the proof. □

Then we have

THEOREM 13.1 see [56]
Let $u \in C^{1,2}(\mathbb{R}_+ \times \mathbb{R})$ be a smooth solution to the one-dimensional nonlinear PDE

$$\partial_t u(t,x) + \left(b(t,x)\partial_x + \frac{1}{2}\sigma^2(t,x)\partial_x^2 \right) u(t,x) + f(u(t,x)) = 0 \qquad (13.8)$$

$$u(T,x) = g(x)$$

with $g \in C^1(\mathbb{R})$ and $\lim_{|x|\to\infty} u(t,x) = 0$. The coefficients b and σ satisfy the assumptions in Proposition 13.1. We have the following representation

$$u(t,x) = -\mathbb{E}\left[\int_{\mathbb{R}} da \, \mathbf{1}_{\{X_T^a > x\}} g'(a) e^{\int_t^T f'(u(T+t-s, X_s^a)) ds} \right] \qquad (13.9)$$

where the Itô process $(X_s^a, s \in [t,T])$ is the solution to

$$dX_s^a = -(b - \sigma\partial_2\sigma)(T+t-s, X_s^a)\, ds + \sigma(T+t-s, X_s^a)\, dB_s \qquad (13.10)$$

where $X_t^a = a$, with B a standard Brownian motion.

PROOF By differentiating Equation (13.8) w.r.t. the variable x, we get

$$\partial_t \Delta(t,x) + \left((b(t,x) + \sigma\partial_x\sigma)\partial_x + \frac{1}{2}\sigma^2(t,x)\partial_x^2 \right) \Delta(t,x)$$

$$+ (\partial_x b(t,x) + f'(u(t,x)))\Delta(t,x) = 0$$

with $\Delta \equiv \partial_x u$ and the terminal condition $\Delta(T,x) = g'(x)$. From Feynman-Kac's theorem, we get

$$\Delta(t,x) = \mathbb{E}\left[g'(Y_T^x) e^{\int_t^T (\partial_x b(s, Y_s^x) + f'(u(s, Y_s^x))) ds} \right]$$

where Y_s^x is solution to the SDE

$$dY_s^x = (b + \sigma\partial_x\sigma)(s, Y_s^x)\, ds + \sigma(s, Y_s^x)\, dW_s, \qquad Y_t^x = x$$

By integration of the function Δ and using the condition $\lim_{|x|\to\infty} u(t,x) = 0$, this leads to

$$-u(t,x) = \int_{\mathbb{R}} \mathbf{1}_{\{y>x\}} \Delta(t,y)\, dy \qquad (13.11)$$

$$= \mathbb{E}\left[\int_{\mathbb{R}} dy \, \mathbf{1}_{\{y>x\}} g'(Y_T^y) e^{\int_t^T (\partial_x b(s, Y_s^y) + f'(u(s, Y_s^y))) ds} \right]$$

By setting $y = \zeta_T^a$ with ζ_s^a the inverse stochastic flow associated to Y_s^x (see Proposition 13.1, Equation (13.7))

$$d\zeta_s^a = -b(T + t - s, \zeta_s^a)\,ds + \sigma(T + t - s, \zeta_s^a)\,d\hat{W}_s \qquad (13.12)$$

we obtain

$$-u(t,x) = \mathbb{E}\left[\int_{\mathbb{R}} da\, \mathbf{1}_{\{\zeta_T^a > x\}} \partial_a \zeta_T^a g'(a) e^{\int_t^T \left(\partial_2 b(s, \zeta_{T+t-s}^a) + f'\left(u(s, \zeta_{T+t-s}^a)\right)\right) ds}\right]$$

where we have used that $Y_T^{\zeta_T^a} = a$ and $Y_s^{\zeta_T^a} = \zeta_{T+t-s}^a$. By using that

$$\partial_a \zeta_T^a = \exp\left(-\int_t^T \left(\partial_2 b + \frac{1}{2}(\partial_2\sigma)^2\right)(T + t - s, \zeta_s^a)\,ds\right.$$
$$\left. + \int_t^T \partial_2\sigma(T + t - s, \zeta_s^a)\,d\hat{W}_s\right)$$

we get

$$-u(t,x) = \mathbb{E}\left[\int_{\mathbb{R}} da\, \mathbf{1}_{\{\zeta_T^a > x\}} \frac{M_T}{M_t} g'(a) e^{\int_t^T f'(u(T+t-s, \zeta_s^a)) ds}\right]$$

with the martingale

$$M_u \equiv \exp\left(-\int_0^u \frac{1}{2}(\partial_2\sigma)^2(T + t - s, \zeta_s^a)\,ds + \int_0^u \partial_2\sigma(T + t - s, \zeta_s^a)\,d\hat{W}_s\right)$$

By switching from the current measure $\mathbb{P}$ to the measure $\mathbb{P}'$ defined by the Radon-Nikodym derivative

$$\frac{d\mathbb{P}'}{d\mathbb{P}}\Big|_{\mathcal{F}_T} = M_T$$

we get our final result. By using Girsanov's theorem, we prove that the dynamics (13.12) reads as (13.10) under $\mathbb{P}'$. □

Monte Carlo algorithm

Let us assume that $g' \geq 0$. The term $\frac{g'(a)}{\int_{\mathbb{R}} g'(y)\,dy}$ can be interpreted as the density of a random variable that we denote A.
The formula (13.9) can then be written as

$$-\frac{u(t,x)}{\int_{\mathbb{R}} g'(y)\,dy} = \mathbb{E}\left[\mathbf{1}_{\{X_T^A > x\}} e^{\int_t^T f'\left(u(T+t-s, X_s^A)\right) ds}\right] \qquad (13.13)$$

The above formula leads to the following algorithm written in the case where $t = 0$:

1. We simulate N independent particles at $t = 0$, $X_0^{a_i} = a_i$, according to the distribution of A.

2. Each independent particle, starting at $(a_i)_{1 \leq i \leq N}$, follows the dynamics (13.10).

3. Let us divide $(0, T)$ into subintervals (t_{i-1}, t_i) and set $\Delta t = t_i - t_{i-1}$. $\frac{u(0,x)}{\int_{\mathbb{R}} g'(y)\,dy}$ is then approximated by the recurrence equations

$$u^{N,\Delta t}(t_n, x) = g(x)$$

$$-\frac{u^{N,\Delta t}(t_{n-1}, x)}{\int g'(y)\,dy} = \frac{1}{N}\sum_{i=1}^{N}\mathbf{1}_{\{X_{t_n}^{a_i} > x\}} e^{f'\left(u(t_n, X_{t_{n-1}}^{a_i})\right)\Delta t}$$

$$-\frac{u^{N,\Delta t}(t_{n-2}, x)}{\int g'(y)\,dy} = \frac{1}{N}\sum_{i=1}^{N}\mathbf{1}_{\{X_{t_n}^{a_i} > x\}}$$

$$e^{f'\left(u(t_n, X_{t_{n-2}}^{a_i})\right)\Delta t + f'\left(u(t_{n-1}, X_{t_{n-1}}^{a_i})\right)\Delta t}$$

$$\vdots$$

$$-\frac{u^{N,\Delta t}(0, x)}{\int_{\mathbb{R}} g'(y)\,dy} = \frac{1}{N}\sum_{i=1}^{N}\mathbf{1}_{\{X_{t_n}^{a_i} > x\}} e^{\sum_{k=0}^{n-1} f'\left(u(t_{n-k}, X_{t_k}^{a_i})\right)\Delta t}$$

Although appealing because it does not involve regression methods, this algorithm is only valid in dimension one (see the link (13.11) between u and Δ) in which case we can simply use a finite difference scheme. In the next section, we introduce another type of purely forward Monte Carlo scheme for nonlinear PDEs, that is valid in high dimensions, and are based on branching diffusions.

REMARK 13.1 Duality   In the case where $f = 0$, $b = 0$, and $g(x) = \mathbf{1}_{\{x \geq K\}}$, Formula (13.9) gives the identity

$$\mathbb{E}[\mathbf{1}_{\{X_T > K\}}] = -\mathbb{E}[\mathbf{1}_{\{X_T^K > X_0\}}]$$

with

$$dX_s = \sigma(s, X_s)\,dW_s$$
$$dX_s^K = \sigma\partial_2\sigma(T - s, X_s^K)\,ds + \sigma(T - s, X_s^K)\,d\hat{W}_s, \qquad X_0^K = K$$

Note that g is not C^1, but the formula remains valid if we derive in the distribution sense. This allows us to get the fair value of a digital option for a strike K and maturity T, $\mathbb{E}[\mathbf{1}_{\{X_T > K\}}]$, as a function of X_0 using a single Monte Carlo simulation:

$$\lim_{N \to \infty} \frac{1}{N}\sum_{i=1}^{N}\mathbf{1}_{\{X_T^{(i)} > K\}} = \lim_{N \to \infty} \frac{1}{N}\sum_{i=1}^{N}\mathbf{1}_{\{X_T^{K,(i)} > X_0\}}$$

We simulate X_T^K and compute the payoff $\mathbf{1}_{\{X_T^{K,(i)}>X_0\}}$ for different values of X_0.

$\square$

13.2   Branching diffusions

13.2.1   Branching diffusions provide a stochastic representation of solutions of some semilinear PDEs

Branching diffusions have been first introduced by McKean and Skorokhod [162, 183] (see also Ikeda, Nagasawa, and Watanabe [140, 141, 142]) to give a probabilistic representation of the Kolmogorov-Petrovskii-Piskunov (KPP) PDE and more generally of semilinear PDEs of the type

$$\partial_t u + \mathcal{L}u + \beta(t)\left(\sum_{k=0}^{\infty} p_k u^k - u\right) = 0 \qquad \text{in } [0,T) \times \mathbb{R}^d \qquad (13.14)$$

$$u(T,x) = g(x) \qquad \text{in } \mathbb{R}^d$$

with $\beta \geq 0$. Here the nonlinearity is a power series in u where the coefficients p_k are required to be nonnegative and sum to one:

$$f(u) \equiv \sum_{k=0}^{\infty} p_k u^k, \qquad \sum_{k=0}^{\infty} p_k = 1, \qquad 0 \leq p_k \leq 1 \qquad (13.15)$$

The probabilistic interpretation of such an equation goes as follows (see [140, 141, 142]). Let a single particle start at the origin, follow an Itô diffusion on $\mathbb{R}^d$ with generator $\mathcal{L}$, after a mean $\beta(\cdot)$ exponential time (independent of X) die and produce k descendants with probability p_k ($k = 0$ means that the particle dies without generating offspring). Then, the descendants perform independent Itô diffusions on $\mathbb{R}^d$ (with the same generator $\mathcal{L}$) from their birth locations, die and produce descendants after a mean $\beta(\cdot)$ exponential times, etc. This process is called a d-dimensional branching diffusion with a branching rate $\beta(\cdot)$. β can also depend spatially on x or be itself stochastic (Cox process). This birth–death process describes a so-called Galton-Watson tree (see Figure 13.2 for examples with 2 and 3 descendants).

REMARK 13.2 $\beta(\cdot)$ **exponential time**   A $\beta(\cdot)$ exponential time can be generated as the first time to default τ of a Poisson process

$$\tau = \inf\left\{t \geq 0 \;\middle|\; \int_0^t \beta(s)\,ds \geq -\ln U\right\}$$

with U a uniform r.v. in $[0,1]$. The probability of survival in the interval $[0,T]$ is $e^{-\int_0^T \beta(s)ds}$ and the probability of default in $[t, t+dt]$ is $e^{-\int_0^t \beta(s)ds}\beta(t)\,dt$.
□

We denote by N_t the number of particles that are alive at time t and by $Z_t \equiv (z_t^1, \ldots, z_t^{N_t}) \in \mathbb{R}^{d \times N_t}$ their locations. We then consider the multiplicative functional defined by[3]

$$\hat{u}(t,x) = \mathbb{E}_{t,x}\Big[\prod_{i=1}^{N_T} g(z_T^i)\Big] \tag{13.16}$$

where $\mathbb{E}_{t,x}[\cdot] = \mathbb{E}[\cdot | N_t = 1, z_t^1 = x]$. Note that N_T can become infinite when $m = \sum_{k=0}^{\infty} k p_k > 1$ (super-critical regime, see [164]). A sufficient condition on g in order to have a well-behaved product is $||g||_\infty \le 1$. We will come back to this assumption later.

THEOREM 13.2 [162]
Let us assume that $||g||_\infty \le 1$. Then $\hat{u}$ is a viscosity solution to the semilinear PDE (13.14).

PROOF By conditioning on $\mathcal{F}_\tau$, with τ the first jump time of a Poisson process with intensity $\beta(t)$, we get from (13.16)

$$\hat{u}(t,x) = \mathbb{E}_{t,x}[\mathbf{1}_{\tau \ge T} g(z_T^1)] + \mathbb{E}_{t,x}\left[\mathbf{1}_{\tau < T}\sum_{k=0}^{\infty} p_k \mathbb{E}_\tau\left[\prod_{j=1}^{k}\prod_{i=1}^{N_T^j(\tau)} g(z_T^{i,j,z_\tau})\right]\right]$$

where z_T^{i,j,z_τ} is the position of the i-th particle at maturity T produced by the j-th particle generated at time τ. By using the independence of the offspring and the strong Markov property, we obtain

$$\hat{u}(t,x) = \mathbb{E}_{t,x}[\mathbf{1}_{\tau \ge T} g(z_T^1)] + \sum_{k=0}^{\infty} p_k \mathbb{E}_{t,x}\left[\mathbf{1}_{\tau < T}\prod_{j=1}^{k}\mathbb{E}_\tau\left[\prod_{i=1}^{N_T^j(\tau)} g(z_T^{i,j,z_\tau})\right]\right]$$

$$= \mathbb{E}_{t,x}[\mathbf{1}_{\tau \ge T} g(z_T^1)] + \sum_{k=0}^{\infty} p_k \mathbb{E}_{t,x}\left[\mathbf{1}_{\tau < T}\prod_{j=1}^{k}\hat{u}(\tau, z_\tau^1)\right]$$

$$= \mathbb{E}_{t,x}[\mathbf{1}_{\tau \ge T} g(z_T^1)] + \sum_{k=0}^{\infty} p_k \mathbb{E}_{t,x}[\hat{u}(\tau, z_\tau^1)^k \mathbf{1}_{\tau < T}]$$

---

[3] $\prod_{i=1}^{0} \equiv 1$ by convention.

$$= \mathbb{E}_{t,x}[e^{-\int_t^T \beta(s)ds}g(z_T^1)] + \sum_{k=0}^{\infty} p_k \int_t^T \mathbb{E}_{t,x}[\beta(s)e^{-\int_t^s \beta(u)du}\hat{u}(s, z_s^1)^k]\, ds$$

$$= \mathbb{E}_{t,x}\left[e^{-\int_t^T \beta(s)ds}g(z_T^1) + \int_t^T \beta(s)e^{-\int_t^s \beta(r)dr}f(\hat{u}(s, z_s^1))\, ds\right] \quad (13.17)$$

Then, by assuming that $||g||_\infty \leq 1$, $\hat{u}$ is uniformly bounded by 1 in $[0, T] \times \mathbb{R}^d$ from (13.16) and from (13.17) we get that $\hat{u}$ is a viscosity solution to PDE (13.14) (see Theorem 4.8). $\qquad \Box$

By assuming that PDE (13.14) satisfies a comparison principle (see Definition 4.6), we conclude that $\hat{u} = u$ is the unique viscosity solution to the semilinear PDE (13.14).

13.2.2 Superdiffusions

13.2.2.1 Definition

A first attempt to obtain a larger class of nonlinearities than those defined by (13.15)—for example $f(u) = u^2 + u$—is to consider superdiffusions, introduced by Dynkin (see [3, 17, 99]). We start by a formal definition:

DEFINITION 13.1 Superdiffusion   *A superdiffusion X_t is a measure-valued continuous Markov process starting at μ_0 satisfying for all positive function g:*

$$e^{-\langle u(t,\cdot),\mu_0\rangle} = \mathbb{E}[e^{-\langle g,X_t\rangle}|X_0 = \mu_0]$$

where $\langle f, \mu\rangle \equiv \int f d\mu$ and u is the nonnegative solution to

$$\partial_t u = \mathcal{L}u - \Psi(u), \qquad u(0, x) = g(x) \quad (13.18)$$

$$\Psi(u) = au + bu^2 + \int_0^{\infty} n(dr)[e^{-ru} - 1 + ru] \quad (13.19)$$

where $a \geq 0$, $b \geq 0$ and n is a measure on $(0, \infty)$ satisfying $\int_0^{\infty}(r \wedge r^2)\, n(dr) < \infty$.

Taking a Dirac mass at x for the initial measure, i.e., $\mu_0 = \delta_x$, the superdiffusion X_t satisfies for a positive payoff g:

$$e^{-u(t,x)} = \mathbb{E}[e^{-\langle g,X_t\rangle}|X_0 = \delta_x]$$

where u is a solution to PDE (13.18). The class of nonlinearities defined by (13.19) is more general than (13.15), in particular it contains $au + bu^2$ with arbitrary positive coefficients a and b. Note that for an appropriate choice of $n(dr)$, this PDE includes KPP-type PDEs. In the following section, we explain how to simulate a superdiffusion and therefore how to solve PDE (13.18).

13.2.2.2 Superdiffusions as infinite collections of branching diffusions

Superdiffusions can be numerically simulated by considering a large collection of branching diffusions leading to a stochastic algorithm for solving PDE (13.18).

We consider a branching diffusion starting from x at $t = 0$ with an intensity β_n and probabilities p_k^n that we define below. We define u_n as

$$e^{-\frac{1}{n}u_n(0,x)} = \mathbb{E}\left[\prod_{i=1}^{N_T} e^{-\frac{1}{n}g(z_T^i)}\right]$$

From Section 13.2.1, $e^{-\frac{1}{n}u_n(t,x)} \equiv U_n(t,x)$ is solution to

$$\partial_t U_n(t,x) + \mathcal{L}U_n(t,x) + \beta_n\left(f_n(U_n(t,x)) - U_n(t,x)\right) = 0, \quad U_n(T,x) = e^{-\frac{g(x)}{n}}$$

with $f_n(u) = \sum_{k=0}^{\infty} p_k^n u^k$. For use below, we define $v_n = n(1 - U_n)$, $g_n = n(1 - e^{-\frac{g}{n}})$ and

$$\Psi_n(u) = n\beta_n\left(f_n\left(1 - \frac{u}{n}\right) - 1 + \frac{u}{n}\right), \quad 0 < \frac{u}{n} < 1 \tag{13.20}$$

v_n is then a solution to

$$\partial_t v_n + \mathcal{L}v_n - \Psi_n(v_n) = 0, \quad v_n(T,x) = g_n(x)$$

By expanding the functions Ψ_n (and f_n) as power series in u and identifying the coefficients in (13.20), we find the intensity β_n and the probabilities p_k^n:

$$\beta_n = \frac{\Psi_n(n)}{np_0^n}, \quad p_1^n = 1 - n\frac{\Psi_n^{(1)}(n)}{\Psi_n(n)}p_0^n,$$

$$p_k^n = \frac{(-1)^k n^k \Psi_n^{(k)}(n)}{k!\Psi_n(n)}p_0^n, \quad k \geq 2 \tag{13.21}$$

The requirement $\sum_{k=0}^{\infty} p_k^n = 1$ gives the condition $\Psi_n(0) = 0$. Then as $n \to \infty$, we have $g_n \to_{n\to\infty} g$, and one can prove that $\Psi_n \to_{n\to\infty} \Psi$ (see Chapter 4 in [3]). Finally, we have $v_n \to_{n\to\infty} v$ with v a solution to (13.18) (after time inversion).

Example 13.1 $\Psi(u) = u^2$
Equations (13.21) with $\Psi_n(u) = \Psi(u) = u^2$ give

$$\beta_n = \frac{n}{p_0^n}, \quad p_1^n = 1 - 2p_0^n, \quad p_2^n = p_0^n, \quad 0 \leq p_0^n \leq \frac{1}{2}$$

By taking $p_0^n = \frac{1}{2}$, the infinite collection of branching diffusions is then characterized by $p_0^n = p_2^n = \frac{1}{2}$ and $\beta_n = 2n$, which explodes when $n \to \infty$. On this simple example, we observe that this method requires simulating a large number of branchings as the default intensity diverges and the nonlinearity is still restrictive.

We leave to the reader as an exercise the characterization of $\{p_k^n\}$ and β_n for a KPP-type PDE. $\square$

Example 13.2 $\Psi(u) = u^\alpha$, $1 < \alpha \le 2$

Proceeding similarly, we obtain

$$\beta_n = \frac{n^{\alpha-1}}{p_0^n}$$

$$p_1^n = 1 - \alpha p_0^n$$

$$p_k^n = \frac{(-1)^k \alpha \cdots (\alpha - k + 1)}{k!} p_0^n$$

By taking $p_0^n = \frac{1}{\alpha}$, we get a branching particle system characterized by $\beta_n = \alpha n^{\alpha-1}$, $p_1^n = 0$, $p_k^n = \frac{(-1)^k \alpha \ldots (\alpha-k+1)}{k!\alpha}$. Note that the nonlinearity u^α with $\alpha > 2$ cannot be reached by the superdiffusion construction as the probabilities p_k^n become negative. $\square$

This leads us to introduce a new class of branching diffusions that can be traced back to Le Jan-Sznitman [153] in the context of stochastic (Fourier) representations of solutions to the incompressible Navier-Stokes equation.

13.3   Marked branching diffusions

13.3.1   Definition and main result

At this stage, PDE (13.14) should be compared with the semilinear PDE (5.55) arising in the pricing of counterparty risk where $f(u) = u^+$. It seems too restrictive and unreasonable to approximate the nonlinearity u^+ by a polynomial of type (13.15) or even (13.19). A natural question is therefore to search if the probabilistic interpretation of the KPP PDE, leading to the forward Monte Carlo scheme described in Section 13.2.1, can be generalized to an arbitrary analytical nonlinearity for which the PDE is

$$\partial_t u + \mathcal{L}u + \beta(t)(F(u) - u) = 0, \qquad u(T, x) = g(x) \qquad (13.22)$$

with $F(u) = \sum_{k=0}^{\infty} a_k u^k$ a power series in u with radius of convergence $R = 1/\limsup_{n\to\infty} |a_n|^{\frac{1}{n}}$. For convenience, we write F as

$$F(u) = \sum_{k=0}^{\infty} \bar{a}_k p_k u^k$$

for some probabilities p_k that we leave unspecified for the moment. We have $p_k \neq 0$ and $\bar{a}_k = \frac{a_k}{p_k}$ for $a_k \neq 0$, and set $p_k = 0$ and $\bar{a}_k = 0$ otherwise. We show below that a probabilistic interpretation of (13.22) can be achieved by counting the number of branchings of each monomial u^k, i.e., the number of branchings where the dying particle gives birth to exactly k descendants.

For each Galton-Watson tree, we denote $\Omega_k \in \mathbb{N}$ as the number of branchings of monomial type u^k. We call this tree, endowed with the number of branchings, a marked Galton-Watson tree. The descendants are drawn with an arbitrary distribution p_k (we discuss how to choose this distribution in Section 13.3.4). In Figure 13.2, we have drawn the diagrams for the nonlinearity $F(u) = (\frac{a}{p_2})p_2 u^2 + (\frac{b}{p_3})p_3 u^3$ up to two defaults. We then define the multiplicative functional:

$$\hat{u}(t,x) = \mathbb{E}_{t,x}\left[\prod_{i=1}^{N_T} g(z_T^i) \prod_{k=0}^{\infty} \bar{a}_k^{\Omega_k}\right], \quad \Omega_k = \sharp \text{ branchings of type } k \qquad (13.23)$$

where we recall that

$$\bar{a}_k = \begin{cases} a_k/p_k & \text{if } a_k \neq 0 \\ 0 & \text{otherwise} \end{cases}$$

and that $\mathbb{E}_{t,x}[\cdot] = \mathbb{E}[\cdot|\tau > t, z_t^1 = x]$; τ stands for the first default time, i.e., the date of the first branching. In general both the payoff function $g(x)$ and the coefficients (a_k) have signs, so $\hat{u}$ is well defined only if

$$\mathbb{E}_{t,x}\left[\prod_{i=1}^{N_T} |g(z_T^i)| \prod_{k=0}^{\infty} |\bar{a}_k|^{\Omega_k}\right] < \infty$$

We now state our main result:

THEOREM 13.3

Let us assume that $\hat{u} \in \mathrm{L}^{\infty}([0,T] \times \mathbb{R}^d)$ with $||\hat{u}||_\infty < R$. Then the function $\hat{u}(t,x)$ is a viscosity solution to (13.22). If PDE (13.22) satisfies a comparison principle for sub- and supersolutions (see Definition 4.6), then $\hat{u}(t,x)$ is the unique viscosity solution to (13.22).

PROOF   The proof proceeds similarly as for Theorem 13.2. By using the independence of descendants and the strong Markov property, we obtain

$$\hat{u}(t,x) = \mathbb{E}_{t,x}[\mathbf{1}_{\tau \geq T} g(z_T^1)]$$

$$+ \sum_{k=0}^{\infty} \mathbb{E}_{t,x}\left[\mathbf{1}_{\tau < T} p_k \bar{a}_k \prod_{j=1}^{k} \mathbb{E}_\tau\left[\prod_{i=1}^{N_T^j(\tau)} g(z_T^{i,j,z_\tau}) \prod_{l=0}^{\infty} \bar{a}_l^{\Omega_l^j}\right]\right]$$

where Ω_l^j is the number of branching of type l produced by the j-th particle generated at time τ. Then we deduce

$$\hat{u}(t,x) = \mathbb{E}_{t,x}[\mathbf{1}_{\tau \geq T} g(z_T^1)] + \sum_{k=0}^{\infty} a_k \mathbb{E}_{t,x}\left[\mathbf{1}_{\tau < T} \prod_{j=1}^{k} \mathbb{E}_\tau\left[\prod_{i=1}^{N_T^j(\tau)} g(z_T^{i,j,z_\tau}) \prod_{l=0}^{\infty} \bar{a}_l^{\Omega_l^j}\right]\right]$$

$$= \mathbb{E}_{t,x}[\mathbf{1}_{\tau \geq T} g(z_T^1)] + \sum_{k=0}^{\infty} a_k \mathbb{E}_{t,x}\left[\mathbf{1}_{\tau < T} \prod_{j=1}^{k} \hat{u}(\tau, z_\tau^1)\right]$$

$$= \mathbb{E}_{t,x}[\mathbf{1}_{\tau \geq T} g(z_T^1)] + \sum_{k=0}^{\infty} a_k \mathbb{E}_{t,x}\left[\mathbf{1}_{\tau < T} \hat{u}(\tau, z_\tau^1)^k\right]$$

$$= \mathbb{E}_{t,x}[\mathbf{1}_{\tau \geq T} g(z_T^1)] + \mathbb{E}_{t,x}[F\left(\hat{u}(\tau, z_\tau^1)\right) \mathbf{1}_{\tau < T}]$$

$$= \mathbb{E}_{t,x}[e^{-\int_t^T \beta(s)ds} g(z_T^1)] + \int_t^T \mathbb{E}_{t,x}[\beta(s) e^{-\int_t^s \beta(r)dr} F\left(\hat{u}(s, z_s^1)\right)] ds$$

From Theorem 4.8, by assuming that $\hat{u} \in \mathrm{L}^\infty([0,T] \times \mathbb{R}^d)$ with $||\hat{u}||_\infty < R$, we deduce that $\hat{u}$ is a viscosity solution to PDE (13.22). If the comparison principle holds for PDE (13.22), it implies uniqueness, hence $u = \hat{u}$.   ☐

REMARK 13.3   The above computation shows that

$$e^{-\int_0^t \beta(s)ds} \hat{u}(t, z_t^1) - \int_t^T \beta(s) e^{-\int_0^s \beta(r)dr} \mathbb{E}\left[F\left(\hat{u}(s, z_s^1)\right)\big|\mathcal{F}_t\right] ds$$

$$= \mathbb{E}\left[e^{-\int_0^T \beta(s)ds} g(z_T^1)\big|\mathcal{F}_t\right]$$

is a martingale, hence has zero drift. For all $s > t$, the drift $\mu_t(s)$ of the martingale process $t \mapsto \mathbb{E}[F(\hat{u}(s, z_s^1))|\mathcal{F}_t]$ vanishes. As a consequence, the drift of the second term on the left hand side is

$$\beta(t) e^{-\int_0^t \beta(r)dr} \mathbb{E}\left[F\left(\hat{u}(t, z_t^1)\right)\big|\mathcal{F}_t\right] - \int_t^T \beta(s) e^{-\int_0^s \beta(r)dr} \mu_t(s) ds$$

$$= \beta(t) e^{-\int_0^t \beta(r)dr} F\left(\hat{u}(t, z_t^1)\right)$$

In the case where $\hat{u}$ is smooth, the Itô lemma implies that the drift of the process $e^{-\int_0^t \beta(s)ds}\hat{u}(t, z_t^1)$ is

$$e^{-\int_0^t \beta(s)ds}\left(\partial_t\hat{u} + \mathcal{L}\hat{u} - \beta(t)\hat{u}\right)(t, z_t^1)$$

so eventually we get

$$\partial_t\hat{u} + \mathcal{L}\hat{u} - \beta(t)\hat{u} + \beta(t)F(\hat{u}) = 0$$

which is precisely PDE (13.22). ☐

REMARK 13.4 When we replace the nonlinearity $F(u(t,x))$ by the term $F(\mathbb{E}[g(X_T)|X_t = x])$, which is independent of u, in (13.22) we get:

$$\partial_t u + \mathcal{L}u + \beta(t)\left(F(\mathbb{E}[g(X_T)|X_t = x]) - u\right) = 0, \quad u(T, x) = g(x) \quad (13.24)$$

Easily adapting the previous proof, we observe that this PDE has a stochastic representation in terms of marked branching diffusions where the particles, generated after a default, become immortal and are not allowed to die anymore.

This boils down to replacing $F(u)$ by $F(v)$ where v is the solution to

$$\partial_t v + \mathcal{L}v = 0, \quad v(T, x) = g(x) \quad (13.25)$$

i.e., the solution to PDE (13.22) with zero default rate: $\beta(t) \equiv 0$. In terms of credit valuation adjustment (CVA), (13.24) is linked to the risk-free close-out convention, while (13.22) corresponds to the risky close-out convention, see Section 13.4. ☐

In the next section, we give an interpretation of Formula (13.23) in terms of Feynman's tree diagrams.

13.3.2 Diagrammatic interpretation

From Feynman-Kac's formula, we have

$$u(t, x) = \mathbb{E}_{t,x}[\mathbf{1}_{\tau \geq T}g(X_T)] + \mathbb{E}_{t,x}[F(u(\tau, X_\tau))\mathbf{1}_{\tau < T}] \quad (13.26)$$

This integral equation can be recursively solved in terms of multiple exponential random times τ_i:

$$u(t, x) = \mathbb{E}_{t,x}[\mathbf{1}_{\tau_0 \geq T}g(X_T)]$$
$$+ \mathbb{E}_{t,x}[F\left(\mathbb{E}_{\tau_0}[\mathbf{1}_{\tau_1 \geq T}g(X_T)] + \mathbb{E}_{\tau_0}[F(\mathbb{E}_{\tau_1}[\mathbf{1}_{\tau_2 \geq T}g(X_T)])\mathbf{1}_{\tau_1 < T}]\right)\mathbf{1}_{\tau_0 < T}]$$
$$+ \cdots \quad (13.27)$$

By expanding F, each term can be interpreted as a Feynman diagram (see Figure 13.2) representing the trajectory of a branching diffusion with a weight

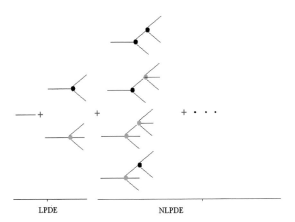

LPDE   NLPDE

Figure 13.2: Marked Galton-Watson random tree for the nonlinearity $F(u) = \frac{a}{p_2} p_2 u^2 + \frac{b}{p_3} p_3 u^3$. The black (resp. grey) vertex corresponds to the weight $\frac{a}{p_2}$ (resp. $\frac{b}{p_3}$). The diagram with two black vertices has the weights $\Omega_2 = 2, \Omega_3 = 0$.

depending on the branching of each monomial. For example, in Figure 13.2, the diagram with two black vertices corresponds to

$$\left(\frac{a_2}{p_2}\right)^2 \mathbb{E}_{t,x}[\mathbf{1}_{\tau_0 < T} \mathbb{E}_{\tau_0}[\mathbf{1}_{\tau_1 \geq T} g(X_T)] \mathbb{E}_{\tau_0}[\mathbf{1}_{\tau_2 < T} \mathbb{E}_{\tau_2}[\mathbf{1}_{\tau_3 \geq T} g(X_T)]^2]]$$

By assuming that the series (13.27) is convergent, one can guess that the solution is given by our multiplicative functional (13.23).

13.3.3   When is $\hat{u}$ bounded?

Theorem 13.3 requires that $\hat{u}$ be bounded, where $\hat{u}$ was defined by (13.23). In this section we derive sufficient conditions under which $\hat{u} \in L^\infty([0,T] \times \mathbb{R}^d)$ if g is bounded. The case where $g \equiv 0$ is trivial so we assume that $||g||_\infty > 0$. We also assume that $||g||_\infty < R$ so that $F(u)$ is well defined at maturity. We have

$$|\hat{u}(t,x)| \leq \mathbb{E}\left[||g||_\infty^{N_T} \prod_{k=0}^{\infty} |\bar{a}_k|^{\Omega_k} \,\Big|\, \tau > t, \, z_t^1 = x\right]$$

$$= \mathbb{E}\left[||g||_\infty^{N_T} \prod_{k=0}^{\infty} |\bar{a}_k|^{\Omega_k} \,\Big|\, \tau > t\right] \equiv \hat{v}(t) \quad (13.28)$$

Hence $\hat{u}$ is guaranteed to be bounded if $\hat{v}$ is bounded. Now, $\hat{v}$ can be easily computed because it corresponds to the *constant* payoff $||g||_\infty$ (a zero-coupon bond). We denote by G the nonlinearity

$$G(v) \equiv \sum_{k=0}^{\infty} |\bar{a}_k| p_k v^k = \sum_{k=0}^{\infty} |a_k| v^k$$

and by η the difference between a rescaled version of G and the identity function:

$$\eta(s) \equiv ||g||_\infty^{-1} G(||g||_\infty s) - s = \sum_{k=0}^{\infty} |a_k| \, ||g||_\infty^{k-1} s^k - s \qquad (13.29)$$

Note that G has same radius of convergence as F, and that η is defined on $\{s \in \mathbb{R} \,|\, |s| < \frac{R}{||g||_\infty}\}$, where $\frac{R}{||g||_\infty} > 1$.

LEMMA 13.1
$\hat{v}$ is solution to the ODE

$$v'(t) + \beta(t)(G(v(t)) - v(t)) = 0, \qquad v(T) = ||g||_\infty \qquad (13.30)$$

As a consequence,

$$\int_{||g||_\infty}^{\hat{v}(t)} \frac{ds}{G(s) - s} = \int_t^T \beta(s)\,ds \qquad \text{if } \eta(1) \neq 0 \qquad (13.31)$$

$$\hat{v}(t) = ||g||_\infty \qquad \text{if } \eta(1) = 0 \qquad (13.32)$$

In particular, $\hat{v} \in \mathrm{L}^\infty([0,T])$ if and only if

(i) $\eta(1) \leq 0$, or

(ii) $\eta(1) > 0$ and there exists $X \in [1, \frac{R}{||g||_\infty})$ such that

$$\int_1^X \frac{ds}{\eta(s)} = \int_0^T \beta(s)\,ds$$

In both cases, $||\hat{v}||_\infty < R$. In Case (i), $||\hat{v}||_\infty \leq ||g||_\infty$.

PROOF Mimicking the proof of Theorem 13.3, with g replaced by $||g||_\infty$ and F replaced by G, we get that

$$\hat{v}(t) = e^{-\int_t^T \beta(s)ds} ||g||_\infty + \int_t^T \beta(s) e^{-\int_t^s \beta(u)du} G\left(\hat{v}(s)\right) ds$$

whence

$$e^{-\int_0^t \beta(s)ds} \hat{v}(t) = e^{-\int_0^T \beta(s)ds} ||g||_\infty + \int_t^T \beta(s) e^{-\int_0^s \beta(u)du} G\left(\hat{v}(s)\right) ds$$

This shows that $\hat{v}$ is differentiable and is a solution of ODE (13.30). Noting that $G(||g||_\infty) = ||g||_\infty \iff \eta(1) = 0$, it is easy to check that the solution of this ODE is given by (13.31)-(13.32). It belongs to $L^\infty([0,T])$ if and only if, going backward from T to 0, the solution of ODE (13.30) has not blown up before reaching $t = 0$. If $\eta(1) = 0$, $\hat{v}$ is constant and hence does not blow up. Otherwise, $\hat{v}$ does not blow up before reaching $t = 0$ if and only if there indeed exists a real number X such that $\int_{||g||_\infty}^X \frac{ds}{G(s)-s} = \int_0^T \beta(s)\,ds$. Since by the change of variable $s \to ||g||_\infty s$,

$$\int_{||g||_\infty}^X \frac{ds}{G(s) - s} = \int_1^{\frac{X}{||g||_\infty}} \frac{ds}{\eta(s)}$$

it is easy to check that this is always verified if $\eta(1) < 0$ because in this case the function η, which is convex and nonnegative at $s = 0$, admits a unique zero $l \in [0,1)$, and $\int_1^l \frac{ds}{\eta(s)} = +\infty$. In this case $0 \le l||g||_\infty < \hat{v}(t) \le ||g||_\infty$ for all $t \le T$.   □

REMARK 13.5   When $\eta(1) > 0$ there might exist no $X \in \frac{R}{||g||_\infty}$ such that $\int_1^X \frac{ds}{\eta(s)} = \int_0^T \beta(s)\,ds$. For instance, if $\eta(s) > 0$ for all $s > 1$ and $\eta(s)$ becomes very large for increasing s, we may have $I \equiv \int_1^{R/||g||_\infty} \frac{ds}{\eta(s)} < \infty$. In this case X does exist if and only if $\int_0^T \beta(s)\,ds < I$, and the solution $\hat{v}$ of ODE (13.30) blows up (backward) at the time t_b defined by $\int_{t_b}^T \beta(s)\,ds = I$ - or it never blows up if $\int_{-\infty}^T \beta(s)\,ds < I$.   □

From (13.28) and Lemma 13.1 we get

PROPOSITION 13.2
Let us assume that $g \in L^\infty(\mathbb{R}^d)$ with $0 < ||g||_\infty < R$. Let $\eta(s)$ be defined by (13.29).

(i) If $\eta(1) \le 0$, $\hat{u} \in L^\infty([0,T] \times \mathbb{R}^d)$ for all T.

(ii) If $\eta(1) > 0$, we have $\hat{u} \in L^\infty([0,T] \times \mathbb{R}^d)$ (as defined by (13.23)) if there exists $X \in [1, \frac{R}{||g||_\infty})$ such that

$$\int_1^X \frac{ds}{\eta(s)} = \int_0^T \beta(s)\,ds$$

In both cases, $||\hat{u}||_\infty < R$. In Case (i), $||\hat{u}||_\infty \le ||g||_\infty$.

Note that, as expected, our blow up criterion does not depend on the probabilities p_k. For $a_k = p_k$ and $||g||_\infty \le 1$, we are in Case (i) and $\hat{u}$ is a viscosity solution of PDE (13.22) as it was proved for McKean KPP branching processes (see Theorem 13.2).

REMARK 13.6 In the particular case of only one branching type $k \neq 1$, i.e., $a_k \neq 0$ and $a_l = 0$ for all $l \neq k$, we have

$$\hat{v}(t) = \left(|a_k| + \left(\frac{1}{||g||_\infty^{k-1}} - |a_k| \right) e^{(k-1) \int_t^T \beta(s) \, ds} \right)^{-\frac{1}{k-1}} \quad (13.33)$$

Indeed, let $w(t) = \hat{v}(t)^{k-1}$. It satisfies the ODE

$$w'(t) + (k-1)\beta(t)(|a_k|w(t)^2 - w(t)) = 0, \qquad w(T) = ||g||_\infty^{k-1}$$

The solution is given by

$$\int_{||g||_\infty^{k-1}}^{w(t)} \frac{ds}{|a_k|s^2 - s} = (k-1) \int_t^T \beta(s) \, ds \quad \text{if } |a_k| \, ||g||_\infty^{k-1} \neq 1$$

$$w(t) = ||g||_\infty^{k-1} \quad \text{if } |a_k| \, ||g||_\infty^{k-1} = 1$$

Using that $s \mapsto \log \left| 1 - \frac{1}{|a_k|s} \right|$ is a primitive function of $s \mapsto \frac{1}{|a_k|s^2 - s}$, we get (13.33).

In this case the sufficient condition for the boundedness of $\hat{u}$ reads as

$$|a_k| \, ||g||_\infty^{k-1} \left(1 - e^{-(k-1) \int_0^T \beta(s) \, ds} \right) < 1 \quad (13.34)$$

☐

Note that the calculations in this section allow to compute the Laplace transform of the distribution of the number of branchings $\Omega = (\Omega_0, \Omega_1, \Omega_2, \ldots)$, see Exercise 13.7.1.

REMARK 13.7 For the linear PDE (13.24), if g is bounded with $||g||_\infty < R$, then $\hat{u}$ is always bounded. Indeed, in this case, from Remark 13.4,

$$\mathbb{P}\left(\forall k \in \mathbb{N}, \ \Omega_k = 0 | \tau > t \right) = e^{-\int_t^T \beta(s) ds}$$

$$\mathbb{P}\left(\forall k \in \mathbb{N}\backslash\{l\}, \ \Omega_k = 0 \quad \text{and} \quad \Omega_l = 1 | \tau > t \right) = \left(1 - e^{-\int_t^T \beta(s) ds} \right) p_l$$

$$\mathbb{P}\left(\exists k \neq l \in \mathbb{N}, \ \Omega_k \geq 1 \quad \text{and} \quad \Omega_l \geq 1 | \tau > t \right) = 0$$

so for all x

$$\hat{u}(t,x) \le \mathbb{E}\left[||g||_\infty^{N_T} \prod_{k=0}^\infty |\bar{a}_k|^{\Omega_k} \Big| \tau > t, z_t^1 = x\right]$$

$$= \mathbb{E}\left[||g||_\infty^{N_T} \prod_{k=0}^\infty |\bar{a}_k|^{\Omega_k} \Big| \tau > t\right]$$

$$= e^{-\int_t^T \beta(s)ds}||g||_\infty + \left(1 - e^{-\int_t^T \beta(s)ds}\right) \sum_{k=0}^\infty p_k ||g||_\infty^k |\bar{a}_k|$$

$$= e^{-\int_t^T \beta(s)ds}||g||_\infty + \left(1 - e^{-\int_t^T \beta(s)ds}\right) \sum_{k=0}^\infty |a_k| ||g||_\infty^k$$

$$\le \max\left(||g||_\infty, \sum_{k=0}^\infty |a_k| ||g||_\infty^k\right)$$

which is finite since $||g||_\infty < R$.   ▯

13.3.4   Optimal probabilities p_k

Let us assume that $\hat{u}$ is bounded. Which probabilities (p_k) should we pick when we use (13.23) to estimate the solution $\hat{u}(0,x)$ of (13.22)? An obvious choice is to choose the probabilities that minimize the variance of the random variable

$$Y \equiv \prod_{i=1}^{N_T} g(z_T^i) \prod_{k=0}^\infty \bar{a}_k^{\Omega_k}$$

For bounded payoffs g, we have

$$\text{Var}_x(Y) \le \mathbb{E}_x[Y^2] \le \mathbb{E}\left[||g||_\infty^{2N_T} \prod_{k=0}^\infty |\bar{a}_k|^{2\Omega_k} \Big| z_0^1 = x\right]$$

$$= \mathbb{E}\left[(||g||_\infty^2)^{N_T} \prod_{k=0}^\infty (|\bar{a}_k|^2)^{\Omega_k}\right] \equiv \hat{w}(0) \quad (13.35)$$

$\hat{w}(0)$ is simply $\hat{v}(0)$ (see (13.28)) with the replacements $||g||_\infty \longrightarrow ||g||_\infty^2$ and $|\bar{a}_k| \longrightarrow |\bar{a}_k|^2$, i.e., $|a_k| \longrightarrow \frac{a_k^2}{p_k}$. Let R_p denote the radius of convergence of the power series $\sum_{k=0}^\infty \frac{a_k^2}{p_k} s^k$. From Proposition 13.2, we have

PROPOSITION 13.3
Let us assume that $g \in \mathrm{L}^\infty(\mathbb{R}^d)$ with $0 < ||g||_\infty^2 < R_p$. For $|s| \le \frac{R_p}{||g||_\infty^2}$, let
$$\theta_p(s) = \sum_{k=0}^\infty \frac{a_k^2}{p_k} ||g||_\infty^{2k-2} s^k - s.$$

(i) If $\theta_p(1) \leq 0$, $\mathrm{Var}_x(Y) \leq ||g||_\infty^2$ for all x and T.

(ii) If $\theta_p(1) > 0$, we have $\mathrm{Var}_x(Y) < \infty$ if there exists $X \in [1, \frac{R_p}{||g||_\infty^2})$ such that

$$\int_1^X \frac{ds}{\theta_p(s)} = \int_0^T \beta(s)\,ds$$

In this case $\mathrm{Var}_x(Y) < R_p$ for all x.

It is then natural to ask the following questions: Under which conditions on (a_k) can we find $p = (p_k)$ such that $\mathrm{Var}_x(Y) < \infty$? Under which conditions on (a_k) do we have $\mathrm{Var}_x(Y) = \infty$ for all (p_k)? The following two propositions give partial results.

PROPOSITION 13.4
Let us assume that $g \in \mathrm{L}^\infty(\mathbb{R}^d)$. If $\sum_{k=0}^\infty |a_k|\,||g||_\infty^{k-1} \leq 1$, then there exists a probability distribution $p = (p_k)$ such that $\mathrm{Var}_x(Y) \leq ||g||_\infty^2$.

PROOF Let us set

$$p_k = \frac{|a_k|\,||g||_\infty^k}{\sum_{i=0}^\infty |a_i|\,||g||_\infty^i} \tag{13.36}$$

This is the probability distribution p that minimizes $\theta_p(1)$. Then

$$\theta_p(1) = \left(\sum_{k=0}^\infty |a_k|\,||g||_\infty^{k-1} \right)^2 - 1 \leq 0$$

so we are in Case (i) of Proposition 13.3: $\hat{w}(0) \leq ||g||_\infty^2$. □

PROPOSITION 13.5
Let us assume that $g \in \mathrm{L}^\infty(\mathbb{R}^d)$ with $0 < R < ||g||_\infty$. For $|s| \leq \left(\frac{R}{||g||_\infty} \right)^2$ let $\kappa(s) = \left(\sum_{k=0}^\infty |a_k|\,||g||_\infty^{k-1} s^{k/2} \right)^2 - s$. If

(i) $\sum_{k=0}^\infty |a_k|\,||g||_\infty^{k-1} > 1$

(ii) $\sum_{k=0}^\infty k|a_k|\,||g||_\infty^{k-1} \geq 1$

(iii) $\int_1^{\left(\frac{R}{||g||_\infty} \right)^2} \frac{ds}{\kappa(s)} \leq \int_0^T \beta(s)\,ds$

then $\hat{w}(0) = \infty$ for all probability distributions $p = (p_k)$. If g^2 is constant, this means that $\mathrm{Var}_x(Y) = \infty$ for all probability distributions $p = (p_k)$.

PROOF   Using Lagrange multipliers, one gets that, for a given $s \geq 1$, the probability distribution $p = (p_k)$ that minimizes $\theta_p(s)$ is

$$p_k = \frac{|a_k| \|g\|_\infty^k s^{k/2}}{\sum_{i=0}^\infty |a_i| \|g\|_\infty^i s^{i/2}}$$

The value of $\theta_p(s)$ is then $\kappa(s)$. As a consequence, for all probability distributions $p = (p_k)$, $\theta_p(s) \geq \kappa(s)$ for all $s \geq 1$. Besides, Conditions (i) and (ii) guarantee that $\kappa(s) > 0$ for all $s \geq 1$. Indeed, Condition (ii) means that $\eta'(1) \geq 0$ (where η was defined in (13.29)), and since η is convex this means that η is nondecreasing on $[1, +\infty)$. Now, $\eta(1) > 0$ because of Condition (i) so $\eta(s) > 0$ for all $s \geq 1$. Equivalently, $(\eta(\sqrt{s}) + \sqrt{s})^2 - s > 0$ for all $s \geq 1$, i.e., $\kappa(s) > 0$ for all $s \geq 1$. As a consequence, for all probability distributions $p = (p_k)$, $\theta_p(s) > 0$ for all $s \geq 1$ so, if it is finite, $\hat{w}(0)$ is given by $\int_1^{\frac{\hat{w}(0)}{\|g\|_\infty^2}} \frac{ds}{\theta_p(s)} = \int_0^T \beta(s)\, ds$. But

$$\int_1^{+\infty} \frac{ds}{\theta_p(s)} \leq \int_1^{+\infty} \frac{ds}{\kappa(s)} \leq \int_0^T \beta(s)\, ds$$

so $\hat{w}(0) = \infty$: the upper bound of $\mathrm{Var}_x(Y)$ is infinite. If besides g^2 is constant, then $\mathbb{E}_x[Y^2] = \hat{w}(0)$ so that $\mathrm{Var}_x(Y) = \infty$ for all probability distributions $p = (p_k)$. □

In the case where there exists a probability distribution (p_k) such that $\hat{w}(0) < \infty$, the distribution that minimizes $\hat{w}(0)$ is a solution to the Lagrange equation: there exists $\lambda \in \mathbb{R}$ such that for all k, $\partial_{p_k} \hat{w}(0) + \lambda = 0$. Since $\int_1^{\frac{\hat{w}(0)}{\|g\|_\infty^2}} \frac{ds}{\theta_p(s)} = \int_0^T \beta(s)\, ds$, we have

$$\frac{\partial_{p_k} \hat{w}(0)}{\|g\|_\infty^2} \frac{1}{\theta_p\left(\frac{\hat{w}(0)}{\|g\|_\infty^2}\right)} - \int_1^{\frac{\hat{w}(0)}{\|g\|_\infty^2}} \frac{\partial_{p_k} \theta_p(s)}{\theta_p(s)^2}\, ds = 0$$

REMARK 13.8   For the linear PDE (13.24), if g is bounded with $\|g\|_\infty < R$, from (13.35) and Remark 13.4, we have that the variance $\mathrm{Var}_x(Y)$ of the Monte Carlo estimator Y is bounded by

$$e^{-\int_0^T \beta(s)ds} \|g\|_\infty^2 + \left(1 - e^{-\int_0^T \beta(s)ds}\right) \sum_{k=0}^\infty \frac{a_k^2}{p_k} \|g\|_\infty^{2k}$$

which is finite if and only if $\|g\|_\infty^2 < R_p$. The probability distribution (13.36) is the one that minimizes this upper bound, which is then equal to

$$e^{-\int_0^T \beta(s)ds} \|g\|_\infty^2 + \left(1 - e^{-\int_0^T \beta(s)ds}\right) \left(\sum_{k=0}^\infty |a_k| \|g\|_\infty^k\right)^2$$

This is finite as $||g||_\infty < R$. ⬜

Note that the probability generating function of N_T, given by $\Pi(z) \equiv \mathbb{E}[z^{N_T}] = \mathbb{E}[\prod_{i=1}^{N_T} z]$, satisfies

$$\int_z^{\Pi(z)} \frac{ds}{\sum_{k=0}^\infty p_k s^k - s} = \int_0^T \beta(t)\, dt$$

By chossing $z = 0$, we get that the probability of extinction at time T, $\mathbb{P}(N_T = 0)$, is given by

$$\int_0^{\mathbb{P}(N_T=0)} \frac{ds}{\sum_{k=0}^\infty p_k s^k - s} = \int_0^T \beta(t)\, dt$$

13.3.5 Marked superdiffusions

Proceeding similarly as in Section 13.2.2, we define marked superdiffusions as limits of marked branching particle systems with the coefficients $\{a_k\}_k$ restricted to take the values $\pm p_k$ with (p_k) a probability. The marked superdiffusion is then linked to a signed measure-valued Markov process (as our coefficients have been restricted to take the values $\pm p_k$) and this leads to more general nonlinearities $\Psi(u)$ than (13.19), which can be obtained as the limit $n \to \infty$ of functions Ψ_n such that:

$$n\beta_n \left(\sum_{k=0}^\infty a_k^n v^k - v \right) = \Psi_n\left(n(1-v)\right) \tag{13.37}$$

By expanding the function Ψ_n as a power series in u and by identifying the coefficients in (13.37), we find

$$\beta_n = \frac{\Psi_n(n)}{na_0^n}, \qquad a_1^n = 1 - n\frac{\Psi_n^{(1)}(n)}{\Psi_n(n)}a_0^n$$

$$a_k^n = \frac{(-1)^k n^k \Psi_n^{(k)}(n)}{k!\Psi_n(n)}a_0^n, \qquad k \geq 2$$

Example 13.3 u^α, $\alpha > 2$
This nonlinearity, which is not reached by the superdiffusion construction (see Example 13.2), can be achieved with the marked superdiffusion with $a_0^n = \frac{1}{\alpha}$ and

$$\beta_n = \alpha n^{\alpha-1}$$
$$a_1^n = 0$$
$$a_k^n = (-1)^k \frac{\alpha \cdots (\alpha-k+1)}{k!\alpha} \equiv (-1)^k p_k^n$$

⬜

Table 13.1: Monte Carlo price quoted in percent as a function of the number of Monte Carlo paths 2^N. PDE pricer (NLPDE) = **21.82**. PDE pricer (LPDE) = **21.50**. Nonlinearity $F(u) = \frac{1}{2}\left(u^3 - u^2\right)$.

N	Fair (NLPDE)	Stdev (NLPDE)	Fair (LPDE)	Stdev (LPDE)
12	20.78	0.78	21.31	0.79
14	22.25	0.39	21.37	0.39
16	21.97	0.19	21.76	0.20
18	21.90	0.10	21.51	0.10
20	21.86	0.05	21.48	0.05
22	**21.81**	0.02	**21.50**	0.02

13.3.6 Numerical experiments

Before applying our marked branching diffusions algorithm to the problem of credit valuation adjustment introduced in Section 5.8, we check it on polynomials which do not belong to the classes defined by (13.15) and (13.19).

13.3.6.1 Experiment 1

We have implemented our algorithm for the two PDE types

$$\text{NLPDE}: \qquad \partial_t u + \mathcal{L}u + \beta(F(u) - u) = 0, \quad u(T, x) = \mathbf{1}_{x>1}$$

(NL stands for nonlinear) and

$$\text{LPDE}: \qquad \partial_t u + \mathcal{L}u + \beta(F(\mathbb{E}_{t,x}[\mathbf{1}_{X_T>1}]) - u) = 0, \quad u(T, x) = \mathbf{1}_{x>1}$$

(L stands for linear) with $F(u) = \frac{1}{2}\left(u^3 - u^2\right)$. $\mathcal{L}$ is the Itô generator of a geometric Brownian motion X with a volatility $\sigma_{\text{BS}} = 0.2$ and the Poisson intensity is $\beta = 0.05$. The maturity is $T = 10$ years. From (13.36), we note that our optimal probability distributions for LPDE and NLPDE coincide with the uniform distribution on $\{2, 3\}$. Moreover Proposition 13.2 gives that the solution to NLPDE does not blow up. For LPDE, this is automatically satisfied as particles become immortal after a default (hence $N_T = 2$ or 3). The numerical method has been checked against a one-dimensional PDE solver with a fully implicit scheme (see Table 13.1) for which we find $u = 21.82\%$ (NLPDE) and $u = 21.50\%$ (LPDE). Note that this algorithm converges as expected and the error is properly indicated by the Monte Carlo standard deviation estimator (see column Stdev).

13.3.6.2 Experiment 2

Same test with $F(u) = \frac{1}{3}\left(u^3 - u^2 - u^4\right)$ (see Table 13.2) and same comments as above.

Table 13.2: Monte Carlo price quoted in percent as a function of the number of Monte Carlo paths 2^N. PDE pricer (NLPDE) = **21.37**. PDE pricer (LPDE) = **20.39**. Nonlinearity $F(u) = \frac{1}{3}\left(u^3 - u^2 - u^4\right)$.

N	Fair (NLPDE)	Stdev (NLPDE)	Fair (LPDE)	Stdev (LPDE)
12	21.14	0.78	20.00	0.78
14	21.56	0.38	19.90	0.39
16	21.62	0.19	20.25	0.20
18	21.31	0.10	20.39	0.10
20	21.38	0.05	20.36	0.05
22	**21.36**	0.02	**20.40**	0.02

Table 13.3: Monte Carlo price quoted in percent as a function of the number of Monte Carlo paths 2^N. PDE pricer (NLPDE, $F(u) = u^2$) = **30.39**. PDE pricer (NLPDE, $F(u) = -u^2$) = **18.36**.

	$F(u) = u^2$		$F(u) = -u^2$	
N	Fair (NLPDE)	Stdev (NLPDE)	Fair (NLPDE)	Stdev (NLPDE)
12	30.20	0.72	17.50	0.81
14	30.75	0.36	18.55	0.41
16	30.47	0.18	18.58	0.20
18	30.55	0.09	18.74	0.10
20	30.51	0.04	18.48	0.05
22	30.39	0.01	18.36	0.01

13.3.6.3   Experiment 3

Same test with $F(u) = u^2$ and $F(u) = -u^2$ (see Table 13.3) for NLPDE and same comments as above. In comparison with the KPP equation $F(u) = u^2$, the replacement of the nonlinearity u^2 by $-u^2$ has added the term $(-1)^{N_T-1}$ in the multiplicative functional (see Equation (13.16)), without changing the complexity of the branching diffusion numerical algorithm:

$$u(0, x) = \mathbb{E}_{0,x}\left[(-1)^{N_T-1}\prod_{i=1}^{N_T}g(z_T^i)\right]$$

13.3.6.4   Experiment 4: Blow up

It is well known that the semilinear PDE in $\mathbb{R}^d \times [0, T]$

$$\partial_t u + \mathcal{L}u + u^2 = 0, \qquad u(T, x) = g(x)$$

Maturity (years)	BBM alg. (Stdev)	PDE
0.5	71.66 (0.09)	71.50
1	157.35 (0.49)	157.17
1.1	$\infty(\infty)$	∞

Table 13.4: Monte Carlo price quoted in percent as a function of the maturity for the nonlinearity $F(u) = u^2 + u$, with $g(x) \equiv \mathbf{1}_{x>1}$ and $\beta = 1$.

blows up in finite time if $d \leq 2$ for any bounded positive payoff g (see [189]). In particular, for $b = \sigma = 0$, i.e., $\mathcal{L} = 0$, the solution

$$u(t,x) = \begin{cases} \dfrac{1}{t - T + \frac{1}{g(x)}} & \text{if } g(x) \neq 0 \\ 0 & \text{if } g(x) = 0 \end{cases}$$

blows up at $t_{\min} = T - \frac{1}{\max_x g(x)}$ if g takes positive values, that is, if $\max_x g(x) > 0$.

We deduce that for bounded positive payoffs the PDE with the nonlinearity $F(u) = u^2 + u$ blows up in finite time $(t_{\min})$ in one dimension. Using Proposition 13.2, our sufficient condition reads as

$$\beta T \|g\|_\infty < 1$$

We have verified this explosion when the maturity T is greater than 1 year (in our case $g = \mathbf{1}_{x>0}$, $\|g\|_\infty = 1$, $\beta = 1$) using our algorithm (and a PDE solver as a benchmark); see Table 13.4. Note that for $T = 1$, the algorithm starts to blow up (Stdev $= 0.49$). A different stochastic representation can be obtained by setting $u = e^{(T-t)}v$. We get

$$\partial_t v + \mathcal{L}v + e^{(T-t)}v^2 - v = 0, \qquad v(T,x) = g(x)$$

and this can be interpreted as a binary tree with a weight $e^{(T-\tau)}$. Our stochastic representation then reads

$$u(t,x) = e^{T-t}\mathbb{E}_{t,x}\left[\prod_{i=1}^{N_T} g(z_T^i) e^{\sum_{i=1}^{\sharp \text{ branchings}}(T-\tau_i)} \right] \tag{13.38}$$

where τ_i is the time of the i-th branching. This representation (13.38) appears in [158] and was used to reproduce Sugitani's blow up criteria [189].

13.3.6.5 Experiment 5

In this example, we consider the degenerate PDEs:

$$\partial_t v_1 + x\partial_a v_1 + \frac{1}{2}\sigma^2 x^2 \partial_x^2 v_1 + \beta(v_1^2 - v_1) = 0, \quad \text{PDE1} \tag{13.39}$$

$$\partial_t v_2 + x\partial_a v_2 + \frac{1}{2}\sigma^2 x^2 \partial_x^2 v_2 + \beta(-v_2^2 - v_2) = 0, \quad \text{PDE2} \tag{13.40}$$

N	Fair (PDE1)	Stdev (PDE1)	Fair (PDE2)	Stdev (PDE2)
12	5.69	0.16	5.36	0.16
14	5.61	0.08	5.23	0.08
16	5.50	0.04	5.15	0.04
18	5.52	0.02	5.16	0.02
20	5.53	0.01	5.16	0.01
22	5.54	0.00	5.17	0.01

Table 13.5: Monte Carlo price quoted in percent as a function of the number of Monte Carlo paths 2^N. PDE pricer (PDE1) = **5.54**. PDE pricer (PDE2) = **5.17** (CPU PDE: 10 seconds). $T = 2$ years. Nonlinearity for PDE1 (resp. PDE2): $F_1(u) = u^2$ (resp. $F_2(u) = -u^2$). For completeness, the price with $\beta = 0$ (which can be obtained using a classical Monte Carlo pricer) is 6.52.

with $v_1(T, x, a) = v_2(T, x, a) = g(x, a) = (a - 1)^+$. These PDEs correspond to the backward SDEs

$$dX_t = \sigma X_t \, dB_t, \qquad X_0 = 1 \tag{13.41}$$
$$dA_t = X_t \, dt, \qquad A_0 = 0 \tag{13.42}$$
$$dY_t = -\beta \left(F(Y_t) - Y_t \right) dt + Z_t \, dB_t, \qquad Y_T = g(X_T, A_T) \tag{13.43}$$

with the nonlinearities $F_1(y) = y^2$ and $F_2(y) = -y^2$. In our numerical experiments, we have taken a diffusion coefficient $\sigma = 0.2$ and a Poisson intensity $\beta = 0.1$, and the maturity $T = 2$ or $T = 5$ years. For $T = 2$ years (resp. 5 years), the probability of default is around 0.18 (resp. 0.39).

Our branching diffusion algorithm has been checked against a two-dimensional PDE solver with an ADI scheme (see Tables 13.5, 13.6). The degenerate PDEs have been converted into elliptic PDEs by introducing the process $\tilde{A}_t = \int_0^t X_s \, ds + (T - t)X_t$, satisfying $d\tilde{A}_t = (T - t) \, dX_t$.

Note that our algorithm converges to the exact PDE result as expected and the error is properly indicated by the Monte Carlo standard deviation estimator (see column Stdev). In order to illustrate the impact of the nonlinearity F on the price v, we have also reported the price corresponding to $\beta = 0$.

13.4 Application: Credit valuation adjustment

13.4.1 Introduction

As a first application, we focus on the computation of counterparty risk, introduced in Section 5.8. Depending on the convention for the mark-to-

N	Fair (PDE1)	Stdev (PDE1)	Fair (PDE2)	Stdev (PDE2)
12	7.40	0.25	5.63	0.26
14	7.28	0.12	5.60	0.13
16	7.20	0.06	5.47	0.07
18	7.24	0.03	5.48	0.03
20	7.24	0.02	5.50	0.02
22	7.24	0.01	5.51	0.01

Table 13.6: Monte Carlo price quoted in percent as a function of the number of Monte Carlo paths 2^N. PDE pricer (PDE1) = **7.24**. PDE pricer (PDE2) = **5.51** (CPU PDE: 25 seconds). $T = 5$ years. Nonlinearity for PDE1 (resp. PDE2): $F_1(u) = u^2$ (resp. $F_2(u) = -u^2$). For completeness, the price with $\beta = 0$ (which can be obtained using a classical Monte Carlo pricer) is 10.24.

market value at default, the pricing equation is (see Equation (5.55))

$$\text{NLPDE}: \qquad \partial_t u + \mathcal{L}u + \beta\left(u^+ - u\right) = 0, \quad u(T, x) = g(x) \qquad (13.44)$$

or (see Equation (5.56))

$$\text{LPDE}: \qquad \partial_t u + \mathcal{L}u + \beta\left(\mathbb{E}_{t,x}[g(X_T)]^+ - u\right) = 0, \quad u(T, x) = g(x) \quad (13.45)$$

For NLPDE, we assume that the payoff is bounded: $g \in \text{L}^\infty$. The solution u of NLPDE can then be rescaled as $v = \frac{u}{||g||_\infty}$ where v satisfies

$$\partial_t v + \mathcal{L}v + \beta\left(v^+ - v\right) = 0, \qquad ||v(T, \cdot)||_\infty = 1$$

Therefore, by rescaling, we can consider that the payoff satisfies the condition $||g|| \leq 1$.

REMARK 13.9 The condition $g \in \text{L}^\infty$, not needed for LPDE, can be easily relaxed as observed in [105], see Remark 3.7. Let g be a payoff with α-exponential growth for some $\alpha > 0$. We scale the solution by an arbitrary smooth positive function ρ given by

$$\rho(x) \equiv e^{\alpha|x|} \quad \text{for } |x| \geq M$$
$$\tilde{v}(t, x) \equiv \rho^{-1}(x)v(t, x)$$

For example, in dimension one, if we write the operator $\mathcal{L}$ as $\mathcal{L}v = \mu(t, x)\partial_x v + \frac{1}{2}\sigma^2(t, x)\partial_x^2 v$, then $\tilde{v}$ satisfies a PDE with the same nonlinearity βv^+:

$$\partial_t \tilde{v} + \tilde{\mathcal{L}}\tilde{v} + \left(\mu\rho^{-1}\partial_x\rho + \frac{1}{2}\rho^{-1}\sigma^2\partial_x^2\rho\right)\tilde{v} + \beta\left(\tilde{v}^+ - \tilde{v}\right) = 0$$

with $\tilde{\mathcal{L}}\tilde{v} = \left(\mu + \sigma^2\rho^{-1}\partial_x\rho\right)\partial_x\tilde{v} + \frac{1}{2}\sigma^2(t, x)\partial_x^2\tilde{v}$. The dimension one plays no particular role here.   ☐

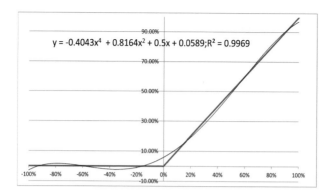

Figure 13.3: $x \mapsto x^+$ versus its polynomial approximation on $[-1, 1]$.

What remains to be done in order to use our marked branching diffusion algorithm is to approximate v^+ by a polynomial $F(v)$:

$$\partial_t v + \mathcal{L}v + \beta\left(F(v) - v\right) = 0, \qquad v(T, x) = g(x) \tag{13.46}$$

In our numerical experiments, we take (see Figure 13.3)

$$F(u) = 0.0589 + 0.5u + 0.8164u^2 - 0.4043u^4 \tag{13.47}$$

We will come back to this choice later (see Section 13.4.6). For such a choice, Proposition 13.2 gives that the solution does not blow up if $\beta T < 0.50829$ (take $X = \infty$ with $||g||_\infty = 1$). Moreover, as a numerical check of the expression (13.1), we have computed the solution to (13.46), using a PDE solver, with $g(x) = 1$, $\tilde{F}(u) = 0.0589 + 0.5u + 0.8164u^2 + 0.4043u^4$, $\beta = 0.05$ and $T = 10$ years. The solution $X = \hat{p}\left(T, -\ln\frac{|a_k|}{p_k}\right)$ coincides with our bound (13.1) and should satisfy

$$\int_1^X \frac{ds}{-s + 0.0589 + 0.5s + 0.8164s^2 + 0.4043s^4} = 0.5 \tag{13.48}$$

We found $X = 4.497$ (PDE solver) and the reader can check that this value satisfies the above identity (13.48) as expected.

13.4.2   Algorithm: Final recipe

The algorithm for solving LPDE (13.45) and NLPDE (13.44) can be described by the following steps:

1. Choose a polynomial approximation of $u^+ \simeq \sum_{k=0}^{M} a_k u^k$ on the domain $[-1, 1]$; see Section 13.4.6.

2. Simulate the assets and the Poisson default time with intensity β. Note that the intensity β can be stochastic (Cox process), usually calibrated to default probabilities implied from CDS market quotes.

3. At each default time, produce k descendants with probability p_k (given by (13.36)). For LPDE, descendants, produced after the first default, become immortal.

4. Evaluate

$$\prod_{i=1}^{N_T} g(z_T^i) \prod_{k=0}^{M} \left(\frac{a_k}{p_k} \right)^{\Omega_k}$$

where Ω_k denotes the number of branchings of type k. For LPDE, $N_T \in \{0, 1, \ldots, M\}$ and $\sum_{k=0}^{M} \Omega_k \in \{0, 1\}$ so the algorithm is always convergent for all T, i.e., the solution of LPDE never blows up.

REMARK 13.10   In the case of collateralized positions, the nonlinearity u_t^+ should be substituted with $(u_t - u_{t-\Delta})^+$ where $\Delta > 0$ is a delay (see Remark 5.3). Using our polynomial approximation, we get $F(u_t - u_{t-\Delta})$. By expanding this function, we get monomials of the form $\{u_t^p u_{t-\Delta}^q\}$. Our algorithm can then be easily extended to handle this case. At each default time τ, we produce p descendants starting at (τ, X_τ) and q descendants starting at $(\tau - \Delta, X_{\tau-\Delta})$. ☐

A natural question is to characterize the error of the algorithm as a function of the approximation error of u^+ by $F(u)$. Using the parabolicity of the semilinear PDE, we can characterize the bias of our algorithm:

PROPOSITION 13.6
*Let us assume that $\underline{F}(v)$ and $\overline{F}(v)$ are two polynomials satisfying (**Comp**), the sufficient condition in Proposition 13.2 for a maturity T, and*

$$\forall x \in \mathbb{R}, \qquad \underline{F}(x) \le x^+ \le \overline{F}(x)$$

We denote $\underline{v}$ and $\overline{v}$ as the corresponding solutions of (13.46) and v as the solution to (13.44). Then

$$\underline{v} \le v \le \overline{v}$$

PROOF   The function $\delta = \bar{v} - v$ satisfies the linear PDE

$$\partial_t \delta + \mathcal{L}\delta - \beta \left(1 - \frac{\bar{v}^+ - v^+}{\bar{v} - v} \mathbf{1}_{v \neq \bar{v}} \right) \delta + \beta \left(\overline{F}(\bar{v}) - \bar{v}^+ \right) = 0, \quad \delta(T, x) = 0$$

Note that the term $r \equiv 1 - \left(\frac{\bar{v}^+ - v^+}{\bar{v} - v} \right) \mathbf{1}_{v \neq \bar{v}}$ is lower bounded. Feynman-Kac's formula (Theorem 1.4) gives

$$\delta(t, x) = \int_t^T \beta \mathbb{E}_{t,x}[(\overline{F}(\bar{v}(s, X_s)) - \bar{v}^+(s, X_s)) \, e^{-\beta \int_t^s r(u, X_u) du}] \, ds$$

where X is the Itô diffusion with infinitesimal generator $\mathcal{L}$, from which we conclude that $\delta \geq 0$, i.e., $v \leq \bar{v}$, as $\overline{F}(x) \geq x^+$ by assumption. The proof of the other inequality is similar. □

A similar result can be found for PDE (13.45). In the case of American options where the nonlinearity is $\mathbf{1}_{v \geq g}$ instead of v^+, our algorithm, based on marked branching diffusions and a polynomial approximation of $\mathbf{1}_{v \geq g}$, gives lower and upper bounds (see Section 6.2).

13.4.3   Complexity

By approximating u^+ with a high order polynomial—say of order N_2—our algorithm can be compared to the brute force nested Monte Carlo method with a complexity at most $O(N_1 \times N_2)$. By comparison, with our choice (13.47), the complexity is at most $O(4N_1)$ for PDE type (13.44). Recall that for this PDE, we do not have multiple branchings.

13.4.4   Numerical examples

We have implemented our algorithm for the two PDE types

$$\partial_t u + \frac{1}{2} x^2 \sigma_{\mathrm{BS}}^2 \partial_x^2 u + \beta \left(u^+ - u \right) = 0, \quad u(T, x) = 1 - 2.1_{x>1}, \quad \text{NLPDE}$$

and

$$\partial_t u + \frac{1}{2} x^2 \sigma_{\mathrm{BS}}^2 \partial_x^2 u + \frac{\beta}{1 - R} \Big((1 - R)\mathbb{E}_{t,x}[1 - 2.1_{X_T>1}]^+$$

$$+ R\mathbb{E}_{t,x}[1 - 2.1_{X_T>1}] - u \Big) = 0, \quad u(T, x) = 1 - 2.1_{x>1}, \quad \text{LPDE}$$

with Poisson intensities $\beta = 1\%$, $\beta = 3\%$ and a recovery rate $R = 0.4$ (see Tables 13.7, 13.8, 13.9, and 13.10). In financial terms, this corresponds to CDS spreads around 100 and 300 basis points.[4] The method has been checked using

---

[4]From PDE (5.54), $\beta = (1 - R)\lambda_C$ with λ_C the intensity of default of the counterparty; β is thus linked to a CDS spread.

Table 13.7: Monte Carlo price quoted in percent as a function of the maturity for LPDE with $\beta = 1\%$; Stdev $= 0.00$ means that Stdev < 0.005.

Maturity (years)	PDE with poly.	BBM alg. (Stdev)	PDE
2	11.62	11.63 (0.00)	11.62
4	16.54	16.53 (0.00)	16.55
6	20.28	20.27 (0.00)	20.30
8	23.39	23.38 (0.00)	23.41
10	26.11	26.09 (0.00)	26.14

a PDE solver with the polynomial approximation (13.47) (see Column "PDE with poly.") In order to justify the validity of (13.47), we have included the PDE price with the true nonlinearity u^+ (see Column "PDE"). As it can be observed, the prices produced by our algorithm converge to the PDE prices with the polynomial approximation and are close to the exact PDE values. We would like to highlight that replacing the Black-Scholes generator $\frac{1}{2}x^2\sigma_{\mathrm{BS}}^2\partial_x^2$ by a multi-dimensional operator $\mathcal{L}$ can be easily handled in our framework by simulating the branching particles with a diffusion process associated to $\mathcal{L}$. This is out of reach with finite difference scheme methods, and not an easy step for the BSDE approach.

Note that in this example the square of the payoff is constant: $g^2 \equiv 1$ so, using the notations of Equation (13.35), $\hat{w}(0) = \mathbb{E}[Y^2]$ which does not depend on x. One can check that conditions (i) and (ii) of Proposition 13.5 are satisfied, and that $\int_1^{+\infty} \frac{ds}{\kappa(s)} \approx 0.2548$. From Proposition 13.5, we get that if $\beta T \geq 0.2548$ the variance of the Monte Carlo estimator is infinite *whatever the probability distribution* (p_k). For $\beta = 3\%$, this reads $T \geq 8.493$ years.

Here we have used the probability distribution (p_k) given by (13.36) for which $\theta_p(s) > 0$ for all $s \geq 1$ (see Proposition 13.3), so that if $\hat{w}(0) = \mathbb{E}[Y^2]$ is finite, it satisfies $\hat{w}(0) \geq 1$ and $\int_1^{\hat{w}(0)} \frac{ds}{\theta_p(s)} = \beta T$. Now, $\int_1^{+\infty} \frac{ds}{\theta_p(s)} \approx 0.2223$ so the variance of the Monte Carlo estimator is finite if and only if $\beta T < 0.2223$. For $\beta = 3\%$, this reads $T < 7.409$ years.

This means that the standard deviations that are reported in Table 13.10 for maturities 8 and 10 years are actually meaningless. They are a numerical artefact. Moreover, there is no way to choose (p_k) so as to get a finite variance for the 10 year maturity.

13.4.5 CVA formulas

We consider a payoff with multiple coupons $(F_1, \ldots, F_n)$ paid at dates $t_1 < \cdots < t_n = T$. Depending on our choice for the mark-to-market of the derivative evaluated at the time of default (with or without provision for counter-

Table 13.8: Monte Carlo price quoted in percent as a function of the maturity for NLPDE with $\beta = 1\%$; Stdev $= 0.00$ means that Stdev < 0.005.

Maturity (years)	PDE with poly.	BBM alg. (Stdev)	PDE
2	11.62	11.64 (0.00)	11.63
4	16.56	16.55 (0.00)	16.57
6	20.32	20.30 (0.00)	20.34
8	23.45	23.45 (0.00)	23.48
10	26.20	26.18 (0.00)	26.24

Table 13.9: Monte Carlo price quoted in percent as a function of the maturity for LPDE with $\beta = 3\%$. The number of Monte Carlo paths used corresponds to a standard deviation $\simeq 0.00$, which means that Stdev < 0.005. In practice, $N \simeq 2^{22}$.

Maturity (years)	PDE with poly.	BBM alg. (Stdev)	PDE
2	12.34	12.35 (0.00)	12.35
4	17.72	17.71 (0.00)	17.75
6	21.77	21.76 (0.00)	21.82
8	25.07	25.06 (0.00)	25.14
10	27.89	27.88 (0.00)	27.98

Table 13.10: Monte Carlo price quoted in percent as a function of the maturity for NLPDE with $\beta = 3\%$. The number of Monte Carlo paths used corresponds to a standard deviation $\simeq 0.00$, which means that Stdev < 0.005. In practice, $N \simeq 2^{22}$.

Maturity (years)	PDE with poly.	BBM alg. (Stdev)	PDE
2	12.38	12.39 (0.00)	12.39
4	17.88	17.86 (0.00)	17.91
6	22.08	22.07 (0.01)	22.14
8	25.58	25.57 (0.01)	25.66
10	28.62	28.60 (0.01)	28.74

party risk), we will get various CVA formulas that we review below. The arbitrage-free price of the derivative u at time t, seen from the point of view of the company facing counterparty risk, can be written as

$$u(t, x) = \mathbb{E}_{t,x}^{\mathbb{Q}} \left[\sum_{i=1}^{n} D_{tt_i} F_i 1_{\tau \geq t_n} \right] + \mathbb{E}_{t,x}^{\mathbb{Q}} \left[D_{t\tau} \tilde{u}_\tau 1_{\tau < t_n} \right] \qquad (13.49)$$

where $\mathbb{Q}$ is a risk-neutral measure, $\mathbb{E}_{t,x}^{\mathbb{Q}}[\cdot] = \mathbb{E}_{t}^{\mathbb{Q}}[\cdot | X_t = x]$, τ is the first time of jump of a Poisson process with intensity β, D_{tT} is the discount factor from T to t, and $\tilde{u}_\tau$ is the mark-to-market value of the derivative just after the default. The CVA is defined by

$$\mathrm{CVA}(F) \equiv \mathbb{E}^{\mathbb{Q}} \left[\sum_{i=1}^{n} D_{0t_i} F_i \right] - u(0, X_0)$$

$$= \mathbb{E}^{\mathbb{Q}} \left[\sum_{i=1}^{n} D_{0t_i} F_i 1_{\tau < t_n} \right] - \mathbb{E}^{\mathbb{Q}} \left[D_{0\tau} \tilde{u}_\tau 1_{\tau < t_n} \right]$$

At the default event, $\tilde{u}_\tau$ is given by

$$\tilde{u}_\tau = \sum_{i \,:\, t_i < \tau} D_{t_i \tau}^{-1} F_i + R M_\tau^+ - M_\tau^-$$

with M_τ the mark-to-market value of the derivative to be used in the unwinding of the position upon default, and R the recovery rate. The first term represents the past coupons capitalized up to the date τ. If the mark-to-market value M_τ is positive, we receive only a fraction amount, i.e., $R M_\tau^+$ with R the recovery rate. Otherwise, we pay M_τ^- if M_τ is negative.

If the mark-to-market of the derivative is evaluated at the time of default without provision for counterparty risk, then $M_\tau = \mathbb{E}_{\tau, X_\tau} [\sum_{i \,:\, t_i \geq \tau} D_{\tau t_i} F_i]$ and

$$\mathrm{CVA}(F) = (1 - R) \mathbb{E}^{\mathbb{Q}} \left[\left(\mathbb{E}_{\tau, X_\tau}^{\mathbb{Q}} \left[\sum_{i \,:\, t_i \geq \tau} D_{0t_i} F_i \right] \right)^+ 1_{\tau < t_n} \right] \qquad (13.50)$$

In the case of collateralized positions, counterparty risk applies to the variation of the mark-to-market value of the corresponding positions experienced over the time it takes to qualify a failure to pay margin as a default event—typically a few days Δ. At the default event, $\tilde{u}_\tau$ is given by

$$\tilde{u}_\tau = \sum_{i \,:\, t_i < \tau} D_{t_i \tau}^{-1} F_i + D_{[\tau]\tau}^{-1} M_{[\tau]} + R \left(M_\tau - D_{[\tau]\tau}^{-1} M_{[\tau]} \right)^+ - \left(M_\tau - D_{[\tau]\tau}^{-1} M_{[\tau]} \right)^-$$

where we denote $[t] = (t - \Delta)^+$. Here $M_\tau = \mathbb{E}^{\mathbb{Q}}_{\tau, X_\tau}[\sum_{i \,:\, t_i \geq \tau} D_{\tau t_i} F_i]$ and $M_{[\tau]} = \mathbb{E}^{\mathbb{Q}}_{[\tau]}\left[\sum_{i \,:\, t_i \geq \tau} D_{[\tau]t_i} F_i\right]$ so that

$$\text{CVA}(F) = (1 - R) \times$$

$$\mathbb{E}^{\mathbb{Q}}\left[\left(\mathbb{E}^{\mathbb{Q}}_{\tau, X_\tau}\left[\sum_{i \,:\, t_i \geq \tau} D_{0t_i} F_i\right] - \mathbb{E}^{\mathbb{Q}}_{[\tau]}\left[\sum_{i \,:\, t_i \geq \tau} D_{0t_i} F_i\right]\right)^+ \mathbf{1}_{\tau < t_n}\right] \quad (13.51)$$

Note that by definition, we do not include the coupon paying between $[\tau]$ and τ in $M_{[\tau]}$, so that the collateralized CVA is always smaller the non-collateralized CVA.

13.4.6 Building polynomial approximations of the positive part

13.4.6.1 Linear duality algorithm

The CVA for LPDE (13.45) requires computing $\mathbb{E}^{\mathbb{Q}}[\mathbf{1}_{\tau < T}\mathbb{E}^{\mathbb{Q}}_\tau[g(X_T)]^+]$ (see Equation (13.50) in the case of a path-dependent payoff). We assume that the defaults are bucketed at dates $(t_j)_{1 \leq j \leq N}$ and we define the random variables

$$Y_j \equiv \left(e^{-\int_0^{t_{j-1}} \beta(u)du} - e^{-\int_0^{t_j} \beta(u)du}\right) \mathbb{E}^{\mathbb{Q}}_{t_j}[g(X_T)]$$

Bucketing default times allows to reduce the variance of the Monte Carlo estimator of $\mathbb{E}^{\mathbb{Q}}[\mathbf{1}_{\tau < T}\mathbb{E}^{\mathbb{Q}}_\tau[g(X_T)]^+]$. The CVA then reads

$$\text{CVA}(g(X_T)) \approx (1 - R) \sum_{j=1}^N \text{CVA}_j, \qquad \text{CVA}_j \equiv \mathbb{E}^{\mathbb{Q}}\left[Y_j^+\right]$$

Let us fix j. By simulating a diffusion that branches at time t_j and produces exactly k independent descendants, we can estimate the moments $m_j^k \equiv \mathbb{E}^{\mathbb{Q}}[Y_j^k]$ for all $k \in \{1, \ldots, M\}$, M being the polynomial order. For use below, we set $v_j \equiv m_j^2 - (m_j^1)^2$. The term CVA_j can then be bounded by

$$\text{LB}_j \equiv \inf_{\mathbb{Q}' \in \mathcal{Q}} \mathbb{E}^{\mathbb{Q}'}[Y_j^+] \leq \text{CVA}_j \leq \text{UB}_j \equiv \sup_{\mathbb{Q}' \in \mathcal{Q}} \mathbb{E}^{\mathbb{Q}'}[Y_j^+]$$

where

$$\mathcal{Q} = \{\mathbb{Q}' \mid \forall k \in \{1, \ldots, M\}, \; \mathbb{E}^{\mathbb{Q}'}[Y_j^k] = \mathbb{E}^{\mathbb{Q}}[Y_j^k]\}$$

Assumption: We will assume that the support of the random variable Y_j is included in the interval $[x_{\min}, x_{\max}] \equiv [m_j^1 - \text{nbstdev}\sqrt{v_j}, m_j^1 + \text{nbstdev}\sqrt{v_j}]$ for some constant nbstdev.

We denote

$$UB_{nbstdev,j} \equiv \sup_{\mathbb{Q}' \in \mathcal{Q} \text{ s.t. supp}(\mathbb{Q}') \subset [x_{min}, x_{max}]} \mathbb{E}^{\mathbb{Q}'}[Y_j^+]$$

$$LB_{nbstdev,j} \equiv \inf_{\mathbb{Q}' \in \mathcal{Q} \text{ s.t. supp}(\mathbb{Q}') \subset [x_{min}, x_{max}]} \mathbb{E}^{\mathbb{Q}'}[Y_j^+]$$

By linear duality, the upper bound $UB_{nbstdev,j}$ can then be written as

$$UB_{nbstdev,j} = \inf_{(a_k)_{0 \le k \le M}} \sum_{k=0}^{M} a_k m_j^k \tag{13.52}$$

where the coefficients $(a_k)_{0 \le k \le M}$ are such that

$$\forall x \in [x_{min}, x_{max}], \qquad \sum_{k=0}^{M} a_k x^k \ge x^+$$

We denote $(a_k^{j,*})_{0 \le k \le M}$ the optimal coefficients. The optimal polynomial P_j^* is then $P_j^*(x) = \sum_{k=0}^{M} a_k^{j,*} x^k$ and the upper bound is

$$UB_{nbstdev,j} = \sum_{k=0}^{M} a_k^* m_j^k$$

Note that by duality this upper bound can be written as

$$UB_{nbstdev,j} = \mathbb{E}^{\mathbb{Q}_j^*}[Y_j^+]$$

with respect to a probability measure $\mathbb{Q}_j^*$ satisfying the moment constraints $\mathbb{E}^{\mathbb{Q}_j^*}[Y_j^k] = m_j^k$. This upper bound can therefore be interpreted as an exact piece of CVA with respect to $\mathbb{Q}_j^*$. Note that in practice, the risk-neutral measure $\mathbb{Q}$ is not perfectly known and $\mathbb{Q}_j^*$ can be seen as a risk provision on $\mathbb{Q}$. A similar construction is achieved for the lower bound $LB_{nbstdev,j}$.

13.4.6.2 Numerical experiments

We have checked our construction of the optimal polynomials with some numerical experiments. We have chosen a Black-Scholes model with a lognormal volatility $\sigma_{BS} = 0.2$ and a spot $X_0 = 1$. The recovery rate R is fixed to a standard value 0.4 and the intensity $\beta = 0.025$ corresponds to a CDS around 150 basis points.[5] Each year we pay a call spread option with payoff $(\frac{X_T}{X_0} - 0.9)^+ - (\frac{X_T}{X_0} - 1.1)^+ - 0.1$ and the maturity of the derivative ranges from 2 years up to 8 years. We have reported the lower and upper bound of the CVA (in percent) as a function of the order M of the polynomial (see Tables 13.11 and 13.12). As expected, as M increases, the upper (resp. lower) bound

---

[5]We have CDS $\approx (1 - R)\beta$.

T (Years)	$M = 3$	$M = 4$	$M = 5$	$M = 6$	Exact
2	0.10	0.10	0.10	0.09	0.09
4	0.31	0.29	0.29	0.28	0.27
6	0.59	0.57	0.54	0.54	0.51
8	0.94	0.92	0.88	0.85	0.81

Table 13.11: Upper CVA. Call spread $\left(\frac{X_T}{X_0} - 0.9\right)^+ - \left(\frac{X_T}{X_0} - 1.1\right)^+ - 0.1$. Black-Scholes model, $\sigma_{\mathrm{BS}} = 0.2$, $\beta = 0.025$, and $R = 0.4$.

T (Years)	$M = 3$	$M = 4$	$M = 5$	$M = 6$	Exact
2	0.05	0.08	0.08	0.08	0.09
4	0.16	0.23	0.23	0.23	0.27
6	0.31	0.43	0.43	0.44	0.51
8	0.51	0.68	0.70	0.70	0.81

Table 13.12: Lower CVA. Call spread $\left(\frac{X_T}{X_0} - 0.9\right)^+ - \left(\frac{X_T}{X_0} - 1.1\right)^+ - 0.1$. Black-Scholes model, $\sigma_{\mathrm{BS}} = 0.2$, $\beta = 0.025$, and $R = 0.4$.

converges toward the exact price, computed using a closed-form formula for the mark-to-market value of the derivative. In Figure 13.4, we have plotted some examples of polynomial approximations (rescaled on $[-1, 1]$) for $M = 2$ and $M = 4$ at two different dates t_j.

13.5 System of semilinear PDEs

13.5.1 Introduction

Before jumping to fully nonlinear PDEs, we first consider systems of semilinear PDEs with polynomial nonlinearities: for all $i \in \{0, \ldots, N\}$,

$$\partial_t u_i + \mathcal{L}u_i + \beta_i(F_i(u_0, \ldots, u_N) - u_i) = 0, \quad u_i(T, x) = g_i(x) \quad (13.53)$$

where

$$F_i(u_0, \ldots, u_N) = \sum_{j=0}^{\infty} M_{ij} \prod_{p=0}^{N} u_p^{\mu_p^i(j)}$$

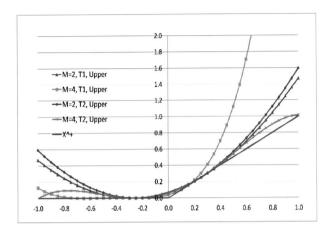

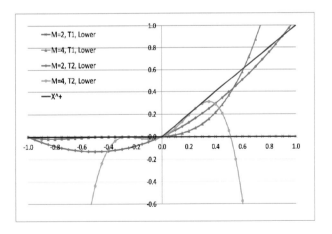

Figure 13.4: Some upper and lower polynomial approximations of $x \mapsto x^+$ (on $[-1, 1]$) as obtained in the examples of Tables 13.11 and 13.12.

with $\mu_p^i(j) \in \mathbb{N}$ and $M_{ij} \in \mathbb{R}$. The coefficients M_{ij} can also depend on (t, x). For use below, we write F_i as

$$F_i(u_0, \ldots, u_N) = \sum_{j=0}^{\infty} \bar{M}_{ij} p_{ij} \prod_{p=0}^{N} u_p^{\mu_p^i(j)}$$

where

$$\bar{M}_{ij} = \begin{cases} M_{ij}/p_{ij} & \text{if } M_{ij} \neq 0 \\ 0 & \text{otherwise} \end{cases}$$

with $p_{ij} \in [0, 1]$, $\sum_{j=0}^{\infty} p_{ij} = 1$, and $p_{ij} = 0$ if $M_{ij} = 0$.

13.5.2 Stochastic representation using multi-species marked branching diffusions

This PDE system has a nice stochastic representation in terms of multi-species marked branching diffusions. We define $N+1$ independent marked branching diffusions with offspring generating function:

$$F_i(u_0, \ldots, u_N) = \sum_{j=0}^{\infty} p_{ij} \prod_{p=0}^{N} u_p^{\mu_p^i(j)}$$

When an individual of the species (type) i dies, it generates with probability p_{ij}, $\mu_0^i(j)$ individuals of type 0, $\mu_1^i(j)$ individuals of type 1, etc., up to $\mu_N^i(j)$ individuals of type N. We denote by N_T^j the number of individuals of the species j that are alive at time T and $(z_T^{1,j}, \ldots, z_T^{N_T^j,j})$ their positions; Ω_k^j denotes the number of branchings of type k for the species j. We have

THEOREM 13.4

$\hat{u}_i$ defined by

$$\hat{u}_i(t, x) = \mathbb{E}\left[\prod_{j=0}^{N} \prod_{i=1}^{N_T^j} g_j(z_T^{i,j}) \prod_{j=0}^{N} \prod_{k=0}^{\infty} (\bar{M}_{jk})^{\Omega_k^j} \,\Big|\, z_t^i = x, N_t^j = \delta_{ij}, \tau^j > t \right] \quad (13.54)$$

is a viscosity solution to the PDE system (13.53).

The proof is not reported here as it duplicates the proof of Theorem 13.3. It is left to the reader as an exercise.

13.6 Fully nonlinear PDEs

In this last section, we briefly explain how our algorithm based on multi-type marked branching diffusion can be extended to fully nonlinear PDEs. Our approach is first illustrated on a classical semilinear PDE, the so-called deterministic Kardar-Parisi-Zhang (KPZ) PDE, which describes the relaxation of an initial rough surface to a flat one.

13.6.1 A toy example: The Kardar-Parisi-Zhang PDE

The one-dimensional (deterministic) Kardar-Parisi-Zhang PDE is given by

$$\partial_t u + \frac{1}{2}\partial_x^2 u + \frac{1}{2}(\partial_x u)^2 = 0, \qquad u(T,x) = g(x) \tag{13.55}$$

A particularly nice property of this PDE is that it can be converted into the (linear) heat kernel equation

$$\partial_t U + \frac{1}{2}\partial_x^2 U = 0, \qquad U(T,x) = e^{g(x)}$$

by applying the transform $u = \ln U$. From Feynman-Kac's formula, we obtain the closed-form solution

$$u(t,x) = \ln \mathbb{E}_{t,x}\big[e^{g(W_T)}\big]$$

with W a Brownian motion.

13.6.2 Bootstrapping

Here we take $g \in C^\infty(\mathbb{R})$. We set $u_0 = u$, $u_i = \partial_x^{(i)} u$ and PDE (13.55) can be written as

$$\partial_t u_0 + \frac{1}{2}\partial_x^2 u_0 + \frac{1}{2}u_1^2 = 0, \qquad u(T,x) = g(x) \tag{13.56}$$

The bootstrapping method, commonly used for proving regularity estimates of solutions of nonlinear PDEs (see e.g., the DeGiorgi-Moser-Nash theorem), consists of differentiating PDE (13.56) iteratively with respect to x:

$$\partial_t u_1 + \frac{1}{2}\partial_x^2 u_1 + u_1 u_2 = 0, \qquad u_1(T,x) = \partial_x g(x)$$

$$\partial_t u_2 + \frac{1}{2}\partial_x^2 u_2 + \big(u_2^2 + u_1 u_3\big) = 0, \qquad u_2(T,x) = \partial_x^2 g(x)$$

and so on. This leads us precisely to an infinite semilinear PDE system with polynomial nonlinearity for which we can use our multi-type marked branching

Table 13.13: Monte Carlo price quoted in percent as a function of the number of Monte Carlo paths 2^N. $T = 1$ year. Exact price $-\frac{\sigma^2}{2} \ln\left(1 - \frac{2}{3}T\right) = \mathbf{2.20}$. $\beta = 1$, $\sigma = 0.2$, $g(x) = x^2/3$. Blow up for $T \geq 1.5$ as expected.

N	Fair	Stdev
12	2.01	0.09
14	2.40	0.08
16	2.14	0.09
18	2.19	0.03
20	**2.20**	0.02

diffusion algorithm described in Section 13.5. From a numerical point of view, our algorithm can deal without any modification with this infinite-dimensional system. Up to the maturity T, we generate species.

This approach can be generalized for a fully nonlinear PDE

$$\partial_t u + \mathcal{L}u + f(t, x, u, Du, D^2u) = 0$$

where f is a polynomial in u, Du and D^2u.

We have numerically illustrated this method in Table 13.13 where we consider a (smooth) payoff $g(x) = x^2/3$ and a maturity $T = 1$ year. $\frac{1}{2}\partial_x^2$ has been replaced by the generator of a geometric Brownian motion with a volatility 0.2. Note that $\partial_x^3 g(x) = 0$, so the infinite semilinear PDE system degenerates into a two-dimensional PDE system involving only 3 species: u_0, u_1, and u_2. Indeed, if species $n \geq 3$ appear, the multiplicative functional will involve a payoff $g_n = 0$ and its contribution will cancel. Stated otherwise, the unique solution to the infinite system

$$\partial_t u_3 + \mathcal{L}u_3 + 3u_2 u_3 + u_1 u_4 = 0, \qquad u_3(T, x) = 0$$
$$\partial_t u_4 + \mathcal{L}u_4 + \left(3u_3^2 + 4u_2 u_4 + u_1 u_5\right) = 0, \qquad u_4(T, x) = 0$$
$$\vdots$$

is $u_n = 0$ for all $n \geq 3$.

13.6.3   The uncertain volatility model

As an example of fully nonlinear PDE, we consider a one-dimensional UVM (possibly depending on path-dependent variables) as described in Chapter 9. Here for the sake of simplicity, the normal volatility (as opposed to the lognormal volatility) is uncertain:

$$\partial_t u + \frac{1}{2}\underline{\sigma}^2 \partial_x^2 u + \frac{1}{2}\left(\overline{\sigma}^2 - \underline{\sigma}^2\right)\left(\partial_x^2 u\right)^+ = 0, \qquad u(T, x) = g(x)$$

Table 13.14: Monte Carlo price quoted in percent as a function of the number of Monte Carlo paths 2^N. $T = 10$ years. Exact price = **20**. "Nonlinearity" $P(\Gamma) = \Gamma$, $\bar{\sigma} = 0.2$, $g(x) = x^2/2$.

N	Fair	Stdev
12	20.18	0.51
14	20.13	0.26
16	19.94	0.13
18	19.94	0.06
20	19.96	0.03

By setting $u = e^{\beta(T-t)}v$ with $\beta = \frac{1}{2}\left(\bar{\sigma}^2 - \underline{\sigma}^2\right)$, we get:

$$\partial_t v + \frac{1}{2}\underline{\sigma}^2 \partial_x^2 v + \beta\left(\left(\partial_x^2 v\right)^+ - v\right) = 0, \qquad v(T,x) = g(x)$$

We approximate the nonlinearity Γ^+ by a polynomial $P(\Gamma)$:

$$\partial_t v + \frac{1}{2}\underline{\sigma}^2 \partial_x^2 v + \beta\left(P\left(\partial_x^2 v\right) - v\right) = 0$$

The bootstrapping method gives the semilinear PDE system

$$\partial_t v_0 + \frac{1}{2}\underline{\sigma}^2 \partial_x^2 v_0 + \beta\left(P\left(v_2\right) - v_0\right) = 0, \quad v_0(T,x) = g(x)$$

$$\partial_t v_1 + \frac{1}{2}\underline{\sigma}^2 \partial_x^2 v_1 + \beta\left(P'\left(v_2\right)v_3 - v_1\right) = 0, \quad v_1(T,x) = g'(x)$$

$$\partial_t v_2 + \frac{1}{2}\underline{\sigma}^2 \partial_x^2 v_2 + \beta\left(P''\left(v_2\right)v_3^2 + P'\left(v_2\right)v_4 - v_2\right) = 0, \quad v_2(T,x) = g^{(2)}(x)$$

and so on. We use the multi-type marked branching diffusion algorithm as described in Section 13.5. We have illustrated this method in Tables 13.14 and 13.15 where we have considered the "nonlinearity" $P(x) = x$ and the nonlinearity $P(x) = x^2/2$ with a payoff $g(x) = x^2/2$ and a maturity $T = 10$ years. In the first case, this is equivalent to pricing in a Black-Scholes model with a volatility $\bar{\sigma}$ by simulating a Brownian motion with a volatility $\underline{\sigma}$.

13.6.4   Exact simulation of one-dimensional SDEs

As a striking application, we consider the exact simulation of a one-dimensional Itô diffusion:

$$dY_t = b(Y_t)\,dt + \sigma(Y_t)\,dB_t$$

This was first considered in [60, 61]. Let us assume that σ is differentiable and positive. By using Lamperti's transformation, i.e., $X_t = \Lambda(Y_t)$ where

Table 13.15: Monte Carlo price quoted in percent as a function of the number of Monte Carlo paths 2^N. $T = 10$ years. Exact price = **11.96**. Nonlinearity $P(\Gamma) = \Gamma^2/2$, $\bar{\sigma} = 0.2$, $g(x) = x^2/2$.

N	Fair	Stdev
12	12.21	0.25
14	12.14	0.13
16	11.99	0.06
18	11.92	0.03
20	11.95	0.02

$\Lambda(y) = \int^y \frac{dz}{\sigma(z)}$ is a primitive of $1/\sigma$, this SDE can be converted into

$$dX_t = A(\Lambda^{-1}(X_t))\, dt + dB_t$$

with $A(y) = \frac{b(y)}{\sigma(y)} - \frac{1}{2}\sigma'(y)$. How can we compute exactly the conditional expectation $u(t, x) = \mathbb{E}_{t,x}[g(X_T)]$, which is the solution to

$$\partial_t u + \frac{1}{2}\partial_x^2 u + A(\Lambda^{-1}(x))\partial_x u = 0, \qquad u(T, x) = g(x)?$$

By a proper rescaling $u(t, x) = e^{\beta(T-t)}v_0(t, x)$ for some positive constant β, we have

$$\partial_t v_0 + \frac{1}{2}\partial_x^2 v_0 + \beta\left(\alpha(x)v_1 - v_0\right) = 0, \qquad v_0(T, x) = g(x)$$

with $\alpha(x) = \frac{A(\Lambda^{-1}(x))}{\beta}$. Proceeding as in the previous section, the bootstrapping method gives the PDE system

$$\partial_t v_1 + \frac{1}{2}\partial_x^2 v_1 + \beta\left(\alpha'(x)v_1 + \alpha(x)v_2 - v_1\right) = 0, \quad v_1(T, x) = g'(x)$$

$$\partial_t v_n + \frac{1}{2}\partial_x^2 v_n + \beta\left(\sum_{k=0}^{n}\binom{n}{k}\alpha^{(n-k)}(x)v_{k+1} - v_n\right) = 0, \quad v_n(T, x) = g^{(n)}(x)$$

where $g^{(n)}$ is the n-th derivative of g. We can then apply our multi-species marked branching diffusion algorithm. This requires only the simulation of a Brownian motion (with generator $\frac{1}{2}\partial_x^2$) and a Poisson process with constant intensity β, which can be done exactly without relying on a rejection method as in [60, 61]. Note that we do not need to introduce a time discretization.

The algorithm can then be described by the following steps:

Algorithm

1. Start at $t = 0$ from X_0 and set $n := 0$, $w := 1$.

2. Simulate a Brownian motion with species n up to the first time of default τ of a Poisson process with a constant intensity β. Generate a new species—say $k + 1$—according to a probability p_{k+1}^n and set $w \to w \times \frac{\binom{n}{k}}{p_{k+1}^n} \alpha^{(n-k)}(X_\tau)$.

3. Iterate Step 2 up to the maturity T. Then compute $W g^{\sharp\text{species}}(X_T)$. Iterative N_{paths} times Steps 1, 2, 3 and then average.

Conclusion

The marked branching diffusion algorithm introduced in this last chapter is a very elegant tool that has the very nice feature of not requiring the computation of conditional expectations, unlike the Tsitsiklis-Van Roy method, the Longstaff-Schwartz method, the method presented in Section 8.7, and methods based on the numerical simulation of BSDEs. This makes it (i) very easy to implement, and (ii) independent of a choice of "good" regressors which, in high dimension, is a very difficult problem. It also has some drawbacks since it only applies to nonlinearities of polynomial type. Our numerical tests show that polynomial approximations of given nonlinearities can lead to accurate results. To the best of our knowledge, the design of a Monte Carlo method that avoids computing conditional expectations and that handles any type of nonlinearity remains a very challenging open problem, of great practical importance. We hope that this book will not only encourage the practitioners to use existing nonlinear Monte Carlo methods and improve them, but also to promote research in this field. There is still so much to do!

13.7   Exercises

13.7.1   The law of the number of branchings

In this exercise we aim at describing the distribution of the number of branchings $\Omega \equiv (\Omega_0, \Omega_1, \Omega_2, \ldots)$ in a branching diffusion. We recall that, for $k \in \mathbb{N}$, Ω_k is the r.v. that gives the number of branchings that produced exactly k descendants, up to maturity T (see Section 13.1.2). To emphasize that the r.v. Ω depends on the maturity T, we use the notation $\Omega(T)$. The distribution of $\Omega(T)$ is characterized by its Laplace transform

$$\hat{p}(T,c) \equiv \mathbb{E}\left[\prod_{k=0}^{\infty} e^{-c_k \Omega_k(T)}\right], \qquad c \in \mathbb{R}^{\mathbb{N}}$$

In Questions 1 and 2, we present two ways to compute $\hat{p}$.

1. By mimicking the proof of Lemma 13.1, show that, if $\hat{p}(T,c)$ is finite, it satisfies

$$\int_{1}^{\hat{p}(T,c)} \frac{ds}{\sum_{k=0}^{\infty} p_k e^{-c_k} s^k - s} = \int_{0}^{T} \beta(t)\, dt \qquad \text{if } \sum_{k=0}^{\infty} p_k e^{-c_k} \neq 1 \qquad (13.57)$$

$$\hat{p}(T,c) = 1 \qquad \text{if } \sum_{k=0}^{\infty} p_k e^{-c_k} = 1 \qquad (13.58)$$

2. In this question we present an alternative way to calculate $\hat{p}$.

   (a) Show that the number $N(\omega)$ of particles produced by the branching $\omega = (\omega_0, \omega_1, \omega_2, \ldots) \in \mathbb{N}^{\mathbb{N}}$ is

   $$N(\omega) = 1 + \sum_{k=0}^{\infty} (k-1)\omega_k = 1 + \sum_{k \in \mathbb{N}\backslash\{1\}} (k-1)\omega_k \qquad (13.59)$$

   Below we use the convention that $N(\omega) = 0$ if at least one of the components of ω is negative.

   (b) Let $p(T,\omega) \equiv \mathbb{P}(\Omega(T) = \omega)$ denote the probability of the configuration ω at time T. We define $\omega^{[k]} = (\omega_0, \omega_1, \ldots, \omega_{k-1}, \omega_k - 1, \omega_{k+1}, \ldots)$. Prove that $p(T,0) = \exp(-\int_0^T \beta(t)\, dt)$ and that, for $\omega \neq 0$, $p(T,\omega)$ satisfies the recurrence equation

$$p(T,\omega) = \sum_{k=0}^{\infty} p_k N(\omega^{[k]}) \int_{0}^{T} \beta(t) p(t,\omega^{[k]}) e^{-(k+N(\omega^{[k]})-1)\int_t^T \beta(s)\, ds}\, dt \quad (13.60)$$

   (Hint: condition on the event $\{\Omega(t) = \omega^{[k]}\}$ for $t \in [0,T]$.)

   (c) Set $q(T,\omega) = e^{N(\omega)\int_0^T \beta(t)\, dt} p(T,\omega)$. Using (13.60), show that

$$\forall \omega \in \mathbb{N}^{\mathbb{N}}, \quad \partial_T q(T,\omega) = \beta(T) \sum_{k=0}^{\infty} p_k N(\omega^{[k]}) q(T,\omega^{[k]}) e^{(k-1)\int_0^T \beta(t)\, dt}$$

   (d) For $c \in \mathbb{R}^{\mathbb{N}\backslash\{1\}}$, let us denote

$$\hat{q}(T,c) \equiv \sum_{\omega \in \mathbb{N}^{\mathbb{N}}} \prod_{k \in \mathbb{N}\backslash\{1\}} e^{-(k-1)c_k \omega_k} q(T,\omega)$$

Show that $\hat{q}(T, c)$ satisfies the first order PDE

$$\partial_T \hat{q}(T, c) = \beta(T) \sum_{k=0}^{\infty} p_k \left(\hat{q}(T, c) - \sum_{q \in \mathbb{N}\backslash\{1\}} \partial_{c_q} \hat{q}(T, c) \right) e^{\left(\int_0^T \beta(t)\, dt - c_k\right)(k-1)}$$

with the initial condition $\hat{q}(0, c) = 1$.

(e) Solve this PDE and recover (13.57)-(13.58).

3. Assuming that $g \in \mathrm{L}^{\infty}(\mathbb{R}^d)$, and using (13.59), show that $\hat{u}$ (defined in (13.23)) can be bounded by

$$|\hat{u}(0, x)| \leq ||g||_{\infty} \hat{p} \left(T, -\ln |\bar{a}_k| - \ln ||g||_{\infty}^{k-1} \right)$$

and give a new proof of Proposition 13.2.

4. Show that in the particular case of only one branching type $k \neq 1$, i.e., $p_k \neq 0$ and $p_l = 0$ for all $l \neq k$, we have

$$\hat{p}(T, c_k) = \frac{e^{\frac{c_k}{k-1}}}{\left(1 - e^{(k-1)\int_0^T \beta(t)\, dt} + e^{c_k + (k-1)\int_0^T \beta(t)\, dt}\right)^{\frac{1}{k-1}}}$$

and recover Condition (13.34).

13.7.2   A high order PDE

In this exercise, we highlight that branching diffusions can also be applied to high order PDEs such as

$$\partial_t u + \frac{1}{2} \partial_x^4 u = 0, \quad u(T, x) = g(x)$$

1. By using the bootstrap method, prove that

$$u(t, x) = \sum_{n=0}^{\infty} \frac{(T - t)^n}{2^n n!} g^{4n}(x) \qquad (13.61)$$

2. We would like to write this series as $u(t, x) = \mathbb{E}[g(x + Z)]$. By identifying this expression with (13.61), deduce the characteristic function of Z:

$$\mathbb{E}_t[e^{sZ}] = e^{\frac{s^4(T-t)}{2}} \qquad (13.62)$$

3. A process with such a characteristic function is obtained in terms of a complex-valued stochastic process $\bar{B}_t$ given by

$$\bar{B}_t = \begin{cases} B_t & \text{if} \quad t > 0 \\ iB_{-t} & \text{otherwise} \end{cases}$$

Prove that $Z = \bar{B}_{W_t}$ with W_t a one-dimensional Brownian motion independent of B_t satisfies the identity (13.62). This representation was obtained by Funaki in [111].

13.7.3 Hyperbolic (nonlinear) PDEs

The branching diffusions can also be applied to second order hyperbolic PDEs as first explained in [148].

1. Consider the PDE system:

$$\partial_t u_1 = v \partial_x u_1 - a u_1 + a u_2$$
$$\partial_t u_2 = -v \partial_x u_2 + a u_1 - a u_2$$

Prove that $u \equiv (u_1 - u_2)/2$ satisfies the hyperbolic PDE (i.e., telegrapher's equation):

$$\partial_t^2 u = v^2 \partial_x^2 u - 2a \partial_t u$$

This is the famous Kac's trick [148]. Note that a similar trick was used by Dirac for writing Dirac's equation as the square root of Klein-Gordon's equation.

2. Design a numerical algorithm for solving the telegrapher's equation.

References

Books and monographs

[1] Brigo, D., Mercurio, F.: *Interest Rate Models – Theory and Practice: with Smile, Inflation and Credit*, Springer-Verlag, 3rd edition, 2007.

[2] Buff, R.: *Uncertain Volatility Models: Theory and Application*, Springer Finance Lecture Notes, 2002.

[3] Dynkin, E.B.: *Diffusions, Superdiffusions and Partial Differential Equations*, American Mathematical Society, 2002.

[4] Flannery, B.P., Press, W.H., Teukolsky, S.A., Vetterling, W.T.: *Numerical Recipes: The Art of Scientific Computing*, 3rd edition, Cambridge University Press, 2007.

[5] Fleming, W.H., Soner, H.M.: *Controlled Markov Processes and Viscosity Solutions*, Springer-Verlag, 1993.

[6] Gatheral, J.: *The Volatility Surface: A Practitioner's Guide*, Wiley, 2006.

[7] Glasserman, P.: *Monte Carlo Methods in Financial Engineering*, Springer-Verlag, 2003.

[8] Graham, C., Kurtz, T., Méléard, S., Protter, P., Pulvirenti, M., Talay, D.: *Probabilistic Models for Non-linear Partial Differential Equations*, Lecture Notes in Mathematics 1627, Springer-Verlag, 1996.

[9] Guyon, J.: *Probabilistic Modeling in Finance and Biology: Limit Theorems and Applications*, Lambert Academic Publishing, 2009.

[10] Henry-Labordère, A.: *Cours de Recherche Opérationnelle*, Presses Ecole des Ponts, 1995.

[11] Henry-Labordère, P.: *Analysis, Geometry and Modeling in Finance: Advanced Methods in Option Pricing*, Chapman & Hall/CRC Financial Mathematics Series, 2008.

[12] Jacod, J., Protter, P.: *Probability Essentials*, Universitext, Springer-Verlag, 2003.

[13] Karatzas, I., Schreve, S.: *Brownian Motion and Stochastic Calculus*, Springer-Verlag, 1991.

[14] Krylov, N.V.: *Controlled Diffusion Processes*, Stochastic Modelling and Applied Probability (14), Springer-Verlag, 1980.

[15] Kunita, H.: *Stochastic Flows and Stochastic Differential Equations*, Cambridge University Press, 1990.

[16] Lapeyre, B., Pardoux, E., Sentis, R.: *Introduction to Monte Carlo Methods for Transport and Diffusion Equations*, Oxford University Press, 2003.

[17] Le Gall, J.-F.: *Spatial Branching Process, Random Snakes, and Partial Differential Equations*, Birkhäuser Verlag, Basel, 1999.

[18] Ma, J., Yong, J.: *Forward-Backward Stochastic Differential Equations and their Applications*, Lecture Notes in Mathematics 1702, Springer-Verlag, 1999.

[19] Malliavin, P., Thalmaier, A.: *Stochastic Calculus of Variations in Mathematical Finance*, Springer-Verlag, 2006.

[20] Øksendal, B.: *Stochastic Differential Equations: An Introduction with Applications*, 5th ed., Springer-Verlag, 1998.

[21] Pham, H.: *Continuous-Time Stochastic Control and Applications with Financial Applications*, Series: Stochastic Modelling and Applied Probability (61), Springer-Verlag, 2009.

[22] Silverman, B.W.: *Density Estimation for Statistics and Data Analysis*, Chapman & Hall, New York, 1986.

[23] Sznitman, A.S.: *Topics in Propagation of Chaos*, Ecole d'été de probabilités de Saint-Flour XIX - 1989, volume 1464 of Lect. Notes in Math. Springer-Verlag, 1991.

[24] Touzi, N.: *Optimal Stochastic Control, Stochastic Target Problems, and Backward SDE*, Fields Institute monographs (29), Springer-Verlag, 2012.

[25] Villani, C.: *Limite de Champ Moyen*, available at http://www.umpa.ens-lyon.fr/~cvillani/Cours/moyen.pdf (in French), 2001.

[26] Villani, C.: *Topics in Optimal Transportation*, Graduate Studies in Mathematics (58), AMS, 2003.

Articles

[27] Abergel, F., Tachet, R.: *A nonlinear partial integrodifferential equation from mathematical finance*, Discrete Cont. Dynamical Systems, Series A, 27(3):907–917, 2010.

[28] Ahdida, A., Alfonsi, A.: *A mean-reverting SDE on correlation matrices*, Stochastic Processes and their Applications, 123(4):1472–1520, 2013.

[29] Alanko, S., Avellaneda, M.: *Reducing variance in the numerical solution of BSDEs*, C. R. Acad. Sci. Paris, Ser. I 340, 2012.

[30] Alfonsi, A., Schied, A., Schulz, A.: *Constrained portfolio liquidation in a limit order book model*, Banach Center Publ. 83:9–25, 2008.

[31] Alvarez, L., Guichard, F., Lions, P.-L., Morel, J.-M.: *Axioms and fundamental equations of image processing*, Arch. Rat. Mech. Anal. 123:199–257, 1993.

[32] Andersen, L., Andreasen, J., Brotherton-Ratcliffe, R.: *The passport option*, The Journal of Computational Finance, 1(3):15–36, 1998.

[33] Andersen, L.: *A simple approach to the pricing of Bermudan swaptions in the multi-factor Libor market model*, Journal of Computational Finance 3(5):5–32, 1999.

[34] Andersen, L., Broadie, M.: *Primal-dual simulation algorithm for pricing multidimensional American options*, Management Science, 50(9):1222–1234, 2004.

[35] Austing, P.: *Repricing the cross smile: An analytic joint density*, Risk magazine, 72–75, July 2011.

[36] Avellaneda, M., Levy, A., Parás, A.: *Pricing and hedging derivative securities in markets with uncertain volatilities*, Applied Mathematical Finance, 2:73–88, 1995.

[37] Avellaneda, M., Friedman, C., Holmes, R., Samperi, D.: *Calibrating volatility surfaces via relative-entropy minimization*, Applied Mathematical Finance, 4(1):37–64, 1997.

[38] Avellaneda, M.: *Minimum-relative-entropy calibration of asset pricing models*, International Journal of Theoretical and Applied Finance, 1998.

[39] Avellaneda, M., Buff, R.: *Combinatorial implications of non-linear uncertain volatility models: The case of barrier options*, Applied Mathematical Finance, 1:1–18, 1999.

[40] Avellaneda, M., Boyer-Olson, D., Busca, J., Friz, P.: *Reconstructing volatility*, Risk magazine, 87–91, Oct. 2002.

[41] Avellaneda, M., Kampen, J.: *On parabolic equations with gauge function term and applications to the multidimensional Leland equation*, Applied Mathematical Finance, 10(3):215–218, 2003.

[42] Azéma, J., Yor, M.: *Une solution simple au problème de Skorokhod*, Séminaire de probabilités de Strasbourg, 13:90–115, 1979.

[43] Balland, P.: *Stoch-vol model with interest rate volatility*, presentation at the Global Derivatives conference, 2005.

[44] Bally, V., Talay, D.: *The law of the Euler scheme for stochastic differential equations : Convergence rate of the distribution function*, Probab. Th. Rel. Fields, 104:43–60, 1996.

[45] Bally, V., Caballero, M.E., Fernandez, B., El Karoui, N.: *Reflected BSDE's, PDE's and variational inequalities*, preprint, 2002.

[46] Bally, V., Pagès, G.: *A quantization algorithm for solving multidimensional discrete-time optimal stopping problems*, Bernoulli, 9(6):1003–1049, 2003.

[47] Bally, V., Pagès, G., Printems, J.: *A quantization tree method for pricing and hedging multi-dimensional American options*, Mathematical Finance, 15(1):119–168, 2005.

[48] Bank, P., Baum, D.: *Hedging and portfolio optimization in financial markets with a large trader*, Mathematical Finance, 14:1–18, 2004.

[49] Barles, G., Souganidis, P.E.: *Convergence of approximation schemes for fully nonlinear second order equations*, Asymptotic Anal., 4(3):271–283, 1991.

[50] Barles, G.: *Solutions de viscosité et équations elliptiques du deuxième ordre*, lecture notes (in French), 1997.

[51] Barraquand, J., Martineau, D.: *Numerical valuation of high dimensiona multivariate American securities*, Journal of Finance and Quantitative Analysis, 30:383–405, 1995.

[52] Beiglböck, M., Henry-Labordère, P., Penkner, F.: *Model-independent bounds for option prices: A mass transport approach*, Finance and Stoch., 17(3):477–501, 2013.

[53] Benhamou, E., Gruz, A., Rivoira, A.: *Stochastic interest rates for local volatility hybrids models*, Wilmott magazine, Mar. 2008.

[54] Benth, F.E., Karlsen, K.H., Rikvam, K.: *On a semilinear Black and Scholes partial differential equation for valuing American options*, Finance and Stoch., 7:277–298, 2003.

[55] Benth, F.E., Karlsen, K.H., Rikvam, K.: *A semilinear Black and Scholes partial differential equation for valuing American options: Approximate*

solutions and convergence, Interfaces and Free Boundaries, 6:379–404, 2004.

[56] Bernard, P., Talay, D., Tubaro, L.: *Rate of convergence of a stochastic particle method for the Kolmogorov equation with variable coefficients*, Math. Comp., 63(208):555–587, 1994.

[57] Bergomi, L.: *Smile dynamics II*, Risk magazine, 67–73, Oct. 2005.

[58] Bergomi, L.: *Smile dynamics III*, Risk magazine, 90–96, Oct. 2008.

[59] Bergomi, L., Guyon, J.: *Stochastic volatility's orderly smiles*, Risk magazine, 60–66, May 2012.

[60] Beskos, A., Roberts, G.O.: *Exact simulation of diffusions*, Ann. Appl. Probab. 15(4):2422–2444, 2005.

[61] Beskos, A., Papaspiliopoulos, O., Roberts, G.O.: *Retrospective exact simulation of diffusion sample paths*, Bernoulli 12(6), 2006.

[62] Bick, A.: *Quadratic-variation-based dynamic strategies*, Management Science, 41:722–732, 1995.

[63] Bokanowski, O., Maroso, S., Zidani, H.: *Some convergence results for Howard's algorithm*, SIAM J. Num. Analysis, 47(4):3001–3026, 2009.

[64] Bos, M., Vandermark, S.: *Finessing fixed dividends*, Risk magazine, 157–158, Sept. 2002.

[65] Bouchard, B., Ekeland, I., Touzi, N.: *On the Malliavin approach to Monte Carlo approximation of conditional expectations*, Finance and Stoch., 8(1):45–71, 2004.

[66] Bouchard, B., Touzi, N.: *Discrete-time approximation and Monte Carlo simulation of backward stochastic differential equations*, Stochastic Processes and their Applications, 111:175–206, 2004.

[67] Brace, A., Gatarek, D., Musiela, M.: *The market model of interest rate dynamics*, Mathematical Finance, 7(2):127–154, 1997.

[68] Brigo, D., Morini, M.: *Close-out convention tensions*, Risk magazine, 74–78, Dec. 2011.

[69] Broadie, M., Glasserman, P.: *A stochastic mesh method for pricing high-dimensional American options*, Journal of Computational Finance, 1997.

[70] Broadie, M., Glasserman, P., Ha, Z.: *Pricing American Options by Simulation Using a Stochastic Mesh with Optimized Weights*, Probabilistic Constrained Optimization: Methodology and Applications (S.P. Uryasev, Editor), pp. 32–50, Kluwer Academic Publishers, 2000.

[71] Bronstein, A.-L., Pagès, G., Wilbertz, B.: *How to speed up the quantization tree algorithm with an application to swing options*, Quantitative Finance, 10(9), 2010.

[72] Carmona, R., Nadtochiy, S.: *Local volatility dynamic models*, Finance and Stoch., 13(1):1–48, 2009.

[73] Carr, P., Jarrow, R., Meyneni, R.: *Alternative characterizations of American put options*, Math. Finance, 2:87–106, 1992.

[74] Carr, P., Lee, R.: *Hedging variance options on continuous semimartingales*, Finance and Stoch., 14(2):179–207, 2010.

[75] Çetin, U., Jarrow, R., Protter, P.: *Liquidity risk and arbitrage pricing theory*, Finance and Stoch., 8(3):311–341, 2004.

[76] Çetin, U., Soner, H.M., Touzi, N.: *Option hedging for small investors under liquidity costs*, Finance and Stoch., 14(3):317–341, 2008.

[77] Chan, T.: *The Wigner semi-circle law and eigenvalues of matrix-valued diffusions*, Probab. Theory Relat. Fields, 93:249–272, 1992.

[78] Chen, N., Glasserman, P.: *Additive and multiplicative duals for American option pricing*, Finance and Stoch., 11(2):153–179, 2007.

[79] Cheridito, P., Soner, H.M., Touzi, N.: *The multi-dimensional super-replication problem under gamma constraints*, Annales de l'institut Henri Poincaré (C) Analyse non linéaire, 22(5):633–666, 2005.

[80] Cheridito, P., Soner, H.M., Touzi, N.: *Hedging under gamma contraints by optimal stopping and face-lifting*, Mathematical Finance, 17(1):59–79, 2007.

[81] Cheridito, P., Soner, H.M., Touzi, N., Victoir, N.: *Second order backward stochastic differential equations and fully non-linear parabolic PDEs*, Comm. on Pure and Appl. Math., 60(7):1081–1110, 2007.

[82] Cherny, A., Dupire, B.: *On certain distributions associated with the range of martingales*, Optimality and Risk: Modern Trends in Mathematical Finance, 29–38, 2009.

[83] Clément, E., Lamberton, D., Protter, P.: *An analysis of a least squares regression method for American option pricing*, Finance and Stoch., 2002.

[84] Cont, R., Deguest, R.: *Equity correlations implied by index options: Estimation and model uncertainty analysis*, Mathematical Finance, 23(3):496–530, 2013.

[85] Corlay, S.: *B-spline techniques for volatility modeling*, available at http://hal.archives-ouvertes.fr/hal-00830378, 2013.

[86] Crandall, M.G., Ishii, H., Lions, P.-L.: *User's guide to viscosity solutions of second order partial differential equations*, Bull. Am. Math. Soc., 27(1):1–67, 1992.

[87] Cvitanić, J., Karatzas, I.: *Convex duality in constrained portfolio optimization*, Annals of Applied Probability, 2:767–818, 1992.

[88] Da Fonseca, J., Grasselli, M., Tebaldi, C.: *Option pricing when correlations are stochastic: An analytical framework*. Review of Derivatives Research, 10:151–180, 2008.

[89] Davis, M.H.A.: *Complete-market models of stochastic volatility*, Royal Society of London Proceedings Series A, 460(2041):11–26, 2004.

[90] De Col, A., Gnoatto, A., Grasselli, M.: *Smiles all around: FX joint calibration in a multi-Heston model*, available at http://arxiv.org/abs/1201.1782, 2012.

[91] Delanoe, P.: *Local correlation with local vol and stochastic vol: Towards correlation dynamics*, presentation at the Global Derivatives conference, April 2013.

[92] Denis, L., Martini, C.: *A theoretical framework for the pricing of contingent claims in the presence of model uncertainty*, Ann. Appl. Probab., 16(2):827–852, 2006.

[93] Derman, E., Kani, I.: *Stochastic implied trees: Arbitrage pricing with stochastic term and strike structure of volatility*, International Journal of Theoretical and Applied Finance, 1:7–22, 1998.

[94] Dupire, B.: *Model art*, Risk magazine, 118–124, Sept. 1993.

[95] Dupire, B.: *Pricing with a smile*, Risk magazine, 7:18–20, 1994.

[96] Dupire, B.: *A Unified theory of volatility*, In Derivatives Pricing: The Classic Collection, edited by Peter Carr, Risk publications, 1996.

[97] Dupire, B.: *A new approach for understanding the impact of volatility on option prices*, presentation at the Risk conference, October 1998.

[98] Durrleman V., El Karoui N.: *Coupling smiles*, Quantitative Finance, 8(6):573–590, 2008.

[99] El Karoui, N., Roelly, S.: *Propriétés de martingales, explosion et représentation de Lévy-Khintchine d'une classe de processus de branchement à valeurs mesures* (in French), Stochastic Processes and their Applications, 38:239–266, 1991.

[100] El Karoui, N., Quenez, M.-C.: *Dynamic programming and pricing of contingent claims in an incomplete market*, SIAM Journal on Control and Optimization archive, 33(1), 1995.

[101] El Karoui, N., Peng, S., Quenez, M.-C.: *Backward stochastic differential equations in finance*, Mathematical Finance, 47(1):1–77, 1997.

[102] El Karoui, N., Peng, S., Quenez, M.-C.: *Reflected solutions of backward SDE's, and related obstacle problems for PDE's*. The Annals of Probability, 25(2):702–737, 1997.

[103] El Karoui, N., Jeanblanc, M., Shreve, S.E.: *Robustness of the Black and Scholes formula*, Math. Finance, 8(2):93–126, 1998.

[104] Evans, L.: Video available at http://www.msri.org/workshops/324/schedules/2313. MSRI workshop for women in mathematics: An introduction to elliptic partial differential equations, Aug. 11–12, 2005.

[105] Fahim, A., Touzi, N., Warin, X. : *A probabilistic numerical method for fully nonlinear parabolic PDEs*, Ann. Appl. Probab., 21(4):1322–1364, 2011.

[106] Föllmer, H., Sondermann, D.: *Hedging of contingent claims under incomplete information*, Applied Stochastic Analysis, eds. M.H.A. Davis and R.J. Elliott, Stochastics Monographs vol. 5, Gordon and Breach, London and New-York, 389–414, 1991.

[107] Föllmer, H., Leukert, P.: *Quantile Hedging*, Finance and Stoch., 3:251–273, 1999.

[108] Forsyth, P.A., Vetzal, K.R.: *Implicit solution of uncertain volatility/transaction cost option pricing with discretely observed barriers*, Appl. Num. Math., 36:427–445, 2001.

[109] Frey, R.: *Perfect option replication for a large trader*, Finance and Stoch., 2:115–149, 1998.

[110] Frey, R., Patie, P.: *Risk management for derivatives in illiquid markets: a simulation study*, In: Sandmann, K., Schönbucher, P. (Eds.), Advances in Finance and Stochastics, 137–159, Springer, 2001.

[111] Funaki, T.: *Probabilistic construction of the solution to some higher order parabolic differential equation*, Proc. Japan Acad. Ser. A Math. Sci., 55(5):176–179, 1979.

[112] Galichon, A., Henry-Labordère, P., Touzi, N.: *A stochastic control approach to no arbitrage bounds given marginals, with an application to lookback options*, to appear in Ann. Appl. Probab., available at http://ssrn.com/abstract=1912477, 2011.

[113] Gatheral, J.: *Developments in volatility derivatives pricing*, slides, presentation at the Global Derivatives conference, 2007.

[114] Gerhold, S.: *The Longstaff-Schwartz algorithm for Lévy models: Results on fast and slow convergence*, Ann. Appl. Probab., 21(2):589–608, 2011.

[115] Glasserman, P., Yu, B.: *Number of paths versus number of basis functions in American option pricing*, Ann. Appl. Probab., 14:2090–2119, 2004.

[116] Gobet, E., Lemor, J.-P., Warin, X.: *A regression-based Monte Carlo method to solve backward stochastic differential equations*, Ann. Appl. Probab., 15(3):2172–2202, 2005.

[117] Gourieroux, C., Sufana, R.: *Wishart quadratic term structure models*, working paper, 2003.

[118] Guyon, J.: *Euler scheme and tempered distributions*, Stochastic Processes and their Applications, 116(6):877–904, 2006.

[119] Guyon, J.: *Numerical methods for nonlinear problems in quantitative finance*, minicourse given at the Research In Options conference, Rio de Janeiro, 2009.

[120] Guyon, J., Henry-Labordère, P.: *Uncertain volatility model: A Monte Carlo approach*, Journal of Computational Finance, 14(3):37–71, 2011.

[121] Guyon, J., Henry-Labordère, P.: *From spot volatilities to implied volatilities*, Risk magazine, 79–84, June 2011.

[122] Guyon, J., Henry-Labordère, P.: *Being particular about calibration*, Risk magazine, 88–93, Jan. 2012. Longer version available at http://ssrn.com/abstract=1885032, 2011. Another longer version published in Post-Crisis Quant Finance, Risk Books, 2013.

[123] Guyon, J.: *A new class of local correlation models*, available at http://ssrn.com/abstract=2283419, 2013.

[124] Gyöngy, I.: *Mimicking the one-dimensional marginal distributions of processes having an Itô differential*, Probability Theory and Related Fields, 71:501–516, 1986.

[125] Haugh, M.B., Kogan, L.: *Pricing American options: A duality approach*, Operations Research, 52(2):258–270, 2004.

[126] Henderson V., Hobson, D.G.: *Local time, coupling and the passport option*, Finance and Stoch., 4(1):69–80, 2000.

[127] Henry-Labordère, P.: *Nonlinear problems in quantitative finance*, presentation at the Numerical Methods in Finance conference, Ecole des Ponts, Paris, 2009.

[128] Henry-Labordère, P.: *Calibration of local stochastic volatility models: A Monte Carlo approach*, Risk magazine, 112–117, Sept. 2009.

[129] Henry-Labordère P.: *Automated option pricing: Numerical methods*, to appear in Intl. Journ. of Theoretical and Applied Finance, available at papers.ssrn.com/sol3/papers.cfm?abstract_id=1968344, 2011.

[130] Henry-Labordère, P.: *Cutting CVA's complexity*, Risk magazine, 67–73, July 2012.

[131] Henry-Labordère, P.: *Forward Monte Carlo schemes for nonlinear PDEs: Multi-type marked branching diffusions*, slides available at http://grozny.maths.univ-evry.fr/pages_perso/crepey/CVA-day/HENRY-LABORDERE_MarkedBMEvry.pdf, 2012

[132] Henry-Labordère, P., Tan, X., Touzi, N.: *A numerical algorithm for a class of BSDE via branching process*, preprint, Arxiv:1302.4624v2, 2013.

[133] Higham N.: *Computing a nearest symmetric correlation matrix - A problem from finance*, IMA Journal of Numerical Analysis, 22(3):329–343, 2002.

[134] Hobson D.G., Rogers, L.C.G.: *Complete models with stochastic volatility*, Mathematical Finance, 8:27–48, 1998.

[135] Hodges, S.D., Neuberger, A.: *Optimal replication of contingent claims under transaction costs*, Review of Futures Markets, 8:222–239, 1989.

[136] Hull, J., White, A.: *The pricing of options on assets with stochastic volatilities*, The Journal of Finance, 42(2):281–300, 1987.

[137] Humez, B.: *2-Factor lognormal stochastic volatility Model*, working paper, Société Générale, 2003.

[138] Hunt, P., Kennedy, J., Pelsser, A.: *Markov-functional interest rate models*, Finance and Stoch., 4(4):391–408, 2000.

[139] Hyer T., Lipton-Lifschitz, A., Pugachevsky D.: *Passport to success*, Risk magazine, 10(9):127–131, 1997.

[140] Ikeda, N., Nagasawa, M., Watanabe, S.: *Branching Markov process I*, J. Math. Kyoto Univ., 8(2):233–278, 1968.

[141] Ikeda, N., Nagasawa, M., Watanabe, S.: *Branching Markov process II*, J. Math. Kyoto Univ., 8:365–410, 1968.

[142] Ikeda, N., Nagasawa, M., Watanabe, S.: *Branching Markov process III*, J. Math. Kyoto Univ., 9:95–160, 1969.

[143] Jamshidian, F.: *Numéraire-invariant option pricing and American, Bermudan, and trigger stream rollover*, working paper, 2004.

[144] Joshi, M., Theis, J.: *Bounding Bermudan swaptions in a swap-rate market model*, Quantitative Finance, 2:370–377, 2002.

[145] Joshi, M.: *A simple derivation of and improvements to Jamshidian and Rogers upper bound methods for Bermudan options*, Applied Mathematical Finance, 14:197–205, 2007.

[146] Jourdain, B.: *Adaptive variance reduction techniques in finance*, Radon Series Comp. Appl. Math 8, 1–18, 2009.

[147] Jourdain, B., Sbai, M.: *Coupling index and stocks*, Quantitative Finance, 12(5):805–818, 2012.

[148] Kac, M.: *A stochastic model related to the telegrapher's equation*, Rocky Mountain J. Math., 4(3):497–510, 1974.

[149] Kholodnyi, V.A.: *The semilinear evolution equation for universal contingent claims: Examples and applications*, preprint, 2006.

[150] Kovrizhkin, O.: *Local volatility + local correlation multicurrency model*, presentation at the Global Derivatives conference, April 2012.

[151] Kramkov, D.: *Optional decomposition of supermartingales and hedging contingent claims in incomplete security markets*, Probab. Theory Related Fields, 105:459–479, 1996.

[152] Langnau, A.: *A dynamic model for correlation*, Risk magazine, 74–78, Apr. 2010.

[153] Le Jan, Y., Sznitman, A.S.: *Stochastic cascades and 3-dimensional Navier Stokes equations*, Prob. Theory Relat. Fields, 109:343–366, 1997.

[154] Leblanc, M., Martini, C.: *Unbounded volatility in the uncertain volatility model*, INRIA Research Report 4065, 2000.

[155] Leland, H.E.: *Option pricing and replication with transaction costs*, The Journal of Finance, 40:1283–1301, 1985.

[156] Lipton, A.: *The vol smile problem*, Risk magazine, 61–65, Feb. 2002.

[157] Longstaff, F.A., Schwartz, E.S.: *Valuing American options by simulation: A simple least-squares approach*, The Review of Financial Studies, 14(1):113–147, 2001.

[158] López-Mimbela, J.A., Wakolbinger, A.: *Length of Galton-Watson trees and blow-up of semilinear systems*, J. Appl. Probab., 35(4):802–811, 1998.

[159] Lyons, T.J.: *Uncertain volatility and the risk-free synthesis of derivatives*, Applied Mathematical Finance, 2(2):117–133, 1995.

[160] Martini, C.: *The Uncertain volatility model and American options*, INRIA Research Report 3697, 1999.

[161] McKean, H.P.: *A class of Markov processes associated with nonlinear parabolic equations*, Proc. Natl. Acad. Sci. U.S.A., 56(6):1907-1911, 1966.

[162] McKean, H.P.: *Application of Brownian motion to the equation of Kolmogorov-Petrovskii-Piskunov*, Communications on Pure and Applied Mathematics, 28(3):323–331, 1975.

[163] Méléard, S.: *Asymptotic behaviour of some interacting particle systems; McKean-Vlasov and Boltzmann models*, in Probabilistic Models for Non-linear Partial Differential Equations, Lecture Notes in Mathematics 1627, Springer-Verlag, 1996.

[164] Méléard, S.: *Modèles aléatoires en écologie et évolution*, Lecture notes (in French), Ecole Polytechnique, 2009.

[165] Meyer, G.: *The Black-Scholes Barenblatt equation for options with uncertain volatility and its application to static hedging*, International Journal of Theoretical and Applied Finance, 9(5):673–703, 2006.

[166] Ninomiya, S., Victoir, N.: *Weak approximation of stochastic differential equations and application to derivative pricing*, Applied Mathematical Finance, 15(2):107–121, 2008.

[167] Obizhaeva, A., Wang, J.: *Optimal trading strategy and supply/demand dynamics*, Journal of Financial Markets, 16(1):1–32, 2013.

[168] Obłój, J.: *Complete characterization of local martingales which are functions of Brownian motion and its maximum*, Bernoulli, 12:955–969, 2006.

[169] Pagès, G.: *Multi-step Richardson-Romberg extrapolation: Remarks on variance control and complexity*, Monte Carlo Methods and Applications, 13(1):37–70, 2007.

[170] Pardoux, E., Peng, S.: *Adapted solution of a backward stochastic differential equation*, Systems Control Lett., 14:55–61, 1990.

[171] Piterbarg, V.: *Smiling hybrids*, Risk magazine, 66–71, May 2006.

[172] Piterbarg, V.: *Markovian projection for volatility calibration*, Risk magazine, 84–89, Apr. 2007.

[173] Pooley, D.M., Forsyth, P.A., Vetzal, K.R.: *Numerical convergence properties of option pricing PDEs with uncertain volatility*, IMA Journal of Numerical Analysis, 23(2):241–267, 2003.

[174] Pooley, D.M., Forsyth, P.A., Vetzal, K.R.: *Two factor option pricing with uncertain volatility*, ICCSA 2003, Lecture Notes in Computer Science 2669, 158–167, Springer, 2003.

[175] Reghai, A.: *Breaking correlation breaks*, Risk magazine, 90–95, Oct. 2010.

[176] Renault, E., Touzi, N.: *Option pricing and implied volatilities in a stochastic volatility model*, Mathematical Finance, 6:279–302, 1996.

[177] Rogers, L.C.G.: *Duality in constrained optimal investment and consumption problems: A synthesis*, Lectures from the CIRANO Workshop on Mathematical Financial Mathematics and Econometrics, June 2001.

[178] Rogers, L.C.G.: *Monte Carlo valuation of American options*, Mathematical Finance 12:271–286, 2002.

[179] Sawyer, N.: *SGCIB launches timer options*, Risk magazine, July 2007.

[180] Schönbucher, P.J.: *A market model for stochastic implied volatility*, Phil. Trans. R. Soc. Lond. A, 357:2071–2092, 1999.

[181] Schweizer, M.: *Variance-optimal hedging in discrete-time*, Math. Oper. Res., 20:1–32, 1995.

[182] Shreve S., Vecer, J.: *Options on a traded account: vacation calls, vacation puts, and passport options*, Finance and Stoch., 4:255–274, 2000.

[183] Skorokhod, A.V.: *Branching diffusion processes*, Theory Probab. Appl., 9(3):445–449, 1965.

[184] Smith, A.T.: *American options under uncertain volatility*, Applied Mathematical Finance, 9(2):123–141, 2002.

[185] Soner, H.M., Touzi, N.: *Super-replication under gamma constraints*, SIAM Journal on Control and Optimization, 39(1):73–96, 2000.

[186] Soner, H.M., Touzi, N.: *The dynamic programming equation for second order stochastic target problems*, SIAM Journal on Control and Optimization, 48(4):2344–2365, 2009.

[187] Soner, H.M., Touzi, N., Zhang, J.: *Well-posedness of second order backward SDEs*, Probability Theory and Related Fields, 153:149–190, 2012.

[188] Stentoft, L.: *Convergence of the least squares Monte Carlo approach to American option valuation*, Manag. Sci., 50:1193–1203, 2004.

[189] Sugitani, S.: *On non-existence of global solutions for some nonlinear integral equations*, Osaka J. Math., 12:45–51, 1975.

[190] Talay, D., Tubaro, L.: *Expansion of the global error for numerical schemes solving stochastic differential equations*, Stoch. Analysis and Appl., 8(4):94–120, 1990.

[191] Tsitsiklis, J.N., Van Roy, B.: *Regression methods for pricing complex American-style options*, IEEE Transactions on Neural Networks, 12(4):694–703, 2001.

[192] Von Petersdorff, T., Schwab, C.: *Numerical solution of parabolic equations in high dimensions*, Modélisation mathématique et analyse numérique, 38(1):93–127, 2004.

[193] Zanger, D.: *Convergence of a least-squares Monte Carlo algorithm for bounded approximating sets*, Appl. Math. Finance, 16(2), 2009.

[194] Zhang, K., Wang, S.: *A computational scheme for uncertain volatility in option pricing*, Applied Numerical Mathematics, 59(8):1754–176, 2009.

Index